OXFORD–WARBURG STUDIES

General Editors
DENYS HAY *and* J. B. TRAPP

OXFORD–WARBURG STUDIES

PROPHETIC ROME

IN THE

HIGH RENAISSANCE
PERIOD

Essays

edited by

MARJORIE REEVES

CLARENDON PRESS · OXFORD

1992

Oxford University Press, Walton Street, Oxford OX2 6DP

Oxford New York Toronto
Delhi Bombay Calcutta Madras Karachi
Petaling Jaya Singapore Hong Kong Tokyo
Nairobi Dar es Salaam Cape Town
Melbourne Auckland
and associated companies in
Berlin Ibadan

Oxford is a trade mark of Oxford University Press

Published in the United States
by Oxford University Press, New York

British Library Cataloguing in Publication Data
Data available

Library of Congress Cataloging-in-Publication Data
Prophetic Rome in the High Renaissance period: essays/edited by
Marjorie Reeves.
— (Oxford–Warburg studies)
Includes bibliographical references (p.   ) and index.
1. Rome—History—Prophecies. 2. Rome—Church history.
I. Reeves, Marjorie. II. Series.
BR878.R7P76 1992
231.7'45'094563209031—dc20   91–24829
ISBN 0–19–920173–0

Typeset by Best-set Typesetter Ltd, Hong Kong
Printed and bound in Great Britain by
Biddles Ltd., Guildford and King's Lynn

PREFACE

The idea of this book originated in the fact that over the last fifteen or twenty years a number of scholars have been working in the field of late Medieval and Renaissance prophecy. The results are scattered in a variety of journals and separate studies. It seemed useful to bring their results together in one publication which would attempt to give a composite picture of a Rome whose features have in the past found little place in the profile of the Eternal City during the high Renaissance period that is usually presented. Most of the essays have been specially written for this book—only two have been previously published, appearing here in a revised form—and we have aimed at producing an integrated work, rather than a collection of separate pieces. The unity of the picture which has resulted is demonstrated in the fact that the individual essays interact at so many points.

There is no central essay on the Popes themselves. This blank, so to say, in the middle of the design emphasizes the key enigma in these studies, that is the attitude of the supreme pontiffs towards the many prophetic voices which clamoured in their ears, and in particular, towards the angelic role continually being set before them. Did any of them, indeed, ever entertain this vision as having any kind of relationship to the political world in which they were called to operate? The distance between the dream and the reality appears to make the question absurd. Yet successive Popes favoured and listened to some, at least, of those declaring the prophetic future. Julius II heard Egidio of Viterbo proclaim the inception of the Golden Age in his pontificate. Leo X allowed a preacher whom he had appointed to point to himself as the expected Angelic Pope. Clement VII turned to Egidio again for the guidance of a prophet in the confused years after the Sack of Rome. In the long run the ambiguity of human motivation darkens counsel and there is no clear answer to the question posed above.

It was tempting to use prophets, but they were also dangerous. A point which emerges strikingly in several of these studies is the extent to which the problem of distinguishing between true and false prophets confronted church leaders at this time. In spite of his condemnation, Savonarola remained the subject of controversy in high places and the decree which was finally promulgated in the Fifth Lateran Council to deal with the vexed question of prophecy in preaching revealed the difficulty of laying down clear criteria for testing the spirits.

Testing the spirits was, indeed, a matter of importance. A clear theme which runs right through these studies is the obsessive concern, not only of intellectuals, but of the people in the market place as well, with what may be called secret sources of knowledge. Everyone paid attention to the clues and signs from which they might read the future and understand the prophetic message. For humanist scholars and theologians this search led them to the hermetic sources of Neoplatonism, to the inner concords of Scripture, to the revelations of mystics and messages of angels, to astrology, to numerology and the cabbala. The messages came to the people in cruder forms, through the hermit-preachers crying in the streets, through natural portents and monstrous births. But the two levels interact in fascinating ways and there is no hard and fast line between the *alto* and the *basso*. From the learned interpretations of an Egidio of Viterbo or a Benigno Salviati to the babblings of the wildest preachers there is a continuous flow of concern with prophetic expectations.

M.E.R.

ACKNOWLEDGEMENTS

My thanks are due in the first place to Professor J. B. Trapp, Director of the Warburg Institute for sponsoring the project which has resulted in this book and for funding the translation of the Italian contributions to it. I am grateful to the British Academy for a grant towards editorial expenses. Among Oxford friends I received generous help over translations from Dr Coke-Enquidados, Mr Richard Jeffery, and Dr Christina Roaf, while Dr R. G. Lewis, Professor George Holmes, and Sir Keith Thomas gave much encouragement. Permission to use published material has been readily given by the Princeton University Press (Professor Chastel's *Sack of Rome*), the Warburg Institute (Professor Jungić's article on San Pietro in Montorio), and the *Burlington Magazine* (Mr Bull's article on the Sistine Chapel Ceiling).

CONTENTS

LIST OF PLATES

Between pages 306 and 307

LIST OF FIGURES

LIST OF CONTRIBUTORS

Malcolm Bull is a Junior Research Fellow at Wolfson College, Oxford.

Thomas Cohen is a professor in the History Department at York University, Downsview, Ontario.

John Headley is a professor in the History Department at the University of North Carolina, Chapel Hill.

William Hudon is an associate professor in the History Department at Bloomsburg University, Pennsylvania.

Josephine Jungić is a professor in the Art History Department at Capilano College, Vancouver.

Aldo Landi is a professor in the Dipartimento di Storia at the Università degli Studi, Florence.

Bernard McGinn is a professor in the Divinity School, University of Chicago.

Angus MacKay is Professor of Medieval History, University of Edinburgh.

Nelson H. Minnich is a professor in the Department of Church History at the Catholic University of America, Washington, DC.

Anna Morisi-Guerra is a professor in the Facoltà di Lettere e Filosofia in the Università 'La Sapienza', Rome.

Ottavia Niccoli is a professor in the Dipartimento di Discipline Storiche at the Università degli Studi, Bologna.

Adriano Prosperi is a professor in the Dipartimento di Storia Moderna e Contemporanea at the Università degli Studi, Pisa.

Marjorie Reeves is an Honorary Fellow of St Anne's and St Hugh's Colleges, Oxford.

Roberto Rusconi is the Director of the Istituto di Filologia e Storia Medievale in the Università degli Studi, Salerno.

Cesare Vasoli is a professor at the Istituto Nazionale di Studi sul Rinascimento, Florence.

ABBREVIATIONS

The following abbreviations are used throughout for journals, works of reference, and collections of source material:

1. *AFH* — *Archivum Franciscanum Historicum*
2. *AFP* — *Archivum Fratrum Praedicatorum*
3. *AHP* — *Archivum Historiae Pontificiae*
4. *AK* — *Archiv für Kulturgeschichte*
5. *Annuarium* — *Annuarium Historiae Conciliorum*
6. *AR* — *Archiv für Reformationsgeschichte*
7. *AS* — *Acta Sanctorum*
8. *ASI* — *Archivio storico italiano*
9. *Bull. ISI* — *Bullettino dell'Istituto Storico Italiano per il Medio Evo e Archivio Muratoriano*
10. *CJC* — *Corpus juris canonici* (ed. E. Friedberg)
11. *CS* — *Cristianesimo nella storia*
12. *DBI* — *Dizionario biografico degli italiani*
13. *DS* — *Dictionnaire de Spiritualité*
14. *DHGE* — *Dictionnaire d'Histoire et de Géographie Ecclésiastiques*
15. *HJ* — *Historisches Jahrbuch*
16. *HZ* — *Historische Zeitschrift*
17. *JWCI* — *Journal of the Warburg and Courtauld Institutes*
18. *Mansi* — *Sacrorum Conciliorum nova et amplissima Collectio*
19. E. Martene and V. Durand,
    Ampl. Coll. — *Veterum Scriptorum et Monumentorum . . . Amplissima Collectio*
20. E. Martene and V. Durand,
    Thesaurus — *Thesaurus novus anecdotorum*
21. *MGHS* — *Monumenta Germaniae Historica Scriptores*
22. *Misc. Franc.* — *Miscellanea Francescana*
23. *PL* — Migne, *Patrologia Latina*
    PG — Migne, *Patrologia Graeca*

I
INTRODUCTION

I

The Medieval Heritage

MARJORIE REEVES

In the scenario of End-Time built up during the later Middle Ages, the city of Rome played a paradoxical double role: on the one hand, it was the focus of Jubilees and prophetically the seat of the Angelic Pope; on the other hand, it was to be the arena for Antichrist and must be chastised, even destroyed, for its wickedness. It was the New Jerusalem (*visio pacis*); it was the New Babylon (*confusio*). The purpose of this chapter is briefly to trace the development of these two roles which formed the ambivalent legacy of Renaissance Rome.

In the early Middle Ages the approach of the millennium had been a focus for religious fears and expectations of the future, although, significantly, these seem to have been distrusted by the official Church.[1] It was not until the later Middle Ages that the centennial, and then the fifty-year milestones, were seized on by the Church as opportunities to build up a powerful image of Rome as well as occasions to draw in needed revenues. Roberto Rusconi has studied the prophetic connotations of the series of jubilees and the following paragraphs draw heavily on his work.[2] The first, and perhaps the most famous, of these occasions was the Jubilee of 1300 proclaimed by Boniface VIII—famous partly because it gave Dante his fictitious date for the Divine Comedy. Rusconi points to the precedent of the Hebrew Year of Jubilee and its eschatological overtones in medieval Judaism as possibly underlying the Indulgence issued by Pope Boniface. The Hebrew vision was of a total

---

[1] For a discussion of this point, see H. Focillon, *L'an Mil* (Paris, 1938), 50–63; M. Reeves, 'The Development of Apocalyptic Thought: Medieval Attitudes', in C. Patrides and J. Wittreich (eds.), *The Apocalypse in English Renaissance Thought and Literature* (Manchester, 1984), 46.

[2] R. Rusconi, 'Millenarismo e centenarismo: tra due fuochi', *Annali della Facoltà di Lettere e Filosofia della Università degli Studi di Perugia*, 22, ns 8 (1984/5), 2. *Studi Storici Antropolici*, 51–64. See also, P. Brezzi, *Storia degli Anni Santi* (Milan, 1975), passim.

pacification of the people of Israel and the apocryphal *Libro dei Giubilei* foresees a final climax to the succession of jubilees when Israel will be purged of all faults and errors and the earth will be pacified forever.[3] The mood of the multitude of pilgrims flocking to Rome in 1300 reflected a sense of extraordinary and final times: 'Omnes ibant securi ad ipsam urbem', writes a chronicler.[4] The momentary illusion of peace in an Italy and a Rome torn by faction and strife created the chimera of a final eschatological tranquillity, and Rusconi points out that, as the prospect of recovering the earthly Jerusalem receded into impossibility, Rome as the New Jerusalem took its place in popular imagination.[5]

The aspirations of the 1350 Jubilee—in the midst of the 'captivity' at Avignon—appear in a much more ambivalent context and are voiced by reformers rather than by the Pope, Clement VI. Rome is seen even more sharply as the longed-for New Jerusalem when juxtaposed to the Avignonese Babylon. The vision is a revolutionary one, threatening the 'establishment' rather than enhancing its glory. Yet Clement responded to pressure by anticipating the Jubilee in a Bull dated as early as 27 January 1343.[6] In this year Cola di Rienzo wrote from Avignon to the Senate and People of Rome proclaiming his hope in the return of the Pope to bring joy and salvation to the whole world:

Let the mountains exult in your circuit, let the hills be clothed with joy . . . Rise again, O Roman city, for so long fallen prostrate; put off your sad vesture of widowhood; clothe yourself in bridal purple; adorn your head with the diadem of freedom . . . take again the sceptre of justice: and encompassed and renewed by all virtues, as a bride adorned to please, show yourself to your spouse. . . . Behold indeed the most clement lamb of God who confounds sin, that is, the most holy Roman pontiff, father of the City, husband and lord of his bride, moved by the cries, complaints and lamentations and having compassion on its misfortunes, calamities and ruin, . . . offers mercy and grace to you for the renewal of the City itself, the glory of its people and the joy and salvation of the whole world, and promises universal redemption.[7]

But Clement's resolution wavered and the Bull remained unpublished. Between 1346 and 1347 Petrarch wrote three passionate sonnets drawing on the imagery of the Apocalypse in a powerful juxtaposition of Babylon/

---

[3] Rusconi, 'Millenarismo', 53–4.
[4] Quoted ibid. 54, from *Annales veteres Mutinenses*.
[5] Ibid. 51–2.
[6] Brezzi, *Anni Santi*, 40.
[7] K. Burdach and P. Piur, *Vom Mittelalter zur Reformation* (Berlin, 1912–29), II. iii. 5–6.

Avignon and Jerusalem/Rome.[8] It is in the New Jerusalem that the imminent age of gold which he announces must dawn. Also in 1347 Cola, by now Tribune of Rome, pressed the claims of the Jubilee on the Pope and dreamed of a universal pacification and an Age of the Spirit, beginning in Rome at the time of the Jubilee.[9] In 1349 St Bridget called on Clement VI to return to Rome and preach the year of salvation there.[10]

The Bull proclaiming the Jubilee was finally published on 18 August 1349. But Clement did not return to Rome. Conditions in the Holy City, stricken by plague and earthquake, were indeed terrible. But the very catastrophes of the time heightened the religious aspirations which were focused on the New Jerusalem. Numbers were certainly exaggerated by chroniclers but the contemporary picture of *tutta la Christianitade* pressing towards Rome *senza impedimento alcuno*[11] conveys the magnetism of the dream which inspired the international throng. The shepherd was absent but the flock gathered of its own accord 'in questo felice tempo del giubileo'[12] and therein lay the threat to Pope and Curia.

That the Avignonese Popes were not impervious to the call of the visionaries and prophets can be traced in several ways in the following period. The Joachite, John of Roquetaillade, although imprisoned at Avignon for his prophesying, was listened to by the Curia there.[13] In 1365 the Infante Pedro of Aragon exhorted Urban V, on the strength of a vision, to return to Rome and reform the Church.[14] The influence of St Bridget's Revelations is evident both in Urban's fleeting return to Rome in 1367 and in the pressure on Gregory XI to remain in Rome in 1377. Most striking is the collection of prophetic writings which can be located in Avignon in the late fourteenth century, and, more particularly, in the papal library of Benedict XIII just after the turn of the century, which included works by Joachim of Fiore and John of Roquetaillade.[15]

The year 1400 also fell at a moment of such confusion for papal Rome that it called forth fervid dreams of a divine eschatological intervention to end the Schism and renew the Church. In 1399 a widespread movement of white-robed penitents, *i Bianchi*, appeared in northern Italy, pressing

[8] F. Petrarca, *Rime*, ed. G. Bezzola (Milan, 1976), 299–301.

[9] Rusconi, 'Millenarismo', 55.

[10] Quoted ibid. 56.

[11] Quoted Brezzi, *Anni Santi*, 43.

[12] Ibid. 44.

[13] On Jean de Roquetaillade, see J. Bignami-Odier, *Études sur Jean de Roquetaillade* (Paris, 1952), rev. edn. *Histoire littéraire de la France*, 41 (Paris, 1981), 75–284.

[14] R. Rusconi, *L'Attesa della fine. Crisi della società, profezia ed Apocalisse in Italia al tempo del' grande scisma d'Occidente (1378–1417)* (Studi storici, 115–18; Rome, 1979), 23.

[15] Ibid. 21–3.

towards Rome and the fulfilment of a vision.[16] It was motivated partly
by a strong penitential appeal but equally, in many cases, by ardent
apocalyptic expectations of a drama to be played out in the Holy City.
Transalpine elements, some perhaps from the Avignonese obedience,
joined with the Italian crowds. According to the chief chronicle sources a
leader emerged, *un santo uomo grande*, and a mysterious book became the
symbol of the movement. Luca Dominici reports thus:

And they say that he who will be the new pope is among them, for they say that
soon, by the grace of God, the present schism will be ended. There is a book
which can only be opened on the altar of St. Peter's at Rome and which is said to
be about life c. 1400. For they say that in 1400 there will be peace in all the
world and the Church of Rome will be in great triumph and, stripped of its
(former) pastor, it will have new pastors and a new style.[17]

Both the *stolae albae* of the *Bianchi* and the theme of the mysterious book
carry overtones of the Apocalypse and the extent to which the movement
had caught up the prophecies which were circulating at that time is
recorded by another chronicler, Sercambi:

And note that in the Apocalypse of St. John . . . are these words which were
posted on the doors of the principal church of Bologna and elsewhere: throughout the world a multitude of peoples, clothed in white and shining garments,
crying grant us peace, and mercy, grant us, O Lord, and then justice and peace
descending from the sky will embrace and truth and peace arise on the earth, the
true pastor will be recognised by all, the just king will arise and there will be
peace and mercy over all the earth.[18]

All this chimes with a prophecy which was circulating, falsely attributed to the Vallombrosan hermit, Giovanni delle Celle: 'The time is at
hand when there will be a revival of my Church. This will be around
1400 or a little after. And there will be a revival in all the world and
increase of my faith.'[19] The concentration of eschatological expectation
on Rome could be highly embarrassing for an insecure Roman pontiff.
The ambivalence of the Curia in Rome over the handling of these
outbursts of apocalyptic fervour is well exemplified in the uncertain
attitude of the Roman Pope in 1400. Boniface IX was at first unwilling
to be involved in any celebration and no Bull to initiate the Jubilee was
officially promulgated. But he did grant the pilgrims pardon and bene-

---

[16] Ibid. 101–8, 204–19; id., 'Millenarismo', 57–8.
[17] Rusconi, 'Millenarismo', 58.
[18] Rusconi, *L'Attesa*, 215.
[19] Rusconi, 'Millenarismo', 56–7.

diction, while seeking to channel their enthusiasm into strictly penitential practices which did not threaten authority.[20]

The 1450 Jubilee proclaimed by Nicholas V was the first since Boniface VIII's to be securely in Papal hands. It was a triumph in pilgrim numbers and financial contributions, despite the tragedy of the crushed mob on the bridge of Sant' Angelo. By this time the medieval image of prophetic Rome could be set alongside the humanist veneration for antique Rome. One diarist records visiting the four main pilgrim churches in the morning and searching out the classical antiquities later in the day.[21] Yet the medieval expectation that the climax of the End-Time would be focused on Rome did not disappear. During the Jubilees of 1475 and 1500, firmly under Papal control as they were, radical prophecies inherited from previous centuries were muted, but they were never far below the surface.

Thus Renaissance Rome inherited a tradition in which high points of celebration at the key milestones of time drew large crowds and built up a tremendous force of popular emotion and visionary aspiration. The confidence and imperial ambition of a Boniface VIII could call forth and control such an emotional tide. In succeeding times of trouble Popes hovered between the desire to harness these high occasions to the glory of their office and fears of the radical implications of eschatological visions. The later fifteenth-century Jubilees appeared to resolve this dilemma satisfactorily but the question still remained: what attitude should the Curia in Rome take towards these popular prophetic forces? They could place the Roman pontiffs at the centre of the eschatological vision of peace or they could threaten them with radical alternatives.

Nowhere is this dilemma more clearly seen than in the emergence of the prophetic figure of the Angelic Pope. This had its origins in Joachim of Fiore's theology of history.[22] Joachim, a Calabrian abbot of the twelfth

[20] Brezzi, *Anni Santi*, 57–8.

[21] Giovanni Rucellai, quoted Brezzi, *Anni Santi*, 72, with the comment: 'Roma antica e Roma Christiana, fede e cultura, coesistevano in parfetta armonia.'

[22] On Joachim of Fiore's theology of history, see M. Reeves, *Influence of Prophecy in the Later Middle Ages: A Study in Joachimism* (Oxford, 1969), 17–21, 139–41, 395–7; id., 'The Abbot Joachim's Sense of History', in *1274 Année Charnière: Mutations et Continuités, Colloques Internationaux du Centre National de la Recherche Scientifique*, 588 (Paris, 1977), 781–96; H. Mottu, *La Manifestation de L'Esprit selon Joachim de Fiore* (Neuchatel and Paris, 1977); B. McGinn, *The Calabrian Abbot: Joachim of Fiore in the History of Western Thought* (New York and London, 1985), 99–203; Joachim of Fiore, *Liber de Concordia Novi ac Veteris Testamenti*, ed. E. R. Daniel, *Transactions of the American Philosophical Society*, 73, pt. 8 (Philadelphia, 1983), xxxv–xlii; B. McGinn, 'Joachim of Fiore's *Tertius Status*: Some Theological Appraisals', *L'Età dello Spirito e la Fine del Tempi in Gioacchino da Fiore e Gioacchinismo Medievale—Atti del II Congresso Internazionale di Studi Gioachimiti* ed. A. Crocco (S. Giovanni in Fiore, 1986), 219–36.

century, believed that the activity of the Triune God could be traced through three main *status* (stages) of history: the work of Father, Son, and Spirit. The fullness of the third *status* of the Spirit was yet to come. First the Church must suffer the tribulation of the worst Antichrist, after which, perhaps only briefly, all Christians would enjoy a period of tranquillity, illumination, and liberty in the Age of the Spirit. I think there is no doubt that for Joachim the reformed and renewed church of the third *status*, the *Ecclesia Spiritualis*, was always the *Ecclesia Romana*, the Latin Church of St Peter.[23] His indications concerning the institutions of the third *status* are not clear-cut, but in at least five passages of his *Liber de Concordia* he points to a renewed and triumphant papacy, figurally prophesied in the Old Testament characters Joseph, David, Zorobabel, Mordecai, and Judas Maccabaeus.[24] But the new life of the Age of the Spirit was symbolized for Joachim in the contemplative figure of St John, rather than in the active life of St Peter.[25] Herein lay the seeds of a revolutionary idea.

The pseudo-Joachimist works of the thirteenth century express a more radical attitude towards the Church of Rome. Disaster, figured in the fall of Eli and his sons, must bring retribution on the wealthy and worldly Church.[26] Some kind of revolutionary action either by the Papacy itself or by others towards it, begins to appear in the programme for the future: Peter must cast off his coat and plunge naked into the waves; the Church must 'fall among thieves'.[27] Yet, although the more extreme Joachite groups of the later thirteenth and fourteenth centuries came to expect a complete revolution in which the Roman Church would be totally superseded by the *Ecclesia Spiritualis*, the remarkable thing is that the Abbot's devotion to Rome was not completely submerged in a tide of eschatological radicalism. Like the Ark of Noah—a figure much used in pseudo-Joachite works—the Church of St Peter must ride the waves.[28] So the strong desire to hold together faithfulness to Rome with the prophetic dream of a radically purified institution is finally crystallized in the extraordinary figure of the Angelic Pope.

Hints of this appear in the pseudo-Joachimist works of the thirteenth

---

[23] For a more radical view, see B. McGinn, 'The Abbot and the Doctors', *Church History*, 40 (1971), 34–5, modified in id., *Calabrian Abbot*, 191–2; Mottu, *La Manifestation*, 47–8, 136–40.

[24] Joachim of Fiore, *Liber Concordie Novi et Veteris Testamenti* (Venice, 1519), fos. 89ʳ, 92ᵛ, 56ʳ, 122ᵛ, 132ᵛ.

[25] See references in Reeves, *Influence*, 139–41, 395–6.

[26] *Super Hieremiam Prophetam* (Venice, 1516), preface.

[27] Ibid., fos. 3ᵛ, 10ʳ.

[28] Ibid., fo. 38ʳ.

century. At one point the *Super Hieremiam* speaks of an expected *pastor bonus* and the later *Oraculum Cyrilli* prophesies his coming under the figure of an *ursus mirabilis*.[29] A prophecy of the Angelic Pope was perhaps already in circulation, for in 1267/8 Roger Bacon, addressing the Pope, speaks of a prophecy forty years old concerning a Pope to come who would purge the Church and bring about the return of the Greeks to the Latin obedience and the conversion or destruction of all infidels: 'et fiet unum ovile et unus pastor.' In 1272 he referred again to 'unus beatissimus papa qui omnes corruptiones tollet ... et renovetur mundus et intret plenitudo gentium et reliquiae Israel convertantur.'[30] We catch another echo from Salimbene who records a circulating verse at the time of Gregory X's death in 1276 which describes a Pope of angelic life who would carry out a programme of reform.[31]

The Angelic Pope, or perhaps a series of angelic popes, appears fully fledged in the *Vaticinia de summis pontificibus*, a sequence of fifteen prophecies with pictures, in the early fourteenth century. The texts and figures are actually drawn from a set of Byzantine prophecies known as the Oracles of Leo the Wise.[32] Whoever were the authors of the Pope Prophecies, the discovery and adaptation of the Leo Oracles was extraordinarily opportune. A succession of emperors leading to a mysterious revolution and transformation of the office provided a strikingly suitable model for a papal succession with imperial overtones which yet reached its climax in a hermit's ascent into the chair of St Peter and a succession of angelic pontiffs. Continuity and revolution are wedded together. The earlier part of the sequence, which begins with Nicholas III, supplies the opportunity for strong denunciations of papal worldliness and corruption, while the 'break', when a voice from heaven calls upon the electors to seek the hermit in the rocks on the most westerly of the seven hills of Rome, is a revolutionary one. Yet the office is not overthrown: a Pope, duly elected, ascends the throne of St Peter.

Associated with the *Vaticinia* is the *Liber de Flore*[33] in which for the

[29] *Oraculum Cyrilli*, ed. P. Piur, in K. Burdach and P. Piur, *Vom Mittelalter zur Reformation*, 2, pt. 4, Appendix (Berlin, 1912).

[30] Roger Bacon, OM, *Opus Tertium*, ed. J. Brewer, *Rolls Series, Opera Inedita*, 86; *Compendium studi philosophiae*, 402.

[31] Salimbene, *Cronica*, *MGHS* 32, 492–3.

[32] On the *Vaticinia de Summis Pontificibus*, see H. Grundmann, 'Die Papstprophecien des Mittelalters', *AK* 19 (1929), 77–159; M. Reeves, 'Some Popular Prophecies from the Fourteenth to the Seventeenth Centuries', in G. J. Cuming and D. Baker (eds.), *Popular Belief and Practice* (Cambridge, 1972), 107–34; R. Lerner, 'On the Origins of the Earliest Latin Prophecies: A Reconsideration', *MGHS* 33 (Hanover, 1988), 611–35.

[33] On the *Liber de Flore*, see H. Grundmann, 'Die Liber de Flore', *HJ* 40 (1929), 33–91; Reeves, 'Popular Prophecies', 115–16.

first time another figure of prophecy, the Second Charlemagne, appears as the partner of a sequence of four Angelic Popes in the work of pacification and renovation. The first angelic, a hermit *pauper et nudus* who will be sought out and crowned by an angel, will form an alliance with the 'generosus rex de posteritate Pipini' and together they will bring the Greek Church back to obedience, recover Jerusalem and purify the Church. The Roman See will renounce all temporal wealth and a General Council will ordain that the clergy shall live on bare necessities. The fourth Angelic Pope will preach throughout the world until in Palestine he is met with palms and songs by the two barbarian peoples, Gog and Magog. Then he will rule the world, as Christendom, until the End.

The political programme of the *Liber de Flore* was developed by the French Joachite, Jean de Roquetaillade (Rupescissa), in the series of prophetic works which he wrote in prison at Avignon in the mid-fourteenth century. Of these the *Vade mecum in tribulatione*,[34] composed in 1356, had a wide circulation in Italy and elsewhere. In brief, the programme for the future which Roquetaillade sets out envisages a period of terrible catastrophes and tribulations to be inflicted on the Church and on the world from 1360 to 1365. An oriental Antichrist, a false Messiah, and an occidental Antichrist, a heretic emperor, will persecute the Church. Only a remnant of the faithful will survive in the Franciscan Order until, by divine intervention, a sovereign Pontiff, *reparator orbis*, appears at the beginning of the third *status*. Aided by a saintly King of France who, contrary to custom, will be elected emperor, the Angelic Pope will exterminate the Antichrists, banish avarice and pride from the clergy, destroy the law of Mahomet, and reform the whole world. The vision of peace which Roquetaillade creates includes the telling detail that the Pope will end the conflict between Guelfs and Ghibellines in Italy and establish the lands of the Church in such manner that in perpetuity the Pope will no more make war.

It is not surprising, therefore, that in 1378 an anonymous Florentine chronicler recorded prophecies by 'uno Frate minore' which are clearly a popularized version of the *Vade mecum*.[35] The prophecy of disaster which Roquetaillade had expected in the 1360s is now applied by the chronicler

---

[34] Pub. in *Fasciculus rerum expetendarum et fugiendarum, prout ab Orthuino Gratio . . . editus est Coloniae . . . 1535 . . . una cum appendice . . . scriptorum veterum . . . qui Ecclesiae romanae errores et abusus, detegunt et damnant . . . opera et studio Edwardi Brown* (London, 1690), ii. 496–507; see also Bignami-Odier, *Roquetaillade* (1952), 157–72.

[35] *Diario d'Anonimo Fiorentino dall'anno 1358 al 1389*, ed. A. Gherardi (Documenti di storia italiana, 6; Florence, 1876), 389–90.

to the uprising of the Ciompi in 1378. But beyond this he sees the drama of the *Riparatore di Christiani*. This Angelic Pope, whom the Florentine identifies with the mystical Elijah who will return to 'restore all things', will reform the Franciscan Order and expel from the Church all worldly and avaricious priests. This prophecy seems to contain echoes of the expectations harboured in Fraticelli circles at that time. In Rusconi's view it represents not part of the popular political programme of that time, so much as a continuing eschatological expectation which interpreted contemporary events as signs of Last Things.[36] This was the mood of the Fraticelli, the remnant of the Spiritual Franciscan movement scattered through Tuscany and Umbria. Their expectations are to be found in the letters or short tracts they wrote.[37] Their radical condemnation of the Roman Church as the New Babylon was uncompromising, but, drawing on the *Vaticinia* and the *Liber de Flore*, they pinned their faith on the advent of a 'papa santo incoronato dagli angioli'.[38]

In the mid-fourteenth century a second set of fifteen *Vaticinia* was produced, probably in the Fraticelli milieu. Like the original set it consisted of 'portraits' starting once more with Nicholas III and containing sharp, recognizable indictments of the Orsini Pope and his successors. But this set ended on a pessimistic note, not with the angelic portraits, but with a fifteenth picture apparently representing Antichrist and bearing the caption: 'Terribilis es et quis resistet tibi?' The continuing attention paid to both sets of *Vaticinia*, as exemplified in the large number of manuscripts, came to a crisis point at the end of the fourteenth century for Urban VI (1378–89) was the fifteenth pope after Nicholas III. Was he therefore, according to the second set, the Antichrist? Had the vision of the Angelic Pope vanished? In the 1370s the Vallombrosan hermit, Giovanni delle Celle, was studying the later set of the *Vaticinia* closely.[39] Attempting to interpret them at the request of his friend Guido del Pallagio, he agonized over the meaning of the fourteenth and fifteenth prophecies. The fourteenth should apply to Gregory XI: was he the last Pope before Antichrist? The hermit was caught between his orthodox loyalty to the Church and his fear that a papal Antichrist was possible. Commenting on the fifteenth prophecy he writes:

---

[36] Rusconi, *L'Attesa*, 67.

[37] On the Fraticelli, see D. Douie, *The Nature and the Effect of the Heresy of the Fraticelli* (Manchester, 1932); Reeves, *Influence*, 212–14. For Giovanni delle Celle's polemic against the Fraticelli, see Rusconi, *L'Attesa*, 57–71,

[38] Rusconi, *L'Attesa*, 67.

[39] Ibid. 57–60; Reeves, *Influence*, 215–16, 253–4.

'This is the last wild beast, of terrible aspect . . . and this beast is Antichrist; and according to what someone said to me this will be another pope. In this matter one of two (alternatives) must be the case: either the book which I saw is corrupt, or someone has added his own opinion; therefore I wait to see.'[40]

The dream, however, could not be quenched. Sercambi records three prophecies[41] the first of which, beginning 'Luna auferetur, quando sol aurietur' is a popularized text from the *Vaticinia* adapted to point to an Angelic Pope expected to follow Pope Benedict XIII (Pedro de Luna). In the second prophecy an angelic voice commands the pope to relinquish his office to 'un uomo umile e quasi ignoto'. Here again there is an echo of the *Vaticinia* and also possibly of the famous election of the hermit Pope Celestine V in 1294, for the expected Angelic Pope is designated *Sextus*. It is therefore likely that this prophecy appeared in 1394 when the death of Clement VII opened up the possibility of ending the Schism. The third prophecy given by Sercambi concerns Antichrist.

Two solutions to the problem of how to extend the *Vaticinia* into the future which were offered in this time of stress bear eloquent testimony to the continued yearning for the ideal that underlay the confused religious attitudes and sharp partisanships of the Great Schism. The first, a short-lived one pointed out by Robert Lerner,[42] consisted of the addition, made in the time of the Avignonese Pope, Benedict XIII, of six prophecies portraying the three Popes after Urban VI (from both obediences), followed by three more predicting the Angelic Pope, Antichrist, and Last Judgement. The second, and more long-term, solution was to merge the two sets into one sequence of thirty. Here the affirmation of the expected *renovatio* emerges strongly. Whoever had the inspiration to put the two sets together placed the second set first so that the fifteenth, the figure of Antichrist, would apply to Urban VI who could be accepted as a menacing figure, though no longer the Antichrist, while the whole series now stretching into the future still culminated in the Angelic Pope portraits. Lerner gives convincing reasons for dating this solution to between 1410 and 1414 and for identifying the author as someone of the Pisan obedience.[43]

---

[40] Letter to Guido del Palagio, in F. Tocco (ed.), *Studii Francescani. Nuova biblioteca di letteratura storia ed arte* (Naples, 1909), iii. 426, quoted Rusconi, *L'Attesa*, 60.

[41] Rusconi, *L'Attesa*, 216–17.

[42] R. Lerner and R. Moynihan, *Weissagungen über die Papste. Vat. Ross. 374* (Zurich, 1985), 59.

[43] Ibid. 59–60.

This became the standard version of the *Vaticinia* in the Renaissance period. Lerner calculates that it is to be found in about fifty-five known fifteenth-century manuscripts, an unsurveyed number of later manuscripts, and at least twenty-four early modern printed editions.[44] Some were probably produced for propaganda purposes in the last days of the Schism and during the Council of Constance. Rusconi notes that Cardinal Pietro Corsini had a copy and also Cardinal Zabarella, a leading figure at Constance.[45] A number of fifteenth-century manuscripts are luxury copies almost certainly owned by important ecclesiastics and Italian aristocrats. Copies were annotated, circulated, and discussed. The humanist Aliotti interpreted the *Papalista* in a letter dated 1464[46] and Zenobio Acciaiuoli discussed them with Delphino, the General of the Camaldolesi who sought to get a copy from the Venetian, Domenico Mauroceno.[47] The persisting relevance of these prophecies, when Schism and Councils were past and the fifteenth-century Popes increasingly secure in their office, testifies to a continuing deep concern for the future of the Papal headship and an unquenchable expectation of a blessed climax to the Roman pontificate.

But Rome was also the doom-laden city. The wicked characters congregated there as well as the blessed ones. The Apocalypse gave Biblical warrant for this; the seven heads of the Beast on which the Great Whore of Babylon sits are 'seven mountains, on which the Woman sitteth' (Apoc. 17: 9). Joachim had expected a double manifestation of Antichrist, one a political tyrant, but the other a pseudo-pope whom he saw figured in the Beast ascending from the earth of Apocalypse 13: 11: 'The beast which comes up out of the earth will have a certain great prelate similar to Simon Magus, like a universal pontiff over the whole earth.'[48] Among later Joachites this figure becomes the *Antichristus mysticus* or *mixtus*[49] who will appear either before or in conjunction with the political tyrant, the 'open' Antichrist. The *Super Hieremiam*, for instance, expects

---

[44] Ibid. 61.

[45] Rusconi, *L'Attesa*, 56.

[46] Lerner, 'Origins', 63.

[47] E. Martène and V. Durand, *Veterum Scriptorum et Monumentorum . . . Amplissima Collectio* (Paris, 1724–33), iii, cols. 1152–3. On Mauroceno, see Reeves, *Influence*, 343.

[48] Joachim of Fiore, *Expositio in Apocalypsim* (Venice, 1527), fo. 168ʳ. On Joachim's Antichrists, see R. Manselli, 'Il problema del doppio Anticristo in Gioacchino da Fiore', in *Geschichts-schreibung u. geistiges Leben im Mittelalter*, ed. K. Hauck and H. Mordek (Cologne and Vienna, 1978), 427–49; B. McGinn, 'Angel Pope and Papal Antichrist', *Church History*, 47 (1978), 155–73; R. Lerner, 'Antichrists and Antichrist in Joachim of Fiore', *Speculum*, 60 (1985), 553–70.

[49] On 'mistico' or 'mixto', see I. Vázanez Janeiro, OFM, 'Anticristo mixto, Anticristo mistico, Varia fortuna de dos expressiones', *Antonianum*, 63 (1988), 522–50.

a pseudo-Pope in the seat of Peter, identified in one passage as the *pseudo-papa futurus Herodes*.[50] The Franciscan, Petrus Ioannis Olivi, feared the advent of the *Antichristus mysticus*, the false Pope who would betray the Franciscan ideal.[51] Although Olivi avoided identifying it with the Roman Church, the 'carnal church' which he denounced must, like the synagogue before it, be rejected and destroyed.[52] Amongst his radical followers, however, this identification was clearly made and Pope John XXII, seen as the persecutor of the Franciscan *zelanti*, was repeatedly identified as the *Antichristus mysticus*.[53] Even in the face of the Inquisition the extremists did not waver in their expectation that the carnal Roman Church would be rejected and the *pauci viri spirituales* would form the foundation of the *ecclesia tercii status*.[54]

Rome becomes the seat of the final cosmic conflict with evil. In the 1250s the Franciscan, Thomas of Pavia, had recorded the vision of a 'certain religious' who dreamt that the Lateran Palace in Rome had collapsed around him and a great voice cried: 'Ecce Nero resuscitatus est et ipse est Antichristus'.[55] At the end of the century Robert D'Uzès, a Dominican, had a vision of Rome thickly covered in dust awaiting its doom.[56] In the first set of *Vaticinia* menacing captions are addressed to Rome: 'Heu, heu miser civitas' and 'Veh tibi civitas septem collis'.[57] In the second set, as we have seen, the final prophecy, probably intended as the *Antichristus mysticus*, bears the caption: 'Terribilis es et quis resistet tibi?'[58] Whoever he might be, Antichrist seems to have been clearly linked with the Papal office. In his commentary on the pseudo-Joachimist *Oraculum Cyrilli* John of Roquetaillade foretold that the political Antichrist would be elected Emperor in Rome and with him the pseudo-Pope, the Beast from the land, would take power.[59] Juxtaposed to this

---

[50] *Super Hieremiam*, fo. 35ʳ. See R. Moynihan, 'The Development of the "Pseudo-Joachim" Commentary "Super Hieremiam": New Manuscript Evidence', *Mélanges de l'École Française de Rome, Série Moyen Âge et Temps Moderne*, 90, 1 (1986), 109–42, for a new examination of the authorship of this work.

[51] Reeves, *Influence*, 198–9; D. Burr, 'The Persecution of Peter Olivi', *Trans. of the American Philosophical Society*, NS 66, pt. 5 (1976), 21–4.

[52] Reeves, *Influence*, 198–9; Burr, *Persecution*, 20–4.

[53] Reeves, *Influence*, 205.

[54] P. a Limborch, *Hist. Inquisitionis, cui subjungitur Liber Sententiarum Inquisitionis Tholosanae 1307–1323* (Amsterdam, 1692), 53.

[55] *AFH* 16 (1923), 25–6.

[56] *AFP* 25 (1955), 274–5.

[57] These are usually numbers eight and ten in the sequence.

[58] See above, p. 11.

[59] See the summary of Roquetaillade's commentary on the *Oraculum* in Bignami-Odier, *Roquetaillade*, 53–109.

pair, of course, were the holy Pope, *corrector et reparator*, and his ally, the King of France. In his final work, the *Vade Mecum* (1356), there are two political Antichrists, as we have seen. The occidental one will be a heretical emperor, a new Nero. He would, according to Roquetaillade, establish his seat in Rome between 1362 and 1370 and the city would tremble at his presence. Finally it would be destroyed.[60]

But the political figure to whom was allotted the role of chastising or destroying Rome was an ambivalent one. He could be the open Antichrist, the new Nero, but if Rome had become the New Babylon there was a part to be played by a just chastiser. The evil chastiser naturally dominates the Italian programmes. He first appears in the *Super Hieremiam* and its associated writings where every kind of biblical symbol for wickedness is applied to the German monarchs who have always, according to these texts, oppressed the Church.[61] The seventh head of the apocalyptic Dragon becomes the Emperor Frederick II and then an expected 'Third Frederick'. One of the figures used to delineate the coming tyrant was taken from the Book of Daniel 8: 23: the *rex impudicus facie*, the king of fierce or shameless countenance.[62] We meet this figure again and again in Renaissance prophecy. But he doubles, as it were, with another character, the just chastiser. We meet the first idea of such a role in the heretical group associated with a Dominican, Arnold of Swabia, whose manifesto, *De correctione Ecclesiae*, called on the Emperor Frederick II to take up this task.[63] After Frederick's death a German prophecy was circulating which described a Third Frederick who would triumph over the papacy.[64] From this a German line of prophecy develops which is exemplified in the prognostications of Gamaleon at the time of the Great Schism.[65] Here the French king is cast for the evil part but in the Rhineland will arise the true King-Chastiser who will destroy Rome, transfer the papal seat to the Patriarchate of Mainz and reduce the clergy to apostolic poverty. This double role is most strikingly illustrated by the inversion of the title *Rex impudicus facie* into *Rex pudicus facie*, the modest or discreet king, yet still a righteous chastiser. This character appeared in a widely circulated Brigittine text.[66] St Bridget herself believed that a terrible chastisement of the Church was imminent. The

---

[60] Ibid. 165, 168.
[61] See references in Reeves, *Influence*, 306–7.
[62] *Super Hieremiam*, fos. 10^r, 18^v, 20^v, 45^v.
[63] Arnold of Swabia, *De correctione Ecclesiae*, ed. E. Winkelmann (Berlin, 1965).
[64] *MGHS* 24, 207.
[65] See the summary and references in Reeves, *Influence*, 332.
[66] See the summary and references in Reeves, *Influence*, 338–9.

executor, variously described as *Dux venturus, Arator, Venator*, appears to be an involuntary agent of God's judgement.[67] Among the popular oracles which began to circulate in her name sometimes the destruction of Rome is emphasized, as in one beginning 'O desolata civitas',[68] but the text of the *Rex pudicus facie* prophecy stresses the positive role of the godly chastiser: he will reform the state of the 'collapsed' Church and threaten the clergy; finally he will rule everywhere.

The role of the future emperor as both chastiser and 'reparator' is seen in several of the popular verse-prophecies circulating in Italy in the second half of the fourteenth and early fifteenth centuries.[69] Attributed to a variety of prophets, among them Joachim, Merlin, Stoppa dei Bostichi, and Tommasuccio da Foligno, they are in part an expression of political propaganda, yet in a land devastated by plague, famine, and war, they are also a haunting expression of a deep longing and a prophetic hope for a better world. In one, after stern chastisement, *un huom che rinnovelle* will make a universal peace.[70] In another, a German emperor must deal with the crisis in Church and State and then establish the blessed regime:

> All infidels will be converted
> And, clad in rough cloth,
> Will live for ever in poverty
> Without private possessions.[71]

Again, in a fifteenth-century prophecy associated with the Emperor Sigismond, the work of the Last World Emperor culminates in a vision of peace:

> And there will be tranquillity, peace and harmony
> as in the days of the good Octavius
> through one single man who will give judgement;
> he will be a most prudent and reasonable man
> and universal lord of Christians. He will maintain
> the Roman empire, ordaining peace to Christians
> and all wars will be forgotten.[72]

---

[67] Ibid. 338.

[68] Telesphorus, *Liber de magnis tribulationibus (Libellus)*, fo. 5ᵛ.

[69] On verse prophecies, see A. Messini, 'Profetismo e profezie ritmiche d'inspirazione gioachimito-franciscana nei secoli XIII, XIV e XV', *Misc. Franc.* 37 (1937), 39–54; 39 (1939), 109–30; Rusconi, *L'Attesa*, 143–63.

[70] Messini, 'Profetismo', 37. 48 n. 1; Rusconi, *L'Attesa*, 154.

[71] Rusconi, *L'Attesa*, 158.

[72] Messini, 'Profetismo', 39. 127–8; Rusconi, *L'Attesa*, 161–2.

In these verses the agent appears most often to be the Imperial one but
through his office the work of renovation lifts the Papacy to a new level
of spirituality:

> Then the Church will be adorned with pastors
> humble and holy as the first founders. . . . [73]

In many of these verses we meet a type of Ghibelline Joachimism which,
despairing of sufficient will power in the papacy to reform itself, looks
for the saviour in Dantesque terms, in the Imperial office.

In the early 1380s there appeared the most famous of the late medieval political prophecies, usually known as the Second Charlemagne
prophecy.[74] It is in essence a revival of the ancient Last World Emperor
prophecy, as appropriated by the French monarchy. In one of its earliest
versions it has already acquired a Joachimist slant, for at the culmination of his victorious career 'Karolus filius Karoli ex natione illustrissimi
Lilii' will be crowned by the Angelic Pope. Here the text follows the
holy partnership developed by Roquetaillade. But this contrasts sharply
with the preceding political programme of triumphant conquests. The
Second Charlemagne will destroy the tyrants of his own realm, subdue
all the neighbouring peoples and then, advancing on Italy, 'Romam cum
Florentia destruet et igne comburet'. The rest of the programme is
couched in the eschatological terms of a Last World Emperor in which
Rome will be superseded by Jerusalem. There, at the heart of the text,
is the uncompromising judgement of Rome by God's appointed 'rod
of anger'. Florence had been associated with Rome in this fate by
Roquetaillade and in this text they are finally linked in this doom.
The ubiquitous appearances of the Second Charlemagne prophecy in
the next two centuries ensured that the future of Rome remained a
menacing one, as it were, on the reverse side of the coin.

The authorship of this prophecy remains in doubt but the early version quoted above is closely associated with the *Libellus* of Telesphorus
of Cosenza.[75] This formed an updating of Joachimist prophecies and
followed Roquetaillade in predicting a schism between a pseudo-Pope
and the Angelic Pope. According to Emil Donckel who studied the
manuscripts of the *Libellus* in detail, Telesphorus' prophecies originated
in the period 1356 to 1365, although the book only took its final form

---

[73] Messini, 'Profetismo', 39. 122; Rusconi, *L'Attesa*, 146.
[74] For the earliest text of the Second Charlemagne prophecy, and its provenance, see
Reeves, *Influence*, 328–31.
[75] Ibid. 328.

between 1378 and 1390.[76] If this dating is correct, Telesphorus' fore-reading of the future was most apt. The last and most monstrous schism is to be God's final judgement on the carnal church. The agents of destroying judgement will be a German king—in Telesphorus' original version a Third Frederick—and a pseudo-Pope, the *Antichristus mysticus*, appointed by him. This antipope will crown the wicked Frederick Emperor and then the stage is set for the final conflict between the cosmic forces of good and evil. This *ultima persecutio* will be heavier and more inhuman than any before. But an angel will summon forth the Angelic Pope and the French king, *generosus rex de posteritate Pipini*, will arise to aid him. Together they will battle with and overcome the forces of evil. The Angelic Pope will crown the French king Emperor and the holy partners will proceed with the work of *renovatio*. In the final climax, however, Telesphorus' emphasis is on the spiritual power: 'Et Romanus pontifex in spiritualibus dominabitur a mari usque ad terminos terrarum et erit reformatio status ecclesie.'[77]

The Schism itself naturally intensified the apocalyptic dream. The stage was now set with two pairs of characters in conflict: the 'bad' and the 'good'. This scenario was peculiarly apt for the terrible period when Christendom was torn apart. Incitement to identify the Antichrists was strong. When the Avignonese Pope, Clement VII, appeared in Naples in 1379 an unknown Florentine chronicler recorded: 'E tutto il popolo di Napoli gridava: "Muoia L'Anticristo".'[78] Petrarch's passionate sonnets had already imprinted the image of Babylon on Avignon, while in 1377 Coluccio Salutati had branded the Avignonese prelates as Babylonish satraps.[79] Salutati was in touch with Giacomo Palladini, Bishop of Florence, to whom he wrote in 1403 concerning eschatological interpretations in which the bishop was deeply interested.[80] At the end of his work, *Consolatio Peccatorum*, Palladini applied Biblical prophecies to the present times.[81] The world is invaded by pseudo-Christs and pseudo-prophets, focused in the papal office—and here, since his work is dedicated to Urban VI, the Avignonese Clement VII is probably intended. Palladini makes a clearly Joachimist statement on the expected third age, but the forces of evil are such that only through a Second

---

[76] See E. Donckel, *AFH* 26 (1933), 29–104, but his dating is now questioned. See also Reeves, *Influence*, 325–8, 423–4.
[77] Telesphorus, *Libellus*, fo. 39ʳ, quoted Reeves, *Influence*, 424.
[78] Rusconi, *L'Attesa*, 54.
[79] Ibid. 99.
[80] Ibid. 120.
[81] Ibid. 122.

Coming of Christ who will Himself be the *rex novus* can the *renovatio mundi* take place. On the other side, the Dominican preacher, Giovanni Dominici, attacks an anonymous writer who identifies the Dragon with seven heads, the Antichrist, with the Roman pontiff.[82]

Once the Great Schism was over the programme for the Last Age was pushed ahead in time. The *Libellus* of Telesphorus is found in many manuscripts of the fifteenth century. With the *Vaticinia* and the Second Charlemagne prophecy it ranked among the most popular prophetic texts and the vivid little pictures with which it was sometimes illuminated brought home to the imagination the still expected scenario for the future: schism, persecution, and conflict, the holy partnership of Angelic Pope and French king, final apotheosis for the Roman pontiff. But alongside it was the alternative future: destruction of Rome and transfer of spiritual authority to Jerusalem or elsewhere. What was Rome to expect? At every hint of political crisis apocalyptic preachers arose to heighten the tension and a *frisson* ran through Italy.

Two occasions in particular brought forth a concatenation of prophecies in the fifteenth century. The first was the appearance of Frederick III in Rome seeking coronation as Emperor in 1452. It was his name rather than his person that caused alarm for, as we know, a Third Frederick had, in the prophetic vocabulary, long been a sinister figure for Rome. We get an amusing light on the reactions of Romans from Aeneas Silvius Piccolomini who tells us that some tried to persuade Pope Nicholas V that this was indeed that future intolerable scourge who would punish Rome.[83] Nicholas, it seems, was in two minds but according to another reporter, the Abbot of Tritheim, decided in a supper-time conversation to raise the question of Frederick's evil role directly with him.[84] Frederick's reply—that his own intentions were good but that he could not evade the role ordained for him by God—shows us how seriously prophecy was taken.

The second occasion was the extraordinary conjunction of the Savonarolan movement in Florence, overpowering the politics of the city with its prophecies, with the invasion of Charles VIII of France in 1494 who came armed with the Second Charlemagne prophecy in which Florence and Rome were together doomed.[85] This crisis forms

[82] Ibid. 110.

[83] Aeneas Silvius Piccolomini, *Historia rerum Friderici III imperatoris*, ed. J. Schilter, *Scriptores rerum Germanicarum* (Strasbourg, 1687), 45. See also Reeves, *Influence*, 334 n. 4.

[84] J. Trithemius, *Chronicon Hirsaugiense* (St Gall, 1690), ii. 423.

[85] For Savonarola, see. P. Villari, *Life and Times of Girolamo Savonarola*, trans. P. Villari (London, 1838); J. Schnitzer, *Savonarola*, trans. E. Rutili (Milan, 1931); R. Ridolfi,

the curtain-raiser to the prophetic dramas studied in this book. Without embarking on a detailed analysis here, four points may be made,
all of which, in one sense or another, foreshadow features which charac-
terize the prophetic themes of the Cinquecento. First, the drama was
played out on two levels. The actual course of events was obviously
determined by political forces. But simultaneously there existed a second level of belief in the force of prophecy. Everyone listened to the
prophetic voices. They had great power. 'Le langage lui-même utilisé
pour décrire la prodigieuse aventure de Charles VIII participe de l'irra-
tional. C'est un langage passionel.'[86] The interworkings of prophecy
and politics in this episode are subtle to trace. Savonarola was defeated
in political terms but the convincing force of prophecy was not overthrown in men's minds. Indeed, the course of events confirmed belief
in prophecy: if God had not chosen the French king he would never
have achieved such success. He was the personification of the *flagellum
Dei*. But he had turned aside from God's purposes as preached by
Savonarola and disaster therefore struck.[87] The prophetic future of
Florence, Rome, and Italy, with all its ambivalence, remained an obsessive concern of the imagination. Secondly, in the Savonarolan movement we meet the confluence of two cultural streams: the medieval
prophetic tradition and the neo-Platonist humanism of Renaissance
Florence. Marsilio Ficino, formally welcoming Charles VIII to the city,
cites the myth of the Second Charlemagne.[88] One also hears a far-off
echo of Dante in his use of biblical messianic language to describe
Charles's mission: in the image of Christ he will be the one who cries
to the people: 'Come unto me all you who are heavy-laden.'[89] Marsilio
later withdrew his support but some of the humanists became *Piagnoni*.
Giovanni Pico della Mirandola was one of these and later Gianfrancesco.
Another disciple, Giovanni Nesi, wrote in 1496 a work, *Oraculo de novo
saeculo*, which is the strangest mixture of Neoplatonism, occult mysteries, and medieval prophecy.[90] In his union of philosophy and Chris-

---

La vita di Girolamo Savonarola (Rome, 1952); R. di Maio, *Savonarola e la curia romana*
(Rome, 1969); D. Weinstein, *Savonarola and Florence: Prophecy and Patriotism in the
Renaissance* (Princeton, NJ, 1970).
   [86] A. Denis, *Charles VIII et les Italiens: Histoire et Mythe* (Geneva, 1979), 151.
   [87] Ibid. 147–8.
   [88] M. Ficino, *Opera* (Basle, 1576), i. 960–1.
   [89] Quoted Denis, *Charles VIII*, 59.
   [90] A. Chastel, 'Antéchrist à la Renaissance', in E. Castelli (ed.), *Christianesimo e ragioni
de Stato, Atti del II Congresso Internazionale di Studi umanistici* (Rome, 1952), 183, describes the climax as 'l'exaltation typique du millénarisme piagnone, où revit en somme
le vieux joachimisme confiant dans le règne prochain de l'Esprit'.

tianity he envisages the eagle of philosophy nesting with Christ the phoenix. The nest is Florence and from it will be born the 'new race of fledglings which will spread throughout the world. Then Man will become heir to the secrets of all religions and mysteries.'[91] Nesi's vision is of the New Jerusalem which is also the Platonic Republic; it is the Joachimist Age of the Spirit and also the classical Golden Age.

Thirdly, the Savonarolan movement inspired at least two visual representations of the prophetic vision. One was Botticelli's *Mystic Nativity*. Its symbolism, with angels embracing men in the foreground and dancing in the sky, while the beaten devil crawls away, carries the Joachimist vision to a high point of ecstasy. In this case the prophetic meaning is made explicit in the well-known inscription on it.[92] Donald Weinstein has interpreted another, less well-known, picture by Botticelli in similar prophetic terms.[93] Here Florence is represented three times: under chastisement on the right, repentant in the centre, and glorious in the light of God's revelation on the left. As Josephine Jungić shows in two studies in this book,[94] the visual statement of prophetic expectation is also found in Renaissance Rome. Fourthly, the persistence with which Savonarola's message of chastisement to be followed by *renovatio* continued to be proclaimed by apocalyptic preachers, by groups of all classes, by self-announced Angelic Popes, kept apocalyptic expectation at fever heat. The wide diffusion of these expectations and the manner in which they gripped the imagination of such varied classes in society are emphasized by Ottavia Niccoli.[95] The focus is on Rome and its prophetic role and the first thirty years of the Cinquecento were characterized by an almost unprecedented wave of apocalyptic emotion.

[91] Giovanni Nesi, *Oracolo de novo saeculo* (Florence, 1497), sig. cvii[v].

[92] The inscription reads: 'I Sandro painted this picture at the end of the year 1500 in the troubles of Italy in the half time after the time according to the xith chapter of St John in the second woe of the Apocalypse in the loosing of the devil for three and a half years. Then he will be chained in the xiith chapter and we shall see him trodden down as in this picture.'

[93] D. Weinstein, *Savonarola*, 336–8.

[94] See below, Jungić, Ch. 17, 18.

[95] See below, Niccoli, Ch. 10.

II

THE PROLIFERATION AND CIRCULATION OF NEW PROPHECIES

Introduction

Research in the last thirty years has brought to light a literary activity which is characteristic equally of the later Middle Ages and the Renaissance period, namely, the collection, circulation, and invention of prophecies. Manuscript 'anthologies', particularly of the fourteenth and fifteenth centuries, have been unearthed in considerable numbers. There was no slackening of interest when print widened the circulation so remarkably. Manuscript anthologies actually survive into the age of printing: there is an example, dated 1500, which is predominantly medieval in its content but is prefaced by a 'Judgement of the learned on the authors'.[1] Even at the end of the sixteenth century three manuscript collections of pope prophecies at least have survived.[2] Among printed anthologies the *Mirabilis Liber* is best known.[3]

Individual prophecies circulated busily. Old favourites turn up again and again but the urge to meet new occasions with solemn judgement or to bring prophetic expectation forward into the immediate future could not be resisted. The manufacture of new prophecies accompanies the collection of old ones. But the authority of new prophecies often depended on the creation of a venerable origin. They must be either sup-

---

[1] The Hague, Bibl. Reg., MS 71 E. 44; see summary in Reeves, *Influence*, 540.
[2] See Reeves, 'Popular Prophecies', 128–31.
[3] For the *Mirabilis Liber*, which went through a number of editions, see J. Britnell and D. Stubbs, 'The *Mirabilis Liber*: Its Compilation and Influence', *JWCI* 49 (1986), 126–49.

posedly dug out of some ancient book, or attributed to a saintly figure, or 'discovered' miraculously as the result of a vision. Roberto Rusconi has studied examples of prophecies produced 'ex quodam antiquissimo libello' in late manuscript and early printed form.[4] One of the most widely circulated of these 'new' prophecies was authenticated both by a saintly attribution and a mysterious discovery. This was the prophecy of S. Cataldo 'quae in libro plumbeo scripta et intra columnam inclusa ut ego vidi Tarenti inventa est A.D. 1402.'[5] It was said to have been revealed in a dream vouchsafed to a deacon. Its instant popularity was no doubt due in part to the timeliness of its strange and solemn warning of calamities shortly to come. By hindsight it seemed to sound the death knell of Italian liberty. Giorgio Tognetti's investigation of its background leads him to the conclusion that its author was a Franciscan, Francesco of Aragon, and that it was designed to warn King Ferrante of Naples of the approaching judgement of God.[6] Reports claimed that its message afflicted the king with such melancholy that it contributed to his death in 1494.[7] Already in 1492 it had appeared in Rome, described by Infessura.[8] It was later commented on and circulated by Pietro Galatino.[9] A second prophecy in circulation in the early sixteenth century claimed an invented author, a non-existent Carthusian, Albert of Trent.[10] In fact it was clearly composed by a Florentine *Piagnone* to justify Savonarola's prophecies as part of a developing divine programme. From *post eventum* evidence it can be dated to 1503. It looks forward to *renovatio* after tribulation and the appearance of an Angelic Pope among the poor.

As Rusconi shows in his essay later in this book, an important circulator of prophecies in Rome was the Franciscan, Pietro Galatino.[11] Whether he was also an inventor it is impossible to say, but he certainly seems to have been responsible for bringing to light at least two pro-

---

[4] R. Rusconi, '"Ex quodam antiquissimo libello"': La tradizione manoscritto delle profezie nell'Italia tardo medioevale: dalle collezioni profetiche alle prime edizioni a stampa', in W. Verbeke, D. Verhelst, and A. Welkenhuysen (eds.), *The Use and Abuse of Eschatology in the Middle Ages* (Louvain, 1988), 441–72.

[5] So wrote Pietro Galatino; see MS. Vat. Lat., 5569, fo. cxliii ᵛ.

[6] G. Tognetti, 'Le Fortune della pretesa profezia di san Cataldo', *Boll. ISI*, 80 (1968), 273–317.

[7] Ibid. 288.

[8] Stefano Infessura, *Diario della Città di Roma*, ed. O. Tommasini (Rome, 1890), 1240.

[9] Tognetti, 'Le Fortune', 308–11.

[10] D. Weinstein, 'The Apocalypse in Sixteenth-Century Florence: The Vision of Albert of Trent', in A. Molho and J. Tedeschi (eds.), *Renaissance Studies in Honor of Hans Baron* (Florence, 1971), 313–31.

[11] See below, Rusconi, Ch. 8.

phecies. The *Vaticinium Romanum* is found among his writings.[12] It purported to have been revealed *c.*1160 and deals with all the Popes from Alexander III down to Clement VII, passing thence from a detailed *post eventum* account to a prophetic mélange on the Angelic Pope and the future glory of the Church. This surely reveals the hand of Galatino, particularly in its citations from the pseudo-Amadeite *Apocalypsis Nova* which is dealt with in the following essay. In 1525 Galatino sent a copy of the *Vaticinium Romanum*, with his own *explicatio*, to Alessandro Spagnuoli of Mantua who, replying, mentions three other prophecies he will send to Galatino to publish.[13] The *Vaticinium Romanum* cites a *Vaticinium Montis Gargano*, which also appears among Galatino's writings, with the names of Popes from Paul II to Clement VII written in the margin, followed once more by the *Pastor Angelicus*.[14]

Some prophets, however, were bold enough to circulate prognostications under their own authority. In 1514 the prophecy of a self-styled Angelic Pope in Rome, Fra Bonaventura, received great publicity and was communicated in a letter to Paris, which was printed in the *Mirabilis Liber*.[15] He was imprisoned for a time but posed no real political threat. Again, in 1527 Bernardino da Parenzo published a letter to Isabella d'Este with a prophecy on the world-shaking events shortly to come which he said he had received in a vision.[16] He also sent this letter to the Pope and other important people.

Among the letters of Benivieni, a disciple of Savonarola, a prophecy has come to light which has a Florentine background and was probably composed *c.*1511.[17] It was supposed to have been taken from a book of oracles by one 'Obetut' and copied by a 'frate Abbias', neither of whom can be identified. But the name of the blessed Amadeus appears prominently on the covering folio and Olivia Pugliese who unearthed this prophecy argues that it has a close connection with the pseudo-Amadeite *Apocalypsis Nova*. This latter was the most famous and substantial of all the new prophetic works which appeared at this time. The mysterious circumstances of its 'discovery', the people associated with

---

[12] See M. Reeves, 'Roma Profetica', in F. Troncarelli (ed.), *La città dei segreti: Magia, astrologia e cultura esoterica a Roma* (Milan, 1985), 294.

[13] Reeves, *Influence*, 445; Tognetti, 'Le Fortune', 311.

[14] Ibid.

[15] See Reeves, 'Roma Profetica', 285–6. On the career of this Bonaventura, see below, Niccoli, Ch. 10.

[16] See G. Tognetti, 'Note sul Profetismo nel Rinascimento e la Lettura Relativa', *Boll. ISI*, 82 (1970), 149.

[17] O. Pugliese, 'Apocalyptic and Dantesque Elements', 127–35.

its circulation, and its strange content all contributed to its fame. It is
the subject of the following essay where its various parts are analysed.
The striking point that emerges is that the famous references to the
coming Angelic Pope occupy such a small portion of the whole. Its
main concern is with the 'secret knowledge' now to be revealed, almost
a 'new theology' which God has willed to be reserved for this age. The
content of this is surprising: angelology and Mariology figure largely,
incorporating and correcting medieval legendary material and drawing
considerably on Duns Scotus, while clearly one of the teasing problems
occupying the mind of the author is that of Predestination and Free
Will. The *Apocalypsis Nova* aroused fierce opposition in some quarters
but was immensely popular and widely disseminated. In the Italian
culture of the high Renaissance, it strikes an unexpected note, revealing theological issues of deep concern which are often ignored by the
historian.

2

The Apocalypsis Nova:
A Plan for Reform

ANNA MORISI-GUERRA

I

Not all prophecies suffer equal fates. Some are consumed in the brief
passage of events and before long fall into oblivion, others survive for
some time unaided, growing and changing according to the times, or are
simply reinterpreted according to the demands of new circumstances;
yet across the centuries they can still inspire both hope and fear. The
Apocalypsis Nova can certainly be counted among these. It has been
handed down to us preserved in dozens of manuscript copies, collected
and annotated over the course of many years—years which bear witness
to an interest in the work and in the problems accompanying its appearance which has not yet been exhausted.[1] Its predictions, although clearly
dated, were extrapolated and circulated even as late as the nineteenth
century, supporting in certain circles the expectation of the *Pastor
Angelicus*.[2] In even more recent times its narrative parts were used to
document an ancient devotion which was then translated into doctrinal
definitions.[3]

It was said that Amadeus, a noble knight who had abandoned his

---

[1] Only a few extracts have been published from the text; for bibliography and analysis
of the sources, see A. Morisi, *Apocalypsis Nova: Richerche sull'origine e la formazione del
testo dello pseudo-Amadeo* (Studi storici, 77; Rome, 1970).

[2] An extract from the work was transcribed at the beginning of the 18th cent. under the
title *Prophetia beati Amadei*, now kept with other prophetic texts in a file of 'Documenti
relativi a Pio VII e alla Francia' (Rome, Museo Centrale del Risorgimento, MS 161). A
Predizione del beato Amadeo, dell'anno 1500 appears in D. Cerri, *I futuri destini degli stati
e delle nazioni ovvero profezie e predizioni riguardanti i rivolgimenti di tutti i regni dell'universo
sino alla fine del mondo*, 7th edn. (Turin, 1871), 182–5. I thank Marina Caffiero for this
information.

[3] C. Balić, *Testimonia de assumptione beatae virginis Mariae* (Rome, 1948); a long extract
is on p. 267.

army career to dedicate himself to prayer, left Spain and set off for
Italy, donning the habit of a Franciscan at the Monastery of Assisi.
Having moved from Milan to join the Minorites, he obtained permission
to embrace a life of meditation in a small, solitary cell, strictly observing
the Rule while remaining, however, directly answerable to the Priests
of the Order. The fame of his saintliness grew from day to day, and
people of all social classes hurried to visit him in his cell: simple com-
moners, the Archbishop of Milan, Duke Francesco Sforza, Duchess
Bianca Maria, who on a number of occasions during the Congress of
Mantua sent him as her messenger to Pope Pius II. In spite of the
hostility and envy of the Observants towards him, the number of those
who followed him grew more and more numerous. They founded other
cells, setting up their own constitutions, and were favoured by the
Superiors of the Order, as well as by the Franciscan Pope Sixtus IV,
who summoned Amadeus to Rome in 1472 to be his confessor. He
spent the last ten years of his life there, often retiring into a cave in the
Janiculum near the Monastery of San Pietro in Montorio. In the late spring
of 1482 he asked the Pope for permission to go to Lombardy to visit
the place where he grew up, and died in Milan on 10 August at the
Monastery of Santa Maria della Pace.

The piety of his devotees at once numbered him among the blessed,
even if it did not suffice to have him canonized.[4] One of his companions,
who a few years later wrote an accurate biography of his life, attributes
prophetic gifts to him, but never presents him as a prophet, nor does
he mention any writings left by him. Yet his fame is, in particular,
linked to the work bearing his name which began to be disseminated
at the beginning of the sixteenth century and which aroused approval,
hostility, and even perplexity. Some welcomed it with ingenuous trust
or with a not altogether disinterested enthusiasm;[5] some criticized it for
its doctrinal approach and its prophetic message;[6] others tried to exonerate Amadeus from the responsibility for a work of doubtful orthodoxy by

---

[4] The earliest biographies have been studied by P. Sevesi, B. Amadeo Menez de Sylva
dei Frati minori, Fondatore degli Amadeiti (Vita inedita di Fra Mariano da Firenze e documenti inediti) (Florence, 1912), and by Morisi, Apocalypsis Nova, 1–6.
[5] On the first reactions, see L. Wadding, Annales Minorum (Rome, 1731–6), Annales
ad a. 1482, 14, XXXVII, 322–3, and AS Aug. II, dies 10, 572 ff.
[6] At a time when the Fifth Lateran Council was condemning the proliferation of prophecies, Tommaso de Vio expressed a harsh opinion of the text, which he interpreted as
doctrinal even in its paradoxical aspects: Secunda secundae sancti Thomae cum comment. car.
Caietani (Lyons, 1540), fo. 287 c. Later Bellarmin drew up his censures with great
precision in 39 articles: X. Le Bachelet, Auctarium Bellarminianum (Paris, 1913), 670–2.
Cf. below, Minnich, Ch. 4, on Tommaso de Vio and prophecy.

suggesting that interpolations had been illicitly made in one of his original writings;[7] still others sought to shed light on the mysterious origins of this writing which, even today, preserves many of its secrets.

The text, which has a fairly homogeneous manuscript tradition, is divided into two parts, of which the first properly merits the title which is usually used to refer to the whole work: *Apocalypsis Nova sensum habens apertum et ea quae in antiqua Apocalypsis erant intus, hic ponuntur foris, hoc est quae erant abscondita sunt hic aperta et manifestata.* This is developed in eight *Raptus* in which the Archangel Gabriel reveals to the friar, rapt in ecstasy in the presence of the heavenly court, the 'arcana fidei', which have yet to be extracted from the Holy Scriptures and therefore have not been made available to all as a truth of faith but are still debated amongst theologians. The second part, referred to in manuscripts as the *Sermones Johannis Baptistae* and the *Sermones domini* or, more rarely, under the singular title of *Revelationes speciales*, is an elaboration, in the form of predictions, of themes which are only mentioned in passing in the Gospels: the exhortations of John the Baptist to the soldiers and to Herod himself, his last words before his death, followed by ten of Jesus' sermons to his disciples and to the multitudes. The two parts are closely linked by frequent cross-references and careful reading reveals that there is a systematic plan; two separate passages appear to supply an index to the subjects dealt with, as if the author had intentionally wished to emphasize the unity of a text which could quite easily be divided up into various parts, each one being of a different literary genre. A second theme closely linked to that of theological revelation, and developed right through the book, is that of the Pastor to whom the message is addressed, because with it he could lead Christendom

---

[7] Against this opinion, which was held by all the Franciscan commentators, the Dominican Bzowski brought to attention the significant uniformity of the manuscripts and the essential unity of the text; he criticized the work severely, censuring its theological, prophetic, and devotional content and attributing the whole to the fancy of an eccentric mind (A. Bzowius, *Annalium ecclesiasticorum . . . continuatio* (Cologne, 1627), xviii. 26–30. Wadding, *Annales*, 14, XXXVII, 322, whilst not wishing to defend a work which he would rather had disappeared 'in suo ortu vel abortu', attempted to show that the Dominican's accusations were without foundation. In fact the latter accepted the contribution that the work was able to offer to the elaboration of the Marian doctrine (*ΠΡΕΣΒΕΙΑ sive legatio Philippi III. et IV . . . de definienda controversia Immaculatae Conceptionis B. Virginis Mariae* (Louvain, 1624), sect. iii, para. iii, no. 13. 343, and *Immaculatae Conceptioni B. Mariae Virginis non adversari eius mortem corporalem Opusculum* (Rome, 1655), 80.) Various Scotist theologians also attempted to revalue the work which offered the authority of a revelation to many of their beliefs (Ildephonsus Brizenus Chilensis, *Prima pars celebriorum controversiarum in Primum Sententiarum Joannis Scoti Doctoris subtilis Theologorum facile Principis* (Madrid, 1642), with an indication of the principal discussions on the doctrines of the *Apocalypsis Nova*).

and, indeed, all humanity to peace and unity. The very beginning of this Apocalypse, which echoes the New Testament Apocalypse (1: 9), seems to be meant as an assurance that the two themes are interwoven:

I Amadeus was snatched from my cave, where I was praying on a certain mountain, into the sphere where the angels and the spirits of the saints whom we worship and venerate stand before God. And when I was there fear and trembling overwhelmed me; I knew not what to say but was like a man without sense or understanding. Then a man stood by me with a fair countenance and shining face and garments.

It is Gabriel, the Angel of the Annunciation, who comforts the bewildered friar and reveals to him the reason for his vision:

Our God wills once more to take pity on the human race, to purge the world from all errors, to lead all men back into the one bosom [gremium] of truth and to appoint over His Church one pastor whom He Himself will choose, so that he may feed his sheep and nourish his people in justice and truth. And he wills to impart these his secrets to you: he ordains that future things which you will not see while in the flesh, now you will see in the spirit and find consolation, gladdening your soul for your many labours and aspirations. And, that you may attend diligently, store up these things in the closet of your heart; write them down, keep them and guard them. Show [them] to no one until God shall send that man who will open the book you have written in the time that God pleases. And because God wills that many vain and superfluous [matters] shall be excised from His faith and that those things which ought to be believed shall be believed purely and simply, He has sent me to you to teach you the several mysteries of the faith, and you will cause them to be written. I have prepared him who, hearing all things as if from your mouth, will write [them] down: in whose time [all things] will be made plain and that Pastor whom God alone knows will publish them to everyone. But you, beware lest you say or reveal these things to anyone, for it is the will of God that in His time, by him whom He Himself wills, these things shall be revealed to men. (R1, fo. 1^(r-v))[8]

This skilfully structured prologue reveals the hand of a person who has a good knowledge of a certain prophetic and theological tradition[9] and who is adept at using a number of devices from the literature of predictions (the cave, the scribe, the simplicity and lack of doctrine of

---

[8] The quotations are taken from MS. Vat. Lat. 3825, one of the earliest codices, and collated with other manuscripts. Hereafter the following abbreviations will be used: Ap.N. (Apocalypsis Nova), R (Raptus), S (Sermones).

[9] Various prophetic texts are quoted either directly or indirectly: in particular, prophecies of St Bridget, for elaborations on the gospel account, and Joachim of Fiore on the doctrine of the Trinity.

the main character, the secrecy of the writing) to make the text itself
tell its story, to suggest to the reader some precise space–time context
and to forestall doubts and objections. The elements form a subtle plot
which runs through the entire text: at the beginning and the end of
each *raptus*, the date of which is not difficult to determine,[10] Amadeus
is in his cave; he is aroused from meditation by members of his brother-
hood who come to interrupt his prolonged fast, by devotees who want
to talk to him, or by a messenger sent by the Pope to summon him. In
all these situations attention to details unusual in prophetic literature
and a meticulous accuracy form characteristics peculiar to this work.
The scribe is summoned immediately at the end of the first *raptus* to
write the initial part of this revelation, and the repeated references to
him make him one of the main characters in the work. He is too real
to be a mere onlooker imposed by tradition. On the other hand, the
presence of a man of letters is made necessary by the *simplicitas*, which
is often stressed, of the modest friar who has been chosen as an intermediary for a message which exceeds his ability. From time to time, in
fact, the angel has to stop to reassure his interlocutor, who is dismayed
by the difficulty of the speech:

Take note, man of God, and although you may be somewhat inept in such
matters, through the grace of God you are made more apt. And where you are
unable to understand fully, write according to what is said to you by me, for
he to whom these things are sent and on account of whom they are written will
understand clearly and teach openly. In this matter you are to God like a horse
to your brother, carrying flour from the mill. And I said: 'May I be a good ass—to
carry something so good to our faithful. You called me a horse out of courtesy
rather than what I deserve but you meant an ass, a rough beast.' (*R3*, fo. 17ʳ)

Sometimes, however, the angel informs him that by divine grace
Amadeus has been granted the right to take an active part in the dialogue
by putting questions and expressing doubts:

I know you are ignorant, not knowing how to declare such uncertain matters.
But God's will is that you will declare them, understanding by His grace vouchsafed to you. I said: 'truly, angel of God, never do I recollect conceiving such

---

[10] Since it was announced to Amadeus that he would shortly die, the ecstasies must all
have taken place in 1482. They occurred at intervals of several days: fifteen, for instance,
between the fourth and the fifth ecstasy; the seventh happened on 'octava resurrectionis'
(14 Apr.), and the eighth on 13 May, three days before the Ascension. In the *Sermones*,
too, there are precise chronological references: in the tenth, for example, the canonization
of St Bonaventura (14 Apr. 1482) is pronounced as imminent.

things and I myself marvel how such unaccustomed thoughts revolve in my mind . . . everyone knows me to be ignorant and foolish.' (*R3*, fo. 15)

Again, in some cases, Amadeus's participation is involuntary:

You seem to me the she-ass of Balaam. I said: 'I speak like a magpie without sense.' And he: 'Thus it is ordained that these things be communicated through a most simple man to any to whom the Lord will open the meaning and understanding; he will immediately understand, transcribe and in his [own] time declare them, that the faithful may not be turned aside among so many opinions but know precisely what to believe in matters of faith.' (*R3*, fo. 23)

Such repeated emphasis reveals a deliberate intention on the part of the writer, since the image given of Amadeus is certainly not that of a theologian. It is possible that this image has become distorted by the usual hagiographic models, and that the friar, born and raised in close proximity to the Spanish court, was not by any means lacking in education. However, it is difficult to attribute to his pen pages which are often patchworks of quotations from Augustine, Jerome, Bede, Anselm, Thomas, Duns Scotus, and from two centuries of the Scotist school, Eastern theologians, and, in particular, the pseudo-Dionysius, John Chrysostom, John of Damascus, and more recent authors, such as Gerson. The angel is aware of this, and sometimes wishes to dispel the reader's puzzlement with an ironic remark: 'I speak in the manner of your philosophers, for I read and understood the works of your Aristotle before he wrote them, and this he said smiling' (*R6*, fo. 85ᵛ).[11]

It may appear surprising that a scholarly, indeed, a most scholarly, prophetic work should have been echoed so widely in popular preaching.[12] Perhaps this was due primarily to the famous prophecy of the Holy Pastor which has fascinated readers and listeners of various cultures and times, and has been seized on as the most striking and authentic part of the *Apocalypsis Nova*, even though, in fact, the actual prediction does not take up more than 2 per cent of the entire text. (This is, of course, only an approximate calculation, since the advent

[11] See also 'Adverte quam subtiliter locutus est dominus Jesus, considera quomodo an etiam posset cathedram theologiae regere et in ecclesiis vestris praedicare. Et hoc dicens subrisit', *S*, fo. 158ʳ.

[12] Wadding, *Annales*, 14, XXXIX, 323, maintains that the work 'apud imperitos et prurientes auribus plurimum obtinuit', and recalls the episodes of two popular preachers, Francesco da Montepulciano and Bonaventura da Subiaco, who may have been inspired by the *Ap.N.* In Dec. 1508, the Amadeite, Antonio da Cremona exhorted the Florentine people to repentance with Savonarolian emphasis (cf. Tognetti, '*Un episodio*', 190–9); his words, however, contain no reference to the text in question.

of a future Pope is prophesied innumerable times 'cum tadeo et nausea legentium'.[13]) There are, however, some longer passages in which both his physical appearance and mission are presented with greater precision, and are accompanied by various details regarding his predecessors and the concomitant political events. These passages, which are found in the fourth *raptus* towards the end of the *Apocalypsis Nova* and at the end of the *Sermones*, show that obvious insertions have been made. The collation of various manuscripts makes it possible to pick out the material common to all of them, to which various revised versions of the prophecy have been added. In the fourth *raptus* the angel complies with the friar's fervent prayers and reveals to him that the chosen Pastor is already in Rome: 'Young, poor, unknown, but God knows him and by degrees will bring him forward and nourish (*irrigat*) him, and in him, when he is an old man, will show the strength of His power' (fo. 29ᵛ). Amadeus is told that he has already met the predestined one without recognizing him but that he will not live long enough to see the Angelic Pastor at the head of the church. The angel then shows Amadeus a portrait of the Holy Pope to comfort him.

Readers of the *Apocalypsis Nova* are supplied with other prophecies which bridge relatively short periods of time. One reads, for example, that Italy will be burdened with a new yoke, that the current Visconti dynasty will fall into irrevocable decline, and that the new Aragonese king will fall from power. Again, when a prince who is very dear to them appears, the Florentines will lose part of the state, they will be tormented by quarrels, suspicion, and fear, and many noble citizens will be forced to leave the city; the 'rex liliorum' will flash through the city like lightning, conquer it, yet return home unvictorious; in his place a new sovereign 'will arise from prison', making many promises and keeping few. The Genoese will change masters, but will always have a master (fo. 30ʳ). These are bits of information—quick and transparent 'predictions', the only purpose of which is to confirm the prophetic time-context. For Florence particularly the prophecy is rich with conjecture and hope: the exiled citizens will gradually return and will be in a stronger position than before; the city will become more devoted to the new pastor than to any other; it will help and encourage him, even though only Rome, the seat of the papacy, is entitled to fulfil the role of the New Jerusalem.

A sequence of papal predictions, couched in the style of the famous

---

[13] Wadding, *Annales*, 14, XXXVII, 322.

Vaticinia de Summis Pontificibus, gives even more precise chronological information. According to the angel:

At present there is the pastor whom you already know and with whom you have spoken many times. He is the originator of much good and evil, but his good-ness and devotions will overcome the evil. After him will come one whom he will raise up, being neither hot in goodness nor cold in evil, but not so luke-warm as to be spewed forth. He will be a man conversing like a man, and sub-ject to the human condition, nor will he do anything above or against men. So God will not abandon him. After him will come an ox in the habit of goring, who is, indeed, not an ox but a bull. He, like Simon Magus, will set up in God's temple those buying and selling and the money-changers' tables, in whom few will be able to trust. (*R4*, fo. 31^r)

This prophecy appears in a limited number of manuscripts, all amongst the oldest. In others one or two passages are found in which additional facts from a later time appear. These may be found in various places in the three prophetic texts, or within a single text, sometimes forming an extension of an unfulfilled prophecy:

Next will come the 'lunated sheep' which by comparison may be thought to be a lamb, and he will be as if he were not. The present one will return under another name which in him will not [be] a lie. He will complete the deeds of this [present] one with greater favour [? from God] and with roaring will prepare the way for his successor in those blessed times. This is hidden: I do not speak all things to you. After him will come the easy-going [*remissus*] one who will displease nobody: next will follow one who is ardent but potent rather for evil [than good] whom a[?] sheep will follow and [? another] sheep will, as it were, not follow. But the successor's voice will follow roaring. It will be heard by all and will spread far and wide; with heavenly pity he will put to flight darkness and black clouds: this man will be the beginning of the happy years.[14]

In the oldest version, this prophetic sequence of Popes does not go beyond the Borgia pontificate: the two additions take it as far as Leo X. Readers of the *Apocalypsis Nova* over a number of years annotated their manuscripts with various interpretations, inserting spaces of decades

[14] There are codices that feature the new parts in the margin, as they have evidently been compared with others. For an examination of these manuscripts, see Morisi, *Apocalypsis Nova*, 20–1. In the appendices of some manuscripts is a long prophetic passage containing a clear allusion to the Sack of Rome, under the title *Extracta novae Apocalypsis beati Amadei, qui liber inventus est Mediolani apud socium eius*. The text incorporates some phrases of the preceding prophecies and seems to have its provenance in Milan, whence also originated a notarial document reporting the opening of the book in that city (J. Mittarelli, *Bibliotheca codicum manuscriptorum monasterii S. Michaelis Venetiarum prope Murianum* (Venice, 1779), 29–30).

between one pope and another in order to lengthen the time span and maintain the validity of the prophecies which vary between precise details and general statements, susceptible to various readings. This is particularly apparent where the prediction becomes most descriptive and places the impious pastor and the saint on opposing sides, in a powerful juxtaposition which, however, lacks the high drama of conflict with the Antichrist:

The Lord will give to Simon Magus the blessing of Esau but to his true elected pastor the blessing of Jacob. But the blessing of Esau was in the dew of heaven and in the fatness of the earth, and so will his [be] also. For he will be strong in strength, healthy in body and lively in mind and what he shall desire will be given to him, so that he will satisfy his appetite with earthly vapours which form the blessing of dew; he shall have many treasures which represent the blessing of earth's fatness; he will plan and carry out worldly designs and occupy his mind with them. He will distribute the goods of the Church (in which he will never be lacking) to his relations, nominated and chosen by himself. He will oppress his subjects with insupportable burdens. At last, on account of his execrable abominations, iniquities, crimes and sins, he will be deposed when the kings come into Italy, for the voice of blood cries against him and his hands are polluted with blood. Yet he will [still] possess and delight in the treasures of the Church. The Pastor whom God loves and chooses, when he comes, will enter into the temple and eject the buyers and sellers, overturn the tables of the money-changers, reconsecrate the temple and reform the Church . . . (fo. 30^{r-v})

A comprehensive reading of this prophetic sequence raises the question of its place in the work. The language is necessarily different from that found in other parts of the *Apocalypsis Nova*. The number of variations in these predictions arouses the suspicion that the sequence developed from a much more concise earlier version. However, within this highly complex and precise text, the reader is primarily led to seek the new Pope's programme for the denunciation of evil and the formulation of proposals for reform capable of meeting the grave crisis in Christian society and culture. This is set out in the final part of the *Apocalypsis Nova*:

Behold, you have heard and seen the great mysteries of the Christian faith which will be made new with the advent of that new pastor of whom I have often spoken to you. And then will be fulfilled that [text] of the Apocalypse: I saw the holy city, the New Jerusalem, descending out of heaven from God (21: 2, 10). This Jerusalem is Rome itself to which God has transferred his kingdom and priesthood and which will again be renewed in those days and will preside over the whole world. And this pastor will be like King David, for, as he renewed that Jerusalem and made his fortress and city in it, so that pastor will reform the new Jerusalem, that is, Rome and the Church. As David was nourished in that old Jerusalem, so this one will be in the new Jerusalem. He will be a true son of the Church and acknowledged by all as Pastor, by God and men. God will give him grace and prudence, and the string of his lips and tongue will be loosed; he will speak of the great things of God openly and all will hear his voice. He will unite the western Church with the eastern in a perpetual union, creating two cardinals amongst eastern [churchmen] and two great patriarchs in the west. Amongst others, seven men, most worthy prelates, will assist him, like the seven angels standing in the presence of God. He will send legates throughout the world to tend God's sheep; he will devote himself to spiritual things and appoint one of the cardinals to attend to temporal affairs. Universal peace will return and reformation. . . . With this great pastor will arise a great king and the will of God will be perfectly fulfilled at this time. (fo. 142^{r-v})

Here traditional elements are not lacking but they are incorporated into a precise plan.[15] The picture painted of a pastor dedicated to spiritual matters, ready to renounce direct intervention in secular affairs and instead to rely in these matters on the co-operation of trustworthy and capable people, was still a valid myth at a time when Popes were failing to respond adequately to the disarray of the Christian world, a disarray and confusion which found expression in anxiety for reform and in pressure for theological and cultural rethinking. This last theme, in particular, is presented here with great insistence. It comes across, not as a common-place, but rather as revealing the true personality and ideals of the main character, author, or person to whom the prophecy is addressed.

II

The *Pastor Angelicus* is thus to be seen as a theologian. The meaning of his message must therefore be sought in the doctrinal part of the text. The first ecstasy opens with a vision:

I looked up and saw a ladder whose top seemed to touch the sky and Christ the Lord rested on the ladder with a shining diadem (Gen. 28: 12–13) and there was a writing which said: Again I will come to you and your heart shall rejoice and no one shall take away your joy from you (John 16: 22). On another part

---

[15] The concord with oriental peoples and the choice of Greek cardinals are elements present in earlier prophecies (cf. Morisi, *Apocalypsis Nova*, 19 n. 40), although here invested with new meanings.

of the ladder was written: He will appear and will not lie; if he appears to you to tarry, wait for him, for he will come (Hab. 2: 3). And I saw the Queen [of heaven], His Mother, supported by that King, from whose mouth came this writing: You have had pity on our race, hasten, do not tarry (1 Kings 20: 38). And another writing was directed to the crowd of people saying: Let not your hearts be troubled (John 14: 1); Lift up your heads, for behold your redemption is at hand (Luke 21: 28). (fo. 2)

The programmatic meaning of this image, which seems to suggest a pictorial representation, is obvious: Jacob's ladder, whether it be a symbol of meditation, contemplation or theology, enables the soul to ascend towards its destination. After this first vision the conversation unfolds gradually in the form of a dialogue between the angel and the friar who, following the path leading to knowledge of God, come to a final discussion, at the end of the *Sermones*, on the significance of *gaudium* and *beatitudo*, in which the vision and fruition are fused in one. The text of the *Apocalypsis Nova* in its entirety reveals a few main themes which do not always correspond with the subdivisions of the ecstasies. Thus, in the first, which comprises two *raptus*, one glimpses a line of thought from the Second Book of Sentences which develops into a broader examination of the first three chapters of Genesis, while the third *raptus* is devoted to a series of open *questiones* about arguments dealt with earlier.

This is the most difficult part of the work, particularly because the tone of the exposition is very varied: demonstrations delivered in impeccable scholarly language alternate with questions and explanations in an intentionally simple style. These prepare the way for two important myths: the Fall of the Angels and the Creation of Man. Here, hidden behind the apparent simplicity, an organized system of thought can be detected, in particular, that of a precise angelology which basically follows Duns Scotus, sometimes correcting and interpreting him with great skill and subtlety. After introducing the vision of the angelic hierarchies in the first ecstasy, the exposition continues with the revelation of the names of the seven higher angels. All these names, which indicate divine operations, are expounded with numerous details of the nature, intellect, and actions of the celestial spirits constituting an infinite process of mediation between divinity and humanity. Having appeared before all other things, before time itself, the primeval beings of creation, they are inferior only to Mary; this is the starting point of the whole work, and introduces the first dramatized story—the myth of the fall of the angels, symbolized by the *proelium magnum* of the Apocalypse (12: 7).

The Scriptures make only a passing reference to it in the Wisdom of
Solomon (2: 24), which states that 'through the envy of the devil death
came into the world'. The story in the *Apocalypsis Nova* seems to start
from the pseudo-Dionysius[16] where the celestial spirits are initiated
into the mystery of the Incarnation. God revealed his purpose to the
angels: he would assume a human form to act as a meeting point between creator and creation; the instrument of this act would be a woman,
who was already foreknown from eternity and whom the angels must
honour as their queen. Put to this test, which is paradigmatic for all
those who follow, some who would like the privilege for themselves did
not accept submission but rebelled. This gave rise to the conflict and
the fall. The good angels were divided from the evil ones, light and
darkness were separated, and human time began. The story, whose
sources both immediate and more remote, can be traced,[17] is told in
a lively, spontaneous manner, with carefully chosen words designed to
be vehicles for theological and philosophical concepts. A myth is thus
created, a translation into narrative form of doctrinal elements. The
second story, about the creation of man, is a development of the biblical
narrative, and provides the opportunity to introduce the mediating role
of the angels. Many of them, wishing to become divine, take part in
this work, and Raphael, Gabriel, and Uriel (*fortitudo*, *medicina*, and
illuminatio dei) in particular are chosen from amongst the spirits of the
first choir. They bestow gifts of original justice on the newborn creature.
The story then continues with episodes of the temptation and Fall of
Man, which have a particularly Scotist ring in their presentation of the
concept of will and responsibility.

In the fourth *raptus*, Amadeus asks his interlocutor a direct question:
'Concerning that controversy which today generates much scandal, I ask
that you will resolve whether the Mother of our Lord Jesus Christ will
have had or contracted that sin' (fo. 32ʳ). The angel affirms that Mary
was kept from sin 'ab aeterno', 'ex voluntate absoluta', not 'ex voluntate
ordinata'. The following exposition embodies controversial points of
the doctrine of the Immaculate Conception which were supported by
the Franciscans. This leads to another narrative section. Woven into the

---

[16] Pseudo-Dionysius, *De caelesti hierarchia*, trans. eds. of the Shrine of Wisdom (London,
1935), iv. 4.
[17] From the *Vangelo di Bartolomeo*, an apocryphal text that was particularly widespread in Slavic lands, to a work that was well known in the circles of Christian cabalists,
the *Corona regia* of Paulus de Heredia, published in Rome in 1486. On the use of these
sources, see A. Morisi, 'Vangeli apocrifi e leggende nella cultura religiosa del tardo
Medioevo', *Bull. ISI* 85 (1974–5), 164–8.

myth here are elements from a vast apocryphal literature which goes
back first to the Pseudo-Matthew, and then in general to early apocryphal gospels. These are combined with new material which, while completing the narrative, conceals a precise theological exposition. The
apocryphal stories concerning Joachim and Anna are replaced in the
Apocalypsis Nova by the proposition that Mary was not exempt from
the laws of nature, but was rather, from the moment of her conception,
in a state of grace, as well as being endowed with innumerable supernatural gifts. To this end she was conceived with the help of the angels
who conferred every favour on her. She is placed half-way between the
natural and the supernatural. From the very first Mary in fact possesses
all the gifts which Adam lost: her will is always upright, her wisdom is
perfect in that, even before birth, she knows grammar and mathematics,
physics and metaphysics, law and theology. The text of the apocryphal
gospel, modified to give a broader doctrinal base, is then utilized more
closely. The description of the Virgin's life in a convent-like community
is a development of chapters 6 and 7 of the Pseudo-Matthew. In the
marriage story a large number of details of the myth are corrected because they are inadequate or inconvenient: Joseph is not old—he is a
man in the full vigour of manhood;[18] he has not contracted any other
marriage, and has made a vow of celibacy. When the priest announces
the marriage to Mary, no uncertainty, refusal, or conflict arises. Even
Duns Scotus had wondered if a marriage contract between two people
who had made a vow of celibacy was valid, and formed the hypothesis
that an angelic revelation to the two marriage partners would resolve
any difficulty. It seems that the author of the *Apocalypsis Nova* bore in
mind this suggestion by the *doctor subtilis* when he invented two revelations, one for Mary and one for Joseph, thus supplying all sufficient
explanations.[19]

In the Annunciation episode another mythical–allegorical element
supplements the evangelic narrative: from the hosts of angels present
in the Virgin's nursery, three come forward as messengers of the entire
Holy Trinity—Gabriel, *fortitudo dei*, for the Father, Euchutiel, *bonum*

---

[18] Gerson had suggested representing the saint as a vigorous character, active in the
events of Jesus' childhood (*Iosephina*, in *Œuvres complètes*, Introduction, ed. M. Glorieux
(Tournai, 1962), iv. 31–100). His example was followed by Franciscan preachers, par-
ticularly Bernardino of Siena ('Sermone intorno a S. Giuseppe, sposo della Vergine Maria',
Misc. Franc. 311/312 (1883), 1–25) and Bernardino of Feltre (*Sermoni di Bernardino da
Feltre* (Milan, 1966), 392–401).

[19] Duns Scotus, *Opus Oxoniense*, IV. xxx. 2, cited in C. Balić, *Joannis Duns Scoti
theologiae marianae elementa* (Sibenik, 1933), 62.

consilium, for the Son, and Barchiel, *benedictio dei*, for the Holy Ghost. Subsequently the Gospel passage relating the meeting between Mary and Elizabeth is retold in a long, rather mannered scene, in which the two mothers and the children who are yet to be born converse, sometimes in human language and sometimes using angelic expressions, inserting five short songs into their conversations.[20]

After the fifth ecstasy the narrative is interrupted to make way for two independent treatises, which occupy the sixth and seventh *raptus*, on the problems of the Eucharist and the Trinity. The problems of divine unity in the dialectics of the Holy Trinity and the appearance of the Word and of its constant presence among men find their place at this stage, and, from the viewpoint of theological proof, the development is coherent. The conversation develops using the same scholastic language as that used in the Scotist tradition. The themes dealt with here were to have damaging consequences in a short space of time. The author returns to problems favoured by Bessarion which had been discussed at the Council of Florence. It is not by chance that passages from the 'Exultate Deo' and 'Cantate Domino' bulls are commented on here.[21] On the question of the Holy Trinity, the long discussion seems to be conducted in order to reconcile the different scholastic positions, classifying them, for the most part, according to differences in language, but the exposition is also designed to include within the doctrine of the Holy Trinity elements which form the basis of Mariology. Not all the questions posed here are answered: this will be the task of the future pastor, who will be given sufficient guidance as to how to resolve them.

In the eighth ecstasy the story of the final years of Mary's life after the Ascension concludes the *Apocalypsis Nova*. If the work, as it now seems, is taken to be a Mariological key comment on the twelfth chapter of the Apocalypse, this final section supplies the exegesis of verses 6 and 14 (the 1,260 'days' the Woman spent in the desert, the 'time and times and half a time') and of verse 1, which portrays the initial vision

---

[20] From the treatment of episodes in the life of Jesus, however, the impression is given that the human dimension of Jesus is avoided. Thus, in the *S*, Amadeo asks for an explanation about Jesus' fear and agony in the Garden, but just as the angel is about to reply, he is interrupted by an imperious voice: only the future shepherd will be able to supply the answer. We are reminded of the discussion between Erasmus and his English friends on this subject so indicative of the religious sensibility of the era (*Opus epistolarum Des. Erasmi Roterodami*, ed. P. S. Allen (Oxford, 1906), 249–53).

[21] H. Denzinger and A. Schönmetzer, *Enchiridion Symbolorum* (Freiburg im Breisgau, 1963), 334–5, 342.

of the 'Woman clothed with the sun', seen in her moment of glorification. The truth has been revealed, but concealed in the words of this most obscure of Christian texts, and hidden until the time when man is able to understand it: 'The evangelists wrote almost nothing of Mary for it was then enough to proclaim those things which pertained to her Son, Christ the Redeemer, lest they should appear to say superfluous things and preach and extol woman, which sex appears to men as base and worthless.'

In the early days of Christianity—continues the angel—it was necessary to announce the beginning of a new age, the age of salvation. Now, however, the Lord wants man to turn to the Mother of all and to know that 'of the whole body which is the Church, whose head is Christ, Mary represents the neck; this is because nothing good, no gift, grace or privilege descends from that Head into the body except through this wonderful neck, which is Mary, the Mother of God.'

The future pastor will be entrusted with the task of propagating these truths which will now be revealed, 'rejecting many popular errors' (fo. 117). This would entail sorting out a collection of legends contained in old apocryphal texts[22] which were beloved of the faithful and documented in iconography, literature, even in doctrinal speculation. The conciliatory intentions of the author are particularly visible in his attempt to settle doctrinal differences. Here, finally, based on the Marian interpretation of the Apocalypse text, Mariology is affirmed as the basis of ecclesiology.[23] Jesus had entrusted his mother to his disciples, who, in turn, depended on her. Mary remained on earth for a further fifteen years, as many as she had lived before the birth of her son, but she had already achieved a state of beatitude because, by taking part in Christ's death and thus becoming a co-redemptrix, she had paid homage to the laws of nature. When she ended her days, the angel Gabriel brought her a palm branch gathered in Paradise which was to be carried in front of her coffin. Here the *Apocalypsis Nova* corrects the legendary tradition in various respects. While the apocryphal texts portray the disciples miraculously assembled and transported on clouds from all the various parts of the world where they have been preaching, the author of the *Apocalypsis Nova* has them all in Palestine. Joined together in the presence of Mary, each of them speaks out in her praise, and each speech contains the proof of theological problems to do with Mariology. The

---

[22] For an examination of these sources, see Morisi, *Apocalypsis Nova*, 73–7.
[23] Cf. A. Prigent, *Apocalypse 12: Histoire de l'exégèse* (Tübingen, 1958).

episode thus depicted gives rise to two interpretations of the 'crown with twelve stars', which both represent the apostles and the virtues of the Virgin, as ecclesiological and Mariological interpretations respectively. Mary takes leave of the disciples, promising always to be close to the faithful, not in the sacrament of the altar, but in images and in the miracles attributed to these images. The rest of the story proceeds along the lines of apocryphal tradition. Once she has been taken up into Heaven, the apostles travel to every part of the world 'announcing the mighty acts of God to all peoples'. At the end of this long revelation the angel entrusts to Amadeus the last chapter of the third Gospel, the one which Luke did not write because the time was not yet right to comprehend it (fos. 141ᵛ–142).[24]

The second part of the work, entitled *Sermones*, brings together further revelations Amadeus received from the angel, while he was in the cave, without being transported into a vision. This is in the form of a partial concordance, in which passages from the Gospels relating to the preaching of John the Baptist and Jesus are rearranged in continuous narrative in such a way as to resolve various problems of agreement, chronology, and even Biblical philology.[25] While the *Apocalypsis Nova* presents the theological programme of the Angelic Pastor, the *Sermones*, through their characteristic moral preaching, enable the eschatological implications of this plan to be grasped more easily. John the Baptist, the prophet who brings to an end the period of expectancy, urges repentance: 'Through your heart you have sinned, through your heart you can abolish all sin and acknowledge the Lamb of God who takes away the sin of the world, who alone is innocent and without sin. He who was to come has now come' (fo. 142ᵛ).

The internalization of the Christian struggle and the expectation of a reign of illumination, when Man will be able to look into his own heart and translate the ancient law into spiritual terms, will create conditions in which the eschatological tension of expecting terrifying events will be replaced by an eager welcome to truth and by a continuous spiritual ascent. There is, indeed, an exhortation to repent, but the language is not that of the 'preachers of penitence'. Freedom, responsibility, grace,

---

[24] On the incompleteness of the gospel account, cf. *Revelationes S. Brigittae . . . a Consalvo Duranto illustratae*, 4, 60–2 (Antwerp, 1611), 489–91.

[25] Compare the debate over the passage in John 21: 20–3: 'Si (or sic) eum volo manere' (fo. 136), which, many years before, had been at the basis of an argument between Bessarion and Trebizond. See L. Mohler, *Aus Bessarions Gelehrtenkreis, Quellen und Forschungen aus dem Gebiete der Geschichte*, 24 (Paderborn, 1942), 70.

the grand themes of salvation theology are expounded, in terms of the 'immense mercy of God'. Yet Man must recognize his negativity, his inability to gain merit, since any good he succeeds in doing is simply God working within him. This exposition touches on the ultimate consequences of absolute predestination, yet the certainty that grace is promised to everyone is confirmed, subject to the willingness to receive it.[26] Unwitting ignorance is not a sin—only obstinacy and arrogance are evil; little old women will not be condemned for their ingenuous and often mistaken faith, and neither will Abbot Joachim for his erroneous doctrine of the Trinity;[27] 'if therefore such are not charged with the crime of heresy, why ought those Christians to be damned who are ignorant of Christ and do good works, if they believe in their simplicity anything contrary to the faith?'[28] The time will come, however, when all will be in agreement, the promised era which will last 'through many ages' (fo. 201ᵛ).

From this perspective of hope and tolerance, the counterpart of God's goodness is the *caritas* of man united in universal solidarity.[29] Everything else is beyond all human comprehension. God is Absolute Will projected into timeless eternity; this eludes definition. He is not bound by any decision; His words, His oaths, His threats are necessarily expressed in human language which cannot, however, reveal the whole meaning.[30] No one, therefore, is certain, before the Final Judgement, of their own status. The damned are in no way certain of what will become of them, neither are the angels and saints who, however, rest at peace because they know that 'God is more prone to spare than to punish, since certainly He is ready to have mercy and to forgive, and through experience we have seen Him liberate some from hell, but eject

---

[26] In this context the vocation of the few chosen to tread the road to perfection is raised and the so-called Osservanti are reproached who, by presuming too much of themselves, err more than others. The Amadeites who, having chosen the narrow way, have not the sufficient resources to traverse it are also criticized, though less severely (fo. 201ʳ).

[27] 'Nota haec et bene signa et bene conscribe, ut audiat universus orbis bonitatem et clementiam dei. Quot enim sunt sanctissimae vetulae, immo homines et mulieres deo gratissimi qui multa fidei credunt contraria; imaginantur enim trinitatem personarum modo contrario, ut abbas Joachim . . . multi simplices adorant imagines sanctorum, non distinguentes inter imaginem et rem cuius est imago' (fos. 199ᵛ–200ʳ).

[28] There is possibly an allusion here to the peoples of the New World.

[29] 'Proximus tuus tu ipse es, odio habere carnem propriam inconsuetum est. Peccata sunt odio habenda, non homines. Iniquitates quibus offenditur deus sunt extirpandae, non proximi, vitia sunt abolenda, non fratres' (fo. 206ᵛ).

[30] 'Disposuit ergo ab aeterno pro diversis temporibus diversa, prout libuit. Non est ergo verum ut quod vult in aeternitate velit pro tota aeternitate. Sed revocat interdum, non revocatione nova, sed aeterna, non tamen pro aeterno' (fo. 213ʳ).

no one from the celestial paradise.' The Day of Judgement should
therefore be awaited without fear. An Origenian motif is interposed
here in the Scotist doctrine of Will. At the end everyone will die and
rise again, but between these two points in time there will be a great
universal fire which will purge the universe of all corruption. The Christ–
Judge will display himself in all his glory. The saved will take their
places around Him, positioned according to whether they are the chosen
people or simply the faithful, unwitting or secret Christians or unbap-
tized children. The damned, those who have refused the offer of true
life are not mentioned, so their presence is meaningless—an almost
non-presence. Through a maze of subtle, quibbling questions on the
times and modes of the Resurrection and the distribution of matter into
separate bodies, the message which comes through is one of reconciliation
and hope.

III

But to whom is this message addressed, and who might the author be?
In the final reading the figure of the Blessed Amadeus seems to become even further removed, while the scant information we have on
the Amadean inheritance does not provide a link between the time of the
founder's death and the time the book appeared or reappeared.[31] The
first readers of the *Apocalypsis Nova* asked themselves the same question, and the information gleaned from them helps to throw some light
on the mysterious origins of the book.

According to one tradition Amadeus carried his secret to the tomb;
the manuscript was buried along with him, and any attempt to salvage
it would have been prevented by terrible omens.[32] The Franciscan
chronicler Mariano from Florence[33] relates, on the other hand, that
Amadeus entrusted the sealed book to his fellow brethren, and that it
was only after many years that Cardinal Domenico Grimani, patron of

---

[31] For the first decades of the congregation, writings by the Amadeites are not known
from which analogies might be drawn with the spirit of the *Ap.N.* In 1516 Mattia da
Milano published a confession written in a tone recognizably similar to that of the *Sermones*
(*Repertorium, seu interrogatorium, sive confessionale . . . per Johannem Angelum Scinzenzeler*
(Milan, 1516). Possibly the author had met Amadeus, who, in this edition is depicted in
the woodcut frontispiece, but it is more likely that he was influenced by a reading of the
text, which at that time was widely known.

[32] *AS*, Aug. II, *dies* 10, 371.

[33] Cited by Wadding, *Annales*, 14, XXXIX, 322–3. On the probable inaccuracy of
Mariano in naming Grimani as present, see below, Minnich, Ch. 4 n. 61.

the Order, and Bernardino Carvajal, Cardinal of Santa Croce, opened it; subsequently, Carvajal, labouring under the illusion that he was the intended pontiff, convened the schismatic Council of Pisa against Julius II. From that time onwards the text would have begun to circulate and become corrupted with frenzied interpolations. However, besides the material on the subject which has got into print, there also exists a mass of information, still very circumstantial, which supplies us with opinions, hearsay, and conjecture annotated by copyists and anonymous readers in the margins of their manuscripts. A Milanese manuscript[34] contains a handful of letters and annotations from which one can glean that in 1502, possibly on Easter Day, Carvajal, patron of the Amadean congregation, the General of the Friars Minor, Brother Egidio Delfini da Amelia, and the guardian of the monastery, Brother Isaia da Varese, met in Rome at the Church of S. Pietro in Montorio; also present was the conventual friar Giorgio Benigno, who was impelled through obedience to keep the book in custody for a year, during which time he was the only person to know its contents, because others, afraid for their own lives, did not even want to touch it. In the course of that year, Friar Giorgio was at liberty to copy it and send the sheets by hand to a Florentine friend, a fervent Savonarolian, Ubertino Risaliti. A somewhat later Florentine manuscript[35] reveals further details which confirm, with few variations, those of the Milanese manuscript and adds the information that copies of the text then in circulation all came from the one transcribed by Giorgio Benigno who, on the point of death in 1520, confessed to having made a number of his own additions. According to all sources, Francesco Biondo wrote down the text dictated to him by the Blessed Amadeus, or copied it in his own hand.

However enigmatic these voices, which reach us like a whisper from across the centuries, may be, they enable us to narrow down the number of main characters involved in the events into which we need to research. The name of Francesco Biondo, one of the ten sons of the better-known Flavio[36] appears consistently in the marginal annotations of several manuscripts, and is also confirmed by Pietro Galatino, who made

---

[34] Milan, Bibl. Civica Trivulziana, MS 402. Written before 1520, it is one of the earliest manuscripts; the last date in the known documents is 1516, cf. Morisi, *Apocalypsis Nova*, 21, 27–32.

[35] Florence, Bibl. Naz. Centrale, MS. Magliab. XXXIX, 11. The known documents do not pre-date 1541.

[36] On Biondo, see Morisi, *Apocalypsis Nova*, 35–6, and Vasoli, *Profezia e ragione*, 126–7.

extensive use of the *Apocalypsis Nova* in his exegetic works.[37] He was
not a stranger to circles in which prophetic and cabbalistic works were
circulated, but we know too little about him to be able to establish what
part he played in the writing and dissemination of the work. A tenuous
link connects the text with two other people: Cardinal Grimani,[38] Patron
of the Order from 1503 (or 1504), supported the plans of General Egidio
Delfini, who sought to unite the large Franciscan family then divided
up into Conventuals, Observants, and other minor congregations—and
unity indeed had a place in the prophetic programme of Amadeus.[39]
However, at the General Chapter Meeting in 1506 the plan failed, and a
disappointed Friar Egidio retired into a Neapolitan monastery. So, if the
details of the part they played in the rediscovery of the book have any
element of truth in them, the event should have taken place before 1506.

As regards Carvajal, who was Spanish like Amadeus, and patron of
the Amadean congregation, everything leads one to suppose that in
1511 he was the first victim of the illusion suggested by the book. However, the image portrayed by the prophecy of the *Pastor Angelicus* does
not seem to be modelled on him. As a leading light in the Council of
Pisa who desired to promote a new wave of renewal to the Church, and
a lover of theological studies, he was, one might say, a man of action
interested in works on political and moral reform.[40] Perhaps he was in
agreement from the beginning with those Marian principles which are
the basic themes of the *Apocalypsis Nova*, and he was probably not a
stranger to prophetic perspectives. It is not by chance that he was so
close to Annio da Viterbo, the Dominican expert on prophecies, who
dared to defy prevailing opinions in his Order by declaring that he
favoured the doctrine of the Immaculate Conception.[41] At the time of

---

[37] A. Morisi, 'Galatino et la Kabbale chrétienne', *Kabbalistes chrétiens* (Cahiers de l'Hermetisme; Paris, 1979), 211–31; C. Vasoli, *Filosofia e religione nella cultura del Rinascimento* (Naples, 1988), 211–29. On Galatino, see also below, Rusconi, Ch. 8.

[38] P. Paschini, *Domenico Grimani cardinale di S. Marco (†1523)* (Storia e letteratura, 4; Rome, 1943), 86–7. A distinguished and urbane humanist more than a theologian, he had occasion to express his admiration for Duns Scotus, and therefore might have read the *Ap.N.* with interest; see the letter of 27 May 1515 introducing an anthology of Scotus's comments on the Sentences: Antonius de Fantis, *Tabula generalis scotice subtilitatis octo sectoribus universam Doctoris Subtilis Peritiam complectens* (Lyons, 1520).

[39] The *S* conclude with this prevision: 'Sicut quatuor sunt evangelistae, ita et quatuor erunt regulae vivendi, neque plures sunt instituendae, et si fuerint institutae, opus est ut tempore illo destruantur. Omnis religiosus aut sub regula Basilii, aut Augustini, aut Benedicti, aut Francisci deo militabit.'

[40] For a discussion of Carvajal's motives, see below, Minnich, Ch. 6.

[41] E. Fumagalli, 'Anecdoti della vita di Annio da Viterbo O.P.', *AFP* 50 (1980), 189–99. Many hoped, for the sake of the unity of Christendom, that the Dominican and Franciscan orders might reach an agreement on the subject of the Immaculate Con-

the Council of Pisa, however, it seems that the Cardinal of Santa Croce was prepared to define the meaning of the doctrine in the sense of the Dominican School in order to gain the support of Savonarolian groups to which he had also promised the canonization of Brother Girolamo.[42]

Because of his Scotist education and his interest in angelological and Mariological themes, the person who, more than anyone else, could be identified with the text of the *Apocalypsis Nova* was Giorgio Benigno Salviati himself, the Bosnian friar who was a pupil of Bessarion.[43] Perhaps through prudence he did not sit with the schismatic prelates at Pisa, but he was at Carvajal's side, when he accompanied him in a legation to the Emperor Maximilian in 1507–8. On this occasion he wrote a work on Trinitarian theology, the *Vexillum Christianae Fidei*,[44] which contains the first evidence of the existence of the *Apocalypsis Nova*. Amongst many quotations from eminent theologians, the treatise quotes ample passages from the *Apocalypsis Nova*. This is never expressly referred to but a reading key is supplied for the *Pastor Angelicus*. In two subsequent works, which were, however, written prior to 1512, the *Libellus de Virginis Matris Assumptione* and the *De Excellentiis et Dignitatibus Virginis Matris*, the use of the text is even more explicit; at every step references are made to the words 'senex' or 'contemplator' contained 'in quadam non contemnenda revelatione' which the author includes in its entirety, openly acknowledging the authority of the doctrinal source.[45] Yet the meaning of these allusions to an unspecified text is not clear: it could be a preface to its circulation or simply a reference

ception. The *Ap.N.* clearly stated that the Virgin protects such preachers as have always shown devotion to her, and criticized the Franciscans for their aggressive polemics (fo. 183ᵛ). Annio was very close to Carvajal (cf. the dedication to the king of Spain in *Antiquitatum variarum volumina XVII* (Paris, 1515), and died in his house on 13 October 1502, bequeathing to him his library (Rome, Archivio del Convento di S. Maria sopra Minerva, MS F. IV. 11, fo. 30ʳ⁻ᵛ: *Cronica della Chiesa, et convento di S.ta Maria in Gradi di Viterbo*; Viterbo, Bibl. Comunale, MS II. C. IV. 380, p. 66: *Cronaca della Chiesa e del convento de Gradi di Viterbo*, written in 1616 by Giacinto Nobile and copied in 1892 by Cesare Pinzi. For Annio's prophetic work, see below, Reeves, Ch. 5 and Headley, Ch. 13.

[42] Tommaso Neri, *Apologia . . . in difesa della dottrina del R.P.F. Girolamo Savonarola da Ferrara* (Florence, 1564), 40.

[43] He probably also harboured the illusion of becoming Pope, see Vasoli, below, Ch. 7. A manuscript of the *Ap.N.*, now in the Escorial Library, seems to be in Benigno's hand, see Pugliese, 'Apocalyptic and Dantesque Elements', 127–35.

[44] We have read the unedited text of the work in the following MSS: Vienna, Öst. Nat. Bibl., MS Pal. 4797 (Theol. 28), which belonged to Maximilian; Milan, Bibl. del Convento dei Cappuccini, MS 16; Rieti, Bibl. Comunale, MS I. 2, 16. For the relationship between Salviati and Carvajal at this time, see below, Minnich, Ch. 6.

[45] For the *Libellus de Virginis Matris Assumptione* the following manuscripts have been consulted, all of which are in Milan: Bibl. Conv. Cappuccini, MS 16; Bibl. Trivulziana, MS 453; Bibl. Ambrosiana, MS A. 30 sup., which contains *De excellentiis*; for a description of these works, see Morisi, *Apocalypsis Nova*, 38–9.

to a work already referred to, but which is still relatively unknown.

It is appropriate to recall here a part of the final prophetic passage at the end of the *Sermones* in which the angel warns:

Close the book you have written and put it in a safe place until God permits it to be opened and gradually come into the report of men. And any who will wish to have the book when it is already partly known will not have it unless they are appointed and ordained to exercise that ministry, which means not only the pastor, but also those whom I see to adhere wholeheartedly to the pastor, of whom the greater part will be found in the city of Florence.

Amadeus obeys:

And I caused the book to be written, closed and sealed, until God should send the one who will open it, and first in secret and afterwards publicly will announce what is contained therein.

The passage seems to indicate that the interpretation of the prophecy was the prerogative of the Pastor and his collaborators, as if this question had been the subject of debate. Again, the hint of a gradual knowledge of the contents of the book seems to indicate that the latter had been concealed for some time, or had required detailed elaboration. A comparative study of various manuscripts has revealed no answer to this, as the passage is found in all the manuscripts so far consulted, except for those which are anthological or fragmentary.

There is no doubt that, amongst the problems which are as yet to be solved, those relating to the time the text was written are fundamental. There were times when the thread of the story unravelled very rapidly; a few years or even months later new perspectives would be discovered and the same person or people, author or authors, would be viewed in a different light. The success of the work so far has clarified many questions, but partial readings, which emphasize some points while abandoning others, have not explained its origins. However, if the information available to us does not supply us with clear answers, it remains for us to concentrate on the text in order to grasp the message contained in it, knowing full well that perhaps this, in itself, is what has in part contributed to its success, and that the same words may assume different meanings in different contexts.

From this perspective, the *Apocalypsis Nova* can be read as a theological manifesto, a plan for reform, which acquires a prophetic value through its universal, holistic character. It is the *summa* of the culture of a time of crisis at the dawning of a new century; it is an attempt to reconcile opposing doctrines, both philosophical and theological, in a unity which alone can save their traditions and be the first step towards harmony, the symbol of the new era. In this light, its relationship with the previous long prophetic tradition can be understood; the prophecies of St Bridget, St Elizabeth, and Joachim, and the prediction on Mount Gargano, all have some element of truth in them. There is a continuity which, without excluding freedom to accept or correct, takes them rather for their exegetic and theological contribution than for their ability to predict the future. It could be said that the message of the *Apocalypsis Nova*, the most authentically Joachimite inheritance, is contained especially in the expectation of a fullness of revelation to be initiated in a 'tempus pacis et innocentiae' when humanity will find the path to a fuller relationship with God. According to the author, this fullness of knowledge will be based on those Marian themes which provide the structure for the entire work. If the stories which are told, taken individually, are myths of mediation, insofar as they express the modes of theophany and the acts of God in this world, they perform a further mediating function, because the author tends to present his material in ways which are profoundly interconnected in every detail. A complete 'ark' of the Holy Story is literally reconstructed in which scriptural elements, apocryphal stories, and revised or newly created myths exist side by side and nourish one another. All this reveals a final moment of mediation between theological research and faith, speculation and devotion, since the story is not only a medium through which theology could address all people in a language which would be universally comprehensible on the level of popular devotion. If that were so, a distinction between *pistis* and *gnosis* as independent moments would become evident in, if not opposed to, the religious experience. The myth, taken in a platonic sense as a figurative expression of intuitive understanding, is also the language of initiation, representing a stage through which speculation must pass to reach higher levels. Thus every plane of experience is linked to others, and the emotional elements of the story become an integral part of the mystic ascent which embraces the entire spectrum from sensibility to pure contemplation. Giorgio Benigno Salviati, who was able to provide a fuller explanation of the *Apocalypsis Nova* also offers, in this case, a reading key. The reconstruction of an entire literal meaning—says the theologian—must be based on three moments of

faith: that which rests on absolute certainty in the case of events testified to by the Scripture; that which the Christian can allow himself
in his meditations; and a third, in which simple devotion leads to
certainty.[46]

[46] In *De Excellentiis*, Contemplatio II, theor. 7, fo. 82[r], Salviati, in relating the episode
of the Annunciation and the events preceding the birth of Christ, expounds that which
Matthew and Luke 'aperte tradunt', and continues: 'quod autem Gabriel sit secundus
seraph, qui cum aliis duobus et exercitu angelorum adventaverit, quod conflictum illum
virgini narraverit, quod eam una cum filio omnes angeli adoraverint, haec pie contemplari
homini christiano licebit, subiciendo semper et captivando intellectum in obsequium
Christi. Illud unum, ut virgo eo momento omnem gratiae gradum sibi competenter con-
secuta fuerit, ultra quam pie citra firmam certitudinem tenere poterimus. Ubi vides tri-
plicem credendi modum: firmiter, pie, supraque pie, id est citra firmiter.'

III

PROPHECY AND THE COUNCILS

Introduction

The schismatic Council of Pisa and its counterblast, the Fifth Lateran
Council, mirror vividly the ambiguities of motive and expectation in
this period. On the one hand, an immediate game of power politics was
being ruthlessly played out; on the other, a sense of grave crisis in the
Church—perhaps the crisis of all time—called for solemn deliberation
and reforming action. 'Without Councils we cannot be saved', proclaimed Egidio of Viterbo.[1] On the one hand, we meet stratagems of
personal ambition; on the other, pursuit of that haunting image from
prophecy, the Angelic Pope. On the one hand, ecclesiastical authority
feared the claims of prophetic inspiration and set out, through conciliar
legislation to suppress its more threatening manifestations; on the other
hand, leading orators in both Councils drew on the whole prophetic
tradition to heighten the sense of momentous days and momentous
decisions. At a deeper level the problem of extra-rational knowledge
and its sources troubled and divided thinkers. 'Testing the spirits,
whether they be of God' is a persistent theme running through the debates and writings of this period. The clash of judgements on Savonarola
which surfaces during the time of the Councils illustrates this problem
strikingly and, as Nelson Minnich shows, the inability to reach a
clear consensus on this most dramatic of cases influenced the final
pronouncement on prophecy in the Fifth Lateran Council.

Fundamentally, it is suggested, these ambiguities must be traced
back to the minds of the actors themselves. Hence arose the conten-

---

[1] Ed. C. O'Reilly, ' "Without Councils we cannot be saved...": Giles of Viterbo
Addresses the Fifth Lateran Council', *Augustiniana*, 27 (1977), 188.

tions concerning true and false prophets and hence the compromise legislation on prophecy which was one of the closing acts of the Fifth Lateran Council.[2]

[2] See Professor Minnich's analysis in Ch. 4.

3

Prophecy at the Time of the Council of Pisa (1511–1513)

ALDO LANDI

The years around the preparation and the actual session of the Council of Pisa[1] saw a large number of preachers and prophets—often friars, but also lay people—who went through street and square announcing either the imminence of divine retribution or the dawn of a new age.

The phenomenon was particularly noticeable in the region of Florence. In fact, it was here, more than anywhere else, that the link between the socio-political crisis and the spread of new religious demands was strongest.[2] Scholars have already been widely interested in this topic,[3] so I shall here limit myself to pointing out certain aspects which are in some way linked to the time and background of the Council of Pisa.

The widespread malaise which the success of those prophets brought to light could only breed suspicion in the Medicean *Signoria*, which consequently took rapid methods of repression (the problem became acute in 1512, precisely the year of the Medici's return to Florence). The Medici Pope, Leo X, found them very dangerous and was therefore very quick to strike at them.[4]

---

[1] For the history of the Council, in the absence of a modern comprehensive study, see the collection of documents edited by A. Renaudet, *Le Concile gallican de Pise-Milan: Documents florentins, 1510–1512* (Paris, 1922).

[2] On the vibrant political debate already alive in Florence at the end of the fifteenth century, see R. von Albertini, *Firenze dalla Republica al Principato. Storia e coscienza politica* (Turin, 1970).

[3] Essential to the study of this subject are works by Marjorie Reeves, esp. *Influence of Prophecy in the Later Middle Ages: A Study in Joachimism* (Oxford, 1969); and 'The Originality and Influence of Joachim of Fiore', *Traditio* 36 (1980), 269–316; also those of C. Vasoli, esp. 'Temi mistici e profetici alla fine del Quattrocento', *Studi sulla cultura del Rinascimento* (Manduria, 1968) 180–240; 'Profezie e profeti nella vita religiosa e politica fiorentina', *Magia, Astrologia e Religione nel Rinascimento* (Wrocław, Warsaw, Krakov, and Gdansk, 1974), 16–29.

[4] Cf. Minnich, 'Concepts of Reform Proposed at the Fifth Lateran Council', *AHP* 7 (1969), 163–251; id., below, Ch. 4.

According to contemporary chroniclers, it was in January 1511 that sermons violently critical of the Roman Curia were given, accusing it of greed,[5] and in the same month, as Roberto Acciaiuoli, writing to *I Dieci*,[6] noted that the Cardinals were already 'plotting something'. A few days later, Louis XII asked the Florentine *Signoria* for the city of Pisa to hold a Council 'pour l'amour de Dieu, paix de la crestienté, utilité et reformacion de l'Église Universal, tant en chef que [s]es membres'.[7] Schism was in the air, and this was only one of the consequences feared from God's anger at a Church governed by a Pope who possessed and demonstrated very few priestly qualities, and who was not even to be trusted at the level of political relations, inasmuch as it was said that he was very able at the dissimulation of 'observing that which he had promised'.[8] As Gregorovius notes, Pope Julius was aptly described by the following judgement of his contemporary Francesco Vettori, who said 'It is assuredly very difficult to be at the same time a secular prince and a priest, for these are two things that have nothing in common. Whoever looks closely into the evangelical law will see that the popes, although calling themselves Vicars of Christ, have introduced a new religion which has nothing of Christ in it but the name. Christ commanded poverty and they seek for wealth; He commanded humility and they desire to rule the world.'[9]

The wrath of God was therefore about to be visited on the world and again in 1511 the anonymous *Memoria delli novi segni* appeared, a collection of wonders announcing the imminence of that ire.[10]

Among the preachers of those years the following are well known: the friar Francesco da Montepulciano,[11] Francesco da Meleto,[12] and the

---

[5] See for example G. Cambi, 'Istorie', in *Delizie degli eruditi toscani* (Florence, 1786), ii. 256.

[6] Letter of 11 Jan. in Renaudet, *Concile de Pise-Milan*, 25–6.

[7] Letter of 27 Jan., ibid. 26.

[8] As expressed by Cardinal Federico Sanseverino, 10 Aug.; see Renaudet, *Concile de Pise-Milan*, 98–9. See also Sir Thomas More's ironic allusion to the 'verye reprochefull thynge, yf in the leagues of them whyche by a peculiare name be called faythfull, faythe should have no place', *Utopia*, ed. J. Lupton (Oxford, 1895), II. vii. 239.

[9] *Sommario della storia d'Italia*, quoted by F. Gregorovius, *History of the City of Rome in the Middle Ages*, trans. A. Hamilton (London, 1902), VIII. i. 116–17.

[10] Cf. O. Niccoli, *Profeti e popolo nell'Italia del Rinascimento* (Rome and Bari, 1987), 44. This work is important for the subject in question; see the illuminating review by C. Vasoli in *Rivista Storica Italiana*, 90 (1987), 3, 796 ff.

[11] See Vasoli, 'Temi mistici', 218–19. See also below, Minnich, Ch. 4, and Index, on all these preachers.

[12] See S. Bongi, 'Francesco da Meleto, una profeta fiorentino a' tempi del Machiavelli', *ASI* 5a ser., 3 (1889)', *ASI* 62–70 ff.; C. Vasoli, 'La profezia di Francesco da Meleto', *Archivio di Filosofia*, 3 (1963), 27–38.

monk Teodoro.[13] These and other figures have already been the subject of scholarly study; however, much research still remains to be done for the systematic study of the themes and content of other preacher–prophets' sermons of the time to be complete.[14] I should like to mention a text here which I came across by chance among the treasures of the libraries in Florence. It is a letter[15] sent to Pope Leo X, in which it is announced to him that 'in the course of the present years 1513 and 1514', because of the conjunction of Saturn, there will be a 'change such as has not occurred for three hundred years'; it further specified that 'this change will take place in the religion of men, so that one sect will destroy another'. It goes on to predict the downfall of the Turks within the space of the decade, exhorting the Pope: 'Hasten, therefore, Holy Father to direct your weapons against them . . .' But it prophesies that 'in the year 1514 in April and May the heavens will reveal a future great misfortune to Christians', and in 1519 and 1520 there would be disturbances in the Church, from which however 'a general reformation will take its beginning . . . ,' and this would reach no less a point than the unification of Rome and the Eastern world. In those years, which fell under the sign of the Scorpion, 'perhaps a pseudo prophet will appear' who, against the Church of Rome 'will spew forth his poison . . . like a scorpion with venom in his tail . . .': the sign of the Scorpion is in fact 'most dangerous to our religion, as the wise say . . .' Such are the subjects revealed by the letter of 'doctor Melchior Scotus Pannonius, humilis servitor Sanctitatis Vestrae', as the author signs himself. I mention this text in a summary manner only because it does not directly concern the Council of Pisa, but I should like to bring it to the notice of interested scholars in that, as far as I know, it is among those which have until now escaped their attention.

On the other hand, what are well known, and bear directly on my topic, are some writings of a Vallombrosan anchorite, by the name of Angelo: five letters from 1511, of which three are addressed to Pope Julius II, one to the King of France, Louis XII, and the last to Cardinal Bernardino Carvajal.[16] Angelo was clearly hostile to the idea of a Council

---

[13] See Prosperi, 'Il monaco Teodoro: note su un processo fiorentino del 1515', *Critica storica*, 11 (1975), 71–101.

[14] Such a study has already been announced by G. Miccoli, 'La storia religiosa', *Storia d'Italia*, ed. G. Einaudi (Turin, 1974), II. i. 955 ff.; cf. also D. Cantimori, 'Le idee religiose del Cinquecento. La storiografia', *Il Seicento, Storia della Letteratura Italiana*, ed. E. Cecchi and N. Sapegno, 5 (Milan, 1967), 37.

[15] I refer to the later copy—perhaps eighteenth-century—in the Bibl. Marucelliana, MS B III 65.

[16] Published *Mansi*, 35 (1902), cols. 1563–73.

convened against the Pope. It was summoned on 16 May by those very same Cardinals who had been 'plotting something': Francesco Borgia, Guillaume Briçonnet, and Carvajal himself. The motive they had adopted for taking their grave decision was the necessity of opposing Papal abuses in Church governance and the need to work for re-establishing peace. The chroniclers do not fail to point out the connection between these actions to bring about the ecclesiastical Council and the raging spread of the plague: in May it was the turn of the city of Rodez, where its Bishop, François d'Estaing—who was to be one of the most influential members of the Council of Pisa—had to rush, to find his See by then 'déserte'.[17] In that same May another leading actor in the Council, Cardinal Francesco Alidosi, was murdered: his assassin, Duke Francesco Maria Della Rovere, was to be envisaged suffering in the flames of Purgatory by another seer.[18]

No alleged sign from the heavens succeeded in stopping either d'Estaing or the other prelates occupied with the Council's preparation. In June of that year, 1511, the Abbot of Monte Subasio, Zaccaria Ferreri, published the text of several lectures given in Milan, in which he claimed the right to summon a Council even without the Pope's assent, and appended to it the proceedings of the Council of Basle.[19] At the same time the Emperor Maximilian I gave the Council his solemn support.[20] On 18 July the King of France wrote to the Cardinals announcing that he had elected 'a large number of prelates, theologians and lawyers' to the Council.[21] The convocation of the Lateran Council by Julius bears the same date: the Pope had in fact decided to oppose the Cardinals' action and take the initiative for the Church's reform in his own hands.[22] However, few succeeded in convincing themselves of this warrior-Pope's conversion to pastoral activity and, in particular, to his genuine adoption of the Council as an instrument to that end: in the same month an informant of Isabella, Marchioness of Mantua, wrote that dissension was worming its way even into the Cardinals who had remained faithful to the Pope, with the comment, 'There will be no peace, because the

[17] C. Belmon, 'Un evêque français à l'assemblée de Tours et au Concile de Pise: François d'Estaing', *Revue des études historiques*, 89 (1923), 305.

[18] C. Vasoli, 'Il notaio e il "Papa angelico". Noterelle su un episodio fiorentino del 1538–1540', *Religione e Civiltà* (Bari, 1982), 636.

[19] See B. Morsolin, *L'abate di Monte Subasio e il concilio di Pisa* (Venice, 1893), 10.

[20] See M. le Glay (ed.), *Négociations diplomatiques entre la France et l'Autriche* (Paris, 1845), i. 416.

[21] Quoted by Renaudet, *Concile de Pise-Milan*, 74.

[22] *Mansi*, 32 (1910), col. 561.

Pope desires too many dishonest things.'[23] So, even if all was not going well for those favouring the Council,[24] it seemed, at least for the time being, that it was the Papalists who were worried: 'Angelus peccator anachorita Vallisumbrose' sent the Pope the text of an 'oratio' he had given as a show of faith towards him. It was 8 September, and Florence had already given the Cardinals Pisa as the seat for the proposed Council, and royal and imperial attorneys had already started to arrive. Angelo declared he was striving 'for the downfall of the Pisan conventicle ... dangers and persecutions notwithstanding', and sought to reassure the Pope 'that the whole ecumenical Church follows, supports and venerates him ... And every creature exclaims "vivat Julius"'.[25] The text of this *oratio* is primarily an attempt to present Julius as an honest and peaceful shepherd of souls,[26] and therefore deny the Cardinals' initiative any legitimacy;[27] it further lists the reforms deemed necessary and looks forward to the summoning of a crusade against the enemies of Christianity. One month later, on 7 October, the anchorite turned on Bernardino Carvajal, 'tunc cardinali':[28] 'in what spirit, with what audacity and temerity do you sacrifice and destroy the Body of Christ? ... Return to your own heart! Give back yourself to yourself! ... Give back our doctor to us!' If he persisted in tearing the seamless robe of Christ, God's judgement would be visited upon him.

In reality the 'former' Cardinal was enjoying success with his and his colleagues' initiative, as many French delegates were arriving in Pisa.[29] Another fortnight passed and the zealous anchorite addressed himself to the Pope once again, to offer him, he said,[30] some comfort, and he related a dream in which it appeared that he had to confront a rabid dog, which, according to the anchorite, represented the King of France; then, at a certain point, a priest arrived, whom he identified as Cardinal Carvajal, to give the dog a helping hand. Three days later in another message to Julius,[31] Angelo da Vallombrosa promised to use his 'mordace calamo' for confuting the arguments of the *doctores* who were

---

[23] Morsolin, *L'abate*, 16.

[24] Cf. ibid. 14–15.

[25] *Mansi*, 35 (1910) col. 1563.

[26] Ibid., col. 1565.

[27] The Dominican Alberto Pasquali was working to a similar end at the same time, cf. J. Mirus, 'On the Deposition of the Pope for Heresy', *AHP* 13 (1975), 233.

[28] *Mansi*, 35 (1910) col. 1570.

[29] As an observer of the Marquis of Mantova writes on the 11th of that month: Morsolin, *L'abate*, 20.

[30] *Mansi*, 35 (1910) cols. 1572–3.

[31] Ibid., col. 1573.

working 'to build up the defence and strength of the gathering at Pisa', and burst out into emphatic lamentations 'O profane outrage! O horrible crime!' God was therefore breeding dogs to safeguard his Church!

Nevertheless at least one of these dogs was not unaware of the spread of apocalyptically prophetic literature: Bernardino Carvajal was an educated man[32] and, as such, gives confirmation—if any were needed—of the fact that such literature was very far from finding an audience only among simple, uneducated people. As Nelson Minnich shows,[33] his concern with prophecy goes back to the early part of his career.[34] He was, of course, a prime mover in the famous episode of the opening of the Amadeite *Apocalypsis Nova* in 1502, and on this subject, the correspondence of Carvajal's friend, Salviati, is illuminating.[35]

On 18 June 1502, Salviati wrote to the Florentine Ubertino Risaliti that the book composed by Amadeus had been 'held by Sixtus [IV], and later by Innocent [VIII], but neither of them had been courageous enough to open it' for the good reason that 'certain friars' who had attempted it had several days later died. But now, however, Cardinal Carvajal, '*uomo doctissimo* in every social grace, begged by certain people', and with the approval of the Minister-General of the Franciscans, ordered the book to be opened; a task which they had entrusted to none other than Salviati himself. The friar goes on to relate that he had 'not a little fear', until a Bishop of the Order 'who was with the Cardinal of S. Croce' (that is, Carvajal) encouraged him to do it in his company. And so during Mass in the church of San Pietro in Montorio, together they read the mysterious pages in which not Amadeus 'sed Gabriel Angelus' predicted the reform of the Church and the conversion of all unbelievers, and even the election of a new shepherd of souls for a Church 'pacatam, puramque ac nitidam'. Ubertino replied to this letter

---

[32] 'Evident from his nobility, his letters and his manners' as defined by F. Guicciardini, *Storia d'Italia*, ed. C. Panigada (Bari, 1929), IX. 49. 'El miembro mas destacado per sus dotes de inteligencia de todo el colegio cardenalicio' says J. Doussinague, *Fernando el Católico y el cisma de Pisa* (Madrid, 1946), 123.

[33] See below, Minnich, Ch. 6.

[34] For the biography of the Cardinal, see G. Fragnito, *DBI* 21 (Rome, 1978), 28–34. Very little of his activity as a theologian and of his thought is known: see H. Rossbach, *Das Leben und die politisch-kirchliche Wirksamkeit des Bernardino Lopez de Caraval, Kardinals von Santa Croce in Gierusalemme in Rom, und das schismatische concilium Pisanum* (Breslau, 1892); see also A. Morisi, *Apocalypsis Nova Richerche sull'origine e la formazione del testo dello pseudo-Amadeo* (Studi Storici, 77; Rome, 1970), 37.

[35] He derived this name from having been adopted by the celebrated Salviati family, see C. Vasoli, *Profezia e ragione* (Naples, 1974), 15–127; see also below, id., Ch. 7. He had also been the theological adviser of Lorenzo the Magnificent, see A. Fabroni, *Laurentii Magnifici vita* (Pisa, 1784), i. 159; ii. 162.

asking for a copy of the text, and Salviati wrote to him again (on 20 July) with the recommendation that he keep it secret, because the Cardinal 'holds this book an oracle . . . *ita et taliter* that we can no longer see it, but—he adds—I have extracted several passages *de verbo ad verbum*'. Several days later (on 13 August), he assured his correspondent that he was endeavouring to live more evangelically, in conformity with the Amadeite prophecy, but he specified however that he did not think he was the future Pope of the prophecy: the Cardinal of S. Croce—he declared—was making the self-same efforts.

The correspondence between the two continued, and in October Salviati repeated that no one but the Cardinal and himself could read the book, which he describes as 'remaining in his [the Cardinal's] room and nowhere else, because it may not be taken outside'. It seems that the Cardinal could not prevent Salviati consulting the book, given that the latter says 'I read it all before he did'; it also seemed likely to the Cardinal that Benigno Salviati, if he was not to become the prophesied Pope, could at least become one of his Cardinals. Finally, quite simply, the Cardinal did not wish people to know the book was in his hands, and to those who asked even the minimum of information, he said: 'believe me, they are things regarding the friars and their devotions!'

We owe the transcription of these letters to one Michele da Trecate, an Amadeite, who also refers to conversations he had with other witnesses of the book's opening, one of whom confirmed the facts in a conversation with him many years later on 8 March 1513.

What can be said of this episode and the role played in it by the Spanish Cardinal?[36] As Cesare Vasoli shows, it seems certain that Salviati's hand can be detected in the extant text of the *Apocalypsis Nova*. The involvement of Carvajal is equally undeniable, if only in the sense that the redrafting of the prophecy happened in his entourage, and that he also intended to utilize it. Another Franciscan, Mariano da Firenze,[37] maintained that the Cardinal was 'too complacent and flattering towards himself, easily believing from certain things contained in the book that he would be the next pope', adding 'therefore, rashly and foolishly he collected together against Julius II the gathering at Pisa, so famous for the audacity of madmen'.[38]

Was the Cardinal of S. Croce using the prophecy as a tool for the

[36] I follow Vasoli, *Profezia e ragione*, 99.
[37] See C. Cannarozzi, 'Ricerche sulla vita di fra Mariano da Firenze', *Studi francescani*, 27 (1930), 31–71.
[38] Morisi, *Apocalypsis Nova*, 27.

advancement of his career?[39] It would be interesting to know how much influence the pseudo-Amadeite text had on the troubled climate of the last months of 1502, which was nearing the end of Alexander VI's pontificate, when the struggle between the opposing Curial factions was becoming more bitter, and later in the next year, at the time of the two conclaves, on the opposing ambitions of the French candidate, the powerful Georges d'Amboise, and the Spanish candidates, Bernardino Carvajal among them, and on how, in the end, the simoniacal agreement on Giuliano Della Rovere was reached. It is a fact that the Spaniard came out of that struggle defeated. But the humiliation he suffered did not quash his hopes: afterwards the evident difference between Julius II's running of the papacy and the dream of a reformist Pope hypothesized by the prophecy was one of the elements that unleashed the rebellion that led to the Council of Pisa. Perhaps Carvajal was only an *arriviste*, as Pastor says;[40] certainly differences of a purely political kind made him opposed to the Pope;[41] however I do not believe there can be any doubts about a sincerely reformist component among the promoters of that Council which has been too quickly stigmatized as a sham and schismatic by current historiographers.[42]

It should not be forgotten that the concept of Conciliarism had a content shared with the apocalyptic-prophetic strand whose defeat by the ecclesiastical hierarchy did not go fortuitously *pari passu* with that of the Conciliarist movement.

There is another aspect in the life of Bernardino Carvajal which not only merits but demands attention by historians, namely his continuing support for Amadeite ideas. In his later years, it appears that the Cardinal, who had by then returned to observing full obedience of the Pope, when he found himself having to face the problem of Martin Luther's condemnation in the preparatory discussion to the Bull of Excommunication (the *Exsurge* which was issued on 15 June 1520), may not have agreed with the idea of inserting the phrase whereby the 'reckless' appeal which the monk had raised to a Council against papal authority was to be considered one of his worst errors.[43] According to a contemporary witness, the decision to insert the phrase was taken 've-

---

[39] Again, the observation is Vasoli's; see *Profezia e ragione*, 100.

[40] Pastor, *History of the Popes*, vi. 387.

[41] But see below, Jungić, Ch. 17, for Carvajal as a reformist.

[42] See W. Ullmann, 'Julius II and the Schismatic Cardinals', in G. Cuming and D. Baker (eds.), *Schism, Heresy and Religious Protest* (Cambridge, 1972), 177–93.

[43] See C. Mirbt and K. Aland, *Quellen zur Geschichte des Papsttums und des Römischen Katholizismus* (Tübingen, 1967), i. 509–10.

hementer obsistente cardinale S. Crucis'.[44] This information, taken as true by P. Kalkoff[45] in his day and later taken up, although with some reservations, by H. Jedin,[46] has recently been placed in doubt by R. Bäumer[47] who attributes it to a tendentious exaggeration ('eine tendenziöse Übertribung') put about in the hopes of obtaining a last-minute revocation of the Bull that threatened Luther with excommunication.

This little *querelle* is destined to remain, because the arguments of Bäumer do not appear to be totally convincing. What is certain is that little more than a year before his death this man, who had cherished the dream of becoming the reformist Pope—to which end, according to the testimony of Salviati, he had made efforts at self-reform—welcomed the pious Adrian of Utrecht to Rome (who, with Carvajal's vote, had become Adrian VI) and in his presence made a vibrant speech in which he invited the Pope to renew the Church on the bases of the ancient Councils and Laws.[48]

---

[44] See Erasmus, *Opuscula* ed. W. Ferguson (The Hague, 1933), 322.

[45] P. Kalkoff, 'Zu Luthers römischen Prozess', *ZKG* 25 (1904), 120 ff.

[46] H. Jedin, *The History of the Council of Trent*, trans. D. Graf (London, 1957) i. 175 n. 2.

[47] *Der Lutherprozess*, in *Lutherprozess und Lutherbann, Vorgeschichte, Ergebnis, Nachwirkung*, ed. R. von Bäumer (Münster, 1972), 37–40. In his *Martino Lutero* (Milan, 1985), i. 680–1, R. Garcia-Villoslada attributes the news to the 'exasperation of the Lutheranists'.

[48] The sermon was edited by C. von Höfler, *Abhandlungen der Münchner Akad., hist. Kl.*, 4, 3 (1846), 57–62. On this sermon, see below, Minnich, Ch. 6.

4

Prophecy and the Fifth Lateran Council (1512–1517)

Nelson H. Minnich

On two occasions the Fifth Lateran Council issued decrees either prohibiting or limiting the foretelling of future events. In the Great Reform Bull of 1514 it severely condemned any attempts to predict the future by recourse to divinations, incantations, or the invocation of demons. In its decree on preaching, *Supernae majestatis praesidio* (1516), it addressed the question of clerics claiming to have a revelation from God. The Council forbade them to predict in their sermons any fixed time of future evils, of Antichrist's coming, or of the day of the Last Judgement. Clerics in their sermons were not to base their predictions on interpretations of Sacred Scripture, nor claim to have their knowledge of the future from the Holy Spirit or through divine revelation, nor seek to prove their statement by foolish divinations. On the other hand, this decree allowed for the possibility of a true revelation from God and set up procedures for testing it before it was announced to the people.[1] These decrees raise a number of questions: how widespread and serious was the problem of prophecy at the time of the Council? why was the Council called upon to issue its decrees? who were their authors? and why were they so formulated?

The practice of divination was widespread in Renaissance society. In educated circles theories of astrological and natural magic were advanced by members of the Florentine Academy and found acceptance among many learned ecclesiastics. Although criticized by humanists and theologians such as Gianfrancesco Pico della Mirandola (1469–1533) and Tommaso de Vio (Cajetan) (1469–1534), who both attended

---

[1] *Concilium Lateranense V generale novissimum sub Julio II et Leone X celebratum*, ed. A. del Monte (Rome, 1521), repr. *Mansi*, 32 (1438–1549), cols. 884DE, 945E, 946C–947B.

the Lateran Council, these theories and the practices based on them were not the targets of the Lateran condemnation.[2]

What attracted the attention of church reformers were the less sophisticated superstitious practices of divination prevalent throughout society. Cajetan, writing at the time of the Council, attacked divination and its various ways of trying to discern future events that use 'neither divine revelation nor the strength of natural light but implicitly or explicitly daemonic vanity', by having recourse to such things as auguries and the invocation of demons.[3] His opposition to the use of lots, dreams, and stars was muted because of their biblical precedents. Other writers of the time also wanted to eliminate these forms of superstition. The two Venetian patricians turned Camaldolese hermits, Paolo Giustiniani (1476–1528) and Pietro Querini (1479–1514), listed some of these practices in their *Libellus*, probably penned in Rome during the summer of 1513 and addressed to Leo X. They urged the Pope to impose harsh penalties on anyone engaging in divination through astrology, marking of days, interpretation of dreams, fortune-telling, palm reading, prestidigitation, magical arts, and recourse to water and fire or lines and figures. Their books on divination were to be destroyed and the practitioners themselves executed if they refused to renounce thoroughly these devilish arts. The hermits' friend, the Venetian patrician and lay-theologian Gasparo Contarini (1483–1542), in his *De officio episcopi* (1517), also attacked the arts of divination, magic, and astrology. These Venetian writers saw these and other superstitious practices as enjoying a wide acceptance in society, despite church legislation to the contrary. While the *Libellus* came to the attention of key persons in the papal circle, it was probably not the immediate source of the 1514 Lateran decree condemning these practices.[4]

[2] See W. Shumaker, *The Occult Sciences in the Renaissance: A Study in Intellectual Patterns* (Berkeley, Calif., 1972); D. Walker, *Spiritual and Demonic Magic from Ficino to Campanella* (Notre Dame, Ind., 1975).

[3] *Sancti Thomas Aquinatis doctoris angelici Opera Omnia iussu impensaque Leonis XIII P. M. edita*, 9: *Secunda secundae Summae Theologiae . . . cum commentariis Thomae de Vio Caietani Ordinis Praedicatorum S.R.E. Cardinalis* (Rome, 1897, hereafter cited as *ST*), 312. For Cajetan's opposition to astrology, magic, and divination, *Tractatus duodecimus De Maleficiis, Opuscula omnia* (Lyons, 1567), 180 and *ST* 9 nn. 312–14, 318–19, 321–2, 325, 332–3.

[4] P. Giustiniani and P. Querini, 'Libellus ad Leonem X Pontificem Maximum', in *Annales Camaldulenses Ordinis Sancti Benedicti*, ed. G. Mittarelli and A. Costadoni (Venice, 1763), vol. ix, cols. 683–4, 686; G. Contarini, *De officio episcopi, Opera*, ed. L. Contarini (Paris, 1571), col. 424; G. Gragnito, 'Cultura umanistica e riforma religiosa: Il *De officio viri boni ac probi episcopi* di Gasparo Contarini', *Studi veneziani*, 2 (1969), 167–8 n. 353, 187–9. For an overview of the extent of and mentality behind popular superstitious practices, see K. Thomas, *Religion and the Decline of Magic* (London, 1971).

The inclusion of this condemnation in the Great Reform Bull is most likely due in good part to another Venetian, Stefano Taleazzi ($c.1445$– 1515). Towards the end of his long career as a curial prelate and noted preacher in the prophetic tradition, Taleazzi composed a work in ten books on the primacy of the Catholic Church that contained many of his ideas on Church reform. On the urgings of Cardinal Lorenzo Pucci, who was an important member of the conciliar reform deputation and, together with Cardinal Pietro Accolti, was entrusted with giving final form to the conciliar reform decree, Taleazzi made a summary of his various proposals for the consideration of the Lateran Council.[5]

On 15 January 1514 he and other clerics in Rome were urged to present their ideas on Church reform to Cardinal Raffaele Riario, the chairman of the conciliar deputation charged with drawing up the Council's reform decrees.[6] The Great Reform Bull of 5 May 1514 is remarkably similar to many of Taleazzi's proposals. Its section condemning any attempts to predict the future by recourse to superstitious practices is very close to the third of Taleazzi's major proposals[7] and it is reasonable to suspect this suggestion as the source of the Bull's

[5] On Taleazzi, see B. Feliciangeli, 'Le proposte per la guerra contro i Turchi presentate de Stefano Taleazzi vescovo di Torcello a papa Alessandro VI', *Archivio della Reale Società Romana di Storia Patria*, 40 (1917), 5–63; for his account of why he penned the summary of his various reform proposals, see his dedicatory letter to Leo X prefacing his 'Summarium de triplici reformatione necessaria pro Concilio Lateranensi', Bibl. Valentiniana di Camerino, MS 78, fo. 1^{r-v}: Eamobrem Reverendissimi Domini Cardinalis Sanctorum quattuor Coronatorum, qui communis quoque salutis zelo maxime afficitur, hortatu ac monitis impulsus, de triplici periclitantium Christianorum reformatione, quam ad succurrendum eorum saluti cito factum iri oportet, summarium perpaucis aedidi a tua beatitudine corrigendum, cui tres etiam tractatus haud quidem inutiles de expeditione adversus infideles castigandos direxi. Nam eo in opere quod in decem libris distinctum de praestantia Christianae fidei ac praeeminentia apostolicae sedis sedentisque in ea, deo favente inscripsi, quod et prope diem tuae divinae monestrati dicatum in lucem prodibit quae et pro universali Ecclesiae reformatione et pro universi orbis ad Christum reductione fieri expediant latissime continentur. On Pucci as a member of the conciliar reform deputation, see Mansi, col. 796E; for his role in putting a reform decree into its final form, see the letter of Francesco Vettori to the Dieci di Balia, Rome, 24 Nov. 1513, Archivio di Stato di Firenze, Dieci di Balia, Carteggi, Responsive, 118, fo. 317^r: 'non si resta per questo [concluding a truce between the Empire and Venice] non si pensi al concilio et di rassectare molte cose circa la religione a che sono proposti li Reverendissimi di Santa Croce et Grimanno et Flischo: rassecteranno anchora li uffici et molte altre cose che sono transcorse et a questo atendono li Reverendissimi Pucci et Accolti.'

[6] *Mansi*, cols. 796E, 847CD.

[7] Taleazzi, *Summarium*, fo. 9^v: Tertio quod episcopi locorum teneantur singulis annis diligenter perquirere et formare processus contra male sentientes in articulis fidei et contra sequentes praestigia, incantationes, sortilegia, et alia prohibita a lege et decretis sanctorum patrum; et non minus contra Judaizantes, marranos et prophetantes de corde suo, ita quod unus quisque purget diocesim suam singulis annis et in necessariis sedem apostolicam consulant.

section condemning fortune-telling, incantations, and other practices forbidden by civil law and church decrees.

The period prior to and during the Fifth Lateran Council was noted for its profusion of prophetic preachers. In the pulpits and piazzas of Italy of the late fifteenth and early sixteenth centuries learned friars, dishevelled hermits, and itinerant troubadours announced the impending punishment of God to be followed by a renovation of Church and society. Men found in the course of the stars, in the events of their day, and in the texts of Sacred Scripture, whose secret meaning was now unlocked with the help of systems of numbers and figures, signs that the End-Times were at hand.[8] Similar in their message of imminent woes, to be followed by restoration, were three key 'prophets' who shaped these predictions: Joachim of Fiore, Amadeus of Portugal, and Girolamo Savonarola. These men enjoyed reputations for personal holiness: the former Cistercian abbot was venerated as a saint in Calabria, the Franciscan hermit was listed among the blessed of his order, and the Dominican friar was proposed for canonization first by Julius II and then by the Council of Pisa—despite the condemnation of Joachim's Trinitarian theology by the Fourth Lateran Council, the attack on Amadeus' revelations by the master general of the Dominicans, and Savonarola's execution for schism and heresy.[9] Similarly, although most of their followers were considered orthodox Christians, at the time of the Fifth Lateran Council or just prior to it, a few were condemned and silenced by Florentine and Roman church authorities. One of these cases may well have become the occasion for the Council itself to issue a decree on prophecy.

The twelve homilists who addressed the Lateran Council often wrapped themselves in the mantle of a prophet. They claimed that the Church was in its last age, that moral corruption was widespread, and that God would punish Christendom if the Council failed to bring about a re-

---

[8] On the general popularity of prophetic preachers in this period, see below, Niccoli, Ch. 10.

[9] On Amadeus, see above, Morisi-Guerra Ch. 2. See also A. Dominques de Sousa Costa, 'Studio critico e documenti inediti sulla vita del beato Amadeo da Silva nel quinto centenario della morte', in I. Vasquez Janeiro (ed.), *Nascere sancta. Miscellanea in memoria di Agostino Amore, OFM.* (+*1982*) (Rome, 1985), 101–360. For Cajetan's condemnation of the revelations attributed to Amadeus, see his commentary on *ST* ix. 400. For works on Savonarola, see above, p. 19. On Savonarola's reputation for holiness and the proposal to Cajetan that the Council of Pisa would canonize him in return for the master general's support, see T. Neri, *Apologia*, 196–7. On Julius II's claim in 1509 that he wanted to canonize Savonarola, see R. Ridolfi, *La vita di Girolamo Savonarola*, 2nd edn. (Rome, 1952), ii. 37.

formation. Continued Moslem victories and schism in the Church were seen as the means with which God would chastise an unrepentant Church. On the horizon, however, was dawning a new age. Let the Council purify the Church and usher in the glorious End-Times.[10]

Perhaps the most famous of these prophetic conciliar homilists was the general of the Augustinian friars Egidio Antonini of Viterbo (1469–1532). In his inaugural address to the conciliar fathers on 3 May 1512 he reminded his listeners that for the past twenty years he had traversed Italy to explain to the people the message of St John's Apocalypse, namely, that the current agitation in the Church would soon lead to its correction. Throughout his speech he pointed to connections between prophecies in the Old Testament, comments and events in the New, and their final fulfilment in the history of the Church. With the Bible as his guide for reading the signs of the times, he proclaimed that the many recent portents and disastrous events such as the battle of Ravenna were warnings from God to the Pope to hold this Council that would reform the Church, end warfare, and restore peace. The opening of the Council was for Egidio the beginning of the long-awaited renewal of the Church, foretold in the Apocalypse. While his listeners found his prophetic message so moving that many wept, the sources for his ideas remain difficult to determine and their complexity is the subject of a separate study.[11]

The Joachimist tradition was represented at the Lateran Council by a number of prominent prelates. Of the sermons given at the Council, the longest discourse, that given at the ninth session by Stefano Taleazzi, may echo the Calabrian abbot. Also present in the Lateran were the reconciled former leaders of the Pisan Council, the Joachimist Zaccaria Ferreri (1479–1524) and Bernardino López de Carvajal (1456–1523).[12]

---

[10] For a summary of their sermons, see N. Minnich, 'Concepts of Reform', 168–207, 234–5. Gianmaria del Monte invoked the imagery of a restoration of justice, a theme from Lactantius (Mansi, col. 779A), Battista de Garghis called for the return of the golden age (ibid. 854C), while Baltassare del Rio cited a Moslem prophecy, confirmed by astrologers and recorded in an Armenian history, that Islam was on the verge of imminent collapse (Mansi, col. 823A). Most other preachers repeated the general message of the Christian prophets of their day.

[11] On Egidio's apocalyptic thought see below, Reeves Ch. 5; on his preaching style, see Niccoli, Ch. 10. The best edn. of his conciliar homily is that by C. O'Reilly, ' "Without Councils we cannot be saved . . .": Giles of Viterbo Addresses the Fifth Lateran Council', *Augustiniana*, 27 (1977), 184–204. For his autobiographical comment, see p. 185; for connections drawn and prophecies fulfilled, see pp. 188–9, 192; for his interpretation of the signs of the times, see pp. 185, 202.

[12] For a listing of the sessions at which they were present, see N. Minnich, 'The Participants at the Fifth Lateran Council', *AHP* 12 (1974), 181 n. 13, 184 n. 91, 186 n. 143, 195 n. 387.

Elements of Joachim's eschatology seem present in the vision of salvation history depicted in the sermon of Taleazzi.[13] This aged prelate
claimed through ecstatic contemplation to find contained in psalm 47
('Great is the Lord . . .') the course of history from creation to the
world's consummation. The turning points for Taleazzi are the two
great Sabbaths.[14] Following the fall of Lucifer and Adam, God brought
about the repair of His creation in Christ. The first Sabbath began with
Peter's confession of faith (Matt. 16: 16) and was built up through the
work of the fishermen (the apostles predicted in Jer. 16: 16). In time
Christianity triumphed in the conversion of Constantine who acknowledged Christ's lordship and surrendered his sceptre to His vicar, Pope
Sylvester. But later rulers of things temporal and spiritual failed to
follow the true meaning of the words of Christ, the prescriptions of the
saints and councils, and the decrees of the pontiffs. They so devoured
the Christian people that the Church could not effectively resist the
attacks of heretics and schismatics and the persecutions of the impious,
especially those of the Sultans and Shahs of Islam. Due to the neglect
of Christian leaders, the Church is much afflicted by three forces of evil:
the jurists representing the knowledge of the world, the Moslems that
of the flesh, and the Averroists that of the devil.[15] By embracing the
knowledge of God, the Lateran Council can initiate the sacrament of
reformation for the second Sabbath. According to the computation of
times this great and final reformation of the Church and world, exaltation of the Apostolic See, and extirpation of the enemies of the Faith
is near at hand and will occur through the agency of the 'spiritual hunters'
(predicted in Jer. 16: 16). Owing to the tearful prayers of those engaged
in both the contemplative and active life (Mary and Martha; John
11: 20–33) and to the work of the glorious Virgin (perhaps a reference
to John 2: 3), Christ will come in spirit to restore life. He will call the

---

[13] On Taleazzi's apocalyptic thought, see his sermon *De futura tribulatione* (1 Dec.
1477), dedicated to Cardinal Giovanni Archimboldi, Milan, Bibl. Ambrosiana, MS J 258
inf., fos. 116^r–128^v, esp. fo. 122. For his *Sermo in materia fidei* (27 Dec. 1480), which
warned of a Moslem conquest of Rome and Italy as a punishment for the sins of Christians,
see J. O'Malley, *Praise and Blame in Renaissance Rome: Rhetoric, Doctrine and Reform
in the Sacred Orators of the Papal Court c.1450–1521* (Durham, NC, 1979), 10 n. 6, 190,
237. In the introduction to his reform proposal for the Lateran Council, Taleazzi claimed
to understand by a clear intuition from reading the Law, Psalms, Prophets, and Apocalypse
and according to the computation of times that the world would soon be converted to
the Catholic Faith and the Church would enter into its seventh *status* with the descent
from heaven upon the Roman Church of the New Jerusalem ('Summarium', fo. 3^{r-v}).

[14] *Mansi*, cols. 916E–917C, 919CD. For Joachim's use of the concept of two Sabbaths,
see M. Reeves and B. Hirsch-Reich, *The Figurae of Joachim of Fiore* (Oxford, 1972),
122–8; B. McGinn, *The Calabrian Abbot: Joachim of Fiore in the History of Western Thought*
(New York and London, 1985), 153–4.

[15] *Mansi*, cols. 919D, 921AB, 923A–925D.

clerical *status* (symbolized in Lazarus, dead for four days; John 11: 17) to the contemplative life. The *status* of the four religious *rotae* (prefigured in the dead daughter of the ruler of the synagogue; Mark 5: 35–43) will be recalled to their first life when the Spirit was in the *rotae*. And the *status* of the laity (foreshadowed in the deceased son of the widow of Nain; Luke 7: 11–17) will be called to an observance of the Catholic Faith and of the Apostolic teaching, as if by remembering the first fruits of our Faith, and will be led to a most dutiful obedience to the Roman Church. Liberation from the persecutions of the Sultan, Shah, and other impious men will be effected not by human power but by God, miraculously and gloriously. By the work of the Holy Spirit the Church will enter the seventh *status* and enjoy true peace. There will be one flock and one shepherd until the consummation of time.[16]

If the preachers at Lateran V did not cite Joachim by name as the source of their inspiration, one of the principal promoters of the Pisan Council, Zaccaria Ferreri, did indirectly in his oration at the anniversary Mass in the cathedral of Lyons on 1 November 1512. This abbot of San Benedetto di Monte Subasio in Umbria referred to Joachim as the *egregius propheta*, *ipse Abbas*, and *praelibatus Abbas* whose allegorical interpretation of the assembly of the sealed elect before the throne of the Lamb (Apoc. 7: 3–9) pointed to a future general council that would gather to renew the faith and morals of both Christians and Moslems and bring together all nations in the Christian religion. When it became clear to Ferreri that his identification of the Pisan Council with the assembly predicted by Joachim was not accurate, he renounced this schismatic council, made new predictions about a glorious pontificate for Leo X, obtained absolution from this Pope, and attended the Lateran Council, being listed at its eighth session as *abbas Sebastiensis* and at the last two congregations and final session as being bishop of Sebaste.[17]

---

[16] *Mansi*, cols. 918C, 919CD, 920A–920C, 921B, 922D–923A, 925AB; the 'spiritual hunters' of Taleazzi may echo Joachim's 'spiritual men', see M. Reeves, *Influence of Prophecy in the Later Middle Ages: A Study of Joachimism* (Oxford, 1969), 135–44; the four religious *rotae* may have as a referent Joachim's use of the four wheels of Ezekiel, see Reeves and Hirsch-Reich, *Figurae*, 224–31; in one of Joachim's concordances the four wheels refer to the four orders of doctors, martyrs, pastors, and contemplatives (ibid. 228). One suspects that Taleazzi is here referring to the four states of religious life, perhaps hermits, monks, canons, and mendicants, or to the four mendicant orders. On the miraculous conversion of the Moslems, see Mansi, cols. 920A, 925A, 929A. Augustinian inspiration may also be inferred from Taleazzi's reference to the City of God and seventh *status* of the Church, see ibid., cols. 920B, 923A, 925B.

[17] On Ferreri's activities at the Pisan Council, see *Acta et decreta sacrosancta secundae generalis Pisani Synodi prout per protonotarios et notarios summarie scripta reperiuntur* (Paris, 1612), 79 (helps to prepare site in Pisa); 91 (appointed conciliar notary); 82–3, 94, 122–30 (preaches at first, second, and fifth sessions); 98 (scrutator of votes); 106 (reads Gospel); For Ferreri's references to Joachim, see *Promotiones et progressus sacrosancti pisani concilii*

Of great interest are the attitudes towards prophecy of Cardinal Bernardino López de Carvajal. His earlier dedication to the causes of a church reform and a crusade against the Moslems was apparently re-energized during the dark days of Alexander VI's reign (1492–1503) by his contact with the prophets who foretold the imminent achievement of these cherished causes. To what extent prophetic ideas may have guided his actions at Pisa and later on in his career is studied elsewhere in this volume.[18] To what extent they may have influenced his role in shaping the Lateran decree on prophecy is addressed later in this article.

The prelate at Lateran V who was most deeply involved in the prophecies of Amadeus was the Bosnian Franciscan known in Italy as Giorgio Benigno Salviati (c.1448–1520).[19] Indeed prophecy had a central position in the thought and actions of this complex person. In his writings over the years, Salviati consistently defended the legitimacy of prophecy after the time of Christ. He insisted that God can and does communicate to angels and man a knowledge of future events: before God acts, He usually first reveals His plans to one of His prophets (Amos 3: 7). As examples of such prophets, Salviati cited St John of the Apocalypse, St Methodius, Joachim of Fiore, and St Bridget of Sweden. He also defended Savonarola for a while and propagated the ideas of Amadeus, his correligionist.[20] On two occasions, thus Salviati recounts, this holy friar had responded with joy to his presence because he knew that God had chosen him for an important role. When they first met in Rome (c.1472) the hermit reportedly embraced him and proclaimed him to be a learned friar. And in Milan in 1482 the body of the recently deceased Amadeus suddenly rose up as if in jubilation from its bier at the very moment Salviati entered that city.[21] In his 1499

---

moderni indicti et incohati anno domini M.D.XI (n.p., n.d.), fo. 44[r–v]; he also preached at the ninth session (fos. 42[v]–43[r]). For Ferreri's presence at the Lateran Council, see Mansi, cols. 830E, 937C, 978B, 982C. For a brief overview of his career, see A. Ferrajoli, Il ruolo della corte di Leone X (1514–1516), ed. V. de Caprio (Rome, 1984), 531–44. For Ferreri's prophecy on the achievements of Leo's pontificate, see his Lugdunense somnium . . . Sylva centesimadecima (n.p., n.d.), sig. Bii[v].

[18] See below, Minnich, Ch. 6.

[19] On Salviati, see below, Vasoli Ch. 7 and Index. See also B. Pandzic, 'Vida y obra de Jorge Dragišić, un humanista filosofo y teologo croato en al Renacimento italiano', Studia Croatica, 11 (1970), 114–31.

[20] C. Vasoli, 'Notizie su Dragišić', Studi storici in onore di Gabriele Pepe (Bari, 1969), 461–73.

[21] Ibid. 497–8; A. Morisi, Apocalypsis Nova: Richerche sull'origine e la formazione del testo dello pseudo-Amadeo (Istituto Storico Italiano per il Medio Evo, Studi Storici, 77; Rome, 1970), 32.

treatise *De natura angelica* he is said to have inserted passages taken from Amadeus' *Apocalypsis Nova*, but this assertion seems suspect, given the injunction that this book was to remain sealed until the election of the Angelic Pope.[22] At the 1502 Mass in S. Pietro in Montorio in Rome, Salviati was the friar ordered by the Minister General Egidio Delfini, on the urgings of Carvajal, formally to break the book's seal, and Salviati seems to have had a major hand in revising it. If the cardinal saw himself as the future Angelic Pope therein announced, the friar may have been convinced that he would be one of the ten oriental cardinals, or perhaps even the Pope. His highest ambitions were supported by prophecy and he certainly did not want to see its authority discredited.

While this Bosnian Franciscan had also been an open admirer and defender of Savonarola in the closing years of the fifteenth century, the person at Lateran V who was best known as a disciple of the Dominican friar was a layman and prince, Gianfrancesco Pico della Mirandola (1469–1533). Soon after attending the eighth session of that Council in 1513, he composed his famous *Ad Leonem Decimum Pontificem Maximum et Concilium Lateranense . . . de reformandis moribus oratio* and sent it to the Pope. This speech is a forceful denunciation of the moral corruption of the clergy and a call for the enforcement of current legislation so that a modicum of decent behaviour is restored. If this reformation is neglected, Pico threatened impending disasters. God Himself will amputate with iron and fire the diseased members and destroy them. The type of cure He has planned is already evident in the signs He has given: pestilence, famine, and the recent bloodiest battles. He has prepared for us a most bitter potion from the hands of the 'perfidious deserters from our religion'—probably a reference to the Moslems rather than a prediction of the Protestants![23] Although many of the distinctive elements of Savonarola's prophetic message are not apparent in this speech, its tone and the lifelong dedication of its author to the cause of the friar and his followers allows one to see it as representative of the Savonarolan message at the Council.[24]

---

[22] Vasoli, 'Notizie', 474–5 n. 109, cites sig. kiiii^{r–v} as containing passages from the *Apocalypsis Nova*: how this section, which addresses the question of desiring proper goods and the intermediate stages towards attaining them, relates to Amadeus's prophecies is unclear to me; Morisi, *Apocalypsis Nova*, 29.

[23] See especially C. Schmitt, 'Gianfrancesco Pico della Mirandola and the Fifth Lateran Council', *AR* 61 (1970), 161–78. Pico's speech went through seven editions, being last reprinted in W. Roscoe's *The Life and Pontificate of Leo the Tenth*, 2nd rev. edn. (London, 1806), vol. 6, appendix 146, pp. 65–77.

[24] Schmitt, 'Pico', 164, 177. On Pico and Savonarola, see ibid. 163 n. 12, 173 n. 44,

The Council father most intimately involved in dealing with the legacy of Savonarola was the master general of the Dominicans, Tommaso de Vio (1469–1534), known as Cajetan because of his birthplace of Gaeta. This famous theologian knew the prophet personally, and is reported to have said that Savonarola had one of the best understandings of Aquinas's theology. In his own writings, Cajetan seems never to have mentioned Savonarola by name, but he implicitly faulted him when he argued that anyone who falsely incriminated himself under torture is guilty of some sin because he retained, despite the pain, some elements of his free will. Cajetan did not openly address the question of the objective merits of Savonarola's case, but he seems to have worked quietly for his rehabilitation, turning down, however, the proposal of the Pisan Council to canonize him if the Dominicans backed that assembly. Indeed, Cajetan and the followers of Savonarola in the Tuscan congregation were ardent opponents of the Pisan Council. The general's concern was to end partisan divisions over Savonarola in that congregation. He intervened to protect his remaining followers and to bring peace, even if that meant dampening their fervour by prohibiting the election of Luca Bettini (c.1489–1527), who was one of Savonarola's ardent promoters, as vicar general of the Tuscan congregation, and supporting his expulsion from that congregation, thus leading Bettini to flee for protection to Gianfrancesco Pico. Cajetan claimed that on numerous occasions he had defended the Tuscan congregation which he 'loved, cherished, and exalted'. The new Roman province promoted by Cajetan incorporated Savonarola's followers in Florence, giving them control of the mother church, S. Maria sopra Minerva, which ensured their strong influence on the headquarters of the generalate in Rome.[25]

At the time of the Lateran Council Cajetan twice addressed the question of prophecy, once in passing, the other at some length. In his sermon before the second session on 17 May 1512, Cajetan argued

---

and D. Weinstein, *Savonarola and Florence: Prophecy and Patriotism in the Renaissance* (Princeton, NJ), 220–6.

[25] A. Cossio, *Il cardinale Gaetano e la Riforma* (Cividale, 1902), i. 119–27; T. Neri, *Apologia . . . in difesa della dottrina del R.P.F. Girolamo Savonarola da Ferrara* (Florence, 1564), 40; A. Giorgetti, 'Luca Bettini', 164–231, esp. 200; G. Benelli, 'Di alcune lettere del Gaetano', *AFP* 5 (1935), 363–75, esp. 364–6, 372–3; R. di Maio, *Savonarola e la curia romana* (Rome, 1969), 17, 126–7, 151–2, 160, 171, 238–9; R. de Roover, 'Cardinal Cajetan on "Cambium" or Exchange Dealings', in E. Mahoney (ed.), *Philosophy and Humanism: Renaissance Essays in Honor of Paul Oskar Kristeller* (Leiden, 1976), 423–33; J. Wicks, *Cajetan und die Anfänge der Reformation*, trans. B. Hallensleben (Münster, 1983), 28–30. For Cajetan's comments on false self-incrimination under torture, see *ST* viii. 135.

that the Roman Church was the new Jerusalem, the perfect city of the Apocalypse (21: 10–22: 5). It is the true Church because it has all its necessary parts: Cajetan lists among its ministers prophets, and among its gifts prophecies, and as if that were not enough, it even has the revelations, illuminations, and protection of angels.[26] He also took up the question of prophecy in his many comments on St Thomas Aquinas's treatment of the same in the *Summa Theologiae*. Cajetan probably wrote on this topic at about the same time that the Council was drafting its decree on prophecy.[27]

According to his commentary, prophecy is the revelation of knowledge known only to God, revealed to the human intellect, directly by God when this involves the supernatural, transitory *habitus* of a prophetic illumination, but mediately through angels who illumine men by strengthening the intellect or by teaching, distinguishing, and proposing perceptibly the thing revealed.[28] That God should choose in each period of human history to reveal His secrets to persons who are to communicate them to others is a procedure befitting man's social nature and is in keeping with the divine plan whereby members of the Church depend on one another.[29] To be a true prophet one need not be morally good, nor spiritual, nor given to ecstatic experiences. In fact, one can even be in a state of mortal sin at the time of receiving the revelation, although sin does make it more difficult to elevate the mind. God freely disposes whomever He chooses to receive this revelation. Even though he may not understand fully the intent of the message he receives, the prophet knows with a clarity of vision that God is revealing this to him and that he speaks from a divine instinct. Anyone who claims that he did not know what he was saying but had been raised to an ecstatic state in which Christ or some spirit spoke through him is either deceived or trying to deceive others, for Scripture clearly states that the spirits of prophets are subject to them (1 Cor. 14: 32). Nonetheless, simple people are often seduced by such sham.[30]

---

[26] *Mansi*, cols. 720E–721A.

[27] On the dating of Cajetan's commentary, see Cossio, *Il cardinale*, 499 nn. 42, 53. Pars II–II contained 189 questions, those dealing with prophecy being qq. 171–4. Cajetan probably began work on Pars II–II soon after completing Pars II on 29 Dec. 1511; he completed work on this second section on 26 Feb. 1517. If he wrote his commentary following the sequence of Aquinas's questions, his reflections on prophecy should have closely coincided with the Council's work on the same. Lateran V issued its decree on 19 Dec. 1516.

[28] *ST* x. 366–7, 379, 383.

[29] Ibid. 374.

[30] Ibid. 375, 378, 380–1, 382, 389–90, 390–1.

Cajetan provides three criteria for determining the truthfulness of the message. First, it must not add to the truths of Christianity. Thus, one should reject any 'revelation' that added new books to the canon of the Bible or introduced new sacramental rites.[31] Second, what is 'revealed' must not teach as licit what the Faith directly or indirectly says is illicit. Prophecies provide no justification for sinning. One cannot invoke the example of Abraham who was ready to sacrifice his son (Gen. 22: 1–19) since Abraham was already a proven prophet and there was no doubt about the message he received. Nor can one point to the adulterous wife of Hosea for she was guilty of her sins (Hos. 1: 2, 3: 1). One must have certitude that his action is good before he engages in it. By conforming to divine and human, especially ecclesiastical, laws one can enjoy this surety and avoid even the appearances of evil.[32] Third, according to the norms in canon law, anyone claiming to have received a mission from God should be denied public trust until he has proved his assertion by a miracle or the special testimony of Scripture. Should someone want to put trust in a private revelation, he may do so as long as he conforms to the universal rules and customs of the Church.[33] Indeed, we should listen to legitimate prophets since their message has been given for the good of the Church (1 Cor. 12: 7) and St Paul explicitly warns us not to spurn prophecy (1 Thess. 5: 20).[34]

On the basis of these norms, Cajetan criticized the claims of some recent prophets. He explicitly rejected the claims of Amadeus to have received revelations from the angel Gabriel because no one can introduce new doctrines to those of the Christian Faith (Gal. 1: 18).[35] He implicitly also attacked the Florentine false prophet Teodoro di Giovanni da Scutari, a Benedictine monk of the order of Monte Oliveto, who had been forced to confess in a public ceremony in 1515 that he had feigned his visions and had falsely presented himself as a holy man, indeed, the prophesied Angelic Pastor. This 'prophet' also admitted that he had introduced new ceremonies and unusual ways of praying and singing during nocturnal gatherings and had seduced women into having sexual relations with him on the grounds that these acts were not sinful. Cajetan had found particularly reprehensible such 'prophets' of his own day who

---

[31] *ST* x. 400.

[32] Ibid. 400–1.

[33] Ibid. 401. Cajetan gives as the reference 'Extra, de Haeret., cap. Cum ex iniuncto'. He is citing the Decretals of Gregory IX, bk. V, tit. VII, Ch. 12, in *CJC* ii. 786.

[34] *ST* x. 400–1.

[35] Ibid. 400.

used their fame to put on pompous airs, introduce 'religious' novelties, and enter into carnal embraces.[36] If the Dominican master general tried to avoid criticizing Savonarola by name, his restraint did not extend to other recent prophets.

This brief survey of how some of the major figures at Lateran V were personally involved with or sympathetic to the possibility of prophecy raises the question of why that Council ever felt a need to legislate on the topic. Indeed, such legislation was not part of its original agenda, but was voted on only in the last four months of a Council that lasted for almost five years.[37] What led to this decree were conditions in Florence and Rome in the years prior to and during the Council, especially an attempt by Leo X to suppress the ideas of Savonarola.

Florentine authorities were nervous whenever a preacher seemed to take up the mantle of the fallen prophet. On 26 December 1508 a Franciscan friar of the Amadeite congregation, Antonio da Cremona, preached a sermon that drew a large crowd of Savonarola's followers. Antonio denounced Florence for having killed the prophet sent to it by God. Because of the sins of the city and of Italy, God would punish the Italians with war, unless they repented. But from this strife would ultimately come good things for the city and a restoration of the Church. Archbishop Cosimo Pazzi was alarmed by this sermon and called in Antonio for questioning on 13 January 1509[38]. The friar insisted that he was not a follower of Savonarola, but refused to divulge the source of his ideas. He denied that they came from mere reason or from Sacred Scripture, but claimed as their source his own heart. (We know now that his ideas were derived probably from Amadeus via Salviati who in 1502 sent in his letters to Ubertino Risalti in Florence excerpts from the *Apocalypsis Nova* that Risalti shared with Antonio.[39]) When the archbishop insisted that he cease preaching such sermons, Antonio agreed to obey, but not before advising the prelate that God had wanted him to preach.[40]

Five years later another Franciscan friar of the Amadeite congrega-

---

[36] Ibid. 400–1. See also A. Prosperi, 'Il monaco Teodoro: note su un processo fiorentino del 1515', *Critica Storica*, 11 (1975), 80–4, 88, 93–5, 99.

[37] N. Minnich, 'Paride de Grassi's Diary of the Fifth Lateran Council', *Annuarium*, 14 (1982), 380–1, 431–3.

[38] Vasoli, 'Notizie', 494, 497.

[39] Morisi, *Apocalypsis Nova*, 29–32 and n. 56 (the 1502 letters of Salviati to Risalti); p. 34 (the Statement of a Fra Antonio); Vasoli, 'Notizie', 494, 497.

[40] G. Tognetti, 'Un episodio inedito di repressione della praedicazione post-Savonaroliana (Firenze, 1509)', *Bibliothèque d'Humanisme et Renaissance*, 24 (1962), 190–9.

tion, Francesco da Montepulciano, stirred up strong reactions among
the followers of Savonarola when he used the *Apocalypsis Nova*, recently
disseminated in a revised version by Salviati, as the basis for his Advent
sermons in 1513 at S. Croce. Francesco predicted the election of an
antipope who would divide Christendom and cause a war that would
result in the near destruction of Florence. Both Duke Lorenzo and
Cardinal Giulio dei Medici were alarmed by this sermon. Leo wrote
from Rome to Florentine officials that this friar was a scandal to the
Faith and a danger to souls. The pope ordered Pietro Andrea Gammaro,
the vicar general of the archdiocese, to take the prophet into custody
immediately and have him sent to Rome under guard, with the help of
the secular arm if necessary. Francesco was ordered to justify his prophecies, but he died suddenly, thus leaving his reputation intact.[41]

About that same time a Florentine layman, Francesco da Meleto, who
engaged in prophecies, also came to the Pope's attention. Leo's personal
adviser, the Camaldolese Pietro Querini (d. 1514), called him to Rome.
His travel expenses were paid by Antonio Zeno of Ferrara, provost
of the bishop of Volterra, Francesco Soderini, who was also cardinal-
protector of the Camaldolese order. Once in Rome Meleto was lodged
for three months in the home of the papal secretary, Pietro Bembo, and
eventually presented to the Pope. Meleto was amazed at the attention
he received, for he was old, poor, of humble origins, and lacking education and connections. What interested the Pope were his prophecies
that he claimed came from reading the Bible with the help of the Holy
Spirit and some mathematical calculations. According to Francesco, this
evidence pointed to the year 1517 as the beginning of the hoped-for
renovation of the Church that would commence with the conversion
of the Jews and finish in 1536 with the extermination of the Moslem
religion. These ideas, which he seems to have derived in part from the
Joachites, from Savonarola, and from his discussions as a youth with
the Jews of Constantinople, he had incorporated into a small book entitled *Convivio de' Segreti della Scriptura Santa* and written in Italian so
as to be read by more people. For Pope Leo X he reworked this same
message into a Latin treatise, *Quadrivium temporum prophetatorum*, which

---

[41] Vasoli, 'Notizie', 482, 489, 492–4; J. Stephens, *The Fall of the Florentine Republic
1512–1530* (Oxford, 1983), 77–8; letter of Leo X to the Florentines, Rome, 22 Dec.
1513, Archivio Segreto Vaticano, Armadio XXXIX, vol. 30, fo. 91ʳ. The Medicis' alarm
should be seen in the context of other predictions earlier that year by friars who preached
an imminent change of government in Florence and spoke favourably of Savonarola, see H.
Butters, *Governors and Government in Early Sixteenth-Century Florence, 1502–1519* (Oxford,
1985), 222–3.

he dedicated to the Pope and presented to him during his audience. With money he received from Zeno, Meleto then had the *Quadrivium* and a small book, the *Enucleatio Psalmi XVIII*, which he dedicated to his patron Zeno, printed, probably in Rome.[42] The initially favourable reception given to Meleto in some quarters of Rome did not survive a closer examination of his writings a few years later in Florence.

Florence was also the home of another prophet who came to the attention of Leo X. But first the case of the Olivetan monk Teodoro, mentioned earlier, was investigated by Florentine church officials: the archdiocesan vicar, Pietro Andrea Gammaro, the local inquisitor Paulo de Fucecchio, and the canons of the cathedral. They established that Teodoro not only led a disreputable private life based on a false doctrine of spiritual espousals that condoned adultery, but he also feigned numerous revelations from God, the angels, and Savonarola, and he claimed to be the Angelic Pastor called by God to renew the Church as was foretold by Savonarola. The sect led by Teodoro held views similar to those of the medieval Fraticelli: evil priests and friars together with the Pope were to be massacred and replaced by clergy who possessed no temporal goods except tithes. From 12 January until 1 February 1515 the Benedictine monk was interrogated and a confession extracted from him which he signed on 2 February.[43]

With this incriminating evidence in hand, Florentine authorities tried to connect the case of Teodoro with that of Savonarola's followers and condemn them both. In their effort to tighten their control over Florence, the Medici wanted to stamp out the remnants of Savonarolan republicanism. On 7 February the youthful Galeotto dei Medici, the local deputy in Florence of Lorenzo dei Medici who was then ruling the city from Rome, wrote to report that Ser Giuliano da Ripa, the last holdout among Savonarola's followers who refused to abjure him, had been accused before the episcopal tribunal of venerating in his house the image and relics of the friar. On 10 February the day before Teodoro was to make his public confession of guilt, Lorenzo wrote to Galeotto that consideration was being given to a papal prohibition on the relics and writings of Savonarola and that Giuliano should be punished if he erred. Lorenzo also wrote that same day to the Gonfaloniere Luigi della

---

[42] S. Bongi, 'Francesco da Meleto, un profeta fiorentino a' tempi del Machiavelli', *ASI* 3 (1889), 64–70; Weinstein, *Savonarola*, 353–7; J. D'Amico, 'A Humanist Response to Martin Luther: Raffaele Maffei's *Apologeticus*', *Sixteenth-Century Journal*, 6 (1975), 40.

[43] Prosperi, 'Teodoro', 80–1, 90–7; Weinstein, *Savonarola*, 357–8; J. Stephens, *Fall of the Florentine Republic 1512–1530* (Oxford, 1983), 78–9.

Stufa ordering him to hand over to Gammaro a copy of the judicial process against Savonarola. The constitution issued by Gammaro in the name of the cardinal–archbishop of Florence, Giulio dei Medici, and read aloud to the ten thousand citizens assembled in the cathedral to hear Teodoro's confession on 11 February not only prohibited prophetic preaching not in keeping with that preached by the first doctors of the Church and condemned as heresy any teaching about a future renovation of the Church, but it also ordered that whoever had in his possession the writings, images, bones, or other relics of someone (that is the Dominican friars Savonarola and his companions Domenico and Silvestro) condemned by the Holy See for heresy and schism should surrender them within fifteen days to the archdiocesan vicar, Gammaro.[44]

This effort to associate the followers of Savonarola with the crimes of Teodoro and condemn both in the same edict was rejected by some theologians. They claimed that diocesan authorities lacked the authority to judge questions of heresy, a power that belonged only to the Holy See. Besides, Teodoro was condemned for holding the ideas of the Fraticelli, views quite different from those espoused by Savonarola.[45] To answer these objections Rome needed to act, and act it did. On 17 April 1515 the archdiocesan decree was formally confirmed by a letter of Leo X addressed to the Archbishop of Florence and his chapter of canons. The Pope praised these Florentine authorities for having condemned Teodoro who usurped the name of Angelic Pope and preached a future renovation of the Church, a teaching identified by the Pope as being the same as that taught by the condemned friars Girolamo Savonarola and Pietro Bernardino. This doctrine had been declared by the Holy See to be heretical and schismatic because the Bible holds that the Church is without spot or wrinkle (Eph. 5: 27). The Pope urged these officials to see that the last spark of this heresy be extinguished and those who hold this erroneous view of the Church be duly punished.[46]

---

[44] Prosperi, 'Teodoro', 98–100; Butters, *Governors and Government*, 204, 236 n. 64, 245–7, 263; Stephens, *Fall of Florentine Republic*, 79, 140; Archivio di Stato-Firenze, *Mediceo avanti il Principato*, Filza 116, fo. 113 (7 Feb. 1515), as cited in Stephens, 79 n. 2, and Filza 141, fos. 96ᵛ–97ʳ (10 Feb. 1515). Fra Gherardo, Franciscan conventual inquisitor at S. Croce in Feb. 1515, is also reported to have ordered, on the authority of the archbishop, that everyone who had the writings and relics of Savonarola should turn them over to Gammaro. Earlier Alexander VI had similarly ordered the surrender of Savonarola's writings to the Florentine archbishop, but these were returned when nothing erroneous was found in them, see Ridolfi, *Savonarola*, 2, 39; A. Giorgetti, 'Fra Luca Bettini e la sua difesa del Savonarola', *ASI* 77 (1919), 187–8.

[45] Giorgetti, 'Fra Luca Bettini', 192, 222; Prosperi, 'Teodoro', 86.

[46] A copy of this papal letter is in Venice, Bibl. Marciana, MS 3049 (formerly Cl. IX. cod. XLI), fos. 35ᵛ–37ʳ, reprinted in D. Moreni, *Memorie istorichi dell'ambrosiana imperial*

This papal letter did not end the debate and on his visit to Florence in 1516 Leo X ordered the archdiocesan vicar and the cathedral chapter to proceed to a new examination of the teaching of Savonarola. The Pope hoped that the friar's message could be shown to be heretical and be formally condemned in a provincial council later that year.[47] The theologians consulted to examine the friar's case tended, however, more to exonerate than to fault him.

One of these was the lay theologian, Gasparo Contarini (1483–1542). Writing from his villa at Pieve di Sacco in the Veneto on 18 September 1516 after having read the writings of Savonarola sent to him, Contarini concluded that there was nothing contrary to the Faith in the friar's works. Reason and an analogy based on circular motion suggest that he may have been correct after all: the Church surely needs a renovation and God has intervened at times to regulate His Church. While Savonarola did disobey Alexander VI, it is also true that papal decrees are not to be obeyed if grave scandal results and a Pope does not intend to do evil.[48]

Others also defended Savonarola. Another lay theologian, Gianfrancesco Pico della Mirandola, insisted that the friar taught no false doctrine but only those of Christ.[49] The Dominican friar, Zaccaria della Lunigiana, showed that Savonarola had not intended to change the Church's universal status, sacraments, or precepts, but sought only the reform of men's morals, and in this sense to renew the Church. Indeed, he asserted, the current Lateran Council and the upcoming Florentine synod have both been convoked for the reformation of the Church.[50] The friar, Lorenzo Macciagnini, also wrote an oration which he hoped to deliver at the provincial council attacking the letter of Leo and its conclusions as false.[51]

The Dominican Luca Bettini (c.1489–1527) was one of Savonarola's principal defenders. Two weeks after Leo's letter of 1515 Bettini published in Bologna two collections of the friar's prophetic sermons: *Sopra i salmi* with his 1495 predictions of a renovation of the Church that would follow a divine chastisement of its sins, and *Sopra Ezechiele* with his exhortations of 1496–7 that Florentines act justly. These publica-

---

Basilica di S. Lorenzo di Firenze (Florence, 1816–18), ii. 511–15.

[47] Ridolfi, *Savonarola*, ii. 39–40; Giorgetti, 'Fra Luca Bettini', 189.

[48] Giorgetti, 'Fra Luca Bettini', 190, 213–14; F. Gilbert, 'Contarini on Savonarola: An Unknown Document of 1516', *AR* 59 (1968), 145–50.

[49] Giorgetti, 'Fra Luca Bettini', 169–70 n. 1, 187.

[50] Ibid. 191, 215–16.

[51] Ibid. 190.

tions were based on the borrowed notes of Ser Lorenzo Violi. Bettini also penned his *Opusculum in defensionem fratris Hieronymi tempore synodi compositum MDXVI*. In this work he argued that Leo's letter did not apply to Savonarola because it was based on false information about the friar. Furthermore, the papacy should condemn only false doctrines and not persons, since it can and has passed erroneous judgements on persons. A more careful examination of the case against Pietro Bernardino would show that he was condemned for sodomy and not heresy. Finally, it is not proper for a decree against Teodoro to contain extraneous material against Savonarola. Leo's letter is thus invalid as regards the Dominican friar.[52]

Probably because of these forceful defences of Savonarola, the Florentine provincial council was postponed until 1517 and it made no mention of the friar. Instead, it proscribed the writings of Francesco da Meleto and ordered him to compose a retraction and to abjure his errors within two months.[53]

If the Pope had difficulty in having Savonarola condemned by a Florentine synod, his determination to have some conciliar legislation restraining prophets could only have been strengthened by the arrival in Rome in August of 1516 of the fanatical figure fra Bonaventura. Claiming to be not only a prophet but the Angelic Pope himself, he warned the faithful to separate themselves from the Roman Church and proceeded to excommunicate the Curia and Leo X whom he predicted would soon die. Indeed, the Pope was gravely ill, but he ordered the prophet to be imprisoned and questioned. Throughout his interrogation Bonaventura insisted that he was correct and if his prophecies did not come true, he should be burnt. While we know that Leo soon recovered, the final fate of the prophet is not clear.[54] It must have become very apparent to the pope that the problem of prophets was not restricted to Florence. If Florentine church authorities had troubles dealing with the legacy of Savonarola, the Pope would have the Lateran Council pass more general legislation that would apply to prophets everywhere.

The task of formulating a conciliar decree on prophecy became the

---

[52] P. Villari, *Life and Times of Girolamo Savonarola*, trans. L. Villari (London, 1838), 331–4, 476–80; Giorgetti, 'Luca Bettini', 201, 217–23.

[53] *Mansi*, cols. 269E, 273E–274B; on the Florentine provincial council of 1517 and its decrees on prophecy, see R. Trexler, *Synodal Law in Florence and Fiesole, 1306–1518* (Studi e testi, 268; Vatican City, 1971), 3–4, 10–13, 112, 132–5, etc.

[54] On Bonaventura, see below, Niccoli, Ch. 10. See also Reeves, *Influence*, 262 n. 4, 438, 448; Paride de Grassi, *Diarium Leonis X*, Bibl. Vat., MS. Vat. Lat. 12275, fo. 178ᵛ.

responsibility of the conciliar deputation on the Faith and Pragmatic Sanction. This deputation as constituted by Leo X on 3 June 1513 consisted of twenty members: eight cardinals chosen by the Pope, eight archbishops and bishops elected by the conciliar fathers, and four other prelates (two curial bishops and two generals of religious orders) selected by Leo X. The Pope also appointed a cardinalatial commission composed of three learned prelates to supervise the final formulation of this deputation's first draft decree.[55] A closer examination of the views of these deputation and commission members helps to explain why the decree was formulated as it was.

The influence of a number of members was probably minimal or non-existent. Two of them can be eliminated on account of poor health, age, or death. Five probably lacked the expertise needed for this deputation. A further three who could have made a contribution did not attend the Council after the ninth session.[56] Given the number of prelates who were no longer active members on the eve of the eleventh session, the Pope may have named some replacements. During the same period that the deputation were probably working on the question of prophecy, another major doctrinal issue was being discussed in Rome with the thought of having the Council make the final decision—namely, the orthodoxy of Johannes Reuchlin's cabbalistic writings. Many of the prelates who were active in the Reuchlin proceedings were members either of the Faith deputation (for example Pietro Accolti and Salviati) or of the cardinalatial commission earlier charged with overseeing the draft of the decree on the soul's immortality (that is Domenico Grimani, Niccolò Fieschi, and Bernardino López de Carvajal). It is thus not unlikely that Cardinals Adriano Castellesi, Leonardo Grosso della Rovere, and Marco

---

[55] *Mansi*, cols. 797B–797D.

[56] Robert Challand Guibé died Nov. 1513 and Jaime Serra (95 at the time of the eleventh session) died within months. Cardinals Sisto della Rovere (who attended only the fourth session), Alfonso Petrucci (who did not attend after the tenth session in 1515), and Bandinello Sauli were not trained theologians, nor had Bishops Girolamo Piccolomini or Gianfrancesco della Rovere the appropriate expertise. Archbishop Rainaldo Graziani OFM, a canon lawyer, Archbishop Zane and Bishop Gianantonio Scotti, with degrees in theology, did not attend after the ninth session (1514). On these participants, see de Grassi, *Diarium Leonis X*, fos. 196ᵛ–197ʳ, 199ᵛ–220ʳ (della Rovere, Serra); C. Eubel, *Hierarchia Catholica Medii Aevi* (Münster, 1898–), iii. 10 n. 6, 212, 309 (Guibé, Piccolomini, della Rovere); Minnich, 'Participants', 187 n. 181, 191 n. 298, 193 nn. 331, 335, 353, 194 n. 360, 196 n. 431 (Graziani, Piccolomini, the della Roveres, Serra, Scotti, Zane); Minnich, 'The Healing of the Pisan Schism (1511–1513)', *Annuarium*, 16 (1984), 86–8, 104, 117, 131–2 (Guibé as involved primarily in French affairs); D. Price, 'The Origins of Lateran V'S *Apostolici Regiminis*', *Annuarium*, 17 (1985), 470 (Graziani, Scotti, Zane); A. Ferrajoli, *La congiura dei cardinale contro Leone X* (Rome, 1920), 8–15, 51–3 (Petrucci, Sauli). For Sauli, see also below, Jungić, Ch. 18.

Vigerio, Bishop Paulus van Middelburg, and the Augustinian Friars' Master General Egidio of Viterbo, who were all involved in the Reuchlin affair, were also consulted or made later members of the conciliar Faith deputation.[57]

Of the still active members of the deputation or of its likely replacements, six were canonists. Four of them were cardinals (Pietro Accolti, Achilles de' Grassi, Niccolò Fieschi, and Leonardo Grosso della Rovere), one a curial bishop (Johannes Staphyleus), and one the minister general of the Franciscans (Bernardino de Prato).[58] When they approached the question of prophecy, they would probably have recalled the classic text of Innocent III's letter of 1199 to the faithful of Metz, a text incorporated into the Decretals of Gregory IX. In his letter Innocent argued that someone who claimed to have an unseen, spiritual mission from God to preach should be required to prove that mission either by the working of a miracle or by providing a special testimony from Scripture, as did Moses with Aaron's rod that turned into a serpent and then back into a rod (Exod. 7: 8–12) and as did John the Baptist who quoted a text of Isaiah (40: 3) to justify his preaching (John 1: 21–3). Indeed God promises to confirm the preaching of Christ's disciples with signs (Mark 16: 20).[59] But surprisingly the Lateran decree did not address specifically the question of how to identify a true prophet, but turned this task over to the Holy See or to the local ordinary and his three or four advisers. Given their jurist tradition the canonists had to allow for some prophecy, even if they strove to restrain it with legal controls.

Difficult to determine with any precision is the possible influence

---

[57] Letter of Martin Gröning to Johann Reuchlin, Rome, 12 Sept. 1516, in J. Schlecht (ed.), 'Briefe aus der Zeit von 1509–1526', *Briefmappe*, pt. 2 (Reformationsgeschichtliche Studien und Texte, 40; Münster, 1922), 65–81, esp. pp. 68, 71, 73–80). Gröning seems to have confused the diocese of Forosemproniensis with Foroliviensis (p. 79) because Pietro Griffi of Forlì was probably then in Spain, while Paulus van Middelburg of Fossombrone was then in Rome actively involved with the Faith deputation; see M. Monaco, *Il 'De Officio collectoris in regno Angliae' di Pietro Griffi da Pisa (1469–1516)* (Rome, 1973), 73–8; D. Marzi, *La questione della riforma del calendario nel Quinto Concilio Lateranense (1512–1517)* (Florence, 1896), 33–8, 48, etc.

[58] On Pietro Accolti, see A. Verde, *Lo studio fiorentino (1473–1503): Ricerche e documenti* (Florence, 1973), iii. 782 n. 1065; P. Litta, *Famiglie celebri italiane* (Milan, 1825), ii. 15, 'Accolti di Arezzo, tavola unica di teste'; B. Ulianich, 'Accolti, Pietro (1455–1532)', *DBI* 1 (Rome, 1960), 106–10; on Achilles de Grassi, see Litta, *Famiglie* (Milan, 1840), vii, 'Grassi di Bologna', tavola 2; on Fieschi, the cardinal-protector of the Dominicans, see I. Crivelli, 'Fieschi di Lavagna (Nicola) (c.1456–1524)', *DHGE* 16 (Paris, 1967), 1440–1; for the legal background of della Rovere, see L. Cardella, *Memorie storiche de' cardinali della santa romana chiesa* (Rome, 1792–7), iii. 313–14; on Johannes Staphyleus, an auditor of the Rota, see Eubel, *Hierarchia*, iii. 299; on Bernardino da Prato, see L. Wadding, *Annales Minorum* (Rome, 1731–6), xv. 532.

[59] Decretal. Gregor. IX, lib. V, tit. VII *De Haereticis*, c.12, *CJC* ii. 786.

exercised by six original members or replacements in the Faith deputation and commission who had some expertise in theology. Cardinal
Castellesi (c.1461–1521?) may have been sympathetic to personal revelation as a way to religious knowledge due to his own fideistic and
sceptical stances.[60] Cardinal Grimani (1461–1523), who was considered
most learned in philosophy and theology, unfortunately left behind no
published writings for fear he might have to repent of something he
had stated. As cardinal-protector of the Franciscan order since 1503,
he may have been attentive to defending these friars' role in the fall of
Savonarola.[61] The Franciscan Cardinal Marco Vigerio, who died on
18 July 1516, may have had a similar concern. His views on Christian
prophecy are not evident, for his *Tractatus de Antichristo* does not seem
to have survived.[62] Archbishop Antonio Trombetta, OFM (1431–1517),
who assisted Vigerio in the earlier direction of the Franciscan Studium
at Padua, had defended the secular priest Gabriele Biondo, accused of
heresy because of various statements in his book *Ricordo*. Among other
things in that book, Biondo attacked Savonarola for hypocrisy, for feign-
ing to be holy. Trombetta felt that those who prophesy are not heretics
unless they pertinaciously adhere to a perverse teaching they know to be
in error. To suggest that the world will end within a certain period of
time is not heretical but temerarious. However, heresy could occur were
one to assert with certainty what is hidden in the mind of God, without having that revealed by the Son through the Holy Spirit. Because
Trombetta returned to Padua after the tenth session in 1515, it is doubtful that his views influenced the conciliar discussions on prophecy.[63]
The same can be said for Bartolomeo da Seniga, OP (d. 1529) from
Bergamo who last attended the Council in 1515. His later career as an
inquisitor general for the area around Brescia in 1520 suggests that he

---

[60] J. d'Amico, *Renaissance Humanism in Papal Rome; Humanists and Churchmen on the Eve of the Reformation* (Baltimore, 1983), 169–83.

[61] P. Paschini, *Domenico Grimani Cardinale di S. Marco (†1523)* (Storia e letteratura, 4; Rome, 1943), 14, 17, 124; J. Moorman, *A History of the Franciscan Order: From its Origins to the Year 1517* (Oxford, 1968), 591; Villari, *Life and Times*, 650–74; Mariano da Firenze's claim that Grimani was present at the opening of Amadeus's book seems to be based on the assumption that Grimani was the cardinal-protector of the Franciscans in 1502, but he assumed that office one year later and was therefore probably not present, see Vasoli, 'Notizie', 492 and Morisi, *Apocalypsis Nova*, 27.

[62] J. Sbaralea, *Supplementum . . . ad Scriptores trium ordinum S. Francisci* (Rome, 1921), pt. 2, 211; Eubel, *Hierarchia*, iii. 10 n. 5.

[63] Paschini, *Grimani*, 79; P. Gios, *L'Attività pastorale del vescovo Pietro Barozzi a Padova (1487–1507)* (Padua, 1977), 305 n. 37; A. Trombetta, 'Quaestio super articulos impositos domino Gabriele sacerdoti', *Opus in Metaphysicam Aristotelis* (Venice, 1502), fo. 106^{r–v} (prophecy and heresy); fos. 107^{r}–108^{r} (Savonarola as hypocrite).

was seen as someone who could act decisively against suspect views. As a Dominican he could also have been protective of his order's reputation and have defended Savonarola. But whatever his views, he seems not to have been a party to these conciliar discussions.[64] The Greek Bishop Alexios Celadenus (1450–1517), while open to the possibility of prophecy, did not personally adopt that style of preaching but did have words of praise for that quality in the sermons of Egidio of Viterbo. Celadenus took an active part in the conciliar proceedings and was probably also very involved in the deputation's discussions.[65]

Of the five remaining members or likely replacements on the Faith deputation, four were deeply involved with prophecy. The Dutchman Paulus van Middelburg (1445–1533) was noted for his prognostications based on the movement of the stars.[66] Egidio of Viterbo had achieved fame by his twenty years of prophetic preaching throughout Italy and by his inaugural sermon at the Council. Cardinal Carvajal and Benigno Salviati[67] were intimately involved in the preparation of the revised prophecies of Amadeus and based their hopes for the highest ecclesiastical promotions in part on these predictions. These prelates were not inclined to let the Lateran decree ban outright all prophecies.

The member of the deputation who probably most influenced the decree's final formulation was Tommaso de Vio. His recently written commentary on Aquinas's treatment of prophecy provided an objective, systematic analysis of the topic. He affirmed the utility for the Church of prophecy and provided criteria for discerning the truthfulness of the revelation. He was also out to protect the reputation of Savonarola and of the Dominican order.

Given a Faith deputation that contained so many important persons favourably disposed toward prophecy, Leo X once again found it difficult to secure a conciliar condemnation of Savonarola or a blanket prohibition on prophesying. The Pope had to settle for a compromise statement.

---

[64] G. Cavalieri, *Galleria de' sommi pontefici, patriarchi, arcivescovi, e vescovi dell'ordine de' predicatori* (Benevento, 1696), 346–7.

[65] N. Minnich, 'Alexios Celadenus: A Disciple of Bessarion in Renaissance Italy', *Historical Reflections*, 15 (1988), 51, 57–63, and his letter to Egidio of Viterbo, Rome, 30 Nov. 1506, Naples, Bibl. Naz., MS 5 F. 20, fos. 171ᵛ–177ʳ.

[66] C. van Leijenhorst, 'Paul of Middelburg', *Contemporaries of Erasmus: A Biographical Register of the Renaissance and Reformation*, 3 (Toronto, 1987), 57–8.

[67] Often unaware of Salviati's membership in the important Faith deputation, scholars have speculated on his role at the Lateran Council in protecting Savonarola. See, for example, G. Benelli, 'Di alcune lettere del Gaetano', *AFP* 5 (1935), 372 n. 1; Ridolfi, *Savonarola*, ii. 182–3; Vasoli, 'Notizie', 481–2.

The decree on preaching, *Supernae majestatis praesidio*, contains significant sections devoted to the correction of abuses. Without mentioning by name the Florentines Girolamo Savonarola, Antonio da Cremona, Francesco da Montepulciano, Francesco da Meleto, or Teodoro da Scutari, or their teachings, the decree censored in general terms the behaviour of some preachers. Instead of basing their sermons on Scripture interpreted according to the common understanding of the approved doctors of the Church, these preachers have dared to claim for themselves an infusion of the Holy Spirit and have twisted the meaning of the text, predicting without reason numerous disasters and making other new and false prophecies. With the help of fabricated miracles, they have disseminated errors and fraudulencies among the simple people whom they lead into deceptions and away from the path of salvation and obedience to the Roman Church. These preachers have inveighed against the prelates of the Church and stirred up the populace.[68] That these abuses are described in such general terms was probably due to the efforts of some on the Faith deputation to protect from condemnation the reputation and distinctive message of their particular prophet.

To put an end to these abuses, specific measures were decreed. Before any cleric can preach he needs to be certified by his ordinary or religious superior regarding his competence and suitability for this ministry. He is to explain the Scriptures in accord with what has been taught over time by those approved by the Church and he is not to add anything at variance with their teachings. Because Scripture tells us that it is not for us to know future events which the Father has predetermined by His power (Acts 1: 7), no cleric is to presume on his own authority to predict in his sermons the fixed time of future evils or the coming of Antichrist, or the particular day of judgement. The decree claimed that those who up to that time have dared to make such predictions have lied and discredited in no small way the authority of those who preach properly. The decree did not say that all prophetic preachers were thus guilty, nor did it explicitly condemn those who avoided predicting precise times. To prevent a continuation of what it considered abuses, the decree forbade any cleric in the future engaged in public preaching to predict firmly (*constanter*) future events on the basis of a

---

[68] *Mansi*, cols. 944D–946A. This decree did not list prognostications based on the movement of the stars as one of the forbidden ways of predicting the future. On this point the influence of Paulus van Middelburg may have been decisive, since Cajetan was somewhat critical of astrology.

scriptural text (*ex literis sacris*) or to claim that he has received knowledge of these things from the Holy Spirit or from a divine revelation or by earnest recourse to strange and inane divinations or by other such means. And finally he is not to defame before the laity and common people the character of bishops, prelates, and other superiors by intemperate and heedless attacks on their names and deeds.[69] While some could have read into this a veiled condemnation of Savonarola and other recent prophets, the decree had once again avoided mentioning anyone by name. What followed in the decree clearly indicates that this prohibition on prophesying was not absolute, but applied only to clerical preachers who had not obtained prior approval for announcing their revelations.

In its concluding section the decree took a surprisingly positive stance towards prophecy. It noted that both the Old and New Testaments teach that God does indeed reveal future events to His servants.[70] Church authorities do not want to impede a true revelation nor label it a fable or lie. But since this is a matter of great importance, trust should not be easily placed in every spirit, but they are to be tested to see if they come from God, as the Apostle testifies.[71] Therefore, as a general rule, before any alleged inspirations are made public or preached to the people they should henceforth be reserved for an examination by the Holy See. But should a delay cause danger or an urgent necessity advise otherwise, they can be referred to the local ordinary who with the advice of three or four learned and prudent men can after a diligent examination of the matter grant his permission for the publication of these revelations, if it seems advisable.[72] The penalties for failing to observe this decree were stipulated to be, besides the normal punishments, an excommunication reserved to the Pope to absolve, unless there is danger of death. In addition, the guilty party is deprived forever of the office of preaching.[73]

---

[69] *Mansi*, col. 946A–946E.

[70] Ibid., cols. 946E–947A. See Amos 3: 7 (The Lord God does nothing unless He has revealed it to His servants the prophets) and I Thess. 5: 19–20 (Do not extinguish the Spirit, do not spurn prophecies).

[71] *Mansi*, col. 947A. The citation from the Apostle could be either that of St Paul (I Thess. 5: 21—Test everything) or that of St John (I Jn. 4: 1—Test the spirits to see if they are from God). It is interesting to note that Aquinas used the quotation from Amos in his treatment of prophecy and his commentator Cajetan cited the texts from the Epistle to the Thessalonians, see *ST* 10, 372, 401. In his reform proposals Taleazzi had urged that bishops on an annual basis should investigate and prosecute those who prophesy *ex corde suo* and, if necessary, consult with the Holy See, see his 'Summarium', fo. 9ᵛ, n. 3.

[72] *Mansi*, col. 947AB.

[73] *Mansi*, col. 947BC.

The decree *Supernae majestatis praesidio* was a compromise statement. Some conciliar fathers wanted a stern condemnation of Savonarola, and in their camp one would have to place Leo X, the Medicean partisans, and persons earlier associated with the friar's disgrace, such as the Franciscans and Savonarola's judge who was now Cardinal Francisco Remolines. But on the deputation that formulated the decree were serious theologians such as Salviati and Cajetan who sought to protect his reputation and would object to labelling his views as heretical, as had been attempted in the papal brief of 1515. The best that the anti-Savonarolans could obtain was a denunciation of abuses that could be read as a veiled reference to the friar's style of preaching. Such leading disciples of Amadeus as Carvajal and Salviati could have rejoiced that neither he nor any other prophet nor their specific message was singled out by name for condemnation. The decree denounced as liars only those who had predicted fixed times for future events. Its prohibitions did not apply to previous prophets but only to clerics henceforth attempting to prophesy. Far from inhibiting all prophecy, the decree openly acknowledged its legitimacy and utility, citing in its support scriptural texts that seem to echo Cajetan's commentary on Aquinas. The decree did, however, mandate the careful scrutiny of prophecies by church authorities before they could be made public. Thus, while some restraint was placed on Renaissance prophecy by the Lateran decree, the Council left intact the reputations and messages of many of its previous practitioners.

Such a compromise decree was easily approved. When a draft of it was read in the general congregation of 15 December 1516 held in the upper chapel of the papal palace, only one prelate protested. Mario Maffei (1463–1537), the humanist bishop of Aquino, felt that preachers should speak according to how the Spirit inspired them. But at the solemn session on the next day in the Lateran Basilica, Maffei joined the other eighty prelates present in approving unanimously this decree.[74] Thus Renaissance prophecy, with its abuses condemned and its preachers controlled, but with its legitimacy formally acknowledged, passed into the legislation of the universal Church on the very eve of the Reformation.

[74] Ibid., cols. 935E–936A, 938C, 941A, 947D; M. Dykmans, 'Le cinquième Concile du Latran d'après le Diarie de Paris de Grassi', *Annuarium*, 14 (1982), 362, 364. On Maffei, see D'Amico, *Renaissance Humanism*, 85–8.

IV

INTERPRETERS OF PROPHECY

Introduction

The five thinkers studied here were not wild preachers or fringe prophets but churchmen with a theological and philosophical education
and a humanist background. Four held influential positions in the ecclesiastical hierarchy.[1] They all exemplify to varying degrees in their
speeches and writings the extraordinary mingling of political and visionary motives. Take Cardinal Carvajal, for instance, passionately striving
to attain the tiara, yet almost certainly a chief participant in the awesome ceremony of opening the *Apocalypsis Nova* and, in his orations,
an ardent pleader for the realization of the *renovatio mundi*. Or again,
consider the colourful Bosnian, Benigno Salviati, pushing himself before every possible patron, yet deeply immersed in a mystical theology
which embraced the role of prophecy. We can picture him as an exile
in Dubrovnik, surrounded by a circle of professional men, lay as well
as ecclesiastical, discussing Free Will, Divine Providence, the validity
of prophecy, and the function of angelic intermediaries. In the midst
of political intrigues at the court of Maximilian I, he dedicated to the
Emperor a work of theology full of mystical Platonic nuances and prophetic overtones. Most influential of all was Cardinal Egidio of Viterbo,
close adviser to successive popes and head of a great order, yet increasingly absorbed in sources of arcane knowledge which he believed held
the key to the providential future.

In the thought of these Churchmen medieval and Renaissance cur-
rents converge. On the one hand through their education they inherited
a biblical and medieval tradition of prophecy; on the other, classical

[1] We may also note the interesting fact that three out of the five came originally from
Eastern Europe.

forms of forecasting the future by portents, astrology, and sibylline oracles influenced them strongly. On the one hand, they were familiar with the debates in medieval scholastic philosophy on free will, the nature of human knowledge, and the possibility of foreknowledge; on the other, they found the neo-Platonist emphasis on mystic and hermetic sources of knowledge immensely attractive. On the one hand, they often exhibited an astonishingly naïve belief in the immediate fulfilment of the prophecies; on the other, they could engage in elaborate philosophical disquisitions on the distinction between true and false prophecy. On the one hand, they were upholders of ecclesiastical authority, with a staunch belief in the hierarchy; on the other, the critical developments in Renaissance thought laid them open to radical ideas. On the one hand, their expected agents to fulfill the prophetic roles were drawn from a medieval world view; on the other, expanding intellectual and geographical horizons were opening up to them entirely new worlds.

5

Cardinal Egidio of Viterbo:
A Prophetic Interpretation of History

MARJORIE REEVES

In his own day Egidio of Viterbo (1469–1532) was acclaimed as an eloquent orator, a 'ciceronian stylist', a humanist poet. He was the 'polished priest of Renaissance circles'.[1] As a philosopher he was an ardent Platonist, regarding himself as a disciple of the Florentine Platonist, Marsilio Ficino. Yet he was also a member of the Order of Augustinian Hermits and from 1507 to 1517 their prior general. He saw the eremitical life as the ideal, was deeply committed to the reform of his order, and during the years of his generalship was constantly active in this cause. However, the evidence for this facet of his life was largely passed over by contemporaries, even by historians of the Augustinian Order itself, and for long Egidio continued to be viewed as the humanist orator and writer, a brilliant member of the papal circle, enjoying the favour of Popes Julius II, Leo X, and Clement VII. He was raised to the cardinalate by Leo X in 1517. It was not until the second half of the nineteenth century that a Frenchman, Léon-Gabriel Pélissier, began to make a critical examination of the manuscripts of Egidio's writings in the Biblioteca Angelica in Rome.[2] Even here, as Father Martin shows, Pélissier's first assessment of Egidio was coloured by the accepted image and it was only in his further essay, published in 1903,[3] that Egidio emerged clearly as one devoted to the religious life, independent in

---

[1] F. X. Martin, OSA, *The Problem of Giles of Viterbo* (Héverlé-Louvain, 1960), 5. I wish to express my gratitude to Professor John O'Malley who provided me with a photocopy of MS 502 (Rome, Bibl. Angelica) and on whose writings I have drawn heavily; also to Professor Martin for help and advice.

[2] L. Pélissier, 'Manuscrits de Gilles de Viterbe à la Bibliothèque Angélique', *Revue des bibliothèques*, 2 (1892), 228–40; id., *De opere historico Aegidii Cardinalis Viterbiensis* (Montpellier, 1896).

[3] Martin, *Problem*, 29–31; L. Pélissier, 'Pour la biographie du Cardinal Gilles de Viterbe', *Misc. di studi critici edita in onore di Arturo Graf* (Bergamo, 1903), 789–815.

his stance towards the papacy and effective as a reformer. Yet his involvement with the humanist world still compromised his position as a reformer in the eyes of those historians, such as Pastor and Boehmer, who 'saw the Renaissance as sharply divided into pagan and Christian'.[4] The most recent work on Egidio[5] has sought to resolve the problem of his different 'faces' by going straight to the main sources, that is, to Egidio's unpublished writings. Now his denunciation of corruption in the Church, his devotion to moral discipline and the life of piety, his zeal for reform in his own order and in the Latin Church can be seen to be closely knit to his enthusiasm for the new learning, his sense of expanding 'worlds', both in knowledge and in geographical exploration, and his expectation of an imminent Golden Age. Perhaps the element which bound all these facets together was Egidio's prophetic view of history in which God's providential purpose for mankind is seen as approaching its climax in the *renovatio mundi* which fuses the Christian and humanist visions of the Golden Age.

One of the striking characteristics of Renaissance Italy—and not only in Italy—was this widespread sense of an imminent crisis in history. Because this stemmed to a considerable extent from the medieval prophetic understanding of history[6] it was interpreted as the final eschatological act in the linear drama of time. But, it must be emphasized, another element in this sense of expectation was the vision of a returning Golden Age looked for by humanists who drew from their classical sources the concept of a cyclical process at work in history. What is emerging clearly in recent studies is the fact that this obsession with prophecy was to be found in all classes, the 'alto' and the 'basso', in Ottavia Niccoli's terms,[7] and the middle citizens who recorded folk oracles and 'mostri'. Characteristic of this mood of imminent crisis was the juxtaposition of a pessimistic and an optimistic expectation. On the one hand, corruption and deterioration, everywhere present, were seen as the signs of an imminent judgement; on the other hand, an approaching *renovatio* was proclaimed by the prophets. In this double

[4] Martin, *Problem*, 35.

[5] E. Massa, 'Egidio da Viterbo e la metodologia del sapere nel cinquecento', in H. Bédarida (ed.), *Pensée humaniste et tradition chrétienne au XVe et XVIe siècles* (Paris, 1950), 185–239; id., 'L'anima e l'uomo in Egidio da Viterbo e nelle fonti classiche e medioevali', *Archivio di Filosofia* (1951), 37–138; id., *I fondamenti metafisici della 'dignitas hominis', e testi inediti di Egidio da Viterbo* (Turin, 1954); J. O'Malley, *Giles of Viterbo on Church and Reform* (Leiden, 1968); id., *Rome and Renaissance: Studies in Culture and Religion* (London, 1981).

[6] Above, Reeves, Ch. 1.

[7] See O. Niccoli, *Profetie e popolo nell'Italia del Rinascimento* (Rome and Bari, 1987).

perception medieval and classical humanist strands mingled. From medieval tradition came the age-old expectation of the final Antichrist: the Church for its sins must suffer a rising scale of tribulation culminating in the last and worst Antichrist, to be followed by the winding up of the historical process in the Last Judgement. In classical tradition Hesiod's four periods symbolized by the four metals, interpreted the time process as a gradual deterioration from gold down to base iron. A similar concept of deteriorating empires appeared in the image of the statue in the Book of Daniel.[8] Decline was also seen mirrored in the ages of Man down to senectude. An aspect of Christian thought—the sense that the further the Christian Church was removed from the primitive purity of the time of Christ and the Apostles, the more its life deteriorated—chimed with this human pessimism.

Yet, on the other hand, medieval tradition also enshrined the hope of a time of blessedness before history was wound up. This was based primarily on certain biblical texts—though their meaning was ambivalent—and can be traced as gathering strength in the eleventh and twelfth centuries, embodied in the idea of a time for the 'refreshment of the saints' between Antichrist and Last Judgement.[9] This concept was given a much wider connotation towards the end of the twelfth century in Joachim of Fiore's vision of an Age of the Holy Spirit which would be ushered in at the death of Antichrist.[10] It was finally crystallized in the idea of the *renovatio mundi* widespread in the later Middle Ages. Although many expressions of this hope have no overt connection with Joachimism, it is perhaps fair to conclude that this was their ultimate source.

In Renaissance thought the medieval linear pattern of the time process met the humanist cyclical concept of a returning sequence of ages. Fundamentally the linear and the cyclical concepts of time were incompatible, but there had always been cyclical or 'returning' elements in Christian thought. The returning cycle of the religious year in Christian (and Jewish) liturgy, the very use of the word *renovatio*, instead of *innovatio*, the vision of a return to apostolic purity, or to the innocence of

[8] See H. Levin, *The Myth of the Golden Age in the Renaissance* (New York, 1969), 14–15. Hesiod actually spoke of five ages, but the fourth is an interlude which breaks the sequence of gold, silver, brass, iron. For the vision of the image in the Book of Daniel, with head of gold, breast and arms of silver, belly and thighs of brass, legs of iron and feet of clay and iron, see Dan. 2: 31–3.

[9] R. Lerner, 'Refreshment of the Saints: The Time after Antichrist as a Station for Earthly Progress in Medieval Thought', *Traditio*, 32 (1976), 97–144.

[10] See Bibliography and Index for Joachim of Fiore.

the Garden of Eden, the use of such texts as 'Elijah truly shall first come, and restore all things'[11]—all suggest, not further light and new creation, but return to the great models or moments of the past. The tension between the two concepts was felt by Joachim whose theology of history was, above all, linear. Yet he combined the linear and cyclical in a spiral figure which expressed at once the liturgical cycle of the Church's year and the ongoing pilgrimage towards a final apotheosis in the Age of the Spirit.[12] So the medieval and classical heritages could, to a certain extent be brought together. Can one determine which was the dominant element in Renaissance prophecy? In so far as imminent judgement is one of the commonest reactions to events, this is clearly medieval, based on the overriding belief in God's control of history. Moreover, it is the most widespread in all classes. The medieval framework of time still obtained. It was among the *literati* that classical ideas of both deterioration and of the returning Age of Gold were most widely cultivated. But at all levels the two strands mingle in fascinating ways.

An obvious question to ask is whether Egidio shows any signs of having been influenced by the *Glosa sive Expositio super Apocalypsim* written by his older fellow-citizen Annio of Viterbo, finished in 1480 and focused partly on the state of the Church from 1481 to the end of the world. Annio glosses most fully chapters 17 onwards of the Apocalypse and this chiefly in terms of recent, present, and future events. Its main themes are the double 'flagellum' of Islam and of Schism, and the glorious triumph of the Roman Church to follow. The destruction of the Turks is symbolized in the fall of Babylon and then Annio argues forcefully that the final monarchy of Christ and his vicar, the Pope, will, by divine right, be a temporal one, even in this present *saeculum*, exclaiming: 'O quanta erit gloria latinorum.'[13] He cites Joachim twice, but only from the pseudo-Joachimist *Super Hieremiam* and only marginally to his main argument.[14] However, his vision of the future beatitude is certainly tinged with Joachimism and at least once he speaks of the 'tercius status in reformatione ecclesie'.[15] Did Annio's work in any

---

[11] See Mal. 4: 5; Matt. 17: 11; Mark 9: 12. For Joachim's use of this text, see references indexed under Elijah in M. Reeves and B. Hirsch-Reich, *The* Figurae *of Joachim of Fiore* (Oxford, 1972).

[12] L. Tondelli, M. Reeves, and B. Hirsch-Reich, *Il Libro delle Figure dell'Abate Gioacchino da Fiore*, vol. ii, 2nd edn. (Turin, 1954), *Mysterium Ecclesiae*, fig. 19; Reeves, and Hirsch-Reich, *Figurae*, 249–61 and pl. 32.

[13] Sig. F in the edition published, according to a later colophon, at Cologne in 1509.

[14] Sig. Eiii (E follows F in this edition).

[15] Sig. Ki.

sense give a lead to Egidio? The evidence seems negative. Apart from the fact that both shared the vision of a coming, this-worldly triumph for the Roman Church, there is little in common between Annio's work and Egidio's *Historia viginti saeculorum* which we shall be studying below. Annio's authorities are conventional medieval ones—the Fathers, Nicholas of Lyra, and so on. There are no traces of Egidio's peculiar use of classical sources, nor of the special prophetic role he gives to the Etruscans. It is significant, however, that Viterbo produced in such a short space of time two writers concerned with the prophetic interpretation of history. The theme of the Etruscan heritage connects the two, for Egidio owned a copy of Annio's *Antiquitates* and apparently intended to write on this theme himself.[16]

Egidio of Viterbo constructs his own prophetic philosophy of history, using, it is true, many of the medieval and classical elements analysed above, but also finding a further and unique key of his own. God's providential working throughout history was the foundation on which he built his structure of the past, present, and future state of the Church. 'Pivotal for understanding Egidio's approach to history is the dominant role he attributes to divine providence. Sacred history is the *providentiae imago*, the earthly fulfilment of a heavenly design.'[17] This theology of history stimulated a search for divine 'signs' which would supply clues to the interpretation of history. 'Search the Scriptures'—'Scrutamini scripturas' (John 5: 39)—was a command which had been followed with particular intent by Joachim and his disciples. Egidio quotes it a number of times in his writings.[18] But, though the key to history was to be sought in the Scriptures, the signs of God's providential purposes at work could be found in all history, including contemporary. This emerges clearly in a celebratory sermon which Egidio delivered in St Peter's before Julius II in 1506, published as a tract in 1507. It was inspired by news of Portuguese conquests in the Far East and the tract was addressed to King Manuel of Portugal.[19] These conquests are a sign of the coming Golden Age of a world-wide Christendom. He called the pamphlet *de aurea aetate* and linked the marvellous expansion of world horizons with the unfolding of the Scriptures. Apostrophizing Pope Julius he exclaims:

---

[16] R. Weiss, 'Traccia per una biografia di Annio da Viterbo', *Italia medioevale e umanistica*, 5 (1962), 437.

[17] O'Malley, *Rome and Renaissance*, 35.

[18] O'Malley, *Church and Reform*, 70; Egidio of Viterbo, *Historia viginti saeculorum*, Rome, Bibl. Angelica, MS 502, fos. 3ᵛ, 194ᵛ, 309ᵛ.

[19] O'Malley, *Rome and Renaissance*, 266–338.

'We have spoken of the golden age. We have spoken of your felicity. We have spoken of the victory and strength of King Emanuel. . . . Therefore now act, most blessed father. See how God calls you by so many voices, so many prophecies, so many deeds well done.'[20] He connects the famous sibylline utterance 'Iam redit et virgo, redeunt Saturnia regna' with Old Testament prophecy.[21] To read correctly the great happenings of the contemporary world one must above all turn to the sacred text. 'Scrutamini scripturas', he urges, and this became increasingly his guiding principle.

Julius II must have found Egidio's type of apocalyptic eloquence very effective in the climate of Rome at that period, for he commissioned him a number of times as a preacher. The two most important occasions were the opening oration at the Lateran Council discussed above by Nelson Minnich[22] and, six months later in 1512, the sermon preached in S. Maria del Popolo to celebrate the treaty between Julius and the Emperor Maximilian.[23] Here Egidio was already anticipating the concords he uses continually in his *Historia* between scriptural and classical images as divine signs of contemporary developments. Italy has lain in the darkness of calamity but light is born from darkness. God created light out of darkness. Beside this Old Testament symbol he places St John's allusion to the light which lightens everyman and Plato's image of the cave as the world of sense and the light as the higher understanding. Referring to the Lateran Council he again uses creation imagery. The creation of light is the inauguration of the Council. The separation of the sky from the waters on the second day signifies the condemnation of the schismatic Council of Pisa. On the third day the Council, like the earth, brings forth its fruits and on the fourth day the sun and moon represent the glorious treaty between Pope and Emperor. Symbolic interpretations of scripture and classical myth are keys to the meaning of history.

The Hebrew cabbala forms another key. Father O'Malley thinks that Egidio probably began to learn Hebrew during his Florentine sojourn in 1494–5; by 1513 he had made great progress in both Hebrew and

[20] Ibid. 337.
[21] Ibid. 337. An early letter of Egidio to Marsilio Ficino is quoted by A. Chastel, 'L'Antéchrist à la Renaissance', in E. Castelli (ed.), *Christianesimo e ragioni di stato* (Rome, 1952), 178: 'Voici le règne de Saturne, l'âge d'or si longtemps célébré par la Sibylle et les devins, l'âge annoncé par Platon'.
[22] See above, Minnich, Ch. 4. On Egidio's preaching style, see below, Niccoli, Ch. 10.
[23] Ed. C. O'Reilly, ' "Maximus Caesar et Pontifex Maximus": Giles of Viterbo Proclaims the Alliance between Emperor Maximilian I and Pope Julius II', *Augustiniana*, 22 (1972), 80–117.

Aramaic. By the time of his final work (the *Scechina*) in 1530 he was 'completely dedicated to investigating the meaning of the scriptures in obedience to the Gospel text'.[24] This emphasis on going back to original languages he shared, of course, with other humanist Biblical scholars. But Egidio's purpose went beyond the scholarly objective of establishing a pure text: he sought the inner meaning of the Scriptures through further clues. The Platonism he derived from the Florentine Academy led him to believe in the secret spiritual knowledge to be sought both in oracular sources of ancient literature and in Christian mysticism. Here he drew on the Florentine Platonists' belief that an ancient Gentile tradition, to be traced back to Noah, constituted a primitive pre-Christian revelation of divine truth and purpose.[25] Like Marsilio Ficino and Pico della Mirandola, he went back to the so-called *prisci theologi* of antiquity, to Orpheus, Pythagoras, Hermes Trismegistus, and others. He cites these many times in his *Historia*, as well as Plato and 'Maro' (Virgil), linking them to the Hebrew oracles and especially to what he terms the secret knowledge of the Aramaeans.[26] Egidio found indications of the doctrine of the Trinity in Plato's philosophy and linked him with St John, his favourite New Testament author.[27] This theology of the ancients pointed to a single divine truth hidden under veils of myth. As an inner core of spiritual understanding it was not available to all: it was a *disciplina arcana* for the few. Here we see how sharply Egidio's interpretation of the sources of true knowledge was in conflict with the Aristotelian philosophy taught, for instance, at Padua. In maintaining that all truth was derived primarily through the senses the Aristotelians, he argued, denied revelation and discounted the inner, mystical sources of knowledge. As early as his examination for the mastership in theology Egidio was attacking the 'Peripatetics' and he continued to do so throughout his life.[28]

Above all Egidio turned to the clues contained in the Hebrew cabbala.[29] He possessed a copy of the great cabbalistic work, the *Zohar*, and collected other cabbalistic writings, including Reuchlin's *De arte*

---

[24] O'Malley, *Church and Reform*, 70.
[25] Ibid. 29.
[26] See, for example, *Historia*, fos. 60^r, 75^v, 80^v, 92^v, 107^v, 108^r, 115^r, 252^v, 254^r, 268^v, 276^r, 277^r, 279^r, 299^r, 366^r.
[27] O'Malley, *Church and Reform*, 24–5, 49–51, 71.
[28] Ibid. 40–3. See, for example, *Historia*, fos. 161^v, 163^r, 190–1^r, 229^r, 235^v.
[29] The correct Hebrew spelling is Kabbalaĥ, see L. Jacob, *Symbols for the Divine in the Kabbalaĥ* (London, 1984).

cabbalistica.[30] Linguistically he saw Hebrew as a holy language, the only language transmitted to men by the Holy Spirit. 'When language and alphabet become a divine code for transmitting sacred doctrine, they become transcendent.'[31] The very letters of the alphabet 'participated in the divine energy in a way which made them not merely symbols of divine reality, but also gave them the power to penetrate into the minds of the pious and humble'.[32] The minutest detail, such as word-order and structure, could be a bearer of divine truth. The two chief requisites for penetrating to the core of meaning were a knowledge of the grammar and syntax of the Hebrew language and the exegetical techniques of the cabbala. Thus, says Egidio, Jerome, lacking the cabbalistic clues, was unable to penetrate to the deepest meanings.[33]

The tradition of the cabbala went back to the belief in an inner spiritual revelation of God to Moses, transmitted in secret books alongside the outer letter of the Law. This mystical wisdom was communicated only to seventy sages and was supposed to have been preserved eventually in seventy books by Ezra. But the cult of the inner secrets really arises in the later Middle Ages, centring in Spanish Judaism where the *Zohar* was produced sometime after 1275 in Castile.[34] The Hebrew word cabbala (kabbalah) literally means tradition, but the emphasis was on methods of reading the inner mysteries by means of the secret 'signs'. One of its main teachings concerned the inner life of the Godhead and its relationship to the created world. Within divinity itself ten emanations, the Sefiroth, embodied the dynamism of divine Being.[35] From these ten manifestations was derived the whole created order. Within creation the self-revelation of God's energy was manifested particularly through language, above all in the Biblical text which contained 'a concentration of energy, with an infinitude of meaning'.[36] Thus for the adepts of the cabbala it provided the final and necessary key to scriptural exegesis.

The method of the cabbala taught Egidio that God revealed Himself in the text of the Scriptures, in His Names, and in the meaning of numbers. The numbers one to ten were particularly significant, since ten was the

[30] O'Malley, *Church and Reform*, 75.
[31] Ibid. 78.
[32] Ibid. 91.
[33] Ibid. 72.
[34] F. Secret, *Les Kabbalistes chrétiens de la Renaissance* (Paris, 1964), 8–21; see also G. Scholem, *Major Trends in Jewish Mysticism*, 3rd edn. (London, 1955), 205–86.
[35] Jacob, *Symbols*, 3–5.
[36] O'Malley, *Church and Reform*, 86.

number of perfection. This led Egidio to frame his theology of history on a sequence of ten ages before and ten following the Incarnation, linked with the ten Sefiroth, that is the ten attributes of God revealed in the cabbala. In adopting this structure it is striking that he departs from the age-old and universally accepted Christian framework of Seven Ages corresponding to the Seven Days of Creation: here his newly found key of the cabbala supersedes medieval tradition.[37]

Egidio's *Historia viginti saeculorum* was written in mid-career during the Fifth Lateran Council.[38] It is a work of more than 300 folios in MS. 502 of the Biblioteca Angelica, which I have used. Its structure is complex, in itself setting up certain tensions in the reading of history. The basic ground plan is an interpretation of the first twenty psalms as providing a 'reading' of the twenty ages, divided into two eras. The history of the first era only occupies the first 26 folios, ending with the age of the Maccabees in the tenth age. The Christian era begins with the eleventh age, that of Christ and the early Church down to Pope Sylvester. It is here that ambiguities in the meaning of history become apparent. Egidio's interpretation of numbers gave him a mounting scale up to ten which again and again he proclaims as the number of perfection. There are innumerable passages in the *Historia* which explore the mystical meaning of numbers, especially the first ten, reaching their climax in the number ten itself, which contains or embraces all things.[39] Yet, by contrast, Egidio also uses the symbolism—derived both from the Book of Daniel and Hesiod—of ages designated in a descending scale of metals from gold down to iron or clay.[40] His instinctive conservatism[41] supported the obvious interpretation of the first Christian age as the Age of Gold. The pure doctrine belonged to the pristine age of Christ and the early Church: the further the Church was removed from its origins, the more corruption set in. So the pattern of the Christian ages appeared to be one

---

[37] On Egidio as a cabbalist, see Secret, *Kabbalistes*, 106–20.

[38] The archetype is MS. Lat. IX B14 in the Bibl. Naz., Naples, with the title 'Aegidii Viterbiensis Hystoria saeculorum per totidem psalmos digesta'. It is dedicated to Leo X. There are marginalia in Egidio's hand and on the title-page is an extended note by Girolamo Seripando stating that the MS was lost at one point but later bought at a stall in Rome by Cardinal Cervini (Pope Marcellus II). These details are drawn from F. X. Martin, 'The Writings of Giles of Viterbo', *Augustiniana*, 29, 1–2 (1979), 170. For Seripando and Cervini, see below, Hudon, Ch. 19.

[39] See especially *Historia*, fos. 239^r, 268^v. On the number symbolism of the *Historia*, see Secret, *Kabbalistes*, 113–14.

[40] *Historia*, fos. 39^v, 42^v, 43^r, 47^r–48^v, 66^r, 75^r, 77^r, 89^r, 91^v, 101^v, 102^v, 154^v, 174 *bis*, 322^r.

[41] O'Malley, *Church and Reform*, 180; id., *Rome and Renaissance*, 5.

of decline. Succeeding the Age of Gold (that is the Age of Christianity and the early Church), the Age of Silver saw the beginnings of decline when the Church became rich under Constantine. Through a lengthy and detailed history of the Christian centuries Egidio chronicles growing corruptions, heresies, and worldly ambitions right down to the ninth age.

But again, this is not the whole story. The very psalms which he is ostensibly interpreting (but often loses sight of) give him clues to God's providential workings in history as well as material for deeply pessimistic interpretations. Above all, the clues lie in the concords between pre- and post-Incarnational eras which reveal a progressive movement towards a final destiny of the Church which is positive, not negative. In the use of concords—which he is never tired of repeating—Egidio's originality lies in the double set of clues he uses, paralleling Old Testament *figurae* with classical legend and history, including the Etruscans. These all prophesy the future glorious role of the Roman Church. Rome is the new Israel; the Vatican is Jerusalem or Sion.[42] Beside the biblical types Egidio sets mysterious and wonderful concords with Etruscan and Roman history. Through divine providence in the pre-Christian era ancient religious teachings were carried to the Etruscans who were a most religious people.[43] This people and their god Janus played an important role in the history of divine revelation. Egidio sets great store by the fact that Viterbo originated as an Etruscan city and he was particularly moved by the archaeological discovery that an Etruscan settlement on the Janiculum hill had preceded the Vatican.[44] All the signs point to the apotheosis of Rome. The promises to Abraham applied ultimately not to the Hebrews but to Etruria and Rome.[45] In Etruria there was always the seat of religion; Janus came first, afterwards Peter.[46] The prophecies point not to Babylon, but Etruria; not to Jerusalem, but the city of Rome; not to Sion, but the Vatican.[47]

The double motifs of decline, on the one hand, and evidences of the fulfilment of God's designs, on the other, run throughout Egidio's history of the Christian ages, down to the ninth, from Celestine V to Julius II.[48] This period was marked by three great tragedies: the Avignonese Captivity, the Great Schism, and the fall of Constantinople

---

[42] *Historia*, fos. 12^{r-v}, 15^v, 16^v, 25^v, 45^r, 46^r, 51^r, 58^v, 156^v, 158^r, 170^r, 172^v, 173^v, 174^v.
[43] Ibid., fo. 3^v.
[44] Ibid., fos. 68^v, 70^v, 91^r, 116^v, 117^r, 166^v, 171^v.
[45] Ibid., fos. 171^v, 220^v.
[46] Ibid., fo. 70^v.
[47] Ibid., fo. 91^r.
[48] Ibid., fos. 85^r, 190^v–191^r, 197^v–198^r.

to the Turks. Moreover, this was the period in which the Aristotelians or Peripatetics gained ascendancy in the Schools, threatening the Church with doctrinal corruption.[49] But psalm nineteen ('The heavens declare the glory of God . . .') prophesies, according to Egidio, the opening out of God's purposes in the ninth (or nineteenth) age.[50] This has become evident in its closing years, first of all in the geographical 'opening up' achieved by Ferdinand, King of Spain, and Emanuel, King of Portugal. 'Their sound has gone forth to all the earth' (verse 4) wonderfully points forward to the discoveries of unknown lands.[51] Secondly, Julius II has promoted the glory of Rome, particularly in the building of St Peter's.[52] Julius is likened to the giant of verse 5 who rejoices to run the course (fo. 266)[53] and the Golden Age is seen to be returning (fo. 223).[54]

So Egidio moves to the threshold of the tenth (or twentieth) age. The Tenth Sefiroth means the revelation of God in majesty and the Tenth Age therefore should see the completion of God's self-revelation. From one point of view this could be seen as a returning Golden Age of the first centuries and this chimed, of course, with the humanist cyclical concept. But the unacknowledged tension emerges again here, for Egidio sees the greatest glory of the Tenth Age as the fullest understanding of the Scriptures finally revealed and at times he appears to envisage this as a developing understanding through the ages. He bases this belief on John 14: 26: 'the Holy Ghost whom the Father will send in my name, he shall teach you all things'—a text used by Joachim of Fiore for precisely the same purpose. This development, proceeding through all the ages of the Church, was given by the direct action of the Holy Spirit, 'an increase of knowledge mysteriously developing in history'.[55] This gradual unfolding of the inner meaning in Scripture appears to contradict his pattern of decline and, indeed, his picture sometimes seems to be one of a sudden dramatic halt to corruption followed by an amazing opening of the Scriptures through the newly acquired key of the cabbala. His orthodox faith taught him that the fullness of doctrine was in Christ; therefore the 'new age' could only be a 'return'. Yet the linear view of history which he inherited from the Middle Ages, now strongly tinged

[49] Ibid., fos. 190^v–191^r, 229^r, 235^v.
[50] Ibid., fo. 223^r.
[51] Ibid., fos. 176^v–177^v, 285^r, 192^v–193^r, 266^{r-v}.
[52] Ibid., fos. 192^v–194^r.
[53] Ibid., fo. 266^r.
[54] Ibid., fo. 223^r.
[55] O'Malley, *Rome and Renaissance*, 450. See also id., *Church and Reform*, 184–6.

with a diffused Joachimist expectation, put the emphasis on the coming age as a triumphant culmination of the time process in new understanding and unity. When Egidio parallels the new opening of the Scriptures with the geographic opening up of new worlds he is implying the expansion of knowledge in both spheres: 'In the tenth age all secrets will be revealed, those of the divine and eternal world through the cabbala and those of the created world by voyages of discovery. Mankind will be brought into an intellectual and religious unity under the Papacy.'[56] Without diminishing the fullness of revelation in Christ he is, none the less, proclaiming prophetically an age in which new and fuller secrets, both of the Scriptures and of the natural world, will be finally opened.

In the *Historia* Leo X is hailed as the first pontiff of the dawning Tenth Age. He is a Medici, with supposedly Etruscan origins, while the very name of his family proclaims them as 'doctors' to cure the wounds of the age.[57] The symbolism of the laurel tree is invoked to laud Leo's father, Lorenzo the Magnificent.[58] The name Leo carries the significances of the Lion of the tribe of Judah and of the mystical *animalia* of Ezekiel and Revelation,[59] while in innumerable passages Egidio proclaims the sacred number ten in the Tenth Age.[60] This happy conjunction of 'signs' is confirmed by achievements in Rome: the new peace and stability in the city, the florescence of arts and humanist learning, the rearing of St Peter's, already begun under Julius II, as the greatest monument to the magnificence of God.[61] In Egidio's vision different aspects of the Golden Age are all coming together. First and foremost there is the opening of the Scriptures to reveal secrets undreamed of before. But this is matched, with no sense of incongruity, by Egidio's enthusiasm for the revival of rhetoric, poetry, and the arts. Leo is, *par excellence*, the great patron in this field and the glorification of St Peter's is his great work.[62] The second facet of this new age, however, concerns Egidio's deeply serious hope for the reform of the Church in doctrine and morals. Here his pessimism with regard to its present state tinges his hope, yet the reading of the signs must prevail. In doctrine it is the purging of philosophy and theology from the poison of the Peripatetics that is urgent.[63]

---

[56] O'Malley, *Church and Reform*, 8.
[57] *Historia*, fos. 287ᵛ, 299ᵛ.
[58] Ibid., fos. 255ʳ⁻ᵛ.
[59] Ibid., fos. 286ᵛ, 296ᵛ.
[60] As examples, see ibid., fos. 293ʳ⁻ᵛ, 300ʳ, 305ʳ.
[61] Ibid., fos. 194ʳ, 286ᵛ, 223ʳ, 314ᵛ–315ᵛ.
[62] See O'Malley, *Rome and Renaissance*, 5–6, 8–9; id., *Church and Reform*, 71.
[63] Id., *Rome and Renaissance*, 8; id., *Church and Reform*, 40–6.

In his aims for moral reform we meet another side to this extraordinary man. As an Augustinian he prized above all the ascetic, hermit life of poverty and contemplation. He praises both Peter the Hermit and Pope Celestine V. From one point of view this aspect of his activities represented his conservatism—the deep desire to return to what he believed to be the model created by St Augustine. His actual reforms while he was General of his Order bear out this austere ideal. While the Church as an institution must portray the magnificence of God and be adorned with all the jewels of art and architecture, for the earnest individual Christian the vision of God must be sought in deep study of the Scriptures and a life of simple piety.[64] Yet the return to a pristine tradition of simplicity is always juxtaposed to the vision of an astonishing new future opening up before the Church. So the third facet of the new age lay in the expansion of the world as exploration opened up vast new territories before amazed eyes. God was revealing the culminating stage of His providential plan in history and Christians were called, with a new sense of the vastness of His purpose, to achieve the conversion of all humanity before the End and to bring all peoples into one flock under one shepherd. Here Egidio's passionate beliefs in the inner harmony of all true knowledge and the unity of all races come together: *plenitudo temporum, plenitude gentium, plenitudo doctrinae*.[65] All things should be nearing their final harmony.

Yet Egidio's realism showed him starkly the present corrupted state of the Church. It is striking that the closing section of the *Historia* is not couched in the euphoric language of an approaching certainty, but is rather in part a call to Leo to fulfill his role in a world where dangers still threaten and in part a prayer to God that the prophecies may be fulfilled. There are many references to the expected Age of Gold throughout the work, but few in this last section. The *Historia* ends on a sober note. The prophetic tradition inherited from the Middle Ages taught Egidio that the Church must suffer great tribulation for its sins before it could enter into its final peace. When Papal Rome suffered the terrible catastrophe of 1527 Egidio saw this as the inevitable judgement of God. The Emperor Charles V, as God's agent in the Sack of Rome, was the new Cyrus, sent to purge the Roman Church of its evils.[65a] But the Tenth

---

[64] Id., *Church and Reform*, 143–4, 186–7; id., *Rome and Renaissance*, 9.
[65] Quoted O'Malley, *Church and Reform*, 185. See also id., *Rome and Renaissance*, 8, 190–1, 296–338.
[65a] Id., *Church and Reform*, 110, 116–17.

Age is still on the threshold of time. In his last years Egidio of Viterbo
hoped that Pope Clement VII and Charles V might be the new messianic
agents of God's final purpose. But the times were ambiguous and might
turn either to disaster or to *renovatio*. His last work, written at the request of Clement VII and dedicated to Charles V, was the *Scechina*,[66]
that is, the habitation of God with men. The dedication may seem strange,
since one of the main purposes of the work was to expound the new
understanding of the Scriptures through the cabbala. By this time, as
Father O'Malley writes, Egidio 'was completely dedicated to investigating the meaning of the Scriptures.'[67] But in his vision the opening
of the inner spiritual world of God's word was still intimately connected
with the opening up of the physical world of His creation. For the outer
world Egidio pins his hope on Charles V as the chosen agent of God. So
the *Scechina* combines in a remarkable way a defence of the new Scriptural exegesis with an eloquent appeal to Charles to take up his role as
world emperor.

François Secret describes Egidio's work as 'sans doute le plus re-
marquable d'assimilation de la Kabbale dans le monde des humanistes
chrétiens'[68] but in the *Scechina* he was also drawing on the late medieval
expectation of a Last World Emperor which had become in many respects
a political form of Joachimism.[69] He writes with great passion, for he
believes that the 'storia assoluta della umanità sub specie aeternitatis'[70]
is indeed reaching its climax. Charles should be the new Moses to lead
the human race into liberty, the new Caesar to conquer the barbarians,
the new David to gather all people into one fold. Egidio even makes a
comparison with the Apostles: 'by them the known world was gained:
by you, the unknown: by them part, by you the whole has been encompassed.'[71] So Egidio calls upon Charles to take up his role, for he is sent
from heaven: 'Now is the propitious time of which the prophets speak;
now you, the new prince, are called to celebrate "the Kalends" and to
renew the earth's orb.'[72] This was the vision of an old man, for Egidio
died two years later.

Egidio of Viterbo's reading of history past, present, and future, in-

---

[66] Egidio of Viterbo, *Scechinae Libellus de litteris hebraicis*, ed. F. Secret (Rome, 1959).
On this work see also O'Malley, *Church and Reform*, 118–20.
[67] O'Malley, *Church and Reform*, 70.
[68] Secret, *Kabbalistes*, 120.
[69] See above, p. 17.
[70] Massa, *Pensée humaniste*, 208.
[71] *Scechina*, 161.
[72] Ibid. 69.

corporated medieval and humanist elements in a prophetic vision which was founded on a particular method of 'searching the Scriptures' derived from Jewish sources. The question we must now ask is how far the specifically Joachimist interpretation of history formed part of Egidio's medieval heritage. Of course the diffused Joachimism of the fourteenth and fifteenth centuries in which a final earthly stage of blessedness was to be inaugurated by an Angelic Pope and a Last World Emperor coloured his expectation. But did Egidio owe anything directly to the thought of Joachim himself? The question is prompted by the evidence of a dedication which Silvestro Meuccio, a member of Egidio's order, wrote to his edition of Joachim's *Expositio in Apocalypsim*, published in Venice in 1527. This was apparently the final publication in a series of Joachimist works produced by a Venetian group of Augustinian hermits. They had begun in 1516 with the *Liber de magnis tribulationibus* of the Joachite Telesphorus of Cosenza and the pseudo-Joachimist work, *Super Hieremiam*, going on to the *Super Esaiam*, also a pseudo-Joachimist work, in 1517 and then turning to Joachim's genuine works with the *Liber de Concordia* in 1519 and finally to the *Expositio* and the *Psalterium decem chordarum*, published together in one volume in 1527. From their prefaces and from other supporting evidence we learn of the undoubted Joachimist interests of this group.[73]

In the 1527 dedication to Egidio,[74] Silvestro Meuccio relates how the revered cardinal came to Venice after a diplomatic mission in Spain. When the Augustinian brother appeared in his presence Egidio asked him to produce the works of the Abbot Joachim which he had published. These pleased Egidio not a little and he exhorted Silvestro to get other works of Joachim into print, especially his commentary on the Apocalypse. Now Silvestro presents this work to the great cardinal with all the usual rhetoric of praise. Egidio excels in learning and in languages, including Hebrew and Arabic. But Silvestro goes on to praise his piety and especially his reforming work in Silvestro's own order. This brings him to the real message he wants to get across to the cardinal. The Abbot Joachim, in his *Expositio in Apocalypsim*, had prophesied in exalted terms the rise of an order of hermits living the life of angels who, in the last age of the world, would, in the spirit and power of Elijah, revive the collapsed church and restore all things.[75] This was a famous passage in Joachim's

---

[73] See B. McGinn, 'Circoli gioachimiti veneziani (1450–1530)', *Cristianesimo nella storia*, 7 (1986), 19–39.
[74] Joachim, *Expos. in Apoc.*, fos. A1–A2ᵛ.
[75] Ibid., fos. 175ᵛ–176ʳ.

works which had already been claimed for the Augustinian hermits[76] and Silvestro cites further texts in support of this claim. It is an order both old and new—thus Silvestro is able to combine his order's tradition of foundation by St Augustine with the great future promised in Joachim's prophecy to an order with a new role in the climactic age of history. The setting is clearly eschatological: the order will arise 'in ipsa tribulatione futura' but after the fall of Babylon, that is, the carnal church, the life of these hermits will be 'quasi ardens in dei amore et zelo ad comburendos tribulos. . . .'

Father O'Malley dates the meeting of Egidio with Silvestro Meuccio to the year 1519 when Egidio spent some time in Venice.[77] This was the year in which the Venetian Augustinians published Joachim's *Liber de Concordia* and it seems most likely that Egidio saw and perhaps read it then. The preface to the *Expositio* certainly suggests that he was genuinely interested in Joachim's writings and he appears to have possessed at least one pseudo-Joachimist manuscript.[78] But Egidio's *Historia* was finished in that year, 1519, and it is hardly likely that he could have made use of the Venetian edition in writing it. There is, however, a possibility that at an earlier stage Egidio was reading Joachim's *Liber de Concordia* in manuscript form. Malcolm Bull's study later in this book draws attention to the striking parallels between the programme of the Sistine Chapel ceiling and the Old Testament concords in Joachim's work.[79] He suggests that an unknown adviser directed Michelangelo in the construction of this programme *c.*1507–8. Egidio of Viterbo would certainly be a candidate for this role, although the case remains unproven.

This is speculation. The case for Joachim's influence on Egidio's theology of history can only rest on clear traces in his writings, especially in the *Historia*. There are enough affinities to support the case for Egidio's interest in Joachim. In the first place both believed in the providential purposes of God revealed through the entire process of history, secular as well as sacred. Secondly, both believed the key to this process lay in a detailed searching of the Scriptures—*scrutamini scripturas* was a favourite text of Joachim and the Joachites as well as Egidio. This searching must be informed by a special exegetical method. Joachim's *spiritualis*

---

[76] M. Reeves, *Influence of Prophecy in the Later Middle Ages: A Study in Joachimism* (Oxford, 1969), 257–8, 261–6.

[77] O'Malley, *Church and Reform*, 61.

[78] Secret, *Kabbalistes*, 115, describes this MS (Paris, Bibl. Nat., 3363) as 'peut-être de sa main'.

[79] See below, Bull, Ch. 16.

intellectus, 'given' in a mystical experience which illumined 'tota veteris ac novi testamenti concordia',[80] here corresponds in a certain sense to Egidio's key of the cabbala, and his belief in a given mystical knowledge beyond reason. Thirdly, numerology gave both their patterns of history. Fourthly, both used the method of concords between the ages in interpreting their patterns. Fifthly, Joachim, like Egidio, though to a more limited extent, found mystical meaning in certain letters and their shapes and arrangement.[81] Alpha, Omega, and the Tetragrammaton embodied for Joachim the deepest mysteries of the Trinity. The shape of the Alpha revealed the mystery of One Person (the Father) sending Two (Son and Spirit) and that of the Omega the corresponding mystery of Two Persons (Father and Son) sending One (the Spirit). The Tetragrammaton, which he transliterated as IEUE gave him a wonderful revelation of Three-in-Oneness, for the letters could be divided into three groups: IE, EU, UE—yet one element—the *E*—is present in all three. He drew these mysteries in diagrams and was never tired of expatiating on them. As we have seen, Egidio also believed that the very form of the letters in the most sacred Hebrew language was an embodiment of divine truth. In an interpretation strikingly similar to Joachim's, he saw the Aleph which was composed of a *Vau* and two *Yod*s as a symbol of the Trinity.[82] The letters themselves 'participated in the divine energy'.[83] He believed that the Tetragrammaton 'had locked within itself the whole of theology'.[84]

In details, too, there are some echoes. Egidio uses the metaphor of cortex and *medulla* for the outer and inner meaning of the Letter, as Joachim did.[85] The symbol of the Tree and the Wheels of Ezekiel, beloved of Joachim, find an important place in Egidio's imagery too.[86] A detail in Egidio's number symbolism suggests more definitely that he had read Joachim's *Liber de Concordia* or at least the little tract *De Septem Sigillis*. Joachim makes a concord between seventy-two elders chosen by Moses and seventy-two disciples sent out by Christ. But the Biblical number in each case is seventy and Joachim gives his own

---

[80] Joachim, *Expos. in Apoc.*, fo. 39[v].

[81] Ibid., fos. 34[r]–39[v]. See Reeves and Hirsch-Reich, *Figurae*, 38, 51, 192–6, for full references.

[82] *Historia*, fos. 18[r–v], 22[r], 115[v]–116[r], 237[r–v]; see also O'Malley, *Church and Reform*, 78–9.

[83] O'Malley, *Church and Reform*, 91.

[84] Ibid. 92.

[85] *Historia*, fos. 277[r], 306[v]. Cf. Joachim, *Liber Concordie Novi et Veteris Testamenti* (Venice, 1519), fos. 65[v], 111[v], 119[r]; *Expos. in Apoc.*, fos. 17[r], 39[r].

[86] *Historia*, fos. 38[v], 48[v], 292[v], 314[r–v]. Cf. Joachim, *Lib. Fig.*, pls. 1, 2, 5, 6, 15, 22, 23. See also Reeves and Hirsch-Reich, *Figurae*, 27–38, 153–83, 224–31.

explanation for changing the number. Egidio uses the same concord with the same variation of seventy-two.[87] Another striking similarity is Egidio's description of St Francis and St Dominic as 'duas olivas et duo candelabra'.[88] This is derived from Apocalypse 11: 14, where the images are used to describe the two witnesses. Joachim saw in this text a prophecy of two future orders of spiritual men who would herald the third age.[89] It is possible, therefore, that we have traces of Egidio's reading of Joachim's two main works. But none of these details are sufficient to establish Joachim as one of Egidio's main sources. The fact remains that his *Historia* is based on entirely different principles from Joachim's theology of history. Although the mystery of the Trinity took a high place in his thought, there is no trace in his interpretation of history of Joachim's Trinitarian structure. Indeed, in his early commentary on the Sentences of Peter Lombard Egidio had opposed Joachim's speculations on the meaning of the Trinity.[90] His structure of 10/10 ages was derived from a totally different source, the 10 Sefiroth of the cabbala. Furthermore, there was, as we have seen, a paradoxical pattern of decline in Egidio's system which is alien to Joachim's thought. The latter could punctuate the periods in both the Old and New Dispensations by a 7/7 series of tribulations, but there is a strong sense of progressive growth in his writings and in his vivid tree figures. There seems to be no trace of Joachim's 'biological' development of history in Egidio's structure: his ages juxtapose elements of evil and good without any sense of organic growth. Again, Egidio's Golden Age is certainly one of illumination like Joachim's third *status*, but it is depicted in his own peculiar blend of humanist, Hebrew and Christian colours. In spite of their affinities Egidio cannot really be classed as a disciple of Joachim.

Yet there remains a speculation in the mind. As I have tried to show, Egidio presents a strange paradox: a man of action who desired the contemplative life; a Renaissance figure, glorifying a magnificent Rome, yet one who promoted simplicity in the spiritual life; a classical humanist and a dedicated Biblical scholar; and finally, a deep conservative who was yet carried away by a vision of newness for the Church, both in

---

[87] *Historia*, fo. 277ᵛ, using the concord between Exod. 24: 1, 9 (or Num. 11: 16, 24), and Luke 10: 1. Cf. Joachim, *Lib. Conc.*, fo. 12ʳ; id., *Psalterium decem Chordarum* (Venice, 1527), fo. 269ᵛ; *De Septem Sigillis*, ed. M. Reeves and B. Hirsch-Reich, 'The Seven Seals in the Writings of Joachim of Fiore', *RTAM* 21 (1954), 240.

[88] *Historia*, fo. 169ᵛ.

[89] Joachim, *Expos. in Apoc.*, fo. 148ᵛ.

[90] O'Malley, *Church and Reform*, 61.

fullness of knowledge and in expansion of the physical horizons. Was it precisely Joachim's vision of the order which was to be both old and yet new which struck a chord in the heart of this Renaissance cardinal? Did the eager fervour of Silvestro for his own order and the clarion note of Joachim's prophecy claimed for it grip his imagination? The 'old, yet new' order encapsulates the quintessence of Joachim's Age of the Spirit. Although in his last work, *Scechina*, Egidio was still thinking in political terms, pinning his faith for the future on the institutions of Empire and Papacy, perhaps he did respond to Joachim's model of new spiritual men and therefore in some sense might be called a Joachite.

6

The Role of Prophecy in the Career of the Enigmatic Bernardino López de Carvajal

NELSON H. MINNICH

The reputation of Cardinal Bernardino López de Carvajal (1456–1523) has rested in large part on the motivations attributed to him in the calling and leadership of the Council of Pisa (1511–12). Both his contemporaries and later historians have disagreed on his intentions, some seeing his actions as the work of one driven by overreaching ambition, others pointing to his longstanding desires for church reform, or yet others to his belief in prophecy.[1]

His contemporary Francesco Guicciardini (1483–1540) claimed in his *Storia d'Italia* that Carvajal was driven by an ardent desire to become Pope and for that reason sowed the seeds of discord that led to the Pisan schism. The cardinal hoped with the help of Emperor Maximilian I to use the Pisan Council to depose Julius II and have himself elected in his stead. While conceding that Carvajal was distinguished for his learning and manner of life, Guicciardini accused him of limitless pride. This view of the man is echoed by papal historians from Odorico Rainaldi to Ludwig von Pastor.[2]

---

[1] For a brief overview of some contemporary writers who mention Carvajal, see J. G. Rossbach, *Das Leben und die politisch-kirchliche Wirksamkeit des Bernardino Lopez de Carvajal, Kardinals von Santa Croce in Gierusalemme in Rom, und das schismatische concilium Pisanum* (Breslau, 1892) 3–11. For the best recent account of his life with ample bibliography, see G. Fragnito, 'Carvajal', *DBI* 21 (Rome, 1978) 28–34.

[2] F. Guicciardini, *Storia d'Italia*, ed. C. Panigada (Bari, 1929), ix. 10; x. 2, 3, 7; xi. 13 (in this edn., pp. 3, 49, 110, 117, 143, 283); O. Rainaldi, *Annales ecclesiastici ab anno MCXCVIII ubi desinit Cardinalis Baronius*, rev. D. Mansi (Lucca, 1747–56), xi. 569, 579, 581–2, *ad annum* 1511, nn. 3, 16, 22, and papal documents condemning Carvajal (e.g. 1511, nn. 13, 26, 36). See also the negative judgements of P. Lehmann, *Das Pisaner Concil von 1511* (Breslau, 1874), 26–30 and L. von Pastor, *History of the Popes*, ed. E. Antrobus (London, 1891–8), vi. 387–8.

That Carvajal sincerely saw in the Pisan Council a vehicle for church reform and the calling of a crusade, goals that both he and the council's supporters frequently affirmed, is a proposition that has gained credibility among scholars from Mandell Creighton to Walter Ullmann and Josephine Jungić.[3] Long before the Pisan schism, it is argued, Carvajal had espoused the causes of church reform and a crusade. At the Mass celebrated by Cardinal Giuliano della Rovere (later Julius II) on 6 August 1492 to begin the conclave that elected Alexander VI, Carvajal vigorously called for a Lateran Council ('in laterano . . . generalis Christianorum conventus') that would reform the Roman curia and organize a defence against Islam.[4] Twenty years later that same desire for church reform and a crusade led him to join in calling the Pisan Council that ironically pressured Julius II into convoking the long-awaited Lateran Council.

Yet a third school of thought sees Carvajal under the influence of prophecy. According to Mariano da Firenze, an Observant Franciscan and contemporary of the cardinal, Carvajal on reading in 1502 the prophecies of Amadeus of Portugal contained in the *Apocalypsis Nova*, saw himself in the role of the Angelic Pope described in that book and therefore dared to call the Pisan Council as a means for fulfilling these prophecies.[5] But another account would seem to contradict this interpretation. Antonio da Cremona, an Amadeite Franciscan, claimed that the cardinal had rejected these prophecies as the work of a devil, not of a saint, and had returned the book to the Amadeite friars with the advice that it be burnt because of its many falsities and heresies.[6] That this account is probably untrue and that Carvajal did indeed believe in Amadeus's prophecies is supported by his reported strategy of publicly

[3] See the cardinals' edict convoking the Pisan Council in *Acta et decreta sacrosancta secundae generalis Pisani Synodi prout per protonotarios et notarios summarie scripta reperiuntur* (Paris, 1612), 35–6. For Carvajal's sermon at Pisa urging church reform, see ibid. 86 and for other conciliar speakers urging the same, see ibid. 94, 107, 128–9, 148, 183, 190. For M. Creighton's view, see *A History of the Papacy from the Great Schism to the Sack of Rome* (London, 1897), v. 151; for W. Ullmann's view, see 'Julius II and the Schismatic Cardinals', in G. Cuming and D. Baker (eds.), *Schism, Heresy and Religious Protest* (Cambridge, 1972), 177–93. See further, J. Jungić's study of Carvajal in Ch. 17 of this book.

[4] Bernardino López de Carvajal, *Oratio de eligendo summo Pontifice* (Rome, c.1492); repr. in E. Martène and V. Durand, *Thesaurus*, vol. ii, cols. 1774–87, esp. 1783. The sermon is studied by P. Paschini, 'Una predica inefficace (propositi di riforma ecclesiastica alla fine del sec. xv)', *Studi romani*, 1 (1953), 31–8 and J. McManamon, 'The Ideal Renaissance Pope: Funeral Oratory from the Papal Court', *AHP* 14 (1976), 16–17, 27, 29–30, 32, 41, 45, 53.

[5] L. Wadding, *Annales Minorum* (Rome, 1731–6), xiv. 371, *ad annum* 1482, n. 39. See also above, pp. 45, 59.

[6] A. Morisi, *Apocalypsis Nova: Richerche sull'origine e la formazione del testo dello pseudo-Amadeo* (Studi Storici, 77; Rome, 1970), 34.

deprecating the book's contents in order to deflect possible papal and royal interest in it, by the fact that he allowed his friend Salviati to study, transcribe, and revise this work, and by his personal retention of the original copy of the *Apocalypsis Nova*, which he claimed he lost during the hasty retreat from Milan of the adherents of the Pisan Council.[7] Other evidence suggesting his support for prophetic ideas can be found in his relations to contemporary prophets other than Amadeus and in his extant speeches subsequent to his reading of Amadeus's book.

In his role as cardinal–protector of various religious groups, Carvajal was familiar with some of the prophets of his day. In addition to protecting the Franciscan Amadeite congregation,[8] Carvajal seems to have taken the place of Cardinal Oliviero Caraffa as protector of the Dominicans when the latter left Rome for fear of being implicated in the affair of Savonarola. Carvajal may have been favourably disposed towards the prophet of Florence. When Savonarola was arrested in 1498, two of his disciples fled to Rome to seek Carvajal's protection.[9] The saintly founder of the Minims, Francesco da Paola, whose prophecies about the future Julius II and Leo X both came true, was probably known to Carvajal. As cardinal–protector of the Minims at the same time that the saint resided in Rome, he secured papal approbation for his revised rule.[10] Carvajal was not ignorant of the prophets of his day.

His views on prophecy can be found in the learned homily he gave on the Feast of the Exaltation of the Holy Cross on 14 September 1508 in the collegiate church of St Rumold in Mechelen before Emperor Maximilian and members of the imperial family. Because the cardinal felt that the arcane material in his talk should not be written down nor ascribed to his name, he initially opposed the suggestion of his friend Salviati that the homily be published, but he eventually yielded. As its editor, Salviati added so much material that the speech grew into a treatise or book, and it is not clear what sections were written by the cardinal and what were added by Salviati. But it seems safe to assume that Carvajal concurred in these additions since the homily was later printed in Rome.[11]

---

[7] Ibid. 31 n. 56, 33–7; C. Vasoli, 'Notizie su Dragišić', *Studi storici in onore di Gabriele Pepe* (Bari, 1969), 491–3, 497.

[8] Ibid. 496.

[9] R. de Maio, *Savonarola e la curia romana* (Rome, 1969), 137 n. 17, 142 n. 40.

[10] G. Roberti, *S. Francesco di Paolo, fondatore dell'ordine dei Minimi (1416–1507): Storia della sua vita*, 2nd edn. rev. (Rome, 1963), 370–4, 523–7.

[11] This single printed edn., apparently published in Rome by J. Besechen of Speyer, probably in 1509 (Carvajal states that seventeen years have passed since the fall of Granada

Carvajal's homily is a commentary on a passage (John 12: 31 'Now is there judgement of this world, now are the princes of this world cast out') taken from the Gospel of the Mass for the feast of the Cross. This verse became the occasion for the cardinal to dare to predict that the Moslem religion would soon, during the sixteenth century, be extirpated by God. He based his prediction on three lines of argument: the nature and present condition of Islam, evidence from Sacred Scripture and its commentaries, and revelations made in his own time by God to many of His elect.[12]

The first type of argument that Carvajal adduced to support his claims about Islam's demise was based on natural reason and organized according to Aristotelian categories. He argued that Islam must fall because it is founded on error and violence. Of all the earlier heresies that afflicted the Church, Islam is the most false, even a form of idolatry. Just as other heresies have come and gone, since they are not of God (as noted by Gamaliel—Acts 5: 38–9), so too will Islam. It is not supported by divine authority, miracles, right reason, or legitimate rights, but only by the sword, and violence cannot last forever. The only reason that the Moslem sect has lasted so long is because God uses it to punish Christians for their sins. Indeed it is on the verge of collapse because of its unjust rule, intellectual decline, incontinent soldiers, and internal divisions into rival camps of the Turkish sultan, Egyptian caliph, and Persian shah.[13]

The strongest natural argument for Islam's imminent ruin was, according to the cardinal, the fullness of time. He rejected, however, speculation based on such arcane calculations as the cumulative number of letters in the Hebrew, Greek, and Latin alphabets, this number 69 supposedly indicating how many centuries there will be in world history,

---

in 1492, sig. d5ᵛ), has no place or date of publication. Further bibliographical confusion arises from differing titles: on sig. aiʳ. *Homelia doctissima Reverendissimi domini Cardinalis sanctae Crucis*; on sig. aiiiʳ. *Omelia habita Mechlinie in collegiata ecclesia sancti Rumoldi Cameracensis diocesis*. In Carvajal's audience were the Emperor, Maximilian I, his daughter, Margaret, Archduchess of Austria, Regent of the Netherlands, his 8-year old grandson, Charles, Duke of Burgundy, Prince of Spain, and his granddaughters, Eleonore (future Queen of Portugal and then of France), Isabella (future Queen of Denmark), and Mary (future Queen of Hungary and Regent of the Netherlands), see K. Brandi, *The Emperor Charles V: The Growth and Destiny of a Man and of a World-Emperor*, trans. C. Wedgewood (London, 1939–67), 41–6. Because Carvajal and Salviati did not return to Rome until Jan. 1509 and they remained at this time on good terms, it is reasonable to assign 1509 as the date of publication and to assume that the cardinal was amenable to Salviati's editorial insertions, see Pastor, *History of the Popes*, vi. 295–6; M. Sanuto, *Diarii* (Venice, 1879–1903), vii. 719; Rossbach, *Leben des Carvajal*, 99; Vasoli, 'Notizie', 494.

[12] Carvajal, *Omelia*, sigs. aiiiʳ, bivʳ.

[13] Ibid., sigs. bivᵛ–cvʳ.

with the end occurring at the beginning of the seventh millennium. Putting credence in such calculations is contrary to Christ's explicit statements and warnings in the Bible (for example Mark 13: 32–3, Acts 1: 7, Matt. 24: 32, Apoc. 16: 15), and there is just as good a reason to reject as to accept systems based on alphabetic centuries, parallels to the days of creation, or balances between periods of time. When Scripture speaks of seventy or a thousand years it means merely many years.[14]

Despite this criticism of speculation based on mathematical calculations, Carvajal later in this speech engaged in it himself. He claimed that just as Jews are people of the seventh day because they keep the Sabbath, and Christians of the eighth day because of Sunday, Moslems are people of the ninth because they keep holy Monday (!), the day of the moon (*lunedi/lunes*), the symbol of their religion. Because Moslems are identified with the imperfect and fatal number 9, their empire cannot long endure. While it has survived for 9 and then 90 years, it cannot last beyond its ninth century or else it could continue for 9,000 years, but such longevity for an erroneous sect would be contrary to the laws of nature, Scripture, and grace. Thus Islam, which has begun its ninth century, must soon come to an end. Indeed seventeen years have already passed since the fall of Granada, which clearly signalled that this long-awaited mystic century has arrived. The conquest of Jerusalem and defeat of the Moslem princes is surely near at hand. This calculation also conforms to the predictions that Islam can last for no more than a millennium.[15]

Other signs indicating the End-Times Carvajal found in the Bible. Some of these he felt had already come to pass, others were just then occurring. The Scriptures repeatedly state that before the final days the Catholic Faith will have been effectively preached through the whole world. The recent victories of the Portuguese have opened up Africa and India, while the Spanish have driven the Moors from Spain and extended their empire to the Fortunate Isles of the Atlantic.[16] Although he claimed to prefer a literal interpretation of the book of Apocalypse, Carvajal identified its 'false prophet' (16: 13, 19: 20, and 20: 10) with Mohammed and his sect with the city of Babylon that will persecute Christians for a

---

[14] Ibid., sigs. cv^r–di^v. Carvajal's rejection of a balance between periods of time as a key to predicting future events seems to place him at odds with Joachim's methodology of concordances, even if he later praised Joachim and agreed with some of his predictions.

[15] Ibid., sigs. dv^{r-v}, dvi^v–ei^r.

[16] Ibid., sigs. di^v–dii^v, eii^r, eiii^v, ev^r. He cites Matt. 24: 14, 31; Mark 13: 27; Apoc. 18: 19.

thousand years, as predicted by Lactantius and those who rearrange the chronology of Apocalypse 20: 2–3. But in the end this city will be destroyed by the angels of God (Apoc. 18: 1–24, 20: 9–10), that is by the Christian princes and servants who will respond to the legates of the Pope sent to proclaim a crusade. The victories of Godfrey of Bouillon in 1097 were but partial and a figure of the full victory to come, which Carvajal felt would involve a great naval battle (Apoc. 18: 17–21). With the fall of Babylon, Moslem princes will be converted to the true Faith and the Church will sing the first Alleluia (Apoc. 19: 1).[17]

Many signs indicate that the great Christian victory is underway. Islam can no longer extend its empire, for God has fixed its limits (Dan. 12: 7). The fall of Granada in 1492 marks the beginning of its reversal. The great victory of the Portuguese in the Arabian Sea has so shifted the spice trade to Lisbon that the merchants of Europe and Africa no longer buy their spices and fragrances from Moslems, thus fulfilling part of the prophecy about Babylon's fall (Apoc. 18: 13). In the person of Charles of Burgundy another prophecy that circulated in Spain during Carvajal's youth has been fulfilled. It predicted the greatest evil for Islam once the hereditary succession of Spain was joined to that of Germany. Pius II at the Congress of Mantua may have known of this prophecy and therefore entrusted the leadership of the crusade to the Duke of Burgundy with the understanding that the future victory would come through such a duke, even though Burgundy was not as yet joined to Spain. With this union now effected, a Christian victory seems assured.[18]

Signs in the heavens also point to major changes. In recent times there have been great conjunctions of stars and lunar eclipses. Moreover every year the moon (a symbol of Islam) has manifested many defects, suggesting the extirpation of the Moslem sect.[19]

Carvajal also cited approvingly the predictions of later prophets. He insisted that while the Holy Spirit abundantly taught the Apostles all that was necessary for salvation, He did not tell them details about the end of the world. Some of these were revealed to Christ's favourite

---

[17] Ibid., Sigs. di^v–div^v, ei^{r-v}. It is not clear from Apoc. 18 on what grounds Carvajal sees the fall of Babylon as the handiwork of God's angels. For Lactantius's prophecies, see his *Divinarum Institutionum Liber Septimus*, *PL* vi, cols. 786–808.

[18] Carvajal, *Omelia*, sigs. dv^v–dvi^r, ei^v–eii^r. On Spanish prophecies about the defeat of Islam, see J. Doussinague, *La politica internacional de Fernando el Católico* (Madrid, 1944); on Pius II, Philip the Good and the Congress of Mantua, see Enea Silvio Piccolomini, *Pii II Commentarii rerum memorabilium que temporibus suis contigerunt*, ed. A. van Heck (Vatican City, 1984), i. 173–239, 461–2; ii. 738–40, 770–3.

[19] Carvajal, *Omelia*, sigs. dv^v, ei^v.

apostle, St John, on Patmos, yet other particular secrets of God were communicated to men even after the death of the Apostles. The ancient Christian writer Lactantius was invoked as an authority on some aspects of the end times, and the cardinal called Joachim of Fiore an 'outstanding teacher in these matters and a great spirit in the concordance of the Scriptures'. He referred to the abbot's interpretation of the text on Peter strengthening his brethren (Luke 22: 32) as indicating the return of the fallen-away patriarchal sees under another Peter. Carvajal claimed that in his own day there are many of God's elect who are full of revelations falling daily from the table of divine wisdom and these allow one to make clear enough conjectures on the imminent demise of Islam.[20]

Carvajal concluded this section of his homily at Mechelen with an exhortation to Maximilian to fight the Moslems. He rehearsed again many of the external signs indicating that God intends to extirpate these infidels soon. Their soldiers are weak while Greek exiles are eager to regain their heredity and Latin princes enjoy military prowess. Egypt lies particularly vulnerable to a three-pronged attack from the Portuguese fleet and the land armies of the Ethiopians and Georgians. A strong war fleet could quickly free the ancient patriarchates that lie on or near the coasts. An alliance with the Shah of Persia against the Turks may even be possible, for this ruler looks kindly on Christians, himself being born of a Christian mother. The signs of the times point to a Christian victory and the destruction of Islam.[21]

Thirteen years later on 29 August 1522, during a ceremony welcoming to Rome Adrian of Utrecht, former tutor of the Hapsburg heir and recently elected Pope, Carvajal's prophecies seemed farther than ever from fulfilment.[22] Instead of Islam, it was Christianity's frontier that seemed on the verge of collapse. In August 1521 Belgrade had fallen to the Turks and in July 1522 the fateful, final siege of Rhodes, the last Latin stronghold in the eastern Aegean, had begun in deadly earnest. The Ottoman Empire under the dynamic leadership of Suleiman the Magnificent (1520–66) had resumed its ominous expansion into Europe.[23]

[20] Ibid., sigs. ci[v], dii[v], ei[r]. (Lactantius); bi[v], div[v] (Joachim); di[r], di[v], dv[r] (Moderns).

[21] Ibid., sigs. eii[r]–fii[r].

[22] For the text of Carvajal's speech, see *Concilium Tridentinum diariorum actorum episto-larum tractatuum nova collectio*, ed. Societas Goerresiana, 12: *Tractatuum pars prior*, ed. V. Schweitzer (Freiburg im Breisgau, 1930), 18–21 (hereafter cited as *CT*). For a description of Adrian's entrance ceremonies, see Pastor, *History of the Popes*, ix. 64–9; for an analysis of Carvajal's sermon, see R. McNally, 'Pope Adrian VI (1522–23) and Church Reform', *AHP* 7 (1969), 261, 269–72.

[23] Pastor, *History of the Popes*, ix. 155; A. Bridge, *Suleiman the Magnificent: The Scourge of Heaven* (New York, 1983), 48–61.

Carvajal's hope that the Amadeite prophecy of an Angelic Pope foretold his own destiny had been repeatedly frustrated. Instead of helping to place him on the papal throne, his leadership of the schismatic Pisan Council had placed him on his knees before Leo X and the College of Cardinals, in which posture he read in a choking voice a humiliating confession of disloyalty.[24] The recent conclave (27 December 1521–9 January 1522) had been a similarly bitter disappointment. On its last day he had tied by fifteen votes with Adrian of Utrecht, but within a matter of minutes a rapid succession of accessions had resulted in his chief competitor's election, despite Carvajal's desperate pleas: 'My most reverend lords, please accede, please accede to me!'[25] Now, welcoming his former rival to Rome in the name of the Sacred College, Carvajal once again looked to prophecy, this time to shed light on the pontificate of Adrian VI.

Carvajal saw in the untainted election of the learned and upright Dutchman a fulfilment of King David's prophecy in ps. 93 (94): 19 'In proportion to the multitude of my previous heartaches so have your consolations cheered my soul.' Because of the many simoniacal elections of previous Popes the Church has suffered cardiac afflictions and the gravest pains. The men thus raised to the pontificate were ignorant and lacking in virtue. But now that God has inspired the free wills of the cardinal electors to choose Adrian, the Church has been blessed with a Pope whose learning is evident for all to see in the scholarly theological books he has published and whose moral character is well known for its constant humility, impartial justice, deep devotion to the divine worship, and frequent celebration of the Mass. In Adrian's election to the papacy has David's prediction come initially true.[26]

So that the consolations of the Lord will fully rejoice the Church's heart, Carvajal urged the pontiff to adopt a programme of reformation. Accordingly simony, ignorance, tyranny, and every vice will be expelled from the Church and it will be reformed according to conciliar and canonical laws. Impartial justice will be handed out by church officials. Provision should be made for supporting members of the hierarchy,

---

[24] For the ceremonies of Carvajal's abjuration and his plea for pardon, see M. Dykmans, 'Le cinquième Concile du Latran d'après le Diarie de Paris de Grassi', *Annuarium*, 14 (1982), 341–7 nn. 989–91; N. Minnich, 'The Healing of the Pisan Schism (1511–13)', *Annuarium*, 16 (1984), 105–8.

[25] Pastor, *History of the Popes*, ix. 23; Sanuto, *Diarii*, xxxii, col. 413; Fragnito, 'Carvajal', 32.

[26] *CT* 12, 18–20.

especially those suffering from poverty. Papal financial support will be given also to members of the nobility and to monasteries in need. Construction work on the new St Peter's Basilica should go forward. And the pope will put his authority and support behind the efforts to rescue the Christians on Rhodes and in Hungary who were oppressed by the victorious Turks. If Adrian were to accomplish these things then would the Church's joy be full and he would attain fame and eternal reward. May God grant to His Church such a favourable, felicitous, and healthy future.[27]

Although Adrian agreed to carry out the reforms urged by Carvajal, his pontificate was too brief to be effective. Rhodes fell in December and the huge debt incurred by his predecessor Leo X greatly limited his largess. Adrian's efforts to reform the Roman Curia found little support among its officials and within thirteen months of his entering Rome, Adrian died on 14 September 1523.[28] Once again Carvajal's prophecy, this time of the Lord's consoling His Church through the effective ministry of Adrian, went unfulfilled. Three months after this Pope's death, Carvajal himself died on 21 December 1523.

That in the later course of his life Carvajal had indeed placed significant confidence in prophecy is supported by the evidence of his association with the prophets of his day, the prophetic message found in some of his major speeches following his reading of the Amadeite *Apocalypsis Nova*, and the unusual actions he undertook seemingly to bring to fulfilment his interpretations of these prophecies.

The credence that Carvajal repeatedly placed in prophecy may have been related to his deep desire for Church reform and a crusade against Islam. It must have been very difficult to remain dedicated to these causes in the distracted Rome of the late fifteenth century, especially during the pontificate of Alexander VI (1492–1503). Those who gave reason to hope for a better day were the prophets. Amadeus, who died peacefully shortly after Carvajal had come to Italy from Spain, may himself have predicted an Angelic Pope and an end to Islam. But Savonarola, who foretold a restored Church, was condemned and executed by Borgia's agents. In the last dark days of Alexander's reign Carvajal turned to the *Apocalypsis Nova* for comfort. The promises he found therein of a reforming Pope and of an extirpation of the Moslems may have helped to

---

[27] Ibid. 20–1.
[28] Pastor, *History of the Popes*, ix. 67–230; McNally, 'Adrian VI and Church Reform', 253–85.

renew his dedication to these causes. Thereafter prophetic themes appear in his speeches, and his own actions may have been influenced by the hope that he himself was destined to play a major role in bringing about this reform and crusade—be it by encouraging the Emperor Maximilian to take up arms against the Turks, by calling and leading the Council of Pisa with its agenda of a reformation in head and members and of a military expedition against the Turks, by seeking his own election to the papacy, or by urging Adrian to reform the Church and call a crusade. Since some of the details of Amadeus's prophecy could easily lead the cardinal–patriarch of Jerusalem into seeing himself as a figure predicted by the Franciscan friar, Carvajal's own hope of further advancement need not be excluded as a contributing factor. But the transfer of belief in his own prophetic role to that of Adrian VI gives undeniable evidence of his genuine hopes and expectations for the future. What some saw as overweening ambition may have been for the most part a long-standing dedication to church reform and the cause of a crusade latterly driven by prophetic visions that often seemed to assign to him a major role in their fulfilment.

7

Giorgio Benigno Salviati (Dragišić)

CESARE VASOLI

I

Giorgio Benigno Salviati[1] is a person of notable importance in Italian religious life of the late fifteenth and early sixteenth centuries. His biography has not yet been completely reconstructed, even though, at this point in time, the essential information about his participation in some of the most important events of that period is well known. He is, perhaps, a complex and elusive character, a participant—undoubtedly— in great prophetic expectations, and yet often implicated in shady political, ecclesiastical, and secular intrigues, dominated by an intense ambition which led him to attempt every possible means to attain the highest ecclesiastical office. He is in particular the man of the Church who played the most decisive role in the elaboration of one of the most famous prophecies of the sixteenth century—the *Apocalypsis Nova*[2]— a text which was to have a wide circulation and to enjoy extensive success, and which, as Secret[3] shows, contributed enormously to feeding Guillaume Postel's eschatological obsession. It is thus understandable that the interest of scholars concerning him and his work should be aroused, as it is symptomatically linked to research into prophetism and esoteric and millenarian Renaissance traditions, and that an attempt should be made to reconstruct his theological doctrines, his relationships with persons of great importance in ecclesiastical history, and his com-

---

[1] For a more extensive biography of Salviati and for the relevant bibliography see C. Vasoli, *Profezia e ragione* (Naples, 1974), 17–120, with recent additions in id., *Filosofia e religione nella cultura del Rinascimento* (Naples, 1988), 141.

[2] For the *Apocalypsis Nova*, see Bibliography and Index. See also A. Morisi, 'Galatino et la Kabbale chrétienne, in *Kabbalistes chrétiens* (Paris, 1979), 212–31.

[3] See F. Secret, *Introduction à Guillaume Postel, Le thrésor des Prophéties de l'Univers; Manuscrit publié avec une introduction et des notes* (The Hague, 1969); id., 'L'émithologie de Guillaume Postel', in E. Castelli (ed.), *Umanesimo e esoterismo. Atti del V Convegno internazionale di studi umanistici. Oberhofen, 16–17 settembre 1960* (Padua, 1960), 381–437, esp. 390–3.

plex personal life, which was always set against a background of some
of the most intellectual circles in Italy at this time.[4]

His forms, therefore, a particularly interesting biography which
affords numerous opportunities to address some of the most important
subjects for those studying Italian religious history, along a 'thread'
leading from the time of Bessarion to the years of the 'querelle' involving
Reuchlin and the early days of the Reform. Salviati was probably born in
Szebrenica, Bosnia, his real name being Juraj Dragišić, about the middle
of the fifteenth century. He must have entered the Minorite Order at a
very young age and lived among those conventual Franciscans who had
an important influence on the Balkan and Dalmatian states which still
remained under Christian domination. The Turkish conquest of Bosnia
(1463) forced him to flee to Dubrovnik (Ragusa), the Adriatic city which
was an open port to Italian culture,[5] and from there he travelled to Italy
as a student of his Order (unconfirmed reports indicate Bologna, Padua,
Pavia, and Ferrara) before going—according to what he himself wrote—
to the great university centres of Europe, Paris, and Oxford. In any case,
whatever his actual course of studies may have been, there is no doubt
that he acquired a good knowledge of scholastic texts and authors (particularly John Duns Scotus and Thomas Aquinas, Henry of Ghent,
Francis de Meyronnes, Godfrey de Fontaines, Landolfo Caracciolo)
often quoted and discussed in his works. Johannes Foxoles or Foxol
(Johannes Anglicus),[6] the theologian and philosopher with a purely
Scotist education who taught in Bologna and possibly in Rome before
completing his career as Bishop of Armagh, also figured amongst his
teachers. It is likely, however, that Dragišić did not, during this time,
confine his interests to the study of theology. He was already attracted to
the eschatological and prophetic traditions and trends which were to be
found in his Order at that time, and which were reviving the older
millenarian vocation of the Friars Minor. Some of the evidence leads

---

[4] For a thorough evaluation of the various studies on Salviati, see G. Ernst's entry in the
DBI.

[5] For his early life, the only source is that left by Salviati himself in his work: *Opus de
Natura Angelica. Impressum cum maxima diligentia Florentiae. XIII Kalendas Augusti* M.
CCCCLXXXXIX, fo. aı[r–v] (hereafter *De Natura Angelica*). (On this incunabulum, printed
by Bartolomeo de' Libri, see *Gesamtkatalog der Wiegendrucke* (Leipzig, 1928), 3 n. 3843. I
have used the copy: Florence, Bibliotèca Nazionale Centrale, C, 6, 23.)

[6] See *De Natura Angelica*, fo. 1 II[r]. On Foxoles, see B. Geyer, *Die Patristische
und Scholastische Philosophie* (Berlin, 1928), 624; also H. Hurter, *Nomenclator literarius
Theologiae catholicae*, ii (Innsbruck, 1906), 994–5; A. Emden, *A Biographical Register of
the University of Oxford* (Oxford, 1958), ii. 726.

us to suppose that he had a close relationship with the congregation, founded in Milan in 1459, of the Portuguese Franciscan João Menezes da Silva, known as the 'Blessed Amadeus',[7] and that he was perhaps familiar with the *vaticinia* attributed to this mystic and his prophecies of the impending advent of a *Papa Angelicus* who would bring about the reformation of Christianity, the decisive destruction of the Muslim enemy, and the final reunification of all peoples in a single Christian body. These were prophecies which were, after all, not in conflict with those which were circulating at that time in the Balkans, nourished by the expectation of an inevitable liberation from the Turkish yoke.

It is not by chance that Dragišić, in a passage from his defence of Savonarola,[8] should refer in particular to these prophecies and hopes of the people expressed in the myth of an impending age of peace and universal harmony under the single eternal 'law' of Christ. His continuing interest in theological discussion relating to the value of prophecy and the nature of prophetic understanding, and to the complex question of 'future contingents' and divine predestination and prescience, are surely linked to those first experiences, nourished and reinforced by the growing spiritual tension of the time and the continual circulation of prophecies, predictions, and horoscopes of every kind. It is also certain that between 1471 and 1472 he was in Rome, where he had connections with Cardinal Bessarion, the great patron of Christians from the 'Eastern' countries.[9] We know, too, that the 'Blessed Amadeus' was also in the Papal City at that time, as confessor of his penitent Cardinal Francesco della Rovere (Sixtus IV) and that he was intimate with Cardinal Bessarion, who treated him as a theological 'adviser'.

Dragišić often harks back to his relationship with Bessarion, recalling, among other things, his participation in the bitter polemics which brought the cardinal into confrontation with George of Trebizond,[10]

---

[7] For Amadeus, see Bibliography and Index; also R. Pratesi, 'De Silva Menesez, Amadeo, beato', *Bibliotheca Sanctorum* iv (Rome, 1969), 587–8. See also above, Morisi-Guerra, Ch. 2 n. 4.

[8] *Georgii Benigni Ordinis Minorum Propheticae Solutiones impressae per Ser Laurentium de Morgianis*, VI idus Aprilis M.cccc.lxxxxvii (hereafter *Propheticae Solutiones*); (cf. also *Gesamtkatalog*, 3 n. 3845. I have used the copy: Florence, BNC, B, 6, 4, fos. bvi^r–bvii^r).

[9] Cf. below, p. 125.

[10] See esp. L. Mohler, *Kardinal Bessarion als Theolog, Humanist und Staatsmann*, i: *Darstellung* (Paderborn, 1923); ii: *Bessarionis in calumniatorem Platonis* (Paderborn, 1927); iii. *Aus Bessarions Gelehrtenkreis* (Paderborn, 1942); also id., *Die Wiederbelebung des Platons: Studium in der Zeit der Renaissance durch Kardinal Bessarion* (Cologne, 1921); P. Loernetz, 'Pour la biographie du Cardinal Bessarion', *Orientalia Christiana periodica*, 10 (1944), 116–49; H. Saffrey, 'Aristoteles, Proclus, Bessarion', *Atti dell' XI Congresso internazionale di filosofia* (Florence, 1960), 153–8; L. Labowsky, 'Bessarion Studies, 1', *Medieval*

and which culminated, between 1469 and 1471, in the publication of the *In calumniatorem Platonis*, and the stern reply of the Cretan humanist, found in the *Annotationes*. His writing was approved by Bessarion, who, on this occasion, bestowed on Dragišić the name of Benigno, which we will use from this point onwards. This tract went astray in England, possibly during a visit made there by the author,[11] but although we know nothing of this work, we have access to another writing preserved in MS. Vat. Lat. 1956, entitled *De libertate et immutabilitate Dei*,[12] which is not only dedicated to Cardinal Bessarion but also refers to a polemic which caused a great stir and strong reactions amongst the majority of the intellectual centres of Europe at that time, from Louvain to Cologne and from Paris to the Roman Curia—the question of the 'future contingents'. This is not the place to give even the broadest outline of a dispute[13] which had its origins in doctrines elaborated by the greatest teachers of the thirteenth and fourteenth centuries, from Thomas Aquinas to Duns Scotus, Peter Aurelius to William of Ockham, Wycliff to Pietro di Candia and Peter Nogent. Neither will I dwell on the reasons for the particular interest shown by Bessarion in this subject, or on Benigno's complex treatment of the question concerning which, in 1471, he declared that he wished to draw a *conclusio* which would, at the same time, save the 'Catholic truth' and enable the various theological schools to 'agree' amongst themselves. It is more significant to note that, following a doctrine similar to that also supported by Francesco della Rovere,[14]

---

and Renaissance Studies, 5 (1961), 108–62; id., 'Il cardinale Bessarione e le origini della Biblioteca Marciana', in A. Pertusi (ed.), *Venezia e l'Oriente tra tardo Medio Evo e Rinascimento* (Venice, 1966), 159–82; id., 'Bessarione, Basilio', *DBI* 9 (Rome, 1967), 686–96; the special issue of *Studi Bizantini e neoellenici* (1968) (dedicated entirely to Bessarion); *Misc. franc.* 73 (1973) *ad ind.*; *Platon et Aristote à la Renaissance. XVIe Colloque international de Tours* (Paris, 1976), *ad ind.*; J. Monfasani, *George of Trebizond: A Biography and a Study of his Rhetoric and Logic* (Leiden, 1976).

[11] See *De Natura angelica*, fos. aI^{r-v}.

[12] MS. Vat. lat. 1056: *Georgii Benigni Macedonis in Francisci de libertate et immutabilitate Dei sententias ad R. patrem et dominum dominum Bessarionem Patriarcam constantinopolitanum sedisque apostolicae episcopum sabiniensem dignissimum* (n.p., n.d.), fos. I^r–96^v. See also *MS Vat. Lat.*, ii. 602–3. A striking aspect of the codex is that of the erased corrections which have been made systematically throughout. Whilst in the beginning Bessarion must have been the principal protagonist in the dialogue and Francesco della Rovere the main interlocutor, later the roles were reversed and Francesco's name, written in conspicuous gold letters, replaces that of the cardinal, whose name now occupies Della Rovere's place. This change was without doubt occasioned by Francesco's election as Pope Sixtus IV and by the death of Bessarion.

[13] See L. Baudry, *La querelle des futurs contingents (Louvain 1465–1475): Textes inédits* (Paris, 1950).

[14] See C. Vasoli, 'Sisto IV professore di teologia e teologo,' *L'età dei Della Rovere. Atti e memorie della Società Savonese di Storia Patria*, NS 24 (1988), 177–207.

he insisted that a complete and absolute knowledge of the 'future contingents' of the mind of God does not limit His omnipotent freedom, just as His predestination does not imply the irrevocability of human destiny. Indeed, Benigno's conclusions always reveal that he made an effort to exclude any suspicion of unorthodoxy from the Scotist texts on the future contingents and to show, by clever adaptation, that a genuine reconciliation existed between the dogma of absolute and divine freedom and the truth of 'prophetic assertions' that Christians could piously approve of.[15]

II

It is likely that Bessarion's death on 18 November 1472 contributed to Benigno's departure from Rome. However, it is also significant that he was immediately offered a position of considerable prestige at the Urbino court of Federico da Montefeltro, who was known to have had a particular liking for and devotion to the Greek cardinal.[16] The Franciscan friar, having been appointed teacher to the young heir of the Duchy of Urbino, Guidubaldo, earned the reputation of being a keen-witted theologian and a clever preacher to 'scholars', a reputation he retained for the rest of his life. We know that he continued to be interested in theological and philosophical problems which were particularly congenial to his Scotist education, but which must also have appealed to the lay scholars at Federico's court. Proof of this is found in the short treatise, *Fridericus, De anima regni principe*,[17] which must have been written between 1474 and 1482. This is a discussion of the supremacy of the will in relation to the intellect, conducted on Scotist lines, but with an interesting stress on the superiority of 'practical' virtues, considered as indispensable for the achievement of a speculative 'beatitude'. It is obviously a subject which must have particularly appealed to a humanistic prince, an old scholar of Vittorino da Feltre, and to the many

---

[15] For a more extensive analysis of the work, see Vasoli, *Profezia e ragione*, 30–4.

[16] See C. Stornajolo, *Alcune ricerche sulla vita del Cardinale Bessarione* (Siena, 1897), 3–8; C. Clough, 'Cardinal Bessarion and Greek at the Court of Urbino', *Manuscripta*, 8 (1964), 160–71. On Dragišić's activity at the court of Urbino, see G. Zannoni, *Scrittori cortigiani del Montefeltro* (Rome, 1894), 1 (taken from *Rendiconti della R. Accademia dei Lincei*, 3).

[17] Cf. P. Sojat OFM, *De voluntate hominis eiusque praeminentia et dominatione in anima secundum Georgium Dragišić (c.1448–1520). Studium historico-doctrinale et editio Tractatus: 'Friedericus, De animae regni principe'* (Rome, 1972). The MS is preserved in MS. Vat. Urb. 995.

men of letters who gathered at his court. Benigno was particularly successful at the Urbino court, however, because of his 'adoption' by one of the most eminent families, the Felici, whose surname he bore for some time before changing it to the far more notable and prestigious name of Salviati.

The friar had to leave Urbino towards the end of 1485, three years after the death of Federico, when Guidubaldo, now advanced in years, had ceased to hold his highly prestigious office. However, he now set off for a new destination: the capital of humanistic Italy—Medicean Florence—where he was immediately enrolled at the College of Theologians[18] before being summoned, from 5 September 1487 onwards, to teach theology at the Studio.[19] Benigno rapidly manœuvred himself into a strong position at Lorenzo the Magnificent's court. He was appointed teacher of philosophy to the young heir, Piero de' Medici, and soon Lorenzo must have been making use of his theological advice at a particularly delicate moment in his dealings with the Church[20] on the eve of the election to the cardinalate of the future Leo X. The Franciscan was certainly favoured by this extremely important patronage, rising rapidly in office and in his position in the Order. In 1488 he was elected Regent of the Franciscan School of Santa Croce, and two years later he became Provincial Father of Tuscany, combining this post with that of Inquisitor.[21] It is said that it was in Florence that he established his reputation as a theologian and sermonizer by his success in a debate of June 1489 with the Dominican teacher Niccolo de Mirabilibus,[22] con-

---

[18] Cf. Arch. del Collegio dei teologi fiorentini, Seminario di Firenze, Reg. B., 1424–1559, fo. 40ʳ: Benigno is designated: 'Georgius Benignus de Feliciis de Urbino.'
[19] For Salviati's 'scholastic' career, see A. Verde, *Lo Studio fiorentino 1473–1503. Ricerche e documenti* (Florence, 1973), 2, 272–83; t. 2 (1985), 740–3, 882–831; t. 3 (1985), 1286–8, 1346–8.
[20] See Vasoli, *Filosofia e religione*, 139–82. See also M. Martelli, 'I Medici e lettere', in C. Vasoli (ed.), *Idee, istituzioni, scienze e arti nella Firenze dei Medici* (Florence, 1980), 113–40; id., 'La politica culturale dello ultimo Lorenzo', *Il Ponte*, 36 (1980), 923–50, 1040–69.
[21] Cf. G. Picotti, 'Un episodio di politica ecclesiastica medicea', *Annali delle Università toscane*, NS 14 (1930), 86 ff.
[22] On this disputation, see F. Fossi, *Novelle letterarie* (Florence, 1790), 20, 609–18; A. della Torre, *Storia dell'Accademia platonica fiorentina* (Florence, 1902, repr. Turin, 1960), 819; C. Dionisotti, 'Umanisti dimenticati?', *Italia medioevale e umanistica*, 4 (1961), 287–321, esp. 315–16. For the theses upheld by the Dominican, see Nicolaus de Mirabilibus ex Septem Castris, *Finis quaestionis disputatae in Domo magnifici Laurentii Medices ultimo die Juni MCCCC-LXXXVIII. Impressum Florentiae per Franciscum Dini Jaccobi civem Florentinum die vigesima septima mensis Julii millesimo quadringentesimo octuagesimo nono* (and cf. H[C] 11221). The printer was Francesco Dini. As regards Niccolò delle Meraviglie (*de mirabilibus*), note that Quétif-Echard (*Scrip. Ord. Praedic.*, i. 878–9) give information only on a certain 'Nicolaus Mirabilis', a Hungarian from Colosvar in Transylvania, master of

ducted in the presence of Lorenzo the Magnificent himself, as well as Ficino, Pico, and Poliziano, on a subject of particular theological importance—*Adae peccatum non est omnium maximum*.[23]

Between 1487 and 1492 he wrote two other equally interesting works, the first dedicated to Lorenzo the Magnificent[24] and the second to his cousin, Lorenzo di Pierfrancesco de' Medici.[25] These are still unedited. In the first, written in the form of a commentary on a sonnet containing a theological argument of Lorenzo the Magnificent's, he again discusses those subjects with which he was already familiar and which most appealed to him, which were predestination and grace, and their relationship to man's freedom. He also quotes with mastery from Averroes, Avicenna, Al-Ghazali, Aristotle, Plato, and Epicurus, not forgetting Bessarion, Ficino, and Pico, along the lines of the Scotist doctrine, but veering, as was his wont, towards a reconciliation with the 'Catholic' schools of thought (particularly the Thomist) which, according to him, if they were really genuine, should in the end be in concord. It is natural that such a method should be well received in an intellectual environment, such as that of Florence, which was particularly responsive to 'harmonization' and attracted by the strong Platonic character, which Benigno attributes to the Scotist doctrine of the *formalitates*. This

theology, apostolic preacher, and general Inquisitor of Hungary in 1483. According to Quétif-Echard, he was author of an *Opus de praedestinatione* and a *Tractatus de foelicitate*. The texts of the disputation were reprinted by Ábel Jenö in *Kiadja a M. T. Akadémia Irodalomtörteneti Birottsága, Irodalomtörteneti Emlékek* (Budapest, 1886), i. 351–426.

[23] *Georgius Benignus de Salviatis Ordinis Minorum sacre Theologiae professor ad Magnanimum Laurentium Medicen Impressum Florentiae* (copy used: Florence, Bibliotecà Riccardiana Ricc., 600², where the brief work is bound together with the *Propheticae solutiones* by the same author). It was printed by the so-called 'stampatore del Benigno', that R. Ridolfi (*La stampa in Firenze nel secolo XV* (Florence, 1958, 23) has renamed 'stampatore del Virgilius C. 6061'. This printer, who was financed by the wealthy Nerli family (cf. ibid., 95–111), also printed the first edn. of the *Dialectica*, also by Salviati (cf. below, n. 33). But cf. *Gesamtkatalog*, 3 n. 3848. The title of the work is *Georgius Benignus de Salviatis mirabilia septem et septuaginta in opuscolo Magistri Nicolai de Mirabilibus reperta mirabili praesenti opera annotavit*.

[24] See Vasoli, *Filosofia e religione*, 139–82. What is evidently the dedication copy is preserved in MS Florence, Bibl. Ricc., 317 (M. III, 18) under the title *Georgii Benigni Fratris Seraphice religionis ad Virum Magnanimum Laurentium Petri Cosmi Patriae Patri In Opus septem quaestionum ab ipso propositarum*. (Cf. A. Lopez, 'Descriptio codicum francescanorum Bibliothecae Riccardianae Florentiae', *AFH* 1 (1908) n. 118; P. D'Ancona, *Le miniature fiorentine* (Florence, 1914), 516; M. Scucirini Greco, *Miniature riccardiane* (Florence, 1958), 79–80). For another later copy, cf. below, p. 149, but see also M. Martelli and C. Vasoli, 'Benigno o Belcari?', *Interpres*, 7 (1987), 206–13.

[25] Florence, Bibl. Laur. Plut. LVIII, MS 16: *Georgi Benigni Salviati Fratris Seraphicae Religionis ad Magnanimum Laurentium Petrifrancisci Medicen in opus de Natura angelica*. The work ought to post-date 1488, a year when Benigno did not yet bear the name 'Salviati'. It is, however, probably earlier than the definitive draft of the previously indicated work, in which it is cited on fo. 7ᵛ.

explains why Ficino held him in such high regard and, perhaps having in mind the close relationship between the Franciscan and Lorenzo, made an urgent request for Benigno to defend his reputation publicly against someone who had accused him of practising magic after the publication of the *De vita coelitus comparanda*.[26]

The other writing, which is equally interesting, is, in contrast, a short treatise on the nature of the angels. It relates closely to current themes of the time, whether in Ficino's work or in Pico's cabbalistic meditations. There is no clear indication yet of the interests Benigno was to display at a later date when, in 1518, he wrote a short introductory eulogy to a classic of Christian cabbalism—Pietro Galatino's *De arcanis Catholicae veritatis*.[27] If, however, as is likely, he had already heard of the role of the angelic revelations in Amadeus's predictions, his propensity for angelological themes would be completely natural; all the more so as, during the years he spent in Urbino, he was already turning towards similar speculations, dealing with the relative superiority of the Archangels Gabriel and Michael.[28] Whilst such interests on the part of a learned theologian are alien to us, they are understandable in the context of the period as a way of conceiving a universal order, consisting of a hierarchy of divine intermediaries, forms of influence and supernatural 'messages' in a disturbing proliferation of 'intelligences' and 'forms', such as fascinated followers of Duns Scotus and adherents of Ficino's *divina Academia*. However, this short *summa de angelis* is, once again, the work of a scholar faithful to Scotus and to his own personal Platonism, which was also recognized by Ficino, who, in more than a hundred papers, discusses the *in corporea* nature of the angels, their creation, and the possibility, or impossibility, of their being, in turn, creators, discussing their immortality and their 'formal' structure, their ways of acquiring knowledge, and their knowledge of God. The predominant tone in Benigno's work is marked by those passages where the philosophical consideration of the nature of 'individual intelligences' gives place to that of the mystical acts of the angelic hosts, dealt with on the lines of the pseudo-Dionysius.[29] In fact, the most typical aspect of Benigno's angelological doctrine is the *communicatio cum angelis* or their function as

---

[26] See Marsilio Ficino, *Opera* (Basel, 1576), i. 574–5.

[27] Cf. below, p. 155.

[28] Cf. *De natura angelica* (as in n. 25), fos. 92ʳ ff. For description of the MS, see A. Bandini, *Catalogus codicum latinorum Bibliothecae Mediceae Laurentianae* (Florence, 1774), i. 468–70.

[29] *De natura angelica* (n. 25), fos. 79ʳ–82ʳ.

'intermediaries' or 'heralds' of the truth to come.[30] In the work we shall discuss later[31] he develops these ideas, which are clearly linked to the extreme prophetic and eschatological tension of those times, with its explicit belief in the indubitable value of prophetic revelations made by the angels. This theme, however, is already present in his writing before 1488–9, and is interwoven with two other 'themes' which already dominate his first writings, namely, the 'contingency' of future events and the 'non-absolute' nature of original sin, with its relationship with predestination and grace.[32]

III

Benigno's days in Florence[33] were, on the whole, rather unfortunate. Apart from gaining Lorenzo's trust, he was 'adopted' by the powerful Salviati family, which was closely related to the Medici. However, after Lorenzo the Magnificent's death he attempted to rise to the position of General of the Order with the full support of Piero de' Medici.[34] This attempt, however, struck a heavy blow at his prestige, involving a series of hard battles and fruitless manœuvres which culminated in physical confrontation and threatening popular demonstrations, probably of an anti-Medicean nature. In the end Salviati had to leave the Florentine Monastery of Santa Croce in 1493 and resign from his positions of responsibility in the Order to take up a post as a teacher at the Studio in Pisa. Then, in November 1494, the fall of Piero de' Medici involved him in the purges against the followers of the former masters of Florence. After a brief period in prison Salviati (as we will now call him) was forced to leave the Florentine Republic.[35]

[30] Ibid., fos. 89ᵛ–92ʳ.

[31] Cf. below, pp. 138–9.

[32] *De natura angelica* (n. 25), fos. 58ᵛ–68ʳ; 95ʳ–101ᵛ.

[33] During his sojourn in Florence, Benigno published a *Dialectica nova secundum mentem Doctoris subtilis. Et beati Thomae Aquinatis aliorumque realistarum. Acutissimi in artibus ac Theologiae doctoris Magistri Georgii Benigni de Salviatis ad praeclarissimos adolescentes: Videlicet Reverendissimum Dominum Joannem sedis apostolicae Cardinalem. Et magnificum Petrum filios optimos Magnanimi Laurentii Medicis* (Impressum Florentiae . . . Die Vero xviii mensis Martii M.CCCC.LXXXVIII) (see *Gesamtkatalog*, 3 n. 3841). The typographer is still the 'stampatore del Benigno'. The work (a clear example of Salviati's 'concordismo') was published again in Rome in 1520 under the title: *Artis dialecticae praecepta vetera et nova.*

[34] See esp. G. Picotti, 'Un episodio di politica ecclesiastica medicea', *Annali delle Università toscane*, NS 14 (1930), 86 ff. and Vasoli, *Profezia e ragione*, 51–7. Salviati attempted, though without success, to overthrow the powerful General of the Order, frate Francesco Sansone.

[35] Cf. Vasoli, *Profezia e ragione*, 57.

He fled to Dubrovnik, the city he considered to be his second home, but it appears that he even went as far as Bosnia, where he was deeply struck by the wide circulation of the prophecies of the imminent destruction of the Turks and the victorious advent of a Christian king who would free the Christians of the East. These were clearly prophecies which were similar to those which had circulated widely, after the initial announcement of the invasion of Charles VIII,[36] combining an intense eschatological expectancy with clever propaganda and forming the culmination of an outbreak of predictions in the 1480s. In exile in Dubrovnik, Salviati would have looked across to Florence, the city which was now dominated by the intransigent prophetic presence of Savonarola and his proposals for a radical reform of the corrupt Church and the establishment of the new Jerusalem which the Dominican from San Marco presented as the only 'ark of salvation' in the impending new 'Flood'. However, he always maintained close contact with a few of his friends and closest disciples, such as Piero Parenti,[37] who soon became a staunch anti-Savonarolian, the Franciscan of Santa Croce, Antonio Sassolini, Ubertino Risaliti, one of the most fervent and enthusiastic *Piagnoni*, and Giovanni Nesi,[38] who was already a follower of Ficino and Pico, and who now became an admirer of Savonarola. He was therefore well informed not only about the turbulent political events taking place in Florence but also about the radical development of Savonarola's predictions, to which was now added the open condemnation of Rome and the Holy See, involving an irreconcilable conflict with Alexander VI, the 'wicked', 'simoniac' Pope who, amongst other things, had left Salviati to the vengeance of the tough General of the Order, Francesco Sansone. In these dramatic and violent years the Franciscan exile must have listened with great interest to these prophecies of Savonarola's which, although they heralded the imminent advent of an age of cruel penance and terrible disasters under the dark figure of the Antichrist, made the promise of an age of peace and universal harmony shine brightly under the leadership of a 'Holy' Pope, the single shepherd of a

---

[36] See M. Reeves, *Influence of Prophecy in the Later Middle Ages: A Study in Joachimism* (Oxford, 1969), 356–8; A. Denis, *Charles VIII et les Italiens: Histoire et Mythe* (Geneva, 1979).

[37] On Piero Parenti, see esp. J. Schnitzer, 'Savonarola nach den Aufzeichnungen des Florentiners Pietro Parenti', *Quellen und Forschungen zur Geschichte Savonarolas* (Leipzig, 1910), 4; for his religious and political attitudes, id., *Savonarola*, trans. E. Rutili (Milan, 1931), ii. 506 ff.; see also A. Matucci, 'Piero Parenti nella storiografia fiorentina', *Studi di filologia e critica offerti dagli allievi a Lanfranco Caretti* (Rome, 1985), i. 149–93.

[38] On Nesi and his relations with Salviati, see C. Vasoli, *I miti e gli astri* (Naples, 1977), 51–128. For further references to Nesi in this book, see Index.

single flock. Furthermore, a man who had always been concerned with calculations for political ambitions might well consider whether the victory of the *Piagnoni*, amongst whom were friends and admirers, might not offer him the opportunity of a propitious return to Florence, allowing him to take up the fight again against his own colleagues who had abandoned him during his hapless attempt to rise to the position of General, and to regain his privileged role in Florence. This may explain why Piero de' Medici's teacher and Lorenzo's theological counsellor in faraway Dubrovnik strongly supported Savonarola, and why Nesi dedicated the first edition of his *Oraculum de novo saeculo* to him, a work which was truly representative in its interpretation of Savonarola as the repository of a secret initiatory truth of a hermetic, cabbalistic nature.

On 8 April 1497 the best-known work of any great length by Salviati, the *Propheticae solutiones*,[39] appeared in Florence, printed by Ser Lorenzo de' Morgiani. The work has justifiably been considered by historians of the Savonarolian period to be a very important polemical document. This was the year in which the excommunication and then the trial and burning at the stake of Savonarola took place. The strong attacks aimed at him by men of the Church, such as the 'hermit of Vallombrosa', Angelo Leonora, the Franciscan Observant Samuele Cassini,[40] the Augustinian Leonardo da Sarzana, and also by anonymous authors whose writings spread throughout Tuscany and Italy, supported with considerable skill the clear condemnation of the highest ecclesiastical authority. On 8 May Savonarola replied to accusations regarding the legitimacy and true nature of his prophetic mission in the *Epistle to All God's Chosen People and Faithful Christians*.[41] Salviati's work appeared at a crucial moment and the fact that it was written by a Franciscan theologian who was a long way away from Florence at the time and therefore not involved in the quarrels taking place in the city, gave the backing of a solid, traditional doctrine which appeared to be free from partisanship to Savonarola's cause.

[39] See Salviati, *Propheticae solutiones*. On the anti-Savonarolan polemic and the reactions of Savonarola and his followers, see esp. Schnitzer, *Savonarola*, i. 485 ff.; but cf. also P. Villari, *Life and Times of Girolamo Savonarola*, trans. L. Villari (London, 1838), ii. 1 ff.; R. Ridolfi, *La vita di Girolamo Savonarola*, 2nd edn. (Rome, 1952), ii. 191 ff. For the relations between Salviati and Savonarola, see Schnitzer, 'Savonarola', i. 340, 496, 501 ff.; ii. 213 ff., 225, 556; Ridolfi, *Savonarola*, ii. 39, 233; Weinstein, *Savonarola*, 317 ff.; C. Vasoli, 'Giorgio Benigno Salviati e la tensione profetica di fine' 400', *Rinascimento*, 2nd ser. 29 (1989), 53–78.
[40] See A. Alessandrini, 'Angelo da Vallombrosa', in *DBI* 3 (Rome, 1961), 238–40; R. Ristori, 'Cassini, Samuele', *DBI* 21 (Rome, 1978), 487–9.
[41] See Ridolfi, *Savonarola*, i. 196; ii. 186.

IV

The *Propheticae solutiones* was a cunning work, conceived by a shrewd theologian, and the result of years of teaching and scholastic debate. This can be seen in the organic and well-thought-out structure. Salviati first addresses the burning question of the legitimacy of prophecy after Christ's advent, and resolves it positively by quoting John's *Apocalypse* as well as the 'new prophets', Methodius, Joachim of Fiore, St Bridget, and many others, who often foretold self-evident truths.[42] Clearly their prophecies were about events which have not yet been verified and which do not deal with Christ's advent, but rather with the fate of members of the Church. According to Salviati, Savonarola is the modern prophet foretelling truths very similar to the predictions of the ancients on the future salvation of all people in Christ and the universal Church. There is no doubt that such revelations arc even more important than the many circulating rumours which deal with the loss and conquest of kingdoms, wars, and other such matters.[43]

As for the accusation that the 'modern' prophets simply predicted 'useless things' and repeated traditional apocalyptic themes, Salviati's reply was quite firm: if Savonarola had only repeated John's prophecy, there is nothing to stop us believing that such a 'repetition' is the Will of God, to convert unbelievers by new evidence, and Florentine prophetic traditions fit perfectly into the mystic pattern of Divine Providence.[44] On the other hand, Savonarola's prophecy had nothing to do with the types of prediction made by observation of the stars, dreams, or other 'natural' means of studying the future. These the theologian recognizes as being valid, at least to a certain extent, but he claims that they are 'dangerous' because they can be 'distorted' by the interference of evil, lying spirits, and are subject to the almost inevitable errors of astrological science (not well understood by human intellect and which, moreover, only studies the 'inclinative' causes and not the 'necessary' or 'coercive' causes of human events).

Disagreeing, at least in part, with the radical Savonarolian condemnation of astrology, Salviati maintains that the Dominican only wanted to admonish the faithful not to believe blindly in the predictions of legal astrologers, which are always based on highly circumstantial, dubious, and inadequate knowledge. However, if it is not permissible for Christians

---

[42] See Salviati, *Propheticae solutiones*, fos. aII^r–aIII^v.
[43] Ibid., fo. aIII^r.
[44] Ibid., fo. aIII^v.

to claim that they know the future 'for certain' through astrological predictions, they must accept those prophecies which God himself has endowed with His knowledge and His will.[45] Such knowledge confirms and perfects poor human knowledge and enables us to receive benefits and overcome difficulties which might arise in the future. Salviati rejects objections which stress the ever-ambiguous and easily deceptive nature of 'visions', the unreliability, obscurity, and enigmatic meaning of the images they transmit and the difficulties which arise in verifying whether their inspiration is divine. Visions, expressions, and mystical words, he maintains, are the natural vehicle for the prophetic mind which is not by chance the matter on which the 'form' of divine inspiration is imprinted.

In propounding an Avicennist doctrine, which was also used by Ficino, Salviati considers God as being the 'active principle' enlightening the minds of the prophets and enabling them to understand 'visions' and to 'sample them'. He is the 'catalyst' for their inner acknowledgement of the certainty of their mission and of the truths to be told. It is not always easy to distinguish true prophets from false, even though there are many ways of ascertaining the genuine nature of their mission. A true prophet, however, has the gift of 'insight' and uses it with the same self-assurance as that which enables us, in everyday life, to distinguish waking from sleeping and real images from illusory ones.[46] If one then objects that 'futures' are unknown to any human mind and are clear only to God, the theologian immediately has recourse to a classical distinction already widely used even in the writings of Salviati's youth: knowledge of the future is God's knowledge, but it goes without saying that He can communicate this power to the angels as well as to man, always using his own 'modus sciendi'. It is precisely for this reason that the prophet, although he is not God and does not identify himself with Divine Nature, is part of the eternal knowledge and knows each and every secret of the Supreme Mind.[47] It is doubtless true that the knowledge of the angels and of the prophets is always infinitely smaller than divine, uncreated knowledge. However, one cannot deny that Christ in his 'human' form prophesied the destruction of Jerusalem, heralded the advent of the pseudo prophets, and predicted the various fates of his disciples. It is also certain that one man, John, knew the whole of the future history of the Church Militant, even though he recounted it

---

[45] Ibid., fo. aIIII^{r-v}.
[46] Ibid., fos. aVI^r–aVII^v.
[47] Ibid., fo. aVIII^{r-v}.

in a way in which few people could understand. But even a mystical obscurity such as this responds to God's design, which does not extend to all knowledge of future contingencies and requires that those who wish to understand prophecy be endowed with the same spirit as that which moves the prophets. God had therefore now sent Savonarola to prepare man for his imminent fate and had endowed him with exceptional gifts as well as the capacity of vision for the very purpose of manifesting divine inspiration.[48]

Salviati does not fail to refute the objections of those *physici* who refute all prophecies and exclude absolutely the possibility of men being able to know the future events which are removed from the control of their will. They are afraid that the acceptance of prophetic knowledge invalidates the ethical principle of free will on which all morality is based. However, such opinions (similar to those expounded by Pietro de Rivo in his polemics on future contingencies, which had already been discussed by Benigno in his *De libertate et immutabilitate Dei*) are not actually valid and can be resolved by a clear theological solution: 'Libertatem quippe voluntatis nostrae confitemur, et ipsum Deum quaecunque volitura est, pro certo cognoscere asserimus.'[49] By indicating various paths followed by different schools of theology to show the complete compatibility between the certain knowledge that God has of the future contingents and the freedom of the human will, this Scotist theologian declares that he prefers the doctrine expounded by his teacher to the Thomist doctrine followed by Savonarola. However, as usual, he does not fail to suggest a 'concordist' solution reconciling the two 'paths'.[50]

V

It is not possible to dwell at length on these and other theoretical pages of the *Propheticae solutiones*, which is devoted to finding a difficult and precarious balance between 'absolute' and 'conditioned' knowledge of the future, and between the truth of 'prescience' and the safeguard of free will. However, as a sample, we will take the most interesting chapters of part two, in which Salviati explains the reasons for his late support of the prophetic truth of Brother Girolamo.

To the question put to him by Risaliti as to whether or not Savonarola

[48] Ibid., fos. bI^r–bII^v.
[49] Ibid., fo. bIII^v.
[50] Ibid., fo. bIIII^v.

had really been sent by God as a 'special prophet of his time', and why Florence had been chosen as the location for the new revelation, he replies not only with a general profession of faith in Savonarola, but also with references to the reformist and liberating French king,[51] which were common and widespread in those days, linked to eschatological experiences which must have had a powerful influence on his personal spiritual experiences. He cites an 'English prophecy' (which may have been one of the many circulated in English Franciscan circles of the Joachimist tradition, or else a well-known politico-ecclesiastical prophecy attributed to Reginald of Oxford),[52] which predicted the victorious advance of the Turks and their invasion of Italy, but followed, after the death of the Sultan, by the reign of his son when the last of their emperors would be more moderate and would live in peace with the Christians. According to Salviati this prophecy (on which Angelo Leonora would also base his), which was circulated in Florence and in Urbino, referred explicitly to the new millennium and the future universal peace. Besides the *prophetia anglica*, Salviati cites other prophecies on the end of the Turkish empire and the unification of all faiths. These are similar to those of Francesco da Meleto and were said to be widely circulated in Constantinople itself.[53] If one reads with an open mind, this long *excursus* raises the question of how a man who had for so long been accustomed to discuss on the intellectual level the *scientia de futuris*, and who was so attracted to all prophetic themes and 'signs', could really maintain that the predictions of the 'prophet of the desperate' were true.

In actual fact, in the same pages he acknowledges Savonarola as the only true contemporary prophet in Christ's Church, the only one who was entitled to speak to people in the name of the Lord, a fact which all the faithful should accept, even if the powerful men of the Church still dared to oppose and condemn him. They—he writes—will most certainly lose their place in the Kingdom of the Church which belongs only to those who know how to gain eternal benefit from it. Indeed, the fame of Savonarola's prediction has become widespread, even in the most

[51] Ibid., fos. bVI^r–bVII^{r-v}.

[52] For the allusion to 'prophetia angelica' and on prophecy in England in general, cf. Reeves, *Influence, ad ind.*; R. Taylor, *The Political Prophecy in England* (New York, 1900). Reeves, *Influence*, 376–8, 391, 500, develops important considerations on the role of prophecy in England at the time of Salviati, giving details on the nature of the prophecy of the pseudo-Telesforus and his relation with the pro-French 'vaticinia'.

[53] On Francesco da Meleto, see C. Vasoli, *Studi sulla cultura del Rinascimento* (Manduria, 1968), *ad ind.*; Weinstein, *Savonarola, ad ind.* See also Index for further references.

far-flung places, even in Dubrovnik itself, where the townspeople asked Salviati persistently for news of the Friar of San Marco and for his opinion of his prophecy and his life. Salviati's reply to them all was to praise the strictness, the poverty, the great power of his preaching, and his firm adherence to the Catholic truth that no slander could belie. He maintained that the fulfilment of Savonarola's predictions was certain and was awaited by all true believers, since he was the standard-bearer of a glorious period for Christianity.[54]

Clearly, when writing especially for the Florentines, Salviati could only be silent about the fact that his appreciation of Savonarola was very recent and suppress his initial doubts about the prophet's sincerity harboured when Zanobi Acciaiuoli, future librarian to Leo X, had told him of the prophecies of an imminent renaissance in Florence and a general *reformatio*. But at the time of Savonarola's Quadragesima sermon of 1491 he was moved by the announcement of the *renovatio*, and when, in the spring of 1494, he was informed of the prophecies of an impending 'Flood', which was approaching *cito et velocior*, he became convinced that Savonarola was guided by a 'non-erring' intelligence, that his prophecy contained nothing contrary to Scripture, the doctrine of the saints, or sound morals, and that the friar was really someone sent by God.[55]

Salviati thus voices no doubt about the absolute legality of Savonarola's prediction, convinced that Decretal texts could not be applied to a true prophet who was so opposed to the introduction of new dogmas and faithful to the old ones, and who was the bearer of God's final admonition to sinners before the Day of Wrath and the Day of Judgement. Neither was it a surprise to him that such a revelation should be made to the Florentines rather than to the Romans, the Milanese, or the Venetians, since the Florentines had always been very religious people and faithful allies to the very Christian king, Charles VIII, chosen by God to reform the Church, expel false priests, and restore Christian unity. The Franciscan thus touched on the most common themes in political prophecies which were popularized in those years by so many restless people desirous of a new ecclesiastical life, even by servants to princes and ambitious ecclesiastics. Nevertheless, although he seemed to be supporting the partisans of Charles VIII, he stresses that Christianity

---

[54] Salviati, *Propheticae solutiones*, fo. bVIII[r-v]. Ridolfi, *Savonarola*, ii. 233, rightly links Salviati's activity with the cult of Savonarola which was widespread at Dubrovnik, and recalls the image of the Dominican, cited by E.-C. Bayonne, *Étude sur Jérôme Savonarola* (Paris, 1879), 356, which still exists in the Adriatic city.

[55] Salviati, *Propheticae solutiones*, fos. cI[r]–cII[r].

must always have its Supreme Pastor who should at all times be responsible for the 'keys' to the highest universal jurisdiction and have the power to guide the entire human 'flock'. Savonarola's invective 'Abiecit te Deus o Roma, et Hierusalem iterum elegit' must be interpreted as a just rebuke for the sins of the Romans, and not as a condemnation of the sovereign power of the Pope.[56] Salviati's prudence as a man of the Church, who was thus always linked with authority and power, perhaps manifested itself in the avoidance of any precise reference to the imminent date for the prophesied reform and to those subjects which were highly criticized by Savonarola's opponents.

VI

Whatever his reasons were for dictating the *Propheticae solutiones*, they produced initially favourable results for Salviati. He was pardoned from exile and was appointed teacher of theology at the Florentine Studio on 1 November 1497.[57] However, whether he wisely evaluated the uncertainty of the times or whether he was still busy with his ecclesiastical and academic duties, Salviati did not leave Dubrovnik. He was undoubtedly right not to return to Florence where political suspicions surrounding him had not, in fact, been dispelled. It is true that even after Savonarola's execution he was again called, on 2 September 1500, to read theology and Aristotelian philosophy, but he never took up this appointment because, as was written in the margin of the document, the Otto di Guardia would banish him again from the Republic.[58] This did not prevent him from maintaining his links with his Florentine friends, particularly Sassolini and Risaliti, and it was the latter who took it upon himself to arrange, and possibly also to finance, the printing of his most ambitious theological work, the thick *tome* known as the *De natura angelica* which expanded the subject matter of Lorenzo di Pier Francesco's work and appeared in Florence in 1499.[59]

Times had, by then, changed a great deal. Savonarola's death at the stake had already brought that brief period of great hope to an end without the prophecies of the *renovatio* being fulfilled. Salviati expressed his disappointment and, perhaps, that of his friends, in a passage from this work in which he wrote that he would never have expected such an

[56] Ibid., fos. cIV^v–cVI^v.
[57] See Verde, *Lo studio fiorentino*, ii. 274.
[58] Ibid. 275.
[59] See above, n. 5.

enormous *mendacio* to come from a man who had lead such a saintly life. However, if his hopes in Friar Girolamo had been dashed, the theologian did not fail to rekindle his Florentine fame by expounding the doctrines which were of most interest to him. He presented his treatise in the form of a series of discussions which must have been conducted in Dubrovnik in May–June 1498 by himself and a few ecclesiastics and influential citizens, men who belonged to the class of lawyers, merchants, and administrators and who, in this quiet Adriatic city, as in restless Florence, were deeply interested in important theological problems concerning Free Will, Divine Providence, the possibility of prophecy and that of understanding the 'future contingents', and the function and nature of angelic 'intermediaries'. These were, in fact, the same themes as those we have constantly met in Salviati's meditations, and which are always discussed through a highly detailed comparison of the theses of Scotus and Thomas Aquinas, Henry of Ghent, Francis de Meyronnes, Landolfo Caracciolo, and Gregory of Rimini, with the express intention of developing a truly genuine *summa de angelis*. Thus, in presenting the intellectual nature of the angels and their function as intermediaries between God and man,[60] Salviati leads the conversation into a discussion on the part angels play in both divine and human knowledge and in knowledge of the future.[61] In book 3 he turns to the distinction between will, nature, and intellect, reaffirming the pre-eminence of the will of the angels over that of man[62] and then, in book 6, with numerous quotations from Avicenna, Cicero, Averroes, Hermes Trismegistus, Plato, and Aristotle, the discussion addresses various theological doctrines on the origin and primary cause for the existence of the angels, concluding that they were created directly by God, and that they have, in a sense, their own part to play in the work of Creation.[63] The question of original state leads into a discussion on guilt and 'angelic sin'. This offers the opportunity to examine the nature of original sin, the fate of the 'evil' angels, and whether or not they are 'ready for Grace'.[64] Book 6 is also important for Salviati's allusion to his first meeting with Bessarion, and for passages which a sixteenth-century reader promptly discovered in the *Apocalypsis Nova*.[65]

---

[60] Salviati, *De natura angelica*, fos. aII^r–bVIII^r.
[61] Ibid., fos. bVIII^v–fI^r.
[62] Ibid., fos. fI^r–gII^v.
[63] Ibid., fos. gIII^r–hIII^r.
[64] Ibid., fos. hIII^r–iVI^r; iVI^v–mIII^r.
[65] Ibid., fo. kIII^{r–v} (the note is in the margin of the incunabulum in the Magliabechiana Collection); for the reference to Bessarion and his own early studies, see fo. 1II^r.

The rest of the work (devoted to the theory of the movement of the angels[66] and, in book 8, to the particular angelic hierarchies and the relationship between man and the angels, especially as media for prophetic truths)[67] does not add a great deal to the concepts already dealt with. The final book makes a distinction between 'eternity' and 'time', which is closely linked to Salviati's meditations on prescience and divine predestination.[68] It is more important to note that the *De natura angelica* is marked by the eschatological tensions of the time reflected in his interest in any visionary or prophetic demonstration, or exceptional divine communication, such as induced even learned and subtle humanists to look to the heavens for signs of the imminent advent of the Antichrist. The friar who wrote this work on the very borders of the Christian world was perhaps not totally aware of this fact. But his exuberant theological imagination, his strong Platonizing Scotism, his faith in angelic revelations, and his belief in the powers of prophetic intellect still echo the myths and esoteric temptations typical of a long tradition later expressed not only in Postel's hallucinatory experience but also in a famous 'confession' of Jean Bodin's.[69]

VII

Salviati's exile to Dubrovnik did not last long. He returned to Italy in 1500 and attended the Franciscan general chapter at Terni. It is probably at this time that he first met Cardinal Bernardino Carvajal[70] patron of the Amadeites, who was very active in both ecclesiastical and political circles through his close relationships with Maximilian I and Ferdinand of

[66] Ibid., fos. mIII^r–nIIII^v.
[67] Ibid., fos. nIIII^v–oVII^v.
[68] Ibid., fos. oVII^v–qVII^r.
[69] I refer to the 'experience' of communing with the angel, which Bodin presents as having happened to a friend, but which was certainly an event in his own life, see Jean Bodin, *De la démonomanie des sorciers* (Paris, 1587), fos. 10^v–15^v. But see F. von Bezold, 'Jean Bodin als Okkultist und seine "Démonomanie"', *HZ* 105 (1910), 1–64; C. Baxter, 'Jean Bodin's Daemon and his Conversion to Judaism', in H. Denzer (ed.), *Verhandlungen der internationale Bodin-Tagung* (Munich, 1973), 1–21; P. Rose, *Bodin and the Great God of Nature* (Geneva, 1980), 164–74.
[70] On Cardinal Carvajal and his ecclesiastical and political activity, see Bibliography and Index; also M. Battlori, 'Bernardino Lopez de Carvajal legado de Alejandro VI en Anagni, 1494', *Saggi storici intorno al Papato*, 21 (Rome, 1959), 171–88. Note that it was Salviati who published the homily delivered by Carvajal in the presence of Maximilian during the Emperor's mission: *Homelia doctissima coram maximo Maximiliano Caesare sempre augusto* (Rome, 1508). For this homily see above, Minnich, Ch. 6. For details of Salviati's career after the return, see Vasoli, *Profezia e ragione*, 83–5.

Aragon, and did not conceal his ambitions for the papacy. In the Roman Monastery of the Twelve Apostles Salviati resumed his ecclesiastical and teaching career, either in the Order or in the Sapienza Romana. However, the most important event in his life at this time was undoubtedly the pre-eminent part he took in the re-elaboration (if not the actual rewriting) of the prophecy attributed to Amadeus.

The *Apocalypsis Nova* is studied elsewhere in this book.[71] I shall confine myself to recalling that it is preserved in a number of manuscripts in Italian and European libraries[72] and is, in its prophetic part, a prediction which was probably created *post factum*, as attested by precise and circumstantial references to the papacies of Sixtus IV, Innocent VIII, Alexander VI, Pius III, and Julius II and to historic events in Italy and Europe at that time, with particular reference to the Aragonese and French monarchies. The author is clearly concerned about the reception of the prophecy in Florence and the assistance this city would give to the future *Papa Angelicus*, the leader of the *reformatio ecclesiae*, and to his struggle against false priests, unbelievers, and the enemies of Christianity. So, Florence, as well as Rome, was the city chosen by God to herald the new reformed Christian world and an age of peace and universal harmony. The *Apocalypsis Nova* is bound up with the success of the myth of the Angelic Pope, and with its use in particular political and ecclesiastical contexts. However, I would like to emphasize its special character which distinguishes it from the other *vaticinia* of the time, in that it is a genuinely theological treatise divided, in most manuscripts, into eight *raptus*. These are analysed by Anna Morisi-Guerra in Chapter 2.

The truly singular character of the *Apocalypsis Nova* did not escape the notice of historians and theologians in subsequent decades and centuries. To allay suspicions expressed by the Blessed Paolo Giustiniani[73] and to

---

[71] See above, Morisi-Guerra, Ch. 2.

[72] I have consulted the texts in the following MSS: Convento dei Cappuccini di Milano, 16; Florence, Bibl. Naz. Cen., Magl. XXXIX, 1, fos. 1r–125v; Arezzo, Bibl. della Fraternità dei Laici, 436. A comparative and systematic study of the various copies of the *Apocalypsis Nova*, is still needed however.

[73] See B. Paolo Giustiniani, 'Trattati, lettere e frammenti dai manoscritti originali dell'Archivio dei Camaldolesi di Monte Corona nell'eremo di Frascati', ed. E. Massa, in E. Massa, *I manoscritti originali custoditi nell'eremo di Frascati* (Rome, 1967), 356–7. At no. 90 of the Tuscan MS. F. II, Giustiniani has assembled a series of writings directed towards the condemnation of personalities, movements, or conventicles of a prophetic character, active or widespread in Florence. At nn. 78 and 79 is a confutation of the 'liber qui dicitur apochalypsis nova Amadei'; at n. 80 an attack upon Savonarola, evidently related to the Florentine Synod of 1517 and its condemnation of all 'prophetic' movements of either Savonarolian or post-Savonarolian origin; at nn. 82, 83, 84, 85, 86, 87, 88, 89 is a group of

counter the radical condemnation pronounced by the Polish Dominican Abraham Bzovius,[74] Luke Wadding had no hesitation in stating that this work was not the product of an honest, learned man, but only the *commentum* of a Scotist *chimicerus* and *maleferiatus* who attributed complex philosophical and theological doctrines which were actually of a Scotist nature to Amadeus's revelatory angel, blending these doctrines with prophecies about the *Pastor Angelicus* and the *renovatio ecclesiae*[75] already discussed. Wadding (who perhaps knew more than he said he did) wrote a revealing phrase: 'Angelus beati Amadei fuit Scotista', suggesting that the original core of the prophecy on the state of the Church, the reform of its morals, and the various changes to take place, which was actually due to Amadeus, underwent radical change and was rendered unrecognizable by the addition of a jumble of *opiniones exoticae*, the suspect nature of which the analyst should not ignore, particularly in the light of the pronouncements at the Council of Trent. For our part we must add that this theologian, whoever he was, expounded doctrines which were either very similar or identical to those proposed in some of Salviati's writings (prior—it should be noted—to the actual circulation of the *Apocalypsis Nova*). The former showed a keen interest in those theological, angelogical, and Mariological themes so often dealt with by the Bosnian teacher, not to mention the ideas on prophecy, the prophetic mind and its relationship to the angelic mind and the Divine Will and Mind, which we have already seen discussed in the *Propheticae solutiones*.

Clearly such considerations would not be sufficient to warrant the

documents relating to the case of the 'prophet' Francesco da Meleto. See Vasoli, *Studi*; id., *Immagini umanistiche* (Naples, 1983), *ad ind.*; Weinstein, *Savonarola*, *ad ind.* At n. 90 is a refutation 'ad quasdam seductas mulieres florentinas in perversissimam heresim et idolatriam G(abrielis) B(iondi)', that is, of Gabriele Biondo, son of Flavio and brother of Francesco Biondo who is said to have written down the original nucleus of the *Apocalypsis Nova*, dictated to him by Amadeus (see Vasoli, *Profezia e ragione*, 121–3; id., *Filosofia e religione*, *ad ind.*). At nn. 91 and 92 are accusations against the Greek monk Theodore who had formed a predominantly female conventicle for the cult of himself, as 'Papa angelico'. Note also that Tommaso De Vio (Cajetan) openly prophesied 'contra novos prophetas et specialiter quendam Amadeum qui, ut aiunt, librum edidit, novam doctrinam earum, quae ad fidei christianae mysteria spectant, introducere conantes et suos sequaces et similes' (see above, Minnich, Ch. 4, on Cajetan's views on prophecy). Ignatius Loyola also, in his famous *Iudicium* of 1549, numbered the book of 'Amadeo' amongst those dangerous sources of false prophetic suggestion of which Savonarola and Postel had been both victims and protagonists (*Iudicium de quibusdam opinionibus quae falso revelationes credebantur, Monumenta ignatiana* (Madrid, 1911), I, xii. 636 ff.).

[74] See A. Bzowius, *Annalium ecclesiasticorum post Illust. et Reverend. D.D. Caesarem Baronium S.R.E. Cardinalem bibliothecarium continuatio, ad an. 1471* (Cologne, 1627), XXX. xviii. 26–30.

[75] See L. Wadding, *Annales Minorum* (Rome, 1731–6), 14, *ad an. 1482*, nos. 16–41, 313–25.

claim that Salviati intervened in the formation and diffusion of the *Apocalypsis Nova*, neither would I hazard this hypothesis if it were not for other pieces of information of notable importance which I believe merit examination. The first is supplied by a manuscript from the Milan Trivulziana[76] which states that Friar Isaia da Varese, the first provincial of the Lombard Amadeites, confided to Friar Michele da Trecate that the sealed volume of the revelations attributed to Amadeus was opened in Rome between Easter Day and Ascension in 1502 by order of the General of the Friars Minor, Egidio Delfini. A later statement says that the same Friar Isaia delivered the parcel 'one Saturday morning to the Cardinal of Santa Croce', that is Bernardino Carvajal himself whose follower was Salviati. That is not all. The contemporary evidence of the Franciscan Florentine Mariano who claimed that the book had been opened by Cardinal Grimani and Bernardino Carvajal has already been cited. It was Mariano who stated that the latter, on reading it, believed himself to be the future Angelic Pope, from which belief sprang the whole design of the Council of Pisa.[77] Interpreting this new information, it would seem that the prophecy of Amadeus was the first 'revelation', subsequently elaborated into its present form in the circle of Carvajal who perhaps intended to use it for his own personal goals as a high-ranking ecclesiastical politician. Whoever worked on the text, only Salviati could have imprinted upon it the doctrinal characteristics which distinguish it.

VIII

Other information which, in my opinion, is more conclusive, comes from letters which an unknown informant added to the *Apocalypsis Nova* in a Magliabechiana manuscript in the National Library of Florence.[78]

---

[76] For a broader treatment of the subject, see Vasoli, *Profezia e ragione*, 89 ff.; also G. Porro, *Catalogo dei manoscritti della Trivulziana* (Turin, 1884), 10; P. Sevesi, 'Il B. Menez de Silva dei Frati Minori, Fondatore degli Amadeiti: Vita inedita di Fra' Mariano da Firenze e documenti inediti', *Luce e Amore*, 8 (1911), 529–42, 586–605, 681–710; id., 'Il beato Amadeo Menezes de Silva e documenti inediti', *Misc. franc.*, 31 (1931), 227–32; A. Morisi, *Apocalypsis Nova: Richerche sull'origine e la formazione del testo pseudo-Amadeo* (Studi Storici, 77; Rome, 1970), 28–32; Vasoli, *Filosofia e religione*, 216–21. For the MSS, see Milan, Bibl. Trivulziana, MS 402.

[77] Wadding, *Annales*, 14, *ad an.* 1482, no. 39, 322–3. See above, Minnich, Ch. 4 n. 61, for the argument that Grimani was unlikely to have taken part.

[78] See Florence, Bibl. Naz. Cen., MS. Magl., XXXIX, I, fos. 219ʳ–294ᵛ (n.n.). At fo. 290ʳ, where the group of letters begins, is clearly legible: 'Copia. duna. letera, auta. da. Slorenzo. violy. delli XVIII. di. marzo. 1540. / Sopra le cose del libro del beato Amadeo! And, before other letters: 'Copia di detta da detto'.

The letters are attributed by the copyist (it is not known whether as author or simply as intermediary) to a person well known to students of Savonarola, the Florentine notary, Ser Lorenzo Violi, the technograph who collected together many of the Dominican's sermons, and who, about 1542, at an advanced age, wrote *L'Apologia per modo di dialogo in defensione delle cose predicata da frate Girolamo Savonarola*.[79] So, this person, who was involved in Florentine religious life at that time, must have taken part in the spring of 1541 in an interesting investigation carried out in order to verify the authenticity of those copies of Amadeus's 'revelation' which were in circulation in Florence, particularly in *Piagnoni* circles and amongst the most tenacious and radical of Savonarola's admirers who were still awaiting the prophesied advent of the *Pastor Angelicus*. The immediate cause of this enquiry may have been the unfortunate case of another Florentine notary, Ser Cristofano da Soci[80] who, after having been sentenced for the first time in May 1538 for divulging dangerous prophecies concerning the recent rise to power of Cosimo I, had joined forces with a hermit, who was proclaimed *Pastor Angelicus* in the mountains of Montefeltro. Both finished up in prison, or rather, in a Florentine lunatic asylum. In the course of the enquiries, an old Amadeite friar, Antonio da Cremona,[81] underwent a particular interrogation. Many years previously he had preached a sermon in Florence which was possibly based on the revelations in the *Apocalypsis Nova*, and had declared that Amadeus's book had actually been opened on the order of Carvajal and the General of the Friars Minor, Delfini, by a certain Friar Giorgio, who had by then been appointed Bishop of Cagli, meaning, without a shadow of a doubt, Benigno Salviati. He added that the revelations had remained in Benigno's possession for almost an entire year, and that during those

---

[79] On Violi, see Vasoli, *I miti*, 129–82; id., 'Un notaio fiorentino del Cinquecento: Ser Lorenzo Violi', *Il notariato nella civiltà toscana. Atti di un Convegno (maggio 1981)* (Rome, 1985), 391–418; G. Garfagnini, 'Ser Lorenzo Violi e le prediche del Savonarola', *Lettere italiane*, 37 (1986), 312–37; Lorenzo Violi, *Le giornate*, ed. G. Garfagnini (Florence, 1986). (*Le giornate* is the vernacular title of the *Apologia*.)

[80] For this event, see Vasoli, *Immagini*, 347–82; L. Polizzotto, 'Confraternities, Conventicles and Political Dissent: The case of the Savonarolian "Capi rossi"', *Memorie domenicane*, NS 16 (1985), 235–83; id., 'Documents', *Memorie domenicane*, NS 17 (1986), 285–300.

[81] For the text of Antonio da Cremona's sermon and of his 'apologia' to archbishop Cosimo de' Pazzi, see Florence, Bibl. Naz. Cen. MS. Conv. soppr., J. 10. 5 (originally from the convent of S. Marco), which also contains a copy of the *Apocalypsis Nova*. Cf. G. Tognetti, 'Un episodio inedito di repressione della praedicazione post-Savonaroliana (Firenze 1509)', *Bibliothèque d'Humanisme et Renaissance*, 24 (1962), 190–9 and see Index for further references to Antonio da Cremona.

months he had recounted them from time to time to Ubertino Risaliti and to Antonio himself. The *Apocalypsis Nova* was then circulated in its revised form by Salviati and Carvajal in copies, all of which came from the 'writing desk' of the Bosnian friar.

The author of the letters was in no doubt that it was to Salviati that the various blemishes and heresies in the current text of the *Apocalypsis Nova* were due. He was also convinced that it would have been difficult, if not impossible, to reconstruct the real core of the prophecy, which had been changed out of all recognition from the original, authentic angelic revelation.[82]

The results of this belated Florentine enquiry (which, by itself, might seem dubious) are confirmed by documents analysed and studied in great detail by Anna Morisi, and by manuscript copies made by Michele da Trecate of a series of letters sent by Salviati himself to Ubertino Risaliti between the spring and autumn of 1502. They have been dealt with by Anna Morisi and by myself elsewhere.[83] I shall therefore confine myself to emphasizing that the following incontestable conclusions may be drawn from the available information: (1) The opening of the book of 'revelations' attributed to Amadeus actually took place in Rome in the spring of 1502 (which did not, however, prevent the original core of the prophecy from having already been circulated before this time, at least in Amadeite circles or amongst those who, like Salviati, had undoubtedly had some contact with them). (2) The prophecy was immediately 'requisitioned' by Carvajal and made available to Salviati who, to put the best construction on it, was allowed to make additions, re-elaborations, and changes, all to conform with the ecclesiastic and political interests of the cardinal. (3) The 'popularized' version of the Pseudo-Amadeus was ready for circulation between 1502 and 1503, in the circle of that powerful man of the Church who was destined to play a very important role in the last days of the Borgia papacy and in the two conclaves of 1503 before becoming one of the promoters of the 'secret meeting of

[82] The letter of 15 April 1541 states that the three characters 'si condussono . . . una matina nella ciesa dimontoro dove era el libro et quivy dica [sic] la messa solene dello Spirito santo volendo dipoy aperire el libro non ardivano niuno diloro metere la mano asciorlo et aprirlo per la paura che ciscuno aveva perche era fama et maxime in questo convento [that of S. Pietro in Montorio] che chi laprisse morisse fra un anno perlo spavento.' It adds that the cardinal ordered Salviati to open it 'in virtute obedientiae' and then left it with him for one year 'stando a vedere quello che seguiva di frate Giorgio in un anno'. For this episode, see below, Jungić, Ch. 18.

[83] See Morisi, *Apocalypsis Nova*, 28–32. Note that P. Sevesi has, strangely, not used these letters in his studies cited in n. 76. See also Vasoli, *Profezia e ragione*, 94–9.

Pisa'. Amongst others, Cardinal Guillaume Briçonnet, who also had a work dedicated to him by Salviati, was involved. (4) The *Apocalypsis Nova* was immediately circulated amongst the Florentine *Piagnoni* to whom Salviati himself had emphasized the convergence between this prophecy, the predictions of Savonarola, and St Bridget's revelations, which were so highly thought of amongst the followers of the Friar of San Marco. (5) The Bosnian Franciscan probably identified himself (or caused others to identify him) as the future *Pastor Angelicus*, or at least thought of himself as being one of the 'Eastern' cardinals who would co-operate with the Pope in achieving a new universal unity in the Church. The hope that he would soon be elected to the Holy College was undoubtedly a constant theme throughout Salviati's ecclesiastical career, from the years of his hapless attempt to become General of the Order right up to virtually the end of his life. (6) There is undoubtedly a very significant agreement between the theological and eschatological doctrines professed, either earlier or later in his life, by Salviati, and the parallel themes in the *Apocalypsis Nova*.

It is therefore very important for a precise evaluation of Salviati's activities during those years to reconstruct the actual extent to which the pseudo-Amadean prophecy was circulated in the turbulent climate of the crumbling Borgia papacy, while the battle between opposing curial fronts became increasingly bitter, and to discover how it was used in the autumn of 1503, when the opposing ambitions of Carvajal and George d'Amboise clashed, right up to the time when an agreement was reached between the two parties through Guiliano della Rovere. One fact is quite certain: the election of Julius II was a fortunate event for Salviati's ecclesiastical career as he was appointed on 21 May 1507 to the episcopal see of Cagli, a diocese which belonged to the Duke of Urbino. From then on his fortunes were linked with those of the Della Rovere family of Julius II.[84]

IX

Salviati, however, still had very close ties with Carvajal. As soon as he was appointed bishop he followed the cardinal in his mission to Innsbruck, where he was to meet Maximilian I, who was on the point of moving south into Italy. Julius II had entrusted Carvajal with the task of dissuading the emperor from invading Italy and of persuading him

---

[84] See C. Eubel, *Hierarchia Catholica Medii Aevi* (Münster, 1898–), 162.

instead to adhere to two alliances: a common Christian alliance against
the Turks, and a special alliance with the Pope against Venice.[85] This
may explain the presence in his retinue of a bishop of Balkan origin
who had a good knowledge of the countries currently under Turkish
domination and of those which were threatened by them, and also of
Venetian forces on the Dalmatian coast. Nevertheless, as can be seen
from two of his letters, Salviati combined this role with that of informer
to the Florentine Gonfalonier, Piero Soderini, whom he kept informed
about the attitude of the emperor towards Florence and the maintenance
of the republican regime, not without making pointed requests of a
financial nature.[86] The Franciscan, therefore, apart from maintaining
his links with *Piagnoni* circles, also had a special friendship with the
unfortunate Gonfalonier, certain that his wish to return to Florence
would be granted very soon. In the mean time, he hastened to spread his
ideas in the Hapsburg court, dedicating a new theological-prophetic
work entitled *Vexillum Christianae victoriae* to Maximilian, and the *Contemplationes Beatae Viriginis Mariae*[87] to his daughter, Maria, Regent of
Burgundy and the Low Countries.

The first is a treatise relating to theological themes beloved of Salviati,

[85] On Carvajal's mission, see L. von Pastor, *History of the Popes*, trans. and ed. E.
Antrobus (London, 1891–8), vi. 295–6. On his Homily to Maximilian I, see above,
Minnich, Ch. 6.
[86] See Florence, Bibl. Naz. Cen. Fondo Ginori-Conti, 29 Carte Michelozzi, n. 112 g.
The letters have been edited in *Rinascimento*, 2nd ser., 19 (1968), 327–30.
[87] The treatise dedicated to Maximilian is in Vienna, Nationalbibliothek, Cod. Palat.,
4797. For the dating and the history of the MS, see T. Gottlieb, *Büchersammlung Kaiser
Maximilians* (Leipzig, 1900), 53, 128 (who also published the dedication and the prefatory
epistle). I have consulted this work (of which there is also a copy in Paris, Bibl. Nat.) in
MS Milan, Bibl. dei Cappuccini di S. Francesco, Cod. 16, fos. 1^r–41^v. The provenance
and history of this MS need further study. It contains also the *De assumptione virginis*
(see below, n. 113), the *Apocalypsis Nova* (fos. 41–148^r), the anonymous *Conclusiones
theologicales disputatae de incarnatione secundum Scotum* (fos. 240^r–241^r), *De sacramento
altaris* (fos. 241^v–244^r), *Sacramenta novae legis* (fo. 245^{r-v}), etc. For a description of the
MS, see C. Varischi, 'Catalogo dei codici della biblioteca dei Minori Cappuccini in Milan',
Aevum, 11 (1937), 259–61. Morisi, who has also seen MS 1. 2. 70., Bibl. comunale, Rieti,
observes that the latter two MSS are later than 1512, because in them Salviati is referred to
as 'archiepiscopus nazarenus', *Apocalypsis Nova*, 39. For the short work for Maria of
Hapsburg see Brussels, Bibl. Royale, MS 10783, entitled in this MS: *Contemplationes
commendationum Virginis Gloriosae*, but in essence the same work as *Libellus de Virginis
Matris Assumptione*, in Milan, Bibl. dei Cappuccini di San Francesco, MS 16, in the
Trivulziano 453—cf. Porro, *Catalogo*, 168—and in Ambrosiano A. 30 sup. The Brussels
text corresponds with the 'short version' found in the MS of the Bibl. dei Cappuccini,
whilst the other two feature extra chapters (31 and 32), consisting mainly of a long passage
from the *Apocalypsis Nova* and a discussion of some extracts from the prophecy of St
Bridget, etc. The opening passage on the symbolism of the pearl (fos. 3^r–5^v) is explained
by the dedication to 'Margarita'.

in which the Platonic and mystical aspects of his particular interpretation of Scotism are stressed, not excluding the use of images and symbols which point to the fact that he frequented Ficinian circles in Florence. The Franciscan stresses, in particular, the theme of the absolute 'simplicity' of the Divine Being, with the obvious purpose of reconciling Scotist with Thomist doctrines, but also developing an interpretation of the Unity/Trinity relationship which is full of Platonic nuances. Subsequent 'contemplations' develop the metaphysical and spiritual meaning of the appearance of the 'divine light' in the figure of a sphere, the reason why it is present in this most perfect of all geometrical figures, and finally, the intimate process by which one reflects on oneself.[88] In the pages devoted to distinguishing and multiplying the Ideas of the Divine Mind, and to the process of spreading the *lumen* derived from it, the principles of the Scotist tradition assume a strongly Platonic character, and are not exempt from shades of gnosticism. Like other Franciscan teachers of the time, Salviati has a tendency to use imaginative and speculative language which resolves the scholastic doctrine of the *formalitates* in a universal cosmic process, and interprets, in neo-Platonic terms, the fundamentals of Scotist metaphysics.[89] Again we meet the incessant recourse to 'angelic' and prophetic motives, derived from the *Apocalypsis Nova*, and to Mariological speculations which are developed in particular in these two writings. Undoubtedly, the insistence on the immaculate conception of the Virgin and on the predestination *ab aeterno* of her motherhood, are typical of Scotist tradition.[90] However, Salviati closely relates these doctrines to his Trinitarian theories and indicates Mary as being not only the indispensable mediator between man and God, but also the co-operator with the Trinity in the incarnation of the Word. Here he inserts other passages from the *Apocalypsis*, which are often repeated word for word but are never quoted as being from the work.[91]

Similarly, the work dedicated to Maria of Hapsburg moves from the apocryphal *De transitu Mariae*[92] to praising all the perfections of the Virgin, her universal knowledge and wisdom, the mystery of her

---

[88] Salviati, *Vexillum*, fos. 1ʳ–3ᵛ.

[89] Ibid., fos. 26ʳ ff., 37ʳ–41ʳ.

[90] Ibid., fos. 30ʳ–37ʳ. See also J. Bonnefoy, *Le vénérable Duns Scot, Docteur de l'Immaculée-Conception; Son milieu, sa doctrine, son influence* (Rome, 1960).

[91] Cf. Morisi, *Apocalypsis Nova*, 40–1.

[92] For Salviati's interest in the apocrypha, see introductory remarks to the text in the MS. Ambrosiano, fo. 1ʳ; also A. Morisi's important observations in: 'Vangeli apocrifi e leggende nella cultura religiosa de tardo Medioevo,' *Bull. ISI*, 85 (1974–5), 151–77.

heavenly ascent and assumption, themes which were also subjects favoured by the Scotist doctrine. This work emphasizes the universal, cosmic role of the Marian 'mediation', and discusses subjects which consistently figured throughout Salviati's theological career: predestination, prescience, the future contingents, the value of prophetic revelations, not to mention further references to the *Apocalypsis Nova*, of which the entire eighth *raptus* and part of the fourth are repeated. It also includes an announcement of the imminent advent of a new age and a glorious future for Christianity and for the Hapsburg Empire. The two works, which were possibly written during his mission with Carvajal at Maximilian's court or not long afterwards, both express, not only the prophetic suggestions and ideas which are so typical of Salviati, but also his desire to spread the *Apocalypsis Nova* in circles which would be particularly sensitive to its use for political ends.

Nevertheless, in his friendship with Carvajal during the years of the grave ecclesiastical-political crisis, which culminated in the Council of Pisa, the newly appointed bishop had to behave with particular prudence to avoid acting in any way which might alienate him from Julius II's favour. Unfortunately we do not have any documents to tell us what his personal situation actually was in 1510 and the first few months of 1511, when the irreparable breach between the French cardinals and the Pope became permanent and precipitated an intricate political web of intrigue which led Louis XII and Maximilian to attempt to bring about a schism and depose Julius II.[93] However, the fact that Salviati supported Della Rovere at the time of the dispute is seen in another of his writings which is also an important document demonstrating his theological boldness: the *Apologeticon seu defensorium ex divinis litteris aggressionis Francisci Mariae de Ruvere*,[94] a work written in defence of the Duke of Urbino who was accused of murdering Francesco Alidosi, the Cardinal of Pavia. The pamphlet, which was probably produced in the summer of 1511, has no great importance in itself. It is, however, a perfect reflection of the special obsequious style of its author who moved from praising Savonarola to praising the Pope, using the prophetic image of the 'Pastor

---

[93] See esp. A. Renaudet, *Le concile gallican de Pise-Milan: Documents florentins, 1510–1512* (Paris, 1922); J. Doussinague, *Fernando el Catolico y el cisma de Pisa* (Madrid, 1946); H. Jedin, *The History of Council of Trent*, trans. D. Graf (London, 1957), 106 ff.; W. Stelzer, 'Neue Beiträge zur Frage des Kaiser-Papstplane Maximilians I im Jahre 1511', *Mitteilungen des Instituts für österreichische Geschichtsforschung*, 71 (1963), 311–32.

[94] Florence, Bibl. Naz. Cen., MS Magl., XXX, 215. On Alidosi's murder and his previous crimes, see L. Frati, 'Il Cardinale Francesco Alidosi e Francesco Maria Della Rovere', *ASI* 47 (1911), 114–58.

Angelicus' and the *Apocalypsis Nova*[95] and other prophetic texts to justify one of the many grim episodes in the ecclesiastical history of those years. Nevertheless, even behind the obvious political intent of the writing, some of Salviati's personal spiritualist convictions are at work: the immediate, direct impulse to be drawn from the enlightenment of the Divine Will and the high value, not subject to any rules or laws, of personal vocation kindled by the supreme, sovereign Spirit. The writing of the *Apologeticon* must have met with the approval of Julius II who, being compelled to put his nephew on trial, sought to obtain a complete absolution. It is no coincidence that Salviati was appointed titular Archbishop of Nazareth in 1512, the final year of the della Rovere papacy (not 1513, as is usually written). This is the only explanation of the fact that he was designated at the opening of the fourth session of the Fifth Lateran Council by the title of his new appointment.[96]

X

Julius II died on the night of 20 February 1513. After a very hard-fought conclave, Giovanni de' Medici, Lorenzo's son, was elected as Leo X. Salviati lost no time in taking action: very soon after the election of the new Pope he had an elaborate copy of his old theological commentary on the Magnificent's sonnet[97] made and sent it to Leo X, adding to it a dedication which was actually an urgent request to be appointed cardinal, based on—so he said—old promises of the Lord of Florence. Naturally, in the letter he referred only to the misfortunes he had suffered after the fall of the Medici, when his loyalty to his former patrons had resulted in his arrest and exile, while he wisely kept quiet about his enthusiasm for Savonarola in 1497 and other connections with circles and people for which the Medici had no particular liking.[98] But the gift of the *Quaestiones septem*, far from stimulating the new Pope to carry out the supposed wishes of his father and of Bessarion, as so clearly remembered by Salviati, caused him rather to take the view that the apparently prestigious title of Archbishop of Nazareth (which actually corresponded to the modest residential diocese of Barletta) was a

---

[95] MS *cit.* (n. 94), fo. 19ᵛ.

[96] For the date of 1513, see Eubel, *Hierarchia*, iii. 162; for that of 1512, see *Mansi*, 32, col. 727.

[97] See Florence, Bibl. Laur., Plut. LXXXIII, MS 18 (and cf. A. Bandini, *Catalogus codicum latinorum Bibliothecae Mediceae Laurentianae* (Florence, 1776), iii. 214–15).

[98] MS *cit.* (n. 97), fo. 2ʳ⁻ᵛ.

more-than-sufficient prize for an old man who was, in fact, a non-too-
trustworthy hanger-on of the Medici family. Nor did Salviati achieve any
better results with the dedication to Leo X of another work entitled
Correctio erroris qui ex equinoctio vernali in Kalendario procedere solet.[99]
Salviati did not achieve his ambition of becoming a member of the
college of cardinals; he did, however, continue to participate in the Fifth
Lateran Council.[100] He saw his last hopes of a common league against
the Turks shipwrecked, the enthusiasm for reform which characterized
the first years of the Leonian papacy fade away, and the political and
religious plans which, at some stage of his life, he had sincerely believed
in fall to the ground, while Germany was beginning to show the first
signs of the great crisis which was to develop in western Christianity. It
is, however, said that, when the most relentless opponents of Savonarola's
memory, anxious to suppress the prophetic revival of the years 1515–16,
induced Leo X to submit the Friar's doctrines to examination by the
Council,[101] Salviati took sides with the defenders of the doctrines.[102] But
the behaviour and activities of Salviati in the last years of Julius II and
the first of Leo X were, perhaps, more intricate and complex than have
been thought up to now, which is why it would be really interesting to
find out whether there was any connection between Salviati and the
Amadeite sermonizers who preached on both old and new prophetic
themes during those years, obviously linking them to specific situations
and political events. We would, in particular, like to know whether
Salviati exerted any influence on the menacing prophetic sermon delivered
by the Amadeite Antonio da Cremona in Florence in 1508, against whom
Archbishop Cosimo de' Pazzi[103] brought charges. Or again, whether he
played a part in the even more controversial sermon of another monastic
Franciscan, Francesco da Montepulciano, in Santa Croce at Advent in
1513, arousing fresh enthusiasm amongst the *Piagnoni*, and causing
considerable anxiety to Lorenzo di Piero dei Medici and to Cardinal
Julius.[104] This is an episode which has recently attracted the attention of

---

[99] MS. Vat. lat., 8226.

[100] See *Mansi*, 32, cols. 679, 709, 728, 744, 764, 786, 807, 829, 860, 902, 940, 981. See
also above, Minnich, Ch. 4.

[101] On the attempt to get Savonarola's doctrines condemned, see above, Minnich, Ch. 4.
See also A. Giorgetti, 'Fra Luca Bettini e la sua difesa del Savonarola', *ASI* 77 (1919),
189 ff.; Ridolfi, *Savonarola*, ii. 39 ff.

[102] See Schnitzer, *Savonarola*, i. 501–2; Ridolfi, *Savonarola*, 2. 39, 182 ff.; Weinstein,
Savonarola, 358 ff.

[103] See above, n. 81.

[104] See Vasoli, *Studi*, 217–19. Salviati appears, however, amongst those, who, in
the eleventh session of the Fifth Lateran Council, approved the severe restrictions on

historians, and a study by Ottavia Niccoli,[105] who has linked it to the impressive revival of prophetic preaching, which was typical of that period, and to the diffusion of both new and old revelations by hermits and itinerant preachers. Research in this area may, perhaps, also throw new light on other aspects of Salviati's life.

One fact is, however, certain: Salviati retained his previous political partiality for the French monarchy, interrupted only by the dramatic period of the 'Conciliabulum' in Pisa. As Franciscan biographers have stated and as documented in a Parisian manuscript, he dedicated a copy of the *Vexillum christianae victoriae* to Francis I (that is after 1 January 1515),[106] for which he wrote an important preface of a purely prophetic nature. In it Salviati salutes Francis as the King of Christendom whose birth was accompanied by omens and miraculous occurrences no different from those which marked the birth of Samuel, the Virgin, and John the Baptist.[107] Linking the King's name with that of Francis of Assisi and Francesco di Paola, he recalls the numerous old and new predictions of the providential duty of the French monarchy which is now called upon to fulfil the 'nova admiranda'. Neither did Salviati fail to mention the Pseudo-Amadeite prophecy (*surget rex magnus cum magno pastore*), the work—he wrote—of a saint, in order to announce the blessed time when the King of France, the Pope, Venice, and Florence would all take up arms to liberate Christian lands and bring back the true faith to the East by defeating the Turks 'et omnes Agarenorum gentes'. According to this prophecy the 'king of the lilies' will not only celebrate his

preaching, the ban on proclaiming the advent of AntiChrist and the 'reformatio Ecclesiae', and the rigorous control of 'revelationes', see above, Minnich, Ch. 4.

[105] See Niccoli, *Profeti e popolo, ad ind.*; id., below, Ch. 10. On the cases of prophetic preaching in Florence, see above, Minnich, Ch. 4.

[106] For the identification of the *Vexillum christianae victoriae* with the *Contemplationes christianae*, dedicated to Francis I, see G. Sbaraglia, *Supplementum et castigatio ad Scriptores Ordinis Minorum* (Rome, 1806), 303; new edn. (Rome, 1908). i. 320. See also *Catalogus manuscriptorum Bibliothecae Regiae* (Paris, 1744), iii. 439, where the MS with the dedication to the King is indicated (no. 3620). For the text of the dedication to Francis I, see F. Secret, 'Aspects oubliés des courants prophétiques au début du XVIe siècle', *Revue de l'histoire des religions*, 173 (1968), 179–201 (esp. 180–2). The text is on pp. 199–201.

[107] On contemporary prophecies relating to Francis I, see F. Secret, 'Paralipomènes de la vie de François I par Guillaume Postel', *Studi Francesi*, 4 (1958), 50–62; see also Brit. Lib., MS. Sloane, 1413, fos. 23^r–53^r; and cf. F. Kvačala, *Postelliana: Urkundliche Beiträge zur Geschichte der Mystick in Reformationszeitalter* (Jurjew, 1915), p. xviii. On Francesco di Paola's prophecy concerning Francis I's assent to the throne, see L. Febvre, *Amour sacré et amour prophane. Autour de l' 'Heptameron'* (Paris, 1944; repr. 1971), 23, 26; G. Roberti, *S. Francesco di Paola, fondatore dell'Ordine dei Minimi (1416–1507): Storia della sua vita*, 2nd edn. rev. (Rome, 1963), 564, 619; Reeves, *Influence*, 380 ff.; G. Vezin, *Saint François de Paule fondateur des Minimes et la France* (Paris, 1971). For general bibliography, see F. Russo, 'Francesco di Paola', *Bibliotheca Sanctorum* (Rome, 1965), v. 116–1175.

splendid triumph over the infidels, but the Holy Pastor of Christianity will bestow upon him, 'ut speramus', the crown of the New Empire.[108] The prophetic enthusiasm of Salviati for the French monarchy is, in my opinion, closely connected to another, still unclear aspect of his life: his relationship with the Briçonnet family of whose profound influence on many of the events of French political and religious history we do not need to be reminded. Research, which unfortunately has not yet been published, carried out by Eugenio Giommi[109] has, however, already shown the considerable influence he exerted on the religious coterie of reformatory and prophetic tendencies and openly pro-French sympathies which gathered around the thaumaturgic nun, Arcangela Panigarola, Arcangela Gabriela Visconti, and Antonio Bellotti at the Milanese Augustinian monastery of Santa Marta. It was known as the *Eterna Sapienza* Society and was considered to be a cradle of the much later Barnabite congregation.[110] We also know that Arcangela Panigarola was regarded as the 'spiritual mother' of the future bishop, Denis Briçonnet, son of the powerful Cardinal of San Malo and brother of the 'evangelist', Guillaume, Bishop of Meaux, and was thus connected to Lefèvre d'Etaples and the circles of the so-called French *préreforme*.[111]

---

[108] This detail would imply that the dedication dates from 1519 when Francis I stood for election as Emperor. It should be remembered that probably already in 1514 the *Liber mirabilis*—a collection of pro-French prophetic texts—had been published with the aim of proving that all the writings contained therein announced the imminent arrival of the Angelic Pope and of the 'last Emperor' of the French dynasty (the 'second Charlemagne'). Note also that in 1516, the Augustinian Silvestro Meuccio published in Venice the 'libellus' on the *Oraculum Cyrilli*, attributed to Telesforus of Cosenza, which was likewise related to the myth of the 'second Charlemagne'. (See Reeves, *Influence*, 262–3, 355–6, 361–2, 374 ff.)

[109] See E. Giommi, *La monaca Arcangela Panigarola, madre spirituale di Denis Briçonnet, L'attesa del 'pastore engelico' annunciato dall' 'Apocalypsis Nova' del Beato Amadeo fra il 1514 e il 1520*. Unpub. doctoral thesis, presented in the academic year 1967–8 at the Università degli Studi di Firenze (Facoltà di Lettere e Filosofia).

[110] On this 'confraternity', see O. Premoli, *Storia dei Barnabiti nel Cinquecento* (Rome, 1913), 7, 13, 80, 408, 410, 411, 414, 457; see also the bibliography in Giommi, *La monaca Arcangela*. But see further, M. Binaghi, 'L'immagine sacra in Luini e il circolo di Santa Marta', *Sacro e profano nella pittura di Bernardino Luini* (Milan, 1975), 50–78; id., 'Bernardino Luini. Affreschi dalla Capella di S. Giuseppe in Santa Maria della Pace a Milano', *Pinacoteca di Brera, Scuole lombarda e piemontese 1300–1535* (Milan, 1988), 234–9.

[111] On the Briçonnet family and their influence on the ecclesiastical and spiritual life of France, see A. Renaudet, *Préréforme et humanisme à Paris pendant les premières guerres d'Italie*, 2nd edn. (Paris, 1953), *ad ind.*; Febvre, *Amour Sacré*, 90 ff.; H. Heller, 'The Briçonnet Case Reconsidered', *Journal of Medieval and Renaissance Studies*, 2 (1972), 223–58; V. Saulnier, 'Marguerite de Navarre au temps de Briçonnet. Étude de la correspondance générale', *Bibliothèque d'Humanisme et Renaissance*, 39 (1977), 437–78; 40 (1978), 7–47, 193–237; M. Veissière and H. Tardif, 'L'emploi de l'écriture par Guillaume

Research has established that Salviati acquainted Briçonnet with the *Apocalypsis Nova* about 1514, and that the expectation of the *Pastor Angelicus* became one of the central themes of spiritual expectation in the *Eterna Sapienza* Society, linked to their hopes for the reforming mission of the French monarchy and its providential duty. Salviati's relationship with the Briçonnets is confirmed by the dedication of another work, the *Virginis matris theoremata*,[112] to the Cardinal of St Malo, dated 18 November 1520 in the manuscript copy of the Ambrosian Library in Milan, but which must have been written before 14 November 1514. The work, which is of a scholarly character and has an 'axiomatic' structure, is close to the other Mariological text dedicated to Margaret of Hapsburg but includes doctrines relating to the *Apocalypsis Nova*, linked as always to the theological arguments favoured by Salviati.

The contents of the work—like that of the *Libellus de assumptione Virginis Dei matris*,[113] preserved in the same manuscript—is thus neither new nor original. The dedication, however, demonstrates the connection between Salviati and the political and ecclesiastical counsellor of Charles VIII and Louis XII. If there is some evidence to show that the *Theoremata* were first dedicated before the crisis of the Council of Pisa, the publication confirms that the Franciscan re-established his relationship with Briçonnet as soon as he could, perhaps in the early days of Leo's pontificate when new political alignments were formed and the circulation of elaborate prophetic texts could be used for particular designs. It would therefore be useful to study the presence of Pseudo-Amadeite prophecy (and perhaps even theological doctrines particularly dear to Salviati's heart) in those French religious circles which were closely linked to Briçonnet, in the years which were to be decisive for French religious history, before the myth of the *Pastor Angelicus* was involved in the eschatological drama of Guillaume Postel.[114]

Briçonnet évêque de Meaux entre 1519 et 1524', *Revue de sciences philosophiques et théologiques*, 63 (1979), 345–64.

[112] See Milan, Bibl. Ambros., MS A. 30 sup.: *Ad reverendissimum et excellentissimum dominum eminentissimum Briconetum cardinalem Episcopum prenestinum Archiepiscopum et primatem Narbonensem Machlovensem vulgo nuncupatum Georgi Benigni de Salviatis virginis matris theoremata incipiunt feliciter* (fos. 53ʳ–98ᵛ). The work cannot be later than 14 December 1514, the date of Guillaume Briçonnet, the cardinal of St Malo's death. Probably it was written prior to 1513, since in his dedication Salviati uses the title 'Bishop of Cagli'.

[113] The short work is located in the same Ambrosiana MS and in MS 16 of the Bibl. dei PP. Cappuccini di Milano. (See also P. Kristeller, *Iter italicum*, i. 327; ii, 537–8).

[114] See n. 3 above, and Vasoli, *Filosofia e religione*, 331 ff.

XI

The most significant episode in the final years of Salviati's life was, however, the part he played in the Reuchlin controversy. He was not greatly familiar with Jewish culture and traditions, although old sources credit him with having a wide knowledge of oriental languages and the 'hebraica veritas'.[115] Neither can it be said that his distant, superficial relationship with Pico had placed him in a position to understand the historical significance of a disruptive 'querelle' which went well beyond his theological range. It was, however, difficult to live so long in Florence without being aware of the interest shown not only by Pico, but also by Savonarola himself, in Jewish traditions. Furthermore, a theologian who was so sensitive to all esoteric and eschatological subjects and so alert to any kind of arcane revelation, could easily have been attracted by the increasing popularity of the cabbala, the texts of which attracted teachers of the Order, from Francesco Giorgio Veneto to Pietro Galatino.[116] Galatino, who lived in Rome during those years, must already have established a close relationship with Salviati, which was probably the source of his knowledge of the *Apocalypsis Nova* and the prophecy of the *Pastor Angelicus*.[117] When the Reuchlin affair reached the Roman See and Salviati was summoned to join the committee that was to pass judgement on the German humanist, he sided with Reuchlin against the sentence of the Sorbonne. Moreover, he intervened directly in the controversy with a *Defensio*[118] of the humanist dedicated to the Emperor. The Count and Canon of Cologne, Hermann von Neunaar, had it printed in his city in September 1517. It is not a great contribution to a debate which was beginning to attract some of the greatest minds in humanist Europe. Nevertheless, his *Defensio* must have been a fair comment, as Reuchlin's main opponent, the Dominican inquisitor, Jakob Hochstraten, replied with a tough *Apologia* dedicated to Leo X and Maximilian, which also appeared in Cologne in 1518.[119]

---

[115] On Salviati's intervention in the Reuchlin 'querelle', see ibid. 183–209.

[116] See Vasoli, *Profezia e ragione*, 117–20; id., *Filosofia e religione*, 211–29; Morisi, 'Galatino'.

[117] Morisi, 'Galatino'; Reeves, *Influence*, 234–8, 366–7, 442–7.

[118] Cf. *Defensio praestantissimi viri Joannis Reuchlin LL. Doctoris a Reverendo patre Georgio Benigno Nazareno archiepiscopo Romae per modum dialogi edita, atque ex opinione decem et octo gravissimorum virorum ad examinandum Oculare speculum a Sanctiss. D. nostro Leone P. M. deputatorum, inter quos ipse primum ex ordine votum emiserat, scripta Divoque Maximiliano Ro. Imp. Augusto dicata . . .*, Anno nativitatis Dei, M. D. mense Septembri. See also Dionisotti, 'Umanisti dimenticati', 319–21.

[119] *Ad sanctissimum dominum nostrum Leonem papam decimum. Ac divum Maximilianum Imperatorem semper augustum. Apologia reverendi patris Jacobi Hochstraten. Artium et sacrae*

In any case, the old theologian had by then decided on the part he was to play in the last of the many debates and disputes he had confronted in the course of his life. In March 1518 he published a short prefatory epistle to a classic in the Christian cabbalistic tradition, the *Opus toti Christianae veritate utile de arcanis Catholicae veritatis* by Pietro Galatino,[120] a work which was mainly a defence of Reuchlin and of his Jewish sources but also, at length, of the western cabbalistic tradition, and the revived interest in the Jewish religious experience. Bearing in mind the various kinds of pressures which prompted the publication of the *De arcanis*, we may speculate as to whether Salviati did not play an important role in the complex ecclesiastical and political manœuvres caused by the Reuchlin 'querelle'. The choice of Galatino as author, sharing, as he did, Salviati's prophetic predilections, is significant.

Destiny, which had robbed Salviati of the brilliant ecclesiastical career he had wished for in his youth, finally enabled him to link his name to one of the greatest cultural and religious crises of his time. This follows the pattern of his early intervention in the Louvain dispute and his even bolder participation in the Savonarola crisis. However, although he sided with the party which represented and predicted a new, more dramatic crisis in Christianity—the dissolution of the old theology and ecclesiastic traditions—it is certain that Salviati had little to do with the causes of the permanent rift between Luther and Rome with its 'damnable synagogue', and opposed any idea of reform which entailed breaking unity with Rome, papal authority, and the Church hierarchy. Nor would he have been able to foresee the realization of the *reformatio ecclesiae* which he had linked with the millenaristic figure of the Angelic Pope.

Ecclesiastical sources all set the year of Salviati's death at 1520, probably at his residential see at Barletta. He thus died, on the other side of the Adriatic, far away from the power he so craved, perhaps in the

---

theologiae professoris eximii. Hereticae pravitatis per Coloniensem Moguntinensem Treverensem provincias Inquisitoris vigilantissimi. Contra dialogum Georgio Benigno Archiepiscopo Nazareno in causa Joannis Reuchlin ascriptum pluribus erroribus scatentem et hic de verbo ad verbum fideliter impressum. In qua quidem Apologia Inquisitor ipse, multis occasionibus eundem coactus, tum catholicam veritatem tum Theologorum honorem, per solidas scripturas verissime tuetur. Opus novum, Anno MCCCCCXVIII Coloniae foeliciter edictum. Impressum Coloniae Anno Mcccccxviii in Februario.

[120] *Opus toti christianae Veritati utile de arcanis catholicae veritatis, contra ostinatissimam Iudeorum nostrae tempestatis perfidiam: ex Talmud genere eleganter congestum. Impressum vero Orthoae maris, summa cum diligentia per Hieronymum Suncinum: Anno christianae nativitatis M D XVIII quintodecimo Kalendas martias.* Salviati's epistle 'ad lectorem' is on fo. 1ʳ. See also Vasoli, *Profezia e ragione*, 119–20; id., *Filosofia e religione*, 183–209; Morisi, 'Galatino', for bibliography.

hope of being Bessarion's true heir in the quest for Christian 'harmony'. The world he left behind was certainly not the one promised by the prophecies he had for so long either acknowledged or propounded. It was, rather, the time when the most radical division in *Christianitas* was to take place, a time of extreme intransigence and irreparable conflict which was to shatter forever the expectation of a return to holy unity.

8

An Angelic Pope Before the Sack of Rome

ROBERTO RUSCONI

Introduction

In the first two decades of the sixteenth century a small group of important people, from the Spanish Cardinal Bernardino Carvajal to the Bosnian theologian Juraj Dragišić (Salviati), from Cardinal Adriano Castellesi to Pope Leo X himself considered, on different grounds, that they had good reason to identify themselves with the Angelic Pope.[1]

Some of these ecclesiastics had no scruples about declaring their conviction in public, even though it had, perhaps, been developed in a casual and superficial manner. Others were like the Apulian Observant Franciscan, Pietro Galatino (1460–c.1540), who placed it at the centre of his writing activity, compiling a truly remarkable number of pamphlets and commentaries.[2] In spite of the fact that some of them were prepared for printing by the author himself, none of them saw the light of day until they were deposited, by express order of Pope Paul III on 11 May 1539, in the library of the Franciscan monastery of Ara Coeli in Rome.[3]

Pietro Galatino was undoubtedly an erudite and cultivated person. Even though the beginnings of his cultural formation are still unknown, they came progressively to include the knowledge of Oriental languages

---

[1] Cf. C. Vasoli, 'A proposito di Gabriele Biondo, Francesco Giorgio Veneto e Giorgio Benigno Salviati', in *Profezia e ragione* (Naples, 1974), 121–7; A. Morisi, *Apocalypsis Nova: Richerche sull'origine e la formazione del testo dello pseudo-Amadeo* (Studi Storici, 77; Rome, 1970); Reeves, *Influence of Prophecy in the Later Middle Ages: A Study in Joachimism* (Oxford, 1969), *ad ind.*; Fragnito, 'Castellesi', *DBI* 21 (1978) 668.

[2] For the first study of Pietro Galatino, see A. Kleinhans, 'De vita et operibus Petri Galatini, O.F.M. scientiarum Biblicarum cultoris', *Antonianum*, 1 (1926), 145–79, 327–56. This gives a brief description of the vast collection of MSS in the Bibl. Vaticana, and a chronology of the writings (which will be corrected in part during the course of the present study). The more important studies which follow will be cited appropriately. It is irrelevant here to establish the true surname of the personage whom we will continue to call 'Pietro Galatino' for reasons of convenience.

[3] L. Wadding, *Annales Minorum* (Rome, 1731–6), 516–17.

and the Hebrew cabbala, Apocryphal literature from the period of the origin of Christianity, Joachimist and pseudo-Joachimist apocalyptic eschatology, and late medieval prophetism—in particular the *Apocalypsis Nova* of the Blessed Amadeus.[4]

As an insider at the highest level of the 'milieu romain', Pietro Galatino expressed, in a totally personal manner, the expectation of a widely spread renewal in Roman ecclesiastic circles.[5] But the roots of his prophetic conviction lay far back in medieval prophecy, now outdated.

1. Self-Identification with the Angelic Pope

In May 1525, the apostolic referendary, Alessandro Spagnoli of Mantua, entrusted to Francesco Calvo, who had succeeded Marcello Silber in 1524 in the role of 'apostolic printer',[6] the text of a 'commentariolum', the *Vaticinii Romani explicatio* by Pietro Galatino—presented by him as an authoritative commentator on prophecies such as those attributed to Methodius, the Bishop of Patara, the mythical Cataldus, Archbishop of Taranto, and Cyril, the Hermit of Mount Carmel.[7]

This pamphlet never saw the light of day, perhaps because it contained a complete, even if not totally explicit identification of a contemporary personage, Pietro Galatino himself, with the awaited Angelic Pastor. This figure had progressively moved to the centre of Galatino's eschatological speculations, particularly from the beginning of the 1520s. He had accumulated numerous, voluminous writings, but it was not until the composition of the *Explicatio*—and thus quite late in his life—that he identified himself with the foretold ecclesiastic reformer of supernatural election.

The stimulus to this move came from two pseudo-prophecies which were otherwise unknown, the *Vaticinium Montis Gargani* and the *Vaticinium Romanum*. These were ready-made to serve as a lexicon of the principal concepts of medieval prophetic literature, giving substance to the expectation of the advent of a reforming Pope. They were probably the product of the Roman ecclesiastic and curial circles in which Galatino

---

[4] Morisi, *Apocalypsis Nova*; Vasoli, *Profezia e ragione*.

[5] F. Secret, 'Guillaume Postel et les courants prophétiques de la Renaissance', *Studi Francesi*, 3 (1957), 375–95, with a series of unconvincing approaches. For a more articulate valuation of these influences, see C. Vasoli, 'Giorgio Benigno Salviati, Pietro Galatino e la edizione di Ortona (1518) del *De arcanis Catholicae veritatis*', in *Filosofia e religione nella cultura del Rinascimento* (Naples, 1988), 188–9.

[6] F. Barberi, 'Calvo, Francesco Giulio', *DBI* 17 (1974), 38–41.

[7] See a brief description of the manuscript in Kleinhans, 'De vita', no. 16, 166–7.

moved rather than his own, appearing in the period of transition between the end of the papacy of Julius II and the beginning of that of Clement VII.[8]

In compiling the 'commentariolum', Pietro Galatino would have been influenced by the political-military events of 1525, and in particular by the military defeats of the King of France, Francis I, which had been anticipated by him in the interpretation of a series of *revelationes* ('Quorum ego vaticinia longe antequam rex ipse caperetur, et vidi et legi, atque statim quum res evenisset, tibi ostendi' (MS. Vat. Lat. 5581, fo. 40ᵛ), attributed to the Provencal visionary Robert d'Uzès and the Minorite Jean de Roquetaillade, whose predictions could be applied exactly to this fatal decade.[9]

In effect, the misfortunes of the French king had caused political prophecy to take on a new strength in Italy, and particularly in Rome itself, through the influence of the hermit Bernardino da Parenzo, who was active again between 1523 and 1525 after a period of silence.[10] The markedly pro-Hapsburg orientation of political prophecy which Galatino had cultivated dates back to an earlier period.[11] When, after 1528, at the exhortation of the general minister of the Observant Franciscans, Francisco de Los Angeles Quiñones, Galatino arranged for a de luxe edition of his *Commentaria in Apocalypsim*, composed before 1524, to be delivered to Charles V on the occasion of a visit to Rome, he took the occasion to remember, with suitable servility, in the dedication that he had already dedicated another work, the *De arcanis Catholicae veritatis* (1518), to Charles's paternal grandfather Maximilian I. Even earlier than that in 1506 Galatino had delivered another work, *De optimi principis diademate*, to his maternal grandfather, Ferdinand the Catholic, who was on a visit to Naples.[12] His strictly pro-imperial point of view is reflected

---

[8] See Reeves, *Influence*, 442 n. 5, whose hypotheses are more likely than those of F. Secret, *Introduction à Guillaume Postel, Le Thrésor des prophéties de l'univers* (The Hague, 1969), 25–6.

[9] 'de iis que ab anno M.D.XX°. usque ad annum M.D.XXXm emersura erant' (MS. Vat. Lat. 5581, fo. 41ʳ).

[10] See R. Rusconi, '"Ex quodam antiquissimo libello": La tradizione manoscritta della profezie nell' Italia tardo medievale: dalle collezioni profetiche alle prime edizione a stampa', in W. Verbeke, D. Verhelst, and A. Welkenhuysen (eds.), *The Use and Abuse of Eschatology in the Middle Ages* (Louvain, 1988), 465–6; id., 'Il collezionismo profetico in Italia alla fine del Medioevo ed agli inizi dell'età moderna', *Florensia*, 2 (1988), 77–8. On prophetism in the years preceding the Sack of Rome in 1527, see also M. Reeves, 'Roma profetica', in F. Troncarelli (ed.), *La città dei segreti: Magia, astrologia e cultura esoterica a Roma* (Milan, 1985), 296; O. Niccoli, *Profeti e popolo nell'Italia del Rinascimento* (Rome and Bari, 1987), *passim*.

[11] On Galatino's pro-Hapsburg orientation, see in particular Reeves, *Influence*, 366–7.

[12] Kleinhans, 'De Vita', no. 2, 152–3.

in the other 'current' predictions within the *Vaticinii Romani explicatio*. These range from a general anti-Lutheran attack ('in Martini Lutheri impiissimo dogmate'), to a prophecy of the outbreak of the Peasants' War of 1524–5,[13] and a reference to the eschatological overtones of the ever present Turkish menace.

Here Galatino's pro-Hapsburg orientation becomes explicit, for Charles V (although not expressly named) is predicted in the role of the Last World Emperor. At the same time, Galatino reduces him to a mere collaborator, purely and simply the secular arm of the Angelic Pope:

From the words of Christ Himself who is truth, it clearly appears that the extermination of the whole Mohammedan sect, the conversion of the whole world, the destruction of abomination in the holy place (that is, simony) and the reformation of the Church must be accomplished by the Angelic Pastor with the aid of the Iberians, that is, the King of Spain who now is Emperor.[14]

In this *Explicatio* the self-identification of Pietro Galatino with the Angelic Pastor who will succeed the current pontiff, Clement VII, is not explicit but made through unequivocal allusions which leave no shadow of a doubt.[15] For the *Vaticinium Romanum* is interpreted by him as prophesying the accession to the Papal throne of a Minorite from Salento—'From this order at length will arise the Angelic Pope who will restore her ancient dominion to the City of Rome and to the Roman Church her pristine dignity'—who entered into the Franciscan order during the papacy of Pope Sixtus IV. He indicates the name by means of an obscure monogram, but there is no doubt that he is alluding to the publication, in Ortona, of his book *De arcanis Catholicae veritatis*, which passes from a first manifestation of the Angelic Pastor to conclude finally that his public revelation is by now imminent: 'from these things it plainly appears that only a very little time can possibly elapse before this

---

[13] MS. Vat. Lat. 5581, fo. 39ᵛ; 'Germani e inobedientia et principum eius ineptia atque lascivia prorogabit tempora foelina' (MS. Vat. Lat. 5581, fo. 43ᵛ).

[14] MS. Vat. Lat. 5581, fo. 50ʳ⁻ᵛ; recurrent in his writings is the idea of a simultaneous intervention by the Angelic Pope and the Last World Emperor; see *De Angelico Pastore* (MS. Vat. Lat. 5578, fo. 30ᵛ); *De septem temporibus* (MS. Vat. Lat. 5579, fo. 10ᵛ), with a Latin poem entitled: 'Authoris carmina de imperatore et pontifice ecclesiam reformaturos'. For both passages cf. Reeves, *Influence*, 366–7.

[15] See F. Secret, 'L'émithologie de Guillaume Postel', in E. Castelli (ed.), *Umanesimo e esoterismo* (Padua, 1960), 390–3. On Galatino's auto-identification with the Angelic Pope, see now on a broader scale G. Vallone, 'Pietro S. detto il Galatino', *Studi offerti ad Aldo Vallone*, i: *Letteratura e storia meridionale* (Florence, 1989), 98–9. Vallone uses Galatino's pseudo prophecies for biographic purposes.

Pastor himself must be made manifest: may God in His mercy quickly grant that we may see him!'[16]

A further clue identifying Pietro Galatino with the eschatological Pope lies in the chronological extent of the predictions, ranging from the implicit ones of the *Vaticinium Monti Gargani*, which embraces the Popes from Paul II to Clement VII and the Angelic Pope who will succeed him, substantially coinciding with the pattern of Galatino's life, to the more explicit ones of the *Vaticinium Romanum*, where the fictitious date of composition, 1160, is designed to fix the date, after a lapse of three hundred years, to fit in with the period which must elapse from it to the birth of the Angelic Pastor (that is 1460, the time of Galatino's birth): 'Qui post trecentos annos nascentur, proximi erunt temporibus illis' (MS. Vat. Lat. 5581, fo. 68ᵛ).

To strengthen his own predictions Pietro Galatino resorts also to astrology. On the one hand he links the awaited events to the revolutions of Saturn which occur every three hundred years, and this allows him to relink the date 1160 with 1460—the year of the birth of the Angelic Pope. On the other hand, referring to a recent event, he connects the great 'mutatio' with the astral conjunction of February 1524: 'Postquam tales rerum ingentium mutationes futuras expectamus.'[17]

In the conclusion to the *Explicatio*, he outlines the eschatological duties of the Angelic Pastor. This is the customary portrait of a reformist pontiff whose eschatological mission includes the final programme which had become, by now, all but a ritual in this type of prediction: 'the reform of the Church, the abolition of the Mohammedan sect, the conversion of all infidels to Christ, one monarch of the world, one Pastor, one sheepfold and the diffusion of divine worship throughout the globe.' (MS. Vat. Lat. 5581, fo. 68ᵛ).[18]

---

[16] The passages referred to are respectively in MS. Vat. Lat. 5581, fos. 12ʳ–13ᵛ; 34ᵛ; 57ᵛ–59ʳ; 37ᵛ–38ʳ; 36ᵛ.

[17] MS. Vat. Lat. 5581, fos. 65ᵛ–67ᵛ; the quotation at 67ᵛ. On the great conjunction, cf. P. Zambelli, 'Fine del mondo o inizio della propaganda? Astrologia, filosofia della storia e propaganda politico-religiosa nel dibattito sulla congiunzione del 1524', *Scienze, credenze occulte, livelli di cultura* (Florence, 1982), 291–368; O. Niccoli, 'Il diluvio del 1524 fra panico collettivo e irrisione carnevalesca', ibid. 369–92. Although the use of the zodiac was not unusual in Galatino's works (see *De septem temporibus*, MS. Vat. Lat. 5579, fos. 43ᵛ–46ᵛ: 'Quod septem tempora ista septem quoque planetis haud incongrue comparentur', also MS. Vat. Lat. 5575, fo. 91ʳ), he had a declared hostility towards astrologers (see A. Morisi, 'Galatino', 229 n. 31).

[18] He also describes the Angelic Pope thus: 'Romam veniet, et serviet, immensi, incredibilis, inauditi, et ingentissimi fati homo. Immutabit in melius omnia, superflua resecabit, moribus, institutis, et legibus novis formabit mundum. Regnabit in posterum

When one reads in the 'commentariolum' regarding the advent and actions of the Angelic Pope: 'Hec enim omnia sub Angelico Pastore adimplenda esse Abbas Ioakim exponit' (MS. Vat. Lat. 5581, fo. 45ᵛ), it would seem that Pietro Galatino has drawn his own inspiration widely from the late medieval prophetic literature and the pseudo-Joachimist works rather than the authentic writings of the Abbot of Fiore.[19]

2. *Pietro Galatino and the Roman Ecclesiastic 'Milieu'*

What were the origins of this sixteenth-century prophet? From a letter by the German humanist, Peter Eberbach, some time after 1515, 'we may deduce that our friar was born at Galatina into an Albanian family who had moved to Puglia from Durazzo in Salento under the Turkish pressure'.[20]

Settled in Puglia and admitted to the Franciscan Order, Galatino yet remains elusive and culturally unknown until the date of his first work, when he was already between forty and fifty years old. It is difficult therefore, if not impossible, to judge to what extent the cultural environment of Puglia may have influenced the Franciscan theologian and writer. The few pieces of biographical information relative to his activities within the Minorite order date back in each case to the period following his entry into Roman ecclesiastic circles. On 9 September 1515, the Papal brief of Pope Leo X approves the nomination by the general minister Bernardino da Chieri of Friar Pietro as provincial vicar to the Observant Franciscans of Puglia. After the breaking of the unity of the order in 1517, Friar Pietro becomes minister of the Provincia S. Nicolai Ordinis Minorum Regularis Observantiae, and is safely re-elected almost twenty years later in 1536.[21]

Pietro Galatino's first two known works are the *De optimi principis diademate*, dedicated to Ferdinand the Catholic on the occasion of his journey to Naples in 1506, and the *Expositio dulcissimi nominis*

---

foelicissimus, et gloriosissimus, homo secundum cor immortalis Dei' (MS. Vat. Lat. 5581, fo. 64ᵛ).

[19] For an analysis of Galatino's Joachimism see above all Reeves, *Influence*, 235–8, 366–7, 442–6, and 'Roma profetica', 291–6.

[20] Vallone, 'Pietro Galatino', 87–105 (with a fairly up-to-date bibliography), citing on p. 105 a note by L. Geiger, *Johann Reuchlin: Briefwechsel* (Tübingen, 1875), 248–9 n. 2: 'patria Dyrachinus, natione Epirota, latine, graece, ebraice chaldaique eruditissimus; iam tum, cum ego adhuc Romae essem, in defensionem Capnionis scribere inceperat, conarique demonstrare omnia nostri dogmatis mysteria in Thalmudicis voluminibus haberi.'

[21] Kleinhans, 'De vita', 147–8; B. Perrone, 'I Frati Minori di Puglia della Serafica Provincia di S. Nicolò (1590–1835)', *Archivi e Biblioteche*, 2 (1977), 156–7.

Tetragrammaton ('Divinis nominis tetragrammaton intepretatio contra Judaeos, ubi divinae trinitatis mysterium distincte continetur'), dedicated to Rutilio Zenone, Bishop of the tiny diocese of St Mark in the ecclesiastical province of Rossano Calabro, who replied to him in a letter dated 31 May 1507 from Naples.[22] Some letters and epigrams which, in the manuscripts, accompany the first of the two works, refer to an intellectual world of ecclesiastical dignitaries holding minor seats in Puglia, and developing cultural activities at a decidedly minor and provincial level.[23]

It is probable that Galatino did not arrive in Rome until the last years of the papacy of Julius II (1503–13). His *Oratio 'Quum ieiunatis'* dates back to this period, and is addressed to the Pope by the author who declares himself at the beginning to be 'sacre theologie professor'.[24]

On 1 January 1515 a short pamphlet appeared in print in Rome, an *Oratio de circumcisione dominica*, dedicated to Pope Leo X by the chaplain to Cardinal Lorenzo Pucci,[25] the Puglian Franciscan Pietro Galatino, 'artium et sacrae theologiae doctor'. It is a traditional enough text divided by the author himself into three points: the redemption of man, the fidelity of God to his own promises, and the necessity of the circumcision of the heart.[26] Not even the final peroration exhorting Christian princes to take up arms against the Turks is couched in the strongly apocalyptic tones which the theme excited in that period,[27] even though it appeared during the Fifth Lateran Council, and came from a personage

---

[22] He was bishop in this seat from 1484 to 1515 (C. Eubel, *Hierarchia Catholica Medii Aevi*, (Münster, 1898–), ii. 185; *Hierarchia Catholica Medii et Recentioris Aevi* (Münster, 1923), iii. 234).

[23] Kleinhans, 'De vita', no. 2, 152–3, was not aware of MS 1366, Bibl. Angelica, Rome, described by H. Narducci, *Catalogus codicum manuscriptorum praeter Graecos et orientales in Bibliotheca Angelica olim Coenobili Sancti Augustini de Urbe* (Rome, 1893), i. 572–3. For other manuscripts of the two works, of which Kleinhans had only an indirect knowledge, cf. now P. Kristeller, *Iter Italicum*, iv: *Alia Itinera* (London and Leiden, 1989), 2, 507a–507b: Granada, Abadia del Sacro Monte, MS. n.n. For *De optimi principis diademate*, see B. Perrone, 'Il *De re publica christiana* nel pensiero filosofico e politico di Pietro Galatino', *Studi storia pugliese in onore di Giuseppe Chiarelli* (Galatina, 1968), ii. 524–9.

[24] The copy in MS 488, Bibl. Angelica, Rome, was described in Narducci, *Catalogus*, 219. It is unknown to O'Malley (see n. 26).

[25] On Pucci, see E. Göller, *Die päpstliche Poenitentiaria* (Rome, 1907–11), 374–5 n. 25.

[26] The only well-known example is conserved at the Bibl. Naz. Vittorio Emanuele II, Rome. See Kleinhans, 'De vita', 172–3, and for the contents J. O'Malley, *Praise and Blame in Renaissance Rome: Rhetoric, Doctrine and Reform in the Sacred Orators of the Papal Court c.1450–1521* (Durham, NC, 1979), 117.

[27] See the long passage from the unedited *Commentaria in Apocalypsim*, reported in Kleinhans, 'De vita', 350–6. Perrone, 'Il *De re publica*', 552 rightly places this *Oratio* of 1515 in the same line of thought as the later *De re publica christiana* of 1519, on the basis of the role assigned to both in the struggle against the Turks.

who, at just 20 years old, would have been present in 1480 at the conquest of Otranto and the massacre of the Christians.

The credit which Pietro Galatino enjoyed in Roman curial and intellectual circles is more than sufficiently demonstrated by the fact that in 1515 he started work at the instigation of Cardinals Lorenzo Pucci and Adriano Castellesi, perhaps by order of Pope Leo X himself, on the defence of Johann Reuchlin and his position on the question of a Christian interpretation of the Hebrew cabbala.[28]

This work was probably finished not later than 4 September 1516, though subsequent interpolations in the text are revealed by a reference to the edition of Reuchlin's *De arte cabalistica* and to the conferment of a cardinalship on Egidio of Viterbo in 1517.[29] It was finally printed at Ortona on 15 February 1518, by the Jewish printer Gershom Soncino, with the title *Opus toti christiane Reipublice maxime utile, de arcanis catholicae veritatis, contra obstinatissimam Judaeorum perfidam*, and as a consequence its author was placed at the centre of one of the greatest intellectual debates of the age.[30]

Galatino's contact with cabbalist circles is demonstrated by his letter to Johann Reuchlin written before 1 July 1515 (the date of Reuchlin's reply), and made public at Hagenau in 1519 in the collection *Illustrium virorum epistolae, hebraicae, graecae et latinae*, dedicated to the German humanist during the course of the controversy of which he was protagonist.[31]

The exact form of his relationships with the persons whom he lists in this letter remains elusive, however, and there is a lack of direct documentation of his relation to the Roman cultural world. For example

---

[28] Morisi, 'Galatino', 212–13. For the dedication of Reuchlin's work, *De accentibus et orthographia linguae hebraicae* (1518), to Cardinal Castellesi, see Fragnito, 'Castellesi', 670. Castellesi was not a simple politician or bureaucrat of the papal curia, but an exponent of an 'Erasmian tendency', as shown by a letter he wrote on 3 Feb. 1516 to Raffaele Maffei da Volterra, in which there is a fervid reminder of the 'hebraica veritas' (an expression which recurs fairly frequently in Galatino's writings) and an undoubted reference to the value of the cabbalist interpretation of the Scripture. See also J. D'Amico, 'Humanism and Theology at Papal Rome, 1480–1520' (Rochester (NY), Ph.D. diss., 1977), 100–3; id., *Renaissance Humanism in Papal Rome; Humanists and Churchmen on the Eve of the Reformation* (Baltimore, 1983), 16–19, 169 ff. For further discussion of the Reuchlin dispute, see above, Vasoli, Ch. 7.

[29] Morisi, 'Galatino', 228 n. 4.

[30] See Kleinhans, 'De vita', 174–9; for the different editions, see Vasoli, *Filosofia e religione*, 183–209, 223–6.

[31] The text of the two letters is edited in Kleinhans, 'De vita', 338–42. See also Morisi, 'Galatino', 213.

there is a possible contact with Antonio Flaminio, Professor of Rhetoric at the University of Rome, who died in 1513, but it is highly improbable that the Franciscan ever taught there.[32] In the *De arcanis* Galatino limits himself to acknowledging his consultation of Flaminio's Hebrew manuscripts, which had been placed in the Apostolic Library immediately following his death in 1513: 'I also found them [the 72 divine names formed cabbalistically on the basis of Exod. 14: 19–21] with their proper punctuation in various places, but particularly in Rome in the Papal library in that place where the books of Flamineus, a man most expert in three languages, were kept.'[33]

The question of Galatino's inclusion, not so much in the Roman curial world, but in its intellectual circles, becomes central when considering the role in the 'circle' of the general minister of Augustinian Friars, later to be Cardinal Egidio of Viterbo. Apart from the presence of his name on the list of cabbalistic *auctoritates* furnished by Galatino himself, the connection between the Franciscan and the Augustinian could be limited to the cross-reference of a series of ideas and concepts which were widely in circulation in that era, especially as it is not possible to ignore the fact that their respective theological orientations were anything but harmonious.[34] In fact confirmation of Galatino's direct contacts with Hebrew-speaking circles lies rather in the personal knowledge he had of Egidio of Viterbo's *Libellus de litteris sanctis*, which had remained unpublished, although a handwritten copy had been dedicated to Cardinal Giulio de Medici in 1517,[35] as well as in the circumstances of the publication of the *De arcanis Catholicae veritatis* in 1518. In the year following Egidio of Viterbo's departure for the Spanish legation (he left Rome on 16 March 1516), the printing of Hebrew books in the city went through a period of inactivity which induced Elia Levita to have one of his works printed at Pesaro by the printer Gershom Soncino, who, in the

---

[32] Kleinhans, 'De vita', 149–50.

[33] Cited in G. Levi Della Vida, *Ricerche sulla formazione del più antico fondo dei manoscritti orientali della Biblioteca Vaticana* (Vatican City, 1939), 161. Cf. F. Secret, 'Flaminio, Antonio', *DHGE* 17 (Paris, 1971), 354 and the bibliography given there.

[34] See J. O'Malley, *Giles of Viterbo on Church and Reform* (Leiden, 1968), 56 n. 2, 71, 76, 115 (where Egidio's reservations with regard to Joachim's theology are underlined; pp. 61 ff., 115 n. 2); id., 'Giles of Viterbo: A Sixteenth-Century Text on Doctrinal Development', *Traditio*, 22 (1966), 448, where Galatino is described as 'Giles' admirer and colleague as expositor of the Christian cabal' (but note that this does not accord with the statement that 'Galatinus' reliance on Joachimite ideas is heavy').

[35] F. Secret, Introduction to Egidio of Viterbo, *Scechina e Libellus de litteris hebraicis*, I (Rome, 1959), i. 9 and n. 3.

same period, had printed the *De arcanis* in Ortona. That text was written in 1515, the year in which Elia Levita arrived in Rome.[36]

A central element in the theological system developed by Pietro Galatino, the progressive revelation of divine mysteries, appears to correspond to the gradual slide from a position which grounded knowledge of the *arcana* in Neoplatonism to one which depended more directly on the contribution of late medieval Hebrew culture. Here the thought of Galatino—and also of Egidio of Viterbo—differed greatly from that of Marsilio Ficino, Pico della Mirandola, or, later, of Reuchlin.[37]

A letter from Johann Reuchlin, dated 12 February 1519 from Stuttgart, shows that Galatino's intervention had caused him to be received directly into the circle of intellectuals who were associated with the German through their letters. Reuchlin promised him to concern himself in the German publication of the 'promissos sex libros contra propositiones Astarothicas', that is against Jacob Hoochstraten.[38] A final indication of the standing acquired by Galatino lies in his inclusion as one of the six proof-readers of the *Salterio tetraglotto* by the Dominican Friar Santi Pagnini, dedicated to Leo X. This had originally gone to press in 1520, but its printing was interrupted for financial reasons in 1521.[39]

3. The Intellectual Pilgrimage of a Cultivator of the Esoteric

After the publication at Ortona in 1518 of the *De arcanis Catholicae veritatis*, and in spite of his vast, solid fame as a student of the cabbala, it does not seem that Pietro Galatino occupied himself with it subsequently to any particular extent. His later works do not display that knowledge of cabbalistic *auctoritates* so abundantly quoted in the published volume.[40] It is probable therefore that the collaboration between Galatino and the Dominican Friar from Genoa, Agostino Giustiniani, in the writing of

[36] Suggested by F. X. Martin, 'Giles of Viterbo as Scripture Scholar', *Egidio da Viterbo, O.S.A. e il suo tempo* (Rome, 1983), 211. On Lévita, see G. Weil, *Elie Lévita, humaniste et massorète (1469–1549)* (Leiden, 1963), and for his influence on Hebrew culture in Rome in the early 16th cent., Martin, ibid. 191–222.

[37] At least according to J. Monfasani, 'Sermons of Giles of Viterbo as Bishop', in *Egidio da Viterbo, O.S.A., e il suo tempo* (Atti del V convegno dell'Istituto Storico Agostiniano; Rome, 1983), 155 and n. 86.

[38] G. Friedländer, *Beiträge zur Reformationsgeschichte* (Berlin, 1837), 87 (Morisi, 'Galatino', 228, n. 18, records that nothing is known of such a work).

[39] T. Centi, 'L'attività letteraria di Santi Pagnini (1470–1536) nel campo delle scienze bibliche', *AFP* 15 (1945), 14.

[40] For the intellectual reconstruction of his personality, see Morisi, 'Galatino', 212–31, summed up by Vasoli, *Filosofia e religione*, 223–6.

the *De arcanis* was a particularly close one,[41] and that in the end, the Franciscan's principal contribution consisted of the utilization of Paulus de Heredia's *Epistola de secretis*, printed in Rome in 1487.[42]

Even if it is difficult fully to grasp the first stages of the author's intellectual development, it is possible, on the evidence from a much later period when a substantial number of his works were written, to argue with some certainty that his interest in cabbalist literature arose originally from the problem of the Hebrew writing of the name of God, the *Tetragrammaton*, because of its possible connections with the Bernardine doctrine of the Name of Jesus. This can be traced back beyond the *De arcanis* to his earlier writing, the *Expositio dulcissimi nominis Tetragrammaton*.[43] Moreover in a treatise which came a little later, the *De ecclesia destituta*, in connection with the marginal rubric 'Nomen Dei tetragrammaton', of which he wishes to furnish the 'allegorica explanatio' or rather the 'mystica interpretatio', Galatino confirms fully that he is dealing, not with 'Christus', but with IESU (MS. Vat. Lat. 5568, fo. 346ʳ)—the name which, in all the manuscripts of his works, is written in capital letters as a Latin version of the Hebrew tetragram IHWH.[44] Pursuing the eschatological potential of the Name of Jesus doctrine which is already present in the writings and preachings of Bernardino of Siena,[45] Galatino, in the treatment of the *De arcanis*,

[41] F. Secret, 'Les Dominicains et la Kabbale chrétienne à la Renaissance', *AFP* 27 (1957), 321–4, on which depend substantially several allusions to Giustiniani in G. Javary, 'Panorama de la kabbale chrétienne en France au XVIe et au XVIIe siècles', *Kabbalistes chrétiens* (Paris, 1979), 69–70. Cf. G. Scholem, 'Considérations sur l'histoire des débuts de la kabbale chrétienne', *ibid.* 34–5, 44–5. The erroneous judgement of A. Berthier, 'Un Maître orientaliste du XIIIe siècle: Raymond Martin O.P.', *AFP* 6 (1936), 284–5, on the 'plagiarism' of Pietro Galatino, re-opens the problem of the Dominican's role. It is not by chance, Secret records regarding the *Gale Razeia* of Paulus de Heredia, that Galatino worked on the *De arcanis* in collaboration with the Dominican Agostino Giustiniani: 'Guillaume Postel', 380.

[42] Cf. Scholem, 'Considérations', 34; F. Secret, 'L'*Ensis Pauli* di Paulus de Heredia', *Sefarad*, 26 (1966), 79–102, 253–72.

[43] Vallone, 'Pietro . . . Galatino', 99 n. 64 observes that in *De arcanis*, bk. ii, Chs. 10–11, 'he recast the epistle in part'. The main theme of the orthography of the name of Jesus in bk. iii, Ch. 8, is derived from Paulus de Heredia, see Secret, *Egidio of Viterbo, Scechina*, 47 n. 55 to the text, 98 ff. The idea that the *Tetragrammaton* coincided with the name of Christ already occurs in the *Pugio Fidei* of Raymond Martin, in the *Conclusiones* of Giovanni Pico della Mirandola and in the *De verbo mirifico* of Johannes Reuchlin, see Monfasani, 'Sermons of Giles of Viterbo', 169 n. 65; Secret, *Le Zohâr dans les Kabbalistes chrétiens de la Renaissance* (Paris and The Hague, 1964), 9, 36.

[44] MS. Vat. Lat. 5568, fo. 346ʳ. In any case he refers to the *De sacra scriptura recte interpretanda*, bk. iii, Ch. 7 (MS. Vat. Lat. 5580; cf. Kleinhans, 'De vita', no. 15, 164–165).

[45] See R. Rusconi, 'Escatologia e povertà nella predicazione di Bernardino da Siena', *Bernardino predicatore nella società del suo tempo* (Todi, 1976), 223–4; id., 'Apocalittica ed escatologia nella predicazione di Bernardino da Siena', *Studi medievali*, 3rd ser. 22 (1981), esp. 109–17.

ascribes a notable importance to the cabbalist texts relevant to the Messiah. This is later confirmed by a passage from the *Commentaria in Apocalypsim* on the 'revelationem quam Cabalam vocant' and on the 'mysteria' relative 'ad Messiam venturum' (MS. Vat. Lat. 5567, fos. CXXIX^v–CXXX^r).[46]

Thus one forms the impression that the centre of his interests lies in the eschatological and millenarian character of his expectations. This is apparent above all in book 12 of the *De arcanis*, 'De secundo Messiae adventu deque in eo futurus'; while a reference in book 3 to Giorgio Benigno Salviati's unpublished *Vexillum Christianae victoriae*, a work full of quotations from the Blessed Amadeus's *Apocalypsis Nova*,[47] indicates the same orientation and points to its roots. The *De arcanis* can, with reason, be held to be a work characterized by 'an intense eschatological perspective' which dated back to the writings and the actual figure of the by now elderly Franciscan theologian, Benigno Salviati.[48]

Perhaps Pietro Galatino 'was obliged to attenuate his cabbalistic enthusiasm',[49] especially after the condemnation of Reuchlin in 1520. The centre of gravity of his speculations appears to be so different that his recourse to the cabbala seems to serve a subordinate purpose: 'like Pico and Reuchlin, the Franciscan considered the cabbalist interpretation of the scripture as a confirmation of all prophetic truths.'[50]

Thus Galatino's interest in the cabbala and his polemic against the Talmud undoubtedly sprang from exegetical and not simply philological concerns. That is to say, he belonged to the category of commentators who concentrated on the 'anagogic' sense when interpreting the Scriptures, rather than the 'literal' sense.[51] This accorded with the 'intellectus spiritalis' which Galatino sought in the 'arcana Dei'.

---

[46] MS. Vat. Lat. 5567, fos. 129^v–130^r. Reproduced in Morisi, 'Galatino', 216.

[47] Vasoli, *Filosofia e religione*, 202 and n. 38. See also F. Secret, 'Aspects oubliés des courants prophétiques au début du XVIe siècle', *Revue de l'histoire des religions*, 173 (1968), 180–2, on the significance of the citation of the unpublished *Vexillum* in bk. iii, Ch. 30. Note that the pseudo-Amadeite prophecy in the dedication is interpreted in a pro-French way by Salviati and Postel, and in favour of Charles V by Galatino.

[48] C. Vasoli, 'Un commento scotista a un sonetto del Magnifico: l'*Opus Septem Questionem* di Giorgio Benigno Salviati', *Filosofia e religione*, 144, where the traces of Salviati's influence on the *De arcanis* are signalled.

[49] Id., 'Giorgio Benigno Salviati', *Filosofia e religione*, 209, with other valuations which we do not completely share.

[50] Id., 'L'Apocalypsis nova', *Filosofia e religione*, 226. See also Morisi, 'Galatino', 220: 'S'il est difficile de prouver que l'influence de la tradition kabbalistique ait été déterminante dans son choix, il est indéniable, en revanche, qu'elle lui a apporté la confirmation de la fécondité de le lecture spirituelle.'

[51] See Morisi, 'Galatino', 214–15, 221, 224. See also Vasoli, 'Giorgio Benigno Salviati', *Filosofia e religione*, 205.

To deal with the question 'de recuperer comme prophéties des écrits nés avant ou même en dehors du christianisme',[52] from the cabbala to the apocrypha writings of Christian origin such as the *Shepherd of Hermas*, is to arrive at the point of connection between the esoteric interests of Pietro Galatino and late medieval prophetism, in particular the apocalyptic eschatology formed in a Joachimist matrix.[53]

4. The Collection of Prophetic Material: Manuscripts and Printed Editions

In early sixteenth-century Rome Pietro Galatino 'collectionait les prophéties';[54] thus the vast assembly of his handwritten works forms a focus for the gathering of prophetic literature accumulated during the course of the late Middle Ages, bringing together not only those of the manuscript tradition but also the first printed editions of prophetic and eschatological-apocalyptic literary works.[55]

Galatino himself acknowledges his debt to this cultural patrimony in the preface to book 8 of the second part of the *De Ecclesia destituta* (MS. Vat. Lat. 5569, fo. XLI[r]): 'If therefore absolutely all the Church in the fifth time was bound to be "deformed" by the corruption of all the *status*, this deformity had to be predicted before it happened, not only by the oracles of divine Scripture, but also indeed by the prophecies of God's saints and servants, in order to expound it perfectly.' His list of prophetic and visionary *auctoritates* is a characteristic one: 'S. Methodius martyr; S. Severus martyr; S. Cataldus arch. Tarentinus; b. Cyrillus Montis Carmeli; Abbas Ioakim; S. Franciscus ord. min. institutor; S. Vincentius o.p.; S. Hildegardis; S. Elisabeth virgo; S. Brigida vidua; b. Robertus o.p.; b. Amadeus o.m.' (MS. Vat. Lat. 5569, fo. XIL[v]). Although a few of the names cited could only have been known to him through the late medieval manuscript tradition, it is the printed prophetic sources which prevail in this review. They begin with Hildegarde of Bingen's *Scivias*, the visions of Elizabeth von Schönau, the works of the Dominican

---

[52] Morisi, 'Galatino', 218.

[53] Ibid. 228.

[54] F. Secret, *Introduction à Guillaume Postel, Le Thrésor des Prophéties de l'univers; Manuscrit publié avec une introduction et des notes* (The Hague, 1969), 13. Cf. id., 'Guillaume Postel', 379.

[55] For a fuller account, see R. Rusconi, 'Circolazione di testi profetici tra '400 e '500: La figura di Pietro Galatino', *Continuità e diversità nel profetismo gioachimita tra '400 e '500 (Atti del 3° Congresso internazionale di studi gioachimiti, San Giovanni in Fiore, 17–21 Sett. 1989*, forthcoming); here only indispensable bibliographical details are given.

Robert d'Uzès, and include the writings of the *Shepherd of Hermas*.[56] These had been published by Jacques Lefèvre d'Étaples in the *Liber trium virorum et trium spiritualium virginum. Hermae liber unus. Uguetini liber unus. Fratris Roberti libri duo. Hildegardis Scivias libri tres. Elizabeth virginis libri sex. Mechtildis virginis libri quinque* (Paris, Estienne, 30 May 1513).[57]

Through this particular printed edition[58] a further link can be forged between Galatino and the cultural circles of Rome, especially that of Egidio of Viterbo, for Jacques Lefèvre d'Étaples had contacted Egidio as a member of the commission charged to pronounce on the disputed Christian interpretation of the cabbala. He wrote to him in Rome in defence of Johann Reuchlin, and Egidio replied to him on 11 July 1516.[59]

This mélange of interests—in the Christian interpretation of the cabbala, in the rediscovery of apocryphal Christian writings, and in visionary, medieval literature—becomes less surprising when seen within its context of aspiration towards religious reform in an ambience in which the interests of modern humanist scholarship mingled with strong elements of medieval forebodings and expectations.[60]

The influence of the *Apocalypsis Nova* on Pietro Galatino must perhaps be considered the most important of all. He always refers to it as the work of its presumed author, the Blessed Amadeus. It is most interesting to note the terms in which the treatise *De ecclesia restituta*—a 'mystica explanatio' focused on the figure of the Pastor Angelicus—affirms the authority of Amadeus: 'Concerning these things, the blessed Amadeus wrote copiously in his book according to the revelation of the angel himself, elucidating the details so precisely, so skilfully, so clearly, that none of the doctors, however learned or great, would have been able to attain to such light of truth' (MS. Vat. Lat. 5576, fo. 313ʳ).

The Bosnian Minorite, Salviati, who played such an important role in

---

[56] Of which he would make ample use also in the *De ecclesia instituta* (MS. Vat. Lat. 5575), see Kleinhans, 'De vita', no. 10, 160–1.

[57] See Secret, 'Guillaume Postel', 379, referring to A. Renaudet, *Préréforme et humanisme à Paris pendant les premières guerres d'Italie (1494–1517)* (Paris, 1916; 2nd edn. 1953). On Lefèvre's edition, see also the *Einleitung* to Hildegarde's *Scivias*, ed. A. Führkötter (Turnhout, 1978) (*Corpus Christianorum, Cont. Med.* xliii), LVI–LIX, with reference to the preceding bibliography.

[58] For a specific copy, used perhaps by Galatino, cf. Rusconi, 'Circolazione', n. 25.

[59] Edited by F. Giacone and G. Bedouelle, 'Une lettre de Gilles de Viterbe (1469–1532) à Jacques Lefèvre d'Étaples (c.1460–1536) au sujet de l'affaire Reuchlin', *Bibliothèque d'Humanisme et Renaissance*, 36 (1974), 335–45.

[60] See Secret, 'Les dominicains', 321; id. *Le Zohâr*, 20.

relation to the *Apocalypsis Nova*,[61] was an exile from the Balkans perhaps like Galatino himself.[62] He died in 1520 in the very period in which Pietro Galatino began the compilation of his prophetic *corpus*. The relationship between Salviati and Galatino was a particularly close one: 'Galatino n'a jamais caché qu'il avait été l'élève et le continuateur de Giorgio Benigno en tant que commentateur de cette *Apocalypsis Nova*.'[63]

A notable stimulus to the distribution of the *Apocalypsis Nova* in ecclesiastic circles was undoubtedly caused by the unrest in the circle of cardinals who had been in a ferment during the last years of the papacy of Julius II (1503–13) and those immediately following. The prophecy was refuted by Paolo Giustiniani and attacked by Cajetan during the repression of post-Savonarolian agitations,[64] but perhaps returned to favour under Clement VII, when, for example, Paolo Angelo incorporated a vernacular version of the fourth *raptus* (on the Angelic Pope) into his *Epistula . . . in Sathan ruinam tyrannidis* (Venice, 1524).[65]

One of Pietro Galatino's contacts with an interest in Amadeus was the Iberian Minorite, Francisco de Los Angeles Quiñones, who became general minister of the new order of Observantines in 1523. As one of his chief protectors, Galatino refers to him in the dedications to the key texts of his own eschatological reflections. Quiñones had assumed the custody of the reformed movement founded by Juan de Puebla. The latter had emphasized the eschatological dimension in the discovery of the New World.[66] Thus, as Adriano Prosperi shows in Chapter 15, when Quiñones became general minister of the Observantine Franciscans in 1523, in his official writings he explicitly set the evangelization of Mexico, begun in 1524, within a prophetic context.[67] This underlines the significance of the fact that he took a copy of the *Apocalypsis Nova* back to Spain from Italy.[68]

[61] See above, Morisi-Guerra, Ch. 2 and Vasoli, Ch. 7. See further, Morisi, *Apocalypsis Nova*, 27–46; Vasoli, *Profezia e ragione*, 85 ff.

[62] See Vallone, 'Pietro . . . Galatino', 105.

[63] Morisi, 'Galatino', 214, 227, with reference to A. Morisi, 'Vangeli apocrifi e leggende nella cultura religiosa del tardo Medioevo. Ricerche sul pensiero teologico di Giorgio Benigno', *Bull. ISI*, 85 (1974–5), 151–77; Vasoli, *Profezia e ragione*, esp. 117–20.

[64] Cf. Vasoli, *Filosofia e religione*, 215–16.

[65] See below, McGinn, Ch. 9. See also Rusconi, 'Ex quodam antiquissimo libello', 467.

[66] See G. Baudot, 'Les missions franciscaines au Mexique au XVIème siècle et les "Douze Premiers"', *Diffusione del Francescanesimo nelle Americhe* (Atti del X Convegno della S.I.S.F.) (Assisi, 1984), 123–52, and bibliography quoted there.

[67] See below, Prosperi, Ch. 15. See also id., 'America e Apocalisse. Note sulla "conquista spirituale" del Nuovo Mondo', *Critica Storica*, 13 (1976), 23–5, quoting the concept of 'the Joachimist alternative to the crusade' (p. 23); St. Lopez Santidrián, 'Quiñones, Francisco de', *DS* 12/2 (Paris, 1986), cols. 2852–3.

[68] Secret, *Introduction à Postel. Le Thresor*, 19 n. 44, relying on the later testimony of Pedro de Alcántara.

There is no doubt, moreover, that Galatino was in close contact with Roman Amadeite circles. This is indicated,[69] in the note on the 'writer' of the *Apocalypsis Nova*, Francesco Biondo, contained in Galatino's *Commentaria in Apocalypsim*: 'As also the blessed Amadeus caused to be written down those marvels, passing the faculty of human intelligence, which the Angel revealed to him in eight *raptus*, only after he came to himself; as Franciscus Blondus, his scribe, prepared by God for him, testified many times.'[70]

'Galatino reveals very clearly the strong impact which the circulation of the *Apocalypsis Nova*, with its vision of the Angelic Pope, produced on those who searched for mystical knowledge in Rome',[71] even if in fact Galatino's announcement that the *Angelicus Pastor* would emerge from the ranks of the Minorites as quoted above, is instead deduced, in the *Vaticinii Romani explicatio*, from the pseudo-Joachimist commentary *super Esaiam* and from the anti-Judaic prophecy attributed to St Cataldus.

Pietro Galatino had further access to the broad prophetic patrimony of the later Middle Ages and the Early Modern Age, through several texts which were available by the first decades of the sixteenth century, not only in manuscript form, but also in print, and sometimes in several editions. This was the case with the *Opusculum de fine mundi* attributed to the Aragonese Dominican Vincent Ferrer,[72] and with the *Revelationes* of Bridget of Sweden—'Brigida vidua'—a text which Galatino cites occasionally at the beginning of the *De arcanis*,[73] and to which he makes a particularly broad reference in the *De ecclesia destituta*. Equally, it is probable that the pseudo-prophecies attributed to 'S. Franciscus ordinis minorum institutor' can be identified with three texts derived from the *libellus* of Telesforus of Cosenza, printed in Venice in 1516, although all three occur in the *De conformitate* of Bartolomeo of Pisa, composed between 1385 and 1390 and published in Milan in 1510 and 1513.[74] In fact a great number of Galatino's quotations from pseudo-Joachimist

---

[69] See Rusconi, 'Circolazione'.

[70] On Francesco Biondo see Morisi, *Apocalypsis Nova*, 32, 35–6; Vasoli, *Profezia e ragione*, 122, and 126–7. For further references here, see Index.

[71] Reeves, 'Roma profetica', 295.

[72] See Rusconi, 'Ex quodam antiquissimo libello', 468; id., 'Il collezionismo', 68–9.

[73] Cf. Vasoli, *Filosofia e religione*, 186 n. 4, and more fully, Rusconi, 'Circolazione'.

[74] For the most recent bibliography on Bartolomeo's compilation, see C. Cargnoni, 'L'immagine di San Francesco nella formazione dell'ordine cappuccino,' *L'immagine di San Francesco nella storiografia dall'Umanesimo all'Ottocento* (Atti del IX Convegno internazionale della S.I.S.F.; Assisi, 1983), 161 n. 105.

literature which had previously been in circulation only in manuscript form are drawn from the *Expositio magni prophete Joachim in librum beati Cyrilli*, the pseudo-Joachimist anthology of 1516 which began with Telesforus's work.[75]

Several cases, however, suggest direct recourse to the manuscript tradition of medieval prophecies, above all where texts had not yet been published. This is the case, probably, with the vernacular prophecies attributed to Tomasuccio da Foligno: 'ut beatus Thomasutius et alii permulti sancti viri praedixerunt', Galatino writes, and links these prophecies to the Turkish menace threatening Rome in the sixth *tempus* (assuming that a presumed printed edition by Gershom Soncino did not exist).[76]

Galatino seems to have had direct access to the manuscript source in one instance at least. This was a 'prophetia mirabilis' recovered 'in quodam libro vetusto'.[77] In the case of several *auctoritates* utilized by Pietro Galatino discrepancies, sometimes considerable, are revealed between Galatino's version and the textual tradition. Therefore it is not possible to determine with certainty which edition—handwritten or printed—he actually used.[78] In the case of the *prophetia S. Severi* transcribed in the treatise *De Ecclesia destituta*, he appears to be quoting from a fuller version than that printed in the *Mirabilis Liber* in 1522 and

[75] On this anthology see B. McGinn, 'Circoli gioachimiti veneziani (1450–1530)', *Cristianesimo nella storia*, 7 (1986), 29–31; Rusconi, 'Ex quodam antiquissimo libello', 465–6. Morisi, 'Galatino', 230 n. 50, notes that Galatino's quotations from the *Vaticinia de summis pontificibus* (in the *De ecclesia restituta* and the *De angelico pastore*) were drawn from this printed edition. To these it is now necessary to add at least the *Liber de Horoscopo* and the *Oraculum Cyrilli cum expositione abbatis Joachim* (in the *De ecclesia destituta*), the *Liber de Flore* (in the *De ecclesia restituta*), and some of the writings of Jean de Roquetaillade.

[76] MS. Vat. Lat. 5579, fo. 38ʳ. See R. Rusconi, *L'attesa della fine; Crisi della società, profezia ed Apocalisse in Italia al tempo del' grande scisma d'Occidente (1378–1417)* (Istituto Storico Italiano per il Medioevo, Studi Storici, 115–18; Rome, 1979), 148–55; id., 'Ex quodam antiquissimo libello', 448, 470; id., 'Il collezionismo', 70. For the presumed Fano edition, see O. Niccoli, 'Profezie in piazza; Note sul profetismo popolare nell'Italia del primo Cinquecento', *Quaderni storici*, 14 (41) (1979), 537 no. 25 (from F. Zambrini, *Le opere volgari a stampa dei secoli XIII e XIV*, 4th edn. (Bologna, 1884), 164–5).

[77] In the treatise *De cognoscendis pestilentibus hominibus* (MS. Vat. Lat. 5579: cf. Kleinhans, 'De vita', no. 14, 163–4), which lacks amongst other things any reference to the *Apocalypsis Nova* or works attributed to Joachim of Fiore. The prophecy begins: 'misericors illis fuit Deus, qui artificiales statuas adorabant' (MS. Vat. Lat. 5579, fos. 73ʳ–74ᵛ).

[78] His quotations from the 'Cyrilli prophetia' do not correspond with the critical edition and therefore cannot be compared with the *Oraculum Cyrilli cum expositione abbatis Joachim* (on which cf. Reeves, *Influence*, 522–3, n. 10, and *ad ind.*). The fact that in this case a MS. may have been used is also suggested by a 'critical note' by Galatino himself: 'Ad horum intelligentia notande sunt quedam glossule interlineares, super quibusdam vocabulis posite, que abbatis Ioakim esse dicunt' (MS. Vat. Lat. 5576, fo. 258ʳ).

earlier still in the Venetian anthology of pseudo-Joachimist texts of 1516.[79] This may be the case elsewhere.

It is interesting, finally, to note that Galatino did not structure his own prophetic compilations entirely in the form of a collection,[80] but that he divided the different sources at his disposal by means of different arguments of his own exposition. This is particularly evident in the case of the *Prophetia Sancti Cataldi*.[81]

Thus we are not here dealing with a mere 'collector' who assembles prophetic texts from different places of origin for his own interpretive ends and who simply arranges them in an anthological manner. Pietro Galatino employed his personal library, which was probably composed mainly of printed texts, in order to extract from it the passages that would fit into his scheme of a prophetic and eschatological interpretation of ecclesiastic history. But his prophetic compilations must be placed after the great period of the Venetian editions of Joachimist works, beginning with the pseudo-Joachimist *Expositio magni prophete Joachim* in 1516 and reaching in 1519 the authentic *Liber concordie Novi et Veteris Testamenti*, headed by a brief extract from the *Expositio in Apocalypsim*.[82] It is possible, therefore, that Galatino was also able to utilize a sort of Joachimist encyclopaedia which contained all the Venetian editions in one volume.[83]

In the general preface to his *Commentaria in Apocalypsim* he makes a long list of his principal Latin exegetic *auctoritates*: St Augustine, Jerome, Bede, Richard of St Victor, Haymo of Auxerre, Albert the Great, Berengar, Pietro di Giovanni Olivi, Hugh of St Cher, Albertinus, Nicholas of Lyra—giving a particular value to the works attributed to Joachim of Fiore: 'Similiter Albertus cognomento magnus, illum mystice et confuse explanat. At Abbas Ioakim ipsum pre ceteris eleganter et spiritualiter elucidat' (MS. Vat. Lat. 5567, fo. III^v).

---

[79] Cf. J. Britnell and D. Stubbs, 'The *Mirabilis Liber*: Its Compilation and Influence', *JWCI* 49 (1986), 134 ff. For the dissemination of prophecies attributed to the mythical archbishop of Ravenna, see Rusconi, 'Ex quodam antiquissimo libello', 451–2, and 'Il collezionismo', 75.

[80] For some examples, see Rusconi, 'Ex quodam antiquissimo libello' and 'Il collezionismo'.

[81] Put into circulation from 1492, widely attested to in the manuscript tradition, printed in Florence in 1497, in Galatino's *De ecclesia destituta* 'the material of the book is divided according to the arguments and not according to sources, and is broken into three parts': G. Tognetti, 'Le fortune della pretesa profezia di san Cataldo', *Bull. ISI*, 80 (1968), 306–11 (quotation on p. 307).

[82] See McGinn, 'Circoli gioachimiti', 19–39; Rusconi, 'Ex quodam antiquissimo libello', esp. 465–6.

[83] For this hypothesis, see Rusconi, 'Circolazione'.

In so doing Galatino was, to a certain extent, limiting the authority of the Calabrian abbot to a purely exegetic level and subordinating his value to that of the 'verum et proprium intellectum', which would coincide with the disclosure of the 'arcana Dei' in the sixth age: 'In the sixth *tempus* of the Church, whose beginning we now witness, all secrets of his book are to be revealed most clearly' (MS. Vat. Lat. 5567, fo. IV[r]).[84]

5. The Years Compiling Prophetic Manuscripts

Pietro Galatino's intensive compilation of prophetic manuscripts took place in the years between 1523 and 1525. This was at the culmination of a particularly concentrated period in the printing of prophetic, cabbalistic, and visionary literature. To mention only the more notable works on the subject, besides the Joachimist editions in Venice,[85] the first edition of the *Mirabilis Liber* was published in Paris in 1522,[86] and the *De harmonia mundi* by the Minorite Francesco Zorzi, was written and published in Venice in 1525.[87] There were also collections of mystic and visionary texts similar in kind to the *Liber* edited by Jacques Lefèvre d'Étaples in 1513. In 1522 a philosopher of the Scotist school, Antonio de Fantis, published in Venice a *Liber gratiae spiritualis visionum et revelationum beate Methildis virginis devotissime ad fidelium instructionem* which, besides the work of the thirteenth-century Rhenish mystic Mechtild von Ackerborn, gives additional short texts: three apocryphal, the *Gospel of Nicodemus*, the *Epistula Lentuli* and the *Visio Isaiae*,[88] an otherwise unknown *Visio sancti Alberti episcopi Agrippinensis*, and finally the pseudo-Joachimist *Vaticinium Sibillae Erithreae* 'in orthodoxe fidei testimonium'.[89]

The theme that seemed to attract Pietro Galatino's interest initially was the reform of the church, in fairly traditional forms. In the period between the death of the Emperor Maximilian I in January 1519 and the election of his successor Charles V in the month of July, the elderly

---

[84] For this, see above all Reeves, *Influence*, 234–8, 442–6, summarized in 'Roma profetica', 291–6.

[85] See Rusconi, 'Ex quodam antiquissimo libello', 459–67.

[86] Britnell and Stubbs, 'The *Mirabilis Liber*', 126–49.

[87] See Vasoli, *Profezia e ragione*, 129 ff.; id., 'Da Marsilio Ficino a Francesco Giorgio Veneto', *Filosofia e Religione*, 233–56.

[88] The title-page reads: 'Visio mirabilis Ysaie prophete in raptu mentis, quae divine Trinitatis archana et lapsi generis humani redemptionis continet.'

[89] See A. Acerbi, 'Antonio de Fantis, editore della *Visio Isaiae*', in his '*Serra Lignea*'. *Studi sulla fortuna della 'Ascensione di Isaia'* (Rome, 1984), esp. 167–75 (also published in *Aevum*, 57 (1983), 396–413).

theologian, now nearly sixty years old, drew up a *Libellus brevissimus de republica christiana pro vera eiusdem reipublicae reformatione, progressu ac felici ad recuperanda Christianorum loca expeditione*, which he would dedicate to Leo X shortly before his death on 1 December 1521, although he later corrected the name of the recipient to that of the new Pope, Adrian VI (1521–3), whose election had been energetically sponsored by the house of Hapsburg.[90] In a text abounding with rhetoric, in which the 'corpus mysticum' of the Church is forcefully presented on the basis of a conception already present in Galatino's early work, the *De optimi principis diademate* and also of more recent debates in the Fifth Lateran Council,[91] the laments of the moralist when confronted with the corruption of the age are resolved in the entirely foreseeable evocation of the *renovatio ecclesiae*. Of this the most auspicious manifestation would be an anti-Turkish crusade led by Charles V, the worthy successor to the Imperium, as prophesied by Joachim of Fiore and the pseudo-Methodius.[92]

Also the original writing of the *De cognoscendis pestilentibus hominibus* probably dates back to the beginning of the 1520s, immediately after the death of Salviati. It was dedicated some years later to Cardinal Andrea della Valle, Protector of the new order of Friars Minor of the Observance.[93] Placed between the *De arcanis* (which he took up again to develop the historic relevance of the model offered by the true Messiah) and the exegetic writings which followed, this text forms a speculative treatise on the two eschatological figures of the Messiah and the Antichrist in a key which is more penitential than millenarian, stressing the need for moral preparations to resist the evil of the final tribulations. It is a warning to the elect to recognize the precursors of Antichrist ('pestilentes homines') and to flee from their sins in the imminence of persecution.

In the attempt to fix the 'tempora de huiusmodi Messia determinata, et simpliciter praenunciata', Galatino embarks on a series of chronological calculations (MS. Vat. Lat. 5579, fo. 91ᵛ–93ʳ), formulated by making reference to the pseudo-Joachimist *Vaticinium Sibillae Erithreae*, and beginning from the conquest and destruction of Jerusalem by Titus (AD

---

[90] MS. Vat. Lat. 5578, fos. 86ʳ–106ᵛ (cf. Kleinhans, 'De vita', no. 13/B, 162), ed. Perrone, 'Il *De re publica christiana*', 609 ff., where it is described as a MS corrected in the author's hand and dated correctly (p. 520).

[91] Thus Perrone, 'Il *De re publica christiana*', 560; cf. D'Amico, *Renaissance Humanism*, 219.

[92] See Reeves, *Influence*, 443; id., 'Roma profetica', 293.

[93] MS. Vat. Lat. 5579 (see Kleinhans, 'De vita', no. 14, 163–4; for the date, see p. 328).

70). First the passing of 'circiter' 1450 or 1452 or 1453 years is indicated and later repeatedly corrected—even erased—to 1457, which in the end would make the 'hodiernus tempus' coincide with 1524, and the date of the awaited events with 1527—after the prophetic three years traditionally foreseen in the apocalyptic tradition.

Without explicitly identifying the 'praecursores' of Antichrist with the Lutherans, Galatino maintains that the relaxation of religious orders and the corruption of customs are contrary to the model put forward by the true Messiah, and characteristic instead of the followers of Antichrist. He speaks in praise of 'voluntaria paupertas', the form of perfection represented by the 'altissima paupertas' of the Minorite religion.[94]

By the middle of the following year Pietro Galatino had drawn up another text, *De Sacra Scriptura recte interpretanda opus*, a *libellus* originally dedicated to the King of England, Henry VIII, after the latter had despatched his pamphlet against the sacramental doctrine of Martin Luther to Pope Leo X in 1521.[95] From the end of the dedication it appears that Galatino planned to proceed to a unitary treatise on the problems of the future of the Church and of its *reformatio* at the hands of the Angelic Pope, with the collaboration of the Last World Emperor. This was to be subdivided into a series of volumes to the intensive writing of which he was now dedicated:

We have given few examples of the aforesaid institutions because in our other books, particularly those on the Angelic Pastor, on the seven times of the Church, on the *ecclesia instituta*, the *ecclesia destituta* and the *ecclesia restituta*, on recognising the pestilent men and unmasking their craftiness, and in commentaries on the Apocalypse of the Apostle John, many other examples of the same things will be given, more suitably placed.

(MS. Vat. Lat. 5580, fo. CXXVII')

The overall key to his exegesis, which was used in all his works, is here clearly laid bare: the secret lies in the revelation of the 'arcana Dei' to the elect, as suggested in the title, *Ostium apertum*.[96] This will take place in the imminent sixth age of the Church, with the coming of the Angelic Pastor: 'In the sixth time of the Church, in which the Church itself must be restored completely to the perfection of its pristine

[94] MS. Vat. Lat. 5579, fos. 100ʳ–101ʳ.
[95] MS. Vat. Lat. 5580 (Kleinhans, 'De vita', no. 15, 164–6). The pamphlet was brought to Henry in 1526 by the Papal Nuncio in England, Gerolamo Ghinucci, Bishop of Worcester, but after the Anglican Schism in 1534 the entire initial rubric was erased from the MS.
[96] See Morisi, 'Galatino', 219.

state, everywhere the door will be opened, so that there will be a full
revelation of all those secrets which lie hidden in the sacred writings'
(MS. Vat. Lat. 5580, fo. XLI^r).

Without dwelling on the chapters of this pamphlet in which Galatino
expatiates on the personal concept of the 'intellectus sive sensus spiritalis',
we must note the topical relevance of his representation of the Precursor:
'Therefore it is sufficiently shown from the above, as I think, that where
the sacred Scriptures recall John the Baptist, besides the principal sense
appropriate to John himself, the *spiritalis intellectus* of the Old Testament
or sacred page is to be understood, as it were the Old Testament itself'
(MS. Vat. Lat. 5580, fo. LXXXV^v).

The Precursor typed as the Baptist would strike a familiar note, for
these were precisely the years in which itinerant hermits journeyed
through the Italian peninsula preaching penitence in the face of the
end of the world and assuming, even outwardly, the 'iconographic'
appearance of the Precursor.[97]

The purpose of the treatise *De ecclesia destituta*, the writing of which
took place between the preceding works and the *Commentaria in
Apocalypsim*, and which was conceived in strict relationship with the
other treatise, *De ecclesia restituta*,[98] is on a different level—that of
delineating a picture of the decadence of the Church and of determining
the real 'deformatio'. Here he uses a 'mystica explanatio' (always contrasted with the 'litteralis interpretatio') of some of the psalms, of
some chapters from the four major prophets—Isaiah, Jeremiah, Ezekiel,
and Daniel—of the eschatological sermon in St Matthew 24, of
Apocalypse 10, and—an important novelty—'ex diversorum servorum
Dei vaticiniis'.[99] Galatino was most preoccupied over the philological
value of his interpretation, and referred back in successive stages to the
Hebrew text of the Scripture ('hebraica littera'), annotating passages of
the Old Testament as the basis of his exegesis. Wherever the Latin
Vulgate text does not appear adequate, it is necessary to turn to the
'hebraica veritas'.[100]

---

[97] See Niccoli, *Profeti e popolo*, esp. 125–32; B. Nobile, '"Romiti" e vita religiosa nella
cronachistica italiana fra '400 e '500', *CS* 5 (1984), 303–40. Reeves, *Influence*, 448, cites a
'John the Baptist Hermit' active in 1525.

[98] MS. Vat. Lat. 5568–9 (Kleinhans, 'De vita', no. 3–4, 153–7) and MS. Vat. Lat.
5576 (Kleinhans, 'De vita', no. 11, 161).

[99] On the significance of this last expression, see Morisi, 'Galatino', 219 and n. 31.

[100] See ibid., 215, for the importance Galatino attached to this expression. His philological views prompted a fairly critical attitude towards Erasmus whom he criticized
explicitly (see, for example, Reeves, 'Roma profetica', 291, and the excursus in MS. Vat.

Previously untapped sources of inspiration appear, moreover, to be central in this text, not so much the *Vaticinium Montis Gargani* as the *Vaticinium Romanum* (this last is only usually quoted in marginal rubrics without ever making clear reference to his own *Explicatio* as already drawn up) and above all the works attributed to Joachim of Fiore ('abbas Ioakim qui prophetiae spiritu claruit'; MS. Vat. Lat. 5568, fo. IX^v), and to the blessed Amadeus (that is, the *Apocalypsis Nova*). These two personages seem at this point to have an equal authority in his eyes (cf. MS. Vat. Lat. 5568, fo. 79^r). But at the end of the *De ecclesia restituta* he states explicitly that between the two unique *auctoritates* on whom he founds his own eschatological ecclesiology, 'qui adeo veridici sunt, ut eorum testimonium nemo inficiari audeat' (MS. Vat. Lat. 5576, fo. CCCXII^r), there exists a difference of notable importance. Of Joachim of Fiore, to whom he refers in attributing to him the composition of the commentary *super Esaiam*, he observes: 'Of [commentators], Ioakim Abbas Florensis, endowed with the spirit of prophecy, foretold many things concerning the Church, as much in its destitute state as when restored'. But as we have seen, he confers a much greater authority on the author of the *Apocalypsis Nova*.[101]

There is no need to dwell further on the dark picture of the ineluctable decadence and corruption of the Church which Galatino traces with the indignant countenance and moralism of all ecclesiastic reformers. The figure of the Angelic Pastor now returns to a central role for Galatino who rereads the texts of the prophet Ezekiel in this light: 'The Angelic Pastor, sent by God Himself to reform the universal Church must be exalted'—the title of one chapter of the treatise (MS. Vat. Lat. 5568, fo. CCCXXXIII^r).

In the other treatise, *De ecclesia restituta*, a natural continuation of the *De ecclesia destituta*, he proposes to furnish a 'mystica explanatio' of texts in the Old and New Testaments: of the psalms of David, of the prophets Isaiah, Jeremiah, and Ezekiel, of Matthew 24, of the Apocalypse, and of the 'testimoniis [. . .] servorum Dei, qui prophetiae spiritu claruerunt', and to focus finally on the eschatological role of the Angelic Pastor, that

---

Lat. 5567, fos. XXVII^v–XXIX^r). His attitude on this subject can be traced back to Egidio of Viterbo, who, 'attributed a special *status* to the Holy Scripture, recognising in it a sort of philological extraterritoriality, not only in its original language but also in the version sanctioned by tradition, such as the Vulgate' (S. Seidel Menchi, *Erasmo in Italia, 1520–1580* (Turin, 1987), 41–72 quotation p. 44). For the philological roots of Galatino's anti-Lutheran attitude, see below, n. 119.

[101] See above, pp. 170, 175.

is 'through a true reformation to restore the Church to its primitive state' (MS. Vat. Lat. 5576, fo. CCLXXXIII[r]).[102]

Galatino gathers together the 'testimonia Servorum Dei' relating to the Angelic Pope in the fifth book in order to determine from them the date and place of his birth, the date of his public appearance, the religious order to which he belongs, his Christian name and surname, the physiognomy of his face, the unusual way in which he is appointed, his personal qualities after his accession to the papacy, and those of his successors and the ecclesiastical prelates of his era.

As well as constituting a first step in the progressive development of the expectation of the Angelic Pastor and in the identification of the supernaturally designated pontiff with his own person,[103] the *De ecclesia restituta* is marked by a notable interest in the eschatological role assigned in emphatic terms to the new Emperor Charles V, identifying him without any hesitation with the Last World Emperor, 'totius ecclesiastice reformationis executore' (MS. Vat. Lat. 5576, fo. CCXCVI[v]). Galatino traces back this prediction to the eighth *raptus* of the *Apocalypsis Nova*,[104] as well as to the 'Cyrilli prophetia' and to a prophecy attributed to Bridget of Sweden.

To Quiñones, who had been elevated to the cardinalate in 1527, Pietro Galatino despatched a sumptuous copy of another work, *De septem ecclesiae tum temporibus tum statibus opus*, which had originally been written in 1523—the year of Pope Clement VII's succession to Adrian VI—and conceived as an 'ostium quoddam, ad mysteriorum eiusdem apocalypseos intelligentiam, patentissimum praebens aditum'. The reference here is to the more voluminous *Commentaria in Apocalypsim*, which Galatino had already sent to Quiñones, imploring the prelate to have both of them printed.[105]

In this treatise Galatino begins by reviewing the subdivision of ecclesiastical history into seven *tempora* (which correspond to the seven *status* and which are assimilated in the seven ages of the world), pausing particularly at the fifth age, held to be 'manifeste deformationis'. To remedy this, God had sent the 'religiosorum mendicantium predicatio',

---

[102] In *De ecclesia instituta* (MS. Vat. Lat. 5575: Kleinhans, 'De vita', no. 10, 160–1), *visiones e similitudines* from the *Shepherd of Hermas*, published by Jacques Lefèvre d'Étaples in 1513, are used to outline the duties of the Angelic Pope and the structure of the Church at the time of the eschatological *reformatio*; cf. Morisi, 'Galatino', 218–19.

[103] Cf. Vallone, 'Pietro . . . Galatino', 101.

[104] MS. Vat. Lat. 5576, fos. CCXCVII[r] and CCCVII[r].

[105] MS. Vat. Lat. 5579 (Kleinhans, 'De vita', no. 14/A, 163. The 'praefatio' is reported there, fo. i[v]). On this text, see Morisi, 'Galatino', 220, 222.

prophesied by Joachim of Fiore (more exactly, by the *super Esaiam* and by the *vaticinium* of the Erithrean Sybil) (MS. Vat. Lat. 5579, fos. 6ᵛ–7ʳ).

Even these orders did not succeed in escaping the ecclesiastic decadence of the fifth *tempus* (MS. Vat. Lat. 5579, fo. 9ᵛ), at least until the beginning of the sixth age, 'perfecte reformationis': 'For at the end of the fifth time . . . all corruption in the carnal Church must be consumed in the fire of great tribulation, so that at the beginning of the sixth time the Church itself, through the Angelic Pastor, like a phoenix, will be made new within, while the whole world will be converted to Christ' (MS. Vat. Lat. 5579, fo. 10ʳ).

In order to disclose the imminent reform of the Church through the work of an Angelic Pope, with the Last World Emperor, Galatino refers to the revelation of the Archangel Gabriel to the blessed Amadeus contained in the *Apocalypsis Nova*. In discussing the time scale of these events Galatino openly parts company with the views of Joachim as expressed in the *Liber de concordia*. For the sixth and seventh *tempus* it is necessary to forecast a 'multo longioris durationis curriculum' than that which the Calabrian abbot had done. A sense of urgency that has not previously been found in Galatino's works appears in the text: 'Thus [the signs] show that we are close to the end of the fifth time, so that now we seem undoubtedly to be at the beginning of the sixth, or even at its gates' (MS. Vat. Lat. 5579, fo. 12ʳ).[106]

The development of the present and future history of the Church is outlined with great clarity. In the fifth *tempus* the prophecy of the Mendicant orders, in particular of the Franciscans, had an eschatological meaning which has been progressively lost on account of their decadence: 'religious mendicants, and especially Minorites, who above all others stood for evangelical perfection, multiplied in the fifth time, of whom, although very many in number, almost all fell into such a multitude of crimes that there were very few among them who did not become worse than secular clergy' (MS. Vat. Lat. 5579, fo. 26ʳ). The sixth *tempus* instead is now the era for the advent of an 'Angelic Pastor', the appearance of an 'evangelicus ordo', and the foundation of the 'ecclesia reformata' (MS. Vat. Lat. 5579, fos. 26ᵛ ff.). The best assets of the

---

[106] There is no clear-cut distinction between *tempora* and *status* in the Church: 'Sic dubio procul et sextum statum, antequam quintus omnino deficiat, et septimum ante sexti desitionem, inchoatum iri, necesse est' (MS. Vat. Lat. 5579, fo. 15ᵛ), but the sixth must be 'plus ceteris duraturus', while the seventh will be 'brevissimum omnium' (MS. Vat. Lat. 5579, fo. 16ᵛ).

Church in this work will be the collaborators of the pontiff, and in particular 'Twelve most holy men, on the pattern of the twelve apostles, called to preach again throughout the world the worship of Christ' (MS. Vat. Lat. 5579, fo. 31ᵛ).

The whole programme is traditional in character including the eschatological collaboration between Pope and Emperor: 'Under the Angelic Pastor and the Emperor whose aid the said Pastor will invoke in the universal reformation, first the Temple will be rebuilt and afterwards the whole city of Jerusalem, for the reformation will begin from the Roman Church as head, by whose example afterwards it will be diffused through the rest of the world's churches' (MS. Vat. Lat. 5579, fo. 30ᵛ).

The defeat and final conversion of the Turks, whose sovereign will be baptized in Rome by the Angelic Pope follows the same pattern (MS. Vat. Lat. 5579, fos. 36ʳ–38ᵛ). Against the last Antichrist, whose persecution will endure for three and a half years, God will send 'duo ordines praedicatorum' and the two 'testes Christi fideles amictos saccis' prophesied in the Apocalypse (MS. Vat. Lat. 5579, fos. 40–41ʳ). The death of Antichrist will be followed, 'haud multo tempore interiecto', by the Parousia, with the final conversion of all peoples. The indeterminate nature of these expectations, however, is underlined by a note on the Angelic Pastor: 'O blessed [shall I be], if before I die, I am worthy to see with my own eyes this most holy one' (MS. Vat. Lat. 5579, fo. 31ᵛ). Here, it would seem, Galatino is not identifying himself with this prophetic figure.

Although Galatino gives pride of place to Joachim among his authorities and can be said to have adopted the Joachimist framework of reference, drastic re-adaptations of the Joachimist eschatological scheme can be found in his works.[107] We have already seen this in his *De septem Ecclesiae temporibus*. We meet the same thing in relation to the Angelic Pope.

In the pamphlet *De Angelico Pastore*, written during the period in which Pietro Galatino was fervidly compiling his own prophetic and eschatological writings, he summarizes, in a sort of 'excerptum', all that has already been stated on this subject in his other works. He himself declares this explicitly in the preface, while in the title it is defined, on the contrary: 'ex sacra veteris et novi testamenti scriptura excerptum'.[108]

---

[107] Reeves, *Influence*, 235–6, where the problem is tackled analytically; also id., 'Roma profetica', 292–3.

[108] MS. Vat. Lat. 5578: Kleinhans, 'De vita', no. 13/A, 162, dating it 'versus finem vitae'. The revision of this chronology had already been suggested by Reeves, *Influence*, 444 n. 2; cf. now Vallone, 'Pietro . . . Galatino', 98.

Set out in sixteen chapters, the pamphlet proposes to illustrate the way in which the *reformatio ecclesiae* will be carried out by an Angelic Pope who has been prefigured by some of the personages from the Old Testament. There would be nothing unusual in this if the influence of the *Apocalypsis Nova*—a text which had become all but canonical on the subject of this eschatological pontiff—did not allow him to make a further deduction in which not only the Old Testament prophecies usually applied to the figure of the Christ/Messiah are referred to the Angelic Pastor, but the author of the *Apocalypsis Nova* himself 'his favourite Amadeus',[109] is held to be a second John the Baptist, the Precursor who must prepare for the coming of the Holy Pope:

But I think the Precursor to have been the blessed Amadeus, who, while [the Pastor himself] was shown by the angel Gabriel in a vision, declared him to be (already) born, educated and now a young man, as it were pointing a finger at him like a second John the Baptist and whatever [the Pastor] will teach, he announced beforehand from the revelation of the angel and caused to be written, leaving his book closed according to the angel's command, to be opened by no one except the Angelic Pastor, that is, to be communicated through a true interpretation. From which we deduce that this most holy Pastor, whom the sacred prophets call angelic, will be manifest very shortly. May I be worthy to see him before I pay the debt of nature and commend my body to the earth. (MS. Vat. Lat. 5576, fo. 137^v)[110]

In appearance, and because of its abridged form, this 'opusculum' lacks the burning actuality which is characteristic of the 'explicatio' of the *Vaticinium Romanum*.[111] This work presupposes a sense of the imminence of the last age, of the 'novissimus dies'. Galatino identifies this with the seventh age of the Church which, after the death of the last Antichrist, will last until the end of the world (MS. Vat. Lat. 5578, fo. 17^r; cf. fo. 18^v). In the imminence of these events, 'circa sexti temporis initium, in quo iam sumus' (MS. Vat. Lat. 5578, fo. 45^v), the announcement of the coming of the Angelic Pope has a purpose which is more eschatological and penitential than apocalyptic: 'Because the Angelic Pastor will announce beforehand precisely and definitely the time when the end of the world will come, so that sinners may be more willing to

---

[109] Reeves, *Influence*, 444.

[110] For this passage from the *De ecclesia restituta*, see Morisi, 'Galatino', 223.

[111] See the passage quoted in Reeves, *Influence*, 445: 'quoniam cito appariturus est, qui non solum (ut in Romano vaticinio praedicitur) orbem universum novis sanctisque legibus moribus et institutis reformabit; verum etiam (ut beato Amadeo angelus revelavit) et mundum ipsum a cunctis purgabit erroribus.'

return to God through repentance, and ought to reform themselves more perfectly, as they expect the Last Day more nearly' (MS. Vat. Lat. 5578, fo. 16ʳ).

To this Angelic Pope, whose advent will be confirmed now 'in sexto ecclesiae tempore', will fall the task of inaugurating the *tertius status* of history, according to Galatino. In the course of this era the *arcana Dei* will be revealed and declared to the whole Christian people: 'That angelic pontiff will not only possess true understanding of the hidden things of God in the sacred texts, but will draw the faithful people also to the study of the sacred words, introducing them by wonderful and unheard of rules' (MS. Vat. Lat. 5578, fo. 54ʳ).[112]

The transformation of the *ecclesia carnalis* into an *ecclesia spiritualis* will take place in Rome with the assistance of a 'princeps saecularis'. The world will be led back to unity by two vicars of Christ, the 'Angelicus Pastor' and the 'Imperator Fidelissimus' (MS. Vat. Lat. 5578, fo. 41ʳ). The Gospel, and with it universal peace, will be preached all over the world, which will include the new territories discovered on the far side of the Atlantic: 'quum multae insulae longe post reperta sint, in quibus diversae habitant gentes, de quibus nulla tunc extabat noticia' (MS. Vat. Lat. 5578, fo. 43ʳ).

As a writer who was, without doubt, profoundly influenced by the Venetian Joachimist editions, Pietro Galatino's great innovation was to transfer the ecclesiastical centre of gravity from the future of the Church to the coming of the Angelic Pastor, which became the central exegetic motive of his eschatological-apocalyptic theology.[113] The circle was completed in 1525, in the *Vaticinii Romani explicatio*, when Pietro Galatino started to identify himself with the figure of the Angelic Pastor, though this constitutes an episode which ultimately appears totally eccentric, and as a consequence almost incidental.[114]

With the passing of time Pietro Galatino has been portrayed in various studies, first as a crude version of the cabbalistic esotericism of Egidio of Viterbo and his circle: 'nous présente, grossis et déformés bien des caractères communs aux esprits du temps'[115]; secondly as the 'perfect' Joachite of the first decades of the sixteenth century: 'one realises with

---

[112] See Reeves, *Influence*, 443, comparing the 'spiritalis intellectus' of Galatino's *Commentaria* (MS. Vat. Lat. 5567, fo. XCVIᵛ), with Joachim of Fiore's *Liber Concordie*.

[113] See Reeves, *Influence*, 442–6, comparing Galatino and Joachim.

[114] Morisi, 'Galatino, 223 ff., for the *De vera theologia* as the culmination of Galatino's speculative reflection.

[115] Secret, Introduction to Egidio of Viterbo, *Scechina*, 13.

astonishment that here is a fully-fledged sixteenth-century Joachite. It is not that he directly quotes Joachim in some works (. . .) but that he has assimilated so much from Joachim, whether he acknowledges the source or not'[116]; and finally almost as a person of thoughtless syncretism: 'Galatino's many works intermingled a variety of themes; Joachimism, the mystical metaphysics of Saint Bonaventure, and a deep study of Hebrew and the cabala.'[117]

The influence of the *Apocalypsis Nova* on Pietro Galatino takes chronological priority and should perhaps be considered the most important of all, particularly on a central theme of his exegetic and theological thought, the temporal progressiveness of the divine revelation.[118] This allows for a sort of extension of the Joachimist-based eschatological-apocalyptic system and helps to explain the transition from a more or less casual interest in the Messianic relevance of the cabbala, to the expectation of the Angelic Pope.

In his polygraphic production, which remains in manuscript form, there appear traces of a training and of interests which are at the same time more coherent and more complex than those of the personages and circles with whom he has from time to time been associated.[119]

6. *The New Hopes and the Twilight*

'The old Franciscan never ceases to await the time of the *renovatio* when the "spiritual" Church will have defeated the "carnal" church',[120] even though the fervid eschatological expectation of the Angelic Pope which he had cultivated in the mid-1520s had not been fulfilled in actuality.

Pietro Galatino continued to be well placed in Roman curial circles in

---

[116] Reeves, *Influence*, 235.

[117] D'Amico, *Renaissance Humanism*, 218.

[118] See the passage from the *De sacra scriptura* (MS. Vat. Lat. 5580, fo. LVIII[r–v]) referred to in Morisi, 'Galatino' 218, and the relevant comment.

[119] See, for example, the extent to which Galatino has still been bound to late 15th-cent. polemics regarding astrology and the Christian doctrine of free will (MS. Vat. Lat. 5579, fo. 51[r], in margin: 'Quod in libro, quem contra astrologos prognosticantes (Deo favente) edituri sumus, latius probabimus.' The reference is in fact to bk. 8 of the third part of *De vera theologia*; see Morisi, 'Galatino', 229 n. 31). Moreover, his writings reveal the rigidly anti-Lutheran attitude characteristic of the circles which, at the beginning of the modern age, were cultivating eschatological prophetism of a medieval type (cf. Niccoli, *Profeti e popolo*, 179–86). However, Galatino's interest in Luther was only marginal compared with that of contemporary Italian polemists (cf. Morisi, 'Galatino', 230 n. 44); the principal root of his dissent is in the rejection of the 'spiritale scripturae intellectum' on the part of the 'lutherani' (see the passage cited by Morisi, 'Galatino', 229 n. 26).

[120] Vasoli, *Filosofia e religione*, 209.

the 1530s, and from the year 1532 refers to himself as 'poenitentiarius apostolicus' both in the dedication of the *De SS. Eucharistiae sacramenti mysterio* to Cardinal Quiñones—to whom, in the mean time, he had become chaplain—and in a series of works which can be dated from the beginning of the pontificate of Paul III.[121] He must have assigned a great significance to the accession of Cardinal Alessandro Farnese to St Peter's Chair in 1534,[122] and this may have caused him to resume his cabbalist interests. In fact he dedicated the third volume of his ecclesiastic eschatology, the *De ecclesia instituta* to Paul III, beginning with a florid eulogy and concluding: 'Therefore we hope greatly that this reformation will be carried out by the work of your Holiness, God willing. For this reason I have dedicated this book to your Holiness' (MS. Vat. Lat. 5575, fo. 5^r).

The pontificate of Paul III (1534–49) was characterized by a definite resumption of expectations for the reform of the Church in an eschatological key, starting with a renewed interest in the prophecy of the Angelic Pope as contained in the *Apocalypsis Nova*.[123] Once again, ten years on, various different hopes for a non-political, non-institutional ecclesiastic *reformatio* start to re-emerge in print, or to be reprinted. For example in 1536, Luca Bettini's *L'Oracolo della renovatione della chiesa secondo la dottrina del Riverendo Padre Frate Hieronimo Savonarola*, first published in 1510, was reprinted in Venice.[124] In Venice in the same year, published by the same printer, there appeared two vernacular editions of Savonarola's *Compendio di rivelazioni*.[125]

In the 1530s Pietro Galatino undertook a sort of frenetic, definitive compilation of all his theological material. In spite of his advanced age— between 70 and 80—he was persuaded by Cardinal Paolo Capizucchi to prepare his cabbalist works for publication: a new edition of Paulus de Heredia's *Epistola de secretis*, which he finished between the end of September and the middle of October 1536, and dedicated to Pope Paul III in 1539, together with an 'index alphabeticus ab ipso authore nuperrime confectus' of the *De arcanis Catholicae veritatis*, which he

---

[121] MS. Vat. Lat. 5580 (Kleinhans, 'De vita', no. 15, 166). Cf. Morisi, 'Galatino', 225. The pamphlet consists of a long extract from the sixth *raptus* of the *Apocalypsis Nova* into which some Talmudic sources have been inserted.

[122] See Reeves, *Influence*, 445; 'Roma profetica', 296.

[123] See Secret, *Introduction à Postel, Le Thrésor*, 17.

[124] See Prosperi, 'Teodoro', 78 n. 19; Reeves, 'Roma profetica', 296 n. 106.

[125] A. Schutte, *Printed Italian Vernacular Religious Books, 1465–1550: A Finding List* (Geneva, 1983), 335.

evidently aimed to have reprinted.[126] In the same year, 1539, the Pope authorized Pietro Galatino to deposit all his voluminous manuscripts in the library of the Franciscan convent of Ara Coeli where they would have remained unpublished and buried forever had it not been for the impassioned reading of Guillaume Postel.[127]

[126] See MS. Ottobon. 2366, fo. 300ʳ–303ᵛ; cf. the brief description by Kleinhans, 'De vita', no. 18, 171–2 and no. 17, 171, for MS. Vat. Lat. 4582, connected to it. For references, see Scholem, 'Considerations', 34–5; Morisi, 'Galatino', 213; Vasoli, *Filosofia e religione*, 209. There is no evidence of the Cardinal's specific interest, see Fragnito, 'Capizucchi, Paolo,' *DBI* 18 (1975), 571–2.

[127] For the influence on Postel of reading Galatino's works and the *Apocalypsis Nova*, see his *Retractationum liber* (1560), quoted by Secret, 'Guillaume Postel et les courants prophétiques', 391. See also id., *Introduction à Postel, Le Thrésor*, 15–17; Vasoli, *Filosofia e religione*, 226–9; id., 'Postel, Galatino e l'*Apocalypsis nova*', *Guillaume Postel, 1581–1981, Actes du Colloque International d'Avranches 5–9 Sept. 1981* (Paris, 1985), 97–108.

9

Notes on a Forgotten Prophet: Paulus Angelus and Rome

Bernard McGinn

Three cities have shared apocalyptic honours in the Christian tradition—Jerusalem, Babylon, and Rome. The roles of the first two were set forth in the Apocalypse of John. Rome, although not mentioned expressly by John, was clearly indicated under the code name of Babylon, the great persecutor, the harlot seated upon the beast at whose downfall kings and merchants will weep but the saints rejoice. The Apocalypse's denunciation of Rome as Babylon, however, was directed against pagan imperial Rome. Need it be true of Christian Rome? The conversion of the empire to Christianity and the growth of papal Rome as the centre of the Western Church provided the city on the Tiber with the possibility of a more positive symbolic value in the history of Christian apocalypticism. Rome came to be perched ambivalently between Jerusalem and Babylon, at times taking on the role of a preliminary or alternate Jerusalem, but often still maintaining a link with evil Babylon. At no time in Western history was this dual view of papal Rome more obvious than in the sixteenth century when Martin Luther and his followers, building upon late medieval identifications of a coming or present false Pope with Antichrist, proclaimed the very institution of the papacy, and by metonymy its capital, as the final enemy of all true Christians, while partly by way of reaction, but also in continuity with late medieval hopes for an imminent final holy Pope or Popes (the *Pastor Angelicus*), papal propagandists emphasized the city's role as the centre of eschatological renewal.[1]

---

[1] On this dual view of the papacy, see above, Reeves, Ch. 1; also B. McGinn, 'Angel Pope and Papal Antichrist', *Church History*, 47 (1978) 155–73. For rich materials on the *Pastor Angelicus*, see M. Reeves, *Influence of Prophecy in the Later Middle Ages: A Study in Joachimism* (Oxford, 1969), *passim*.

In the first quarter of the sixteenth century, late medieval apocalyptic ideas were decisively tested by their confrontation with a rapidly changing religious and political situation. The discovery of America and the outbreak of the Reformation were just two of the major new factors that cried out for apocalyptic interpretation. The decade of the 1520s saw a pitch of excitement, as hopes and fears for the final days grew strong. The elevated but still ambiguous role of Rome in these expectations is an important part of sixteenth-century apocalypticism. Prophetic warnings of judgement on the Popes and their city were overtaken by events in 1527 when the city was captured and cruelly sacked by the imperial army under Charles de Bourbon.[2] But Rome survived; its ambivalent role in Christian apocalypticism was to continue to elicit varied reactions among prophets and pundits for centuries to come.

Many famous sixteenth-century writers, both Catholic and Protestant, have left us their reflections on the role of Rome and the Popes in the final events. Some, like Martin Luther, are well known; others, like Egidio of Viterbo, cardinal, reformer, and polymath, though less famous today than in their own time, still merit detailed study. There are other figures of more restricted appeal and influence, but whose careers and ideas may still have something to tell us about the apocalyptic atmosphere of the times. Paulus Angelus is one of these.[3]

Paulus Angelus claimed to be a descendant of the Byzantine imperial family of the Angeli.[4] His father, Peter, an exile from the 'turcharum accerbissima rabies', fled his native Albania for Venice where he fought for the Maritime Republic against the Turks and raised his family in deep piety.[5] We do not know when Paulus was born, but according to G. Tognetti, following F. Pall, he obtained a parochial benefice in

---

[2] See A. Chastel, *The Sack of Rome*, trans. B. Archer (Princeton, NJ), Ch. 2.

[3] Little has been written on Paulus in modern times. There are several references in Reeves, *Influence*, see index. More extensive accounts are given in F. Secret, 'Paulus Angelus descendant des empéreurs de Byzance et la prophétie du Pape Angélique', *Rinascimento*, 2nd ser. 2 (1962), 211–24; G. Tognetti, 'Note sul profetismo nel Rinascimento e la letteratura relativa', *Boll. ISI*, 82 (1970), 15–54, 129–57; B. McGinn, 'Circoli gioachimiti veneziani (1450–1530)', *Cristianesimo nella storia*, 7 (1986), 36–7. Unavailable to me, though summarized by Tognetti, are the biographical and bibliographical details found in F. Pall, 'Marino Barlezio. Uno storico umanista', *Mélanges d'histoire générale* (dell'università di Cluj), 2 (1938), 135–315.

[4] Paulus Angelus, *In Sathan ruinam tyrannidis* (Venice, 1524), fo. 3ʳ 'Paulum cognomento Angelum natione Romanum, Patria Drivastensem, ac Brianensem vel Raquilinum, et profugum de stirpe olim antiquiorum imperatorum Byzantinorum commorantem adhuc sic in Dierum Tarvisina' (my transcription and numeration of folios follow the Vatican copy; I have expanded abbreviations and modernized punctuation, leaving spelling as in the text).

[5] *In Sathan*, fos. 15ᵛ–16ʳ. Cf. fo. 23ʳ for another notice.

Treviso in 1513, so his birth was probably prior to 1490. These same authorities say that he died in 1568 at what must have been an advanced age.[6] Though a priest, he was no university man, describing himself as 'that pitiable little man . . . not a theologian, a philosopher, not a lawyer, nor a canonist, but a very simple little known student.'[7] This did not stop him from having a considerable literary career, one largely devoted to importuning the Popes of his time about many matters, not least ones apocalyptic.[8]

There is a good deal of confusion about the titles, dates, and number of editions of his various writings, caused in large part by their extreme rarity.[9] I propose to investigate here only one work, his major apocalyptic writing, the *In Sathan ruinam tyrannidis*, published in Venice in 1524.[10] This odd book, written in a forced and turgid style of numbing prolixity, might be described as a prophetic miscellany, or even prophetic mishmash.[11] Despite its confusing arrangement and bizarre style, it is not without interest for students of sixteenth-century apocalypticism.

The *In Sathan* is a mixture of letters written to three Popes, Leo X (1513–22), Hadrian VI (1522–3), and Clement VII (1523–34), along with a variety of other documents. Rather than following the confusing order of the printed book, we may get a better sense of Paulus's role as prophet by beginning with a chronological analysis of the texts contained

---

[6] Tognetti, 'Note', 15–51.

[7] *In Sathan*, fo. 7ᵛ.

[8] Tognetti, 'Note', 152, puts it well: 'ma l'Angelo non era uomo che potesse restare troppo a lungo senza far pervenire la sua voce al papa regnante.'

[9] On the basis of the literature cited in n. 3, it seems there are at least four works besides the *In Sathan*: (a) *Epistola ad Saracenos cum Libello contra Alcoranum*, a translation of Ricoldo of Montecroce's *Confutatio Alcorani* dedicated to Hadrian VI and therefore presumably published in 1522 or 1523. (b) *Mirabile interpretatione di prophetie del fine del mondo*, a translation of the *Opusculum de fine mundi* attributed to St Vincent Ferrer, published in Venice in 1527, dedicated to the Doge Andrea Gritti and presented to him in a special audience. In 1530 Paulus issued a revised version under the title *Profetie certissime . . . dell'Anticristo*, adding material from Telesphorus of Cosenza's *Liber de magnis tribulationibus* printed in Venice in 1516. (See Reeves, *Influence*, 264). (c) *Expositio novissima . . . supra nonum capitulum Apocalypsis*. According to Tognetti, 'Note', 152, this was published after 1526. (d) *Apologia*, an anti-Lutheran treatise, which, according to the British Museum Catalogue, was published in Rome in 1537 and again in 1544.

[10] Besides the Vatican copy, which I am using, there is one in Paris in the Bibliothèque Nationale which Secret used in his article. This appears to differ in some details from the Vatican copy, if his transcriptions are correct.

[11] Tognetti, 'Note', 153–4, summarizes his character and style thus: 'resta l'immagine di un uomo che dava al gruppo cui si era associato solo l'apporto di una sforzata magniloquenza, priva di vigore e di originalità, e di una convulsa laboriosità di traduttore, editore, divulgatore, mentre tentava di posi in evidenza con ogni mezzo e nessun senso della misura.'

in it. This procedure will also serve to highlight the prophet's relations with the Popes and prelates of Renaissance Rome.

A note on fo. 22^v informs us that the combination dialogue and meditation under the title 'Quintus Angelus tuba canens' contained on fos. 18^r–22^v was gradually revealed to the prophet beginning in 1518 and was fully revised by the Holy Spirit in 1522.[12] This generalized apocalyptic work threatens divine judgement on the sins of the wicked,[13] but in typical fashion also promises a coming renewal of the Church— 'I have sworn very soon to renew my Spouse, like an eagle or phoenix, in your sight and that of my other designated faithful.'[14] It is not always clear in this text (though evident from the later materials in which it is embedded) that the Fifth Angel is not Paulus Angelus himself, but rather the message he bears.

The other early piece contained in the *In Sathan* is a letter 'Usque modo sanctissime Pater' (fo. 26^r–26^v), which had been first addressed to Leo X and was later sent to Hadrian VI in July 1523. A note introducing this brief missive concerning the cure for the illnesses of the Church tells us that it contains in implicit fashion matters that had been openly shown to the Pope, to Bernardino Cardinal Carvajal, and to his friend Lawrence, the 'Archiepiscopus Antibariensis'.[15] If this is true, it implies that Paulus had moved in very high circles in the time of Leo X.[16]

The *In Sathan* contains three letters written and sent to Hadrian VI, the unfortunate Dutch reformer whose brief pontificate stretched from 9 January 1522 to 14 September 1523. These, all written at Rome in the spring of 1523, are entitled 'Quoniam lucem' (fos. 23^r–26^r), 'Sapientes seculi huius' (fos. 26^v–28^v) and 'Cum omne bonum' (fos. 29^r–31^r).[17] Paulus says that he had come to Rome to obtain the Pope's permission for his work, but that a meeting did not take place.[18] The first letter is interesting for its critique of some of Hadrian's predecessors on the

---

[12] On fo. 17^r, Paulus, following his apocalyptic fiction, claimed that the work was composed by the mysterious 'servus Dei' to be considered below.

[13] E.g. fo. 22^r: 'Virgam virtutis mee porro ferream, virgam dirrectionis, equitatis, ac regni mei, aeterni ex alto emittam corruscantem; insuper, et arcum gladiumque meum, iamdudum vibratum super terram evaginabo cito.'

[14] Fo. 21^v.

[15] Fo. 26^r: 'Epistola parva iam olim exhibita Domini Leoni X. In qua modus implicite continetur qui iam explicitae sibi et Reverende Domino Bernardino Cardinali Sancte Crucis et suo familiari Laurentio Archiepiscopo Antibariense palam monstratus est.'

[16] For Bernardino (or Bernardo) Carvajal (at that time Dean of the Sacred College), see above, Minnich, Ch. 6, and Index. Secret, 'Paulus Angelus', 213 n. 2, cites texts showing his interest in prophecy.

[17] The final letter, according to a note on fo. 31^r, was delivered to the 'Episcopus Abulensis', who promised to give it to Pope Hadrian.

[18] Fos. 23^v–24^r.

throne of Peter.[19] In it we also begin to see the emergence of several of the main themes of Paulus Angelus's full apocalyptic programme. There is a veiled reference to a coming flood,[20] a theme that becomes more explicit in the second letter where the deluge is interpreted as symbolizing impending war: 'This deluge is not simply of material waters from the harmony of the planets that have converged . . . , but [is] of the waters of many peoples meeting in battle and slaughtering each other with warlike instruments and diverse implements.'[21] Given the state of Italy at this time as the cockpit of the power struggle between Francis I and Charles V this seems a relatively safe prediction. In the same letter, Paul calls on the Pope to open the 'new Apocalypse' sealed with seven seals, apparently a reference to the apocalyptic writings he was preparing for the pontiff.[22] Finally, in the letter 'Cum omne bonum' the prophet not only calls for the cleansing of the Church in a general way, but also offers specific suggestions, among them the summons of a crusade against the Turk,[23] and the more daring call to crown the emperor and convoke a general council as the best means of initiating reform.[24]

The remainder of the *In Sathan* appears to date from 1524, though it may include earlier materials brought up to date for inclusion. There are two letters to the new Medici Pope, Clement VII, who had begun his reign on 19 November 1523. The first (fos. 1^v–3^r), 'Dum medium fere silentio', was written on 10 June 1524. An important second letter to Clement, 'Altera autem die' (fo. 17^{r-v}) summarizes Paulus's apocalyptic message. It is said to have been written at Treviso near the Vigil of the Nativity in 1524.[25] Between the two is a major apocalyptic section of the newer material under the title 'Praefatio praefati angelici in angelicam

---

[19] Fo. 25^v.
[20] Cf. fo. 25^v: 'Sic in corde cito pessime adultere terre flagellum sanguinis horridissimum, necisque multiplicis horrende diluvium in ianuis iam iam appropinquans futurum est.'
[21] Fo. 27^{r-v}.
[22] Fo. 28^r.
[23] Cf. fo. 30^r.
[24] Fo. 29^v: 'Ergo sanctissime Pater, cum quocumque vero christicola nullactenus unquam ambigimus sanctitatem prudentissimam tuam velle quam citius poterit coronare sacrosanctum christianissimum catholicissimumque nostrum Imperatorem legitimum. . . . Interimque, per viscera misericordiae Dei nostri in quibus visitavit nos oriens ex alto in tua electione Pontificii summi, caeteros christianorum potestates, vel oratores eorum, et precipue veriorum Christi fidelium convocare, secumque preparare, ac celebrare auxilio Divi Michaelis et fortitudine Angelorum cunctorum tam excellens concilium generalissimum.' As an appendix, on fo. 31^{r-v}, the *In Sathan* includes the 'Forma Consensus' of the Council of Basle for a newly elected Pope, which includes a promise to summon a general council.
[25] If this is the true date, it would seem to indicate that the *In Sathan* was not actually published until 1525.

reformationem' (fos. 3ʳ–16ʳ). Several other brief apocalyptic texts of uncertain date are scattered through the book,[26] and the final folios contain a miscellany of texts and translations, including one important apocalyptic piece, a translation of the prophecy of the coming *Pastor Angelicus* ascribed to the Franciscan Frater Amadeus, together with Latin comments by Paulus.[27]

The message that Paulus tried to convey in his meandering and laboured prose was an unexceptional one by apocalyptic standards, involving a sense of present crisis leading to judgement on the wicked and vindication for the just in the form of a renewal of the Church. His apocalypticism is what we might call an interim one, since he sees the coming judgement and renewal not as the final act of history, but as a preliminary stage whose temporal relation to the eschaton is not commented upon.

Many of his comments about the signs of the coming proleptic judgement and renewal are vague animadversions on the corruptions and sins of the time, the stock in trade of prophets over the millennia. More interesting are the signs specific to the situation of Rome in the 1520s. In the 'Praefatio' he summarizes three tasks that he expects Clement VII to effect in order to usher in the renewal:

First, he will silence and close the impudent mouth of Luther, who, I think, is the deadly new 'Antichristus mixtus' (or his predecessor). Second, he will trample on the Mohammedan religion, putting a stop to its insane fury. Third, the very evil little foxes who ruined the vineyard of the Lord Jesus Christ that was taken out of Egypt, and who destroyed the gatherers of grapes, will now be eradicated, submerged and will perish at the sight of the face of Jesus himself and by his rebuke.[28]

The identification of Luther with the *Antichristus mixtus* (a term used by the Franciscan Spirituals and their followers to indicate the coming false pope of the last days) is interesting, though in this work, at least, Paulus Angelus shows little other concern about Antichrist and his identity.[29]

---

[26] These include: (a) the 'Canticum fidelium Dei viventium' on fo. 16ᵛ, mentioning an experience of rapture that Paulus Angelus enjoyed on Good Friday of 1524 at Padua; and (b) the 'Sententia diffinitiva . . . Domini nostri Iesu triumphantis', followed by a 'Dialogue' and a 'Gratiarum actio' (fos. 31ᵛ–33ᵛ).

[27] Under the title 'Prophetia fratris Amadei ordinis minorum' on fos. 33ᵛ–34ᵛ. On the pseudo-Amadeus, see above, Morisi-Guerra, Ch. 2.

[28] Fo. 7ʳ.

[29] The only other reference appears to be a brief mention of the final Antichrist, called the *Antichristus purus*, on fo. 5ʳ. See above, Reeves, Ch. 1, for the *Antichristus mixtus*.

There are several other references to Luther's career as an apocalyptic sign.[30] The dread Turk, of course, was an apocalyptic menace for both Lutheran reformers and papal loyalists. Paulus Angelus mentions the Mohammedan threat in a number of places.[31] The third evil of the day, the 'pessime vulpecule', are difficult to identify precisely. Foxes were traditional symbols for heretics, but a different kind of dissimulation may be hinted at here, that of the false converts from Judaism forced to adopt Christianity by the Spanish decrees of 1492.[32]

In the letters and treatises addressed to Clement VII, Paulus returns to the theme of the coming deluge mentioned in his earlier letters, but now the deluge seems to be more concretely taken as an imminent inundation, especially due to its connection with a planetary conjunction. The 'Praefatio' expresses part of its task as '(To show) by a truly uncovered, open and revealed declaration what kind of flood the conjunction of planets in the coming year 1524 is meant to signify.'[33] Perhaps Paulus came to mingle a symbolic with a real flood in his list of apocalyptic signs.

Like most apocalyptic propagandists, Paulus Angelus was not content to lament the evils of the present, but he also issued a call or summons to action. His invitation was directed not so much to the Christian community at large (though it did not exclude it), as to the Pope in his capacity as the supreme head of the Church. Paulus's naïve confidence in the capacity of the impolitic and misguided reformer Hadrian VI and the devious and indecisive Clement VII to cope with the great events that had overtaken the Church says little for his judgement, but does witness to the power of the apocalyptic view of the papacy that had been evolving since the twelfth century.

The first thing that the Pope must do is to read and grasp the meaning of the present crisis as revealed in the 'Quintus Angelus tuba canens seu foetus sacrosanctae Eucharistie', that is, the book sent to the pontiffs through Paulus Angelus, God's inspired servant.[34] The message contains a number of proposals for concrete action. The summons to put down

---

[30] E.g. fos. 1^v and 7^v.

[31] E.g. fos. 1^v, 7^v, 30^r, and 30^v.

[32] This interpretation is based on a passage in the second description of the three enemies found on fo. 7^{r-v} which begins, 'primus draco diabolicus est falsitas christianorum admixta iudeis vita consimili'. For the presence of such in Rome, see below, MacKay, Ch. 11.

[33] Fo. 3^v. For other references, see fos. 1^r, 4^r, and 32^r.

[34] That the 'Quintus Angelus' is actually the book and not the prophet is clear from texts like fos. 17^r and 32^r.

Luther and crush the Turks was scarcely exceptional. The repeated encouragement to convoke a general council, however, was something more problematic, especially to Clement VII who resisted many such calls. To these major politico-religious issues, Paulus Angelus added some more particular suggestions of his own for reform within an apocalyptic context that will seem bizarre to the modern reader. The first, expounded on fos. 2^r–3^r, is his view that the establishment of Eucharistic Confraternities throughout the Church could be a potent instrument of universal renewal (Paulus had a profound Eucharistic piety evident throughout the *In Sathan*). A second proposal is even stranger. In writing to Clement VII, the prophet introduces a plan for revising the order of the various *summae* of moral instruction, especially the very popular *Summa angelica* of the Franciscan Angelo of Chivasso (*c.*1411–95). This was apparently to be effected according to two indices of materials that Paulus included in his 'Praefatio' (fos. 9^r–15^v).[35] Paulus obviously thought that some kind of revision of moral teaching was a vital element in the coming reform of the Church, but the kind of change he had in mind seems scarcely adequate to the task.

Apart from these particulars, Paulus's message is too vague to be of much interest. The few visionary accounts that he presents, such as the mirror vision (fos. 4^v–5^r), or the allegorical vision of the 'Agiro' bird (fo. 6^{r-v}), are not particularly striking. One element, however, in the presentation of Paulus's apocalyptic message is worthy of more detailed study—his dealings with the 'senex sacerdos', the holy old man who acts as the mediator of the divine revelation.

The early apocalypses of Intertestamental Judaism and nascent Christianity always involved a mediating figure, usually an angel, who conveyed the divine message to the earthly seer.[36] In later Western Christianity, revelations concerning the imminent last events were often presented through other forms of literature, especially commentary on scriptural texts, though the visionary apocalyptic genre, and even mediated visions, never died out. Early in his 'Praefatio', Paulus Angelus introduces a mediating figure whose presence is central in the 'Clementine' portions of the *In Sathan*.

The mediator is initially presented as a real historical figure:

---

[35] The reform of the *summae*, generally referred to as a 'transmutatio capitulorum', is discussed on fos. 1^r, 7^v, and esp. 8^{r-v}.

[36] On apocalypse as a genre, see J. J. Collins *et al.*, *Apocalypse. The Morphology of a Genre* (Missoula, Mont., 1979).

I consulted a servant and priest of God Himself (our Lord Jesus Christ), one who truly keeps God's commandments and whom I know is completely free of mortal sin. First of all, I asked him and humbly petitioned that he might deign to pray and ask the Lord our Father in heaven on my behalf for the saving, blessed and supersubstantial daily bread, [that is] for the Lord Jesus Christ, omnipotent creator of all, who descends from heaven for the salvation of the true and not of the false Christians.[37]

Although his initial appearance is merely as one who prays that Paulus may be worthy of the special divine grace that he has been given, the holy priest soon takes on a more important role. After a week's fast, Paulus returns to the 'praefatum virum Dei sacerdotem' and receives 'omnia optata responsa' to his questions.[38] Paulus is instructed by the priest to the point where he becomes worthy to behold the angels and to convey the message of coming judgement and renewal embodied in the 'Quintus Angelus tuba canens'. At this point, a transition to a transcendental realm takes place:

Here the joyful man of God lifted me up rejoicing in the same spirit and led me apart to a high mountain. He transfigured himself into the Word of God and into our Lord, and by the name of the same Jesus Christ he gave me these commands, especially not to reveal the vision openly, so that it could be seen and beheld by anyone unless I humbly ask for express permission from him or from Christ.[39]

Then the 'sanctus vir' is given a book written within and without and sealed with seven seals, that is, a new version of the sealed book of the Apocalypse of John (Apoc. 5: 1), which the priest, now functioning fully as a Christ figure, begins to open (cf. Apoc. 5: 5). Paulus is allowed to read part of the book's message of vengeance and to witness the opening of the fifth seal.[40]

This 'sanctus ipse vir Dei', who is also described as an 'angelus paradisi Dei mei', is the intermediary for Paulus's visionary message throughout the 'Praefatio'. Who is he? We are given a hint about his identity towards the end of this text when he is sent by God to give the prophet the interpretation of the allegory of the 'Agiro' bird. Paulus begs the holy one to take up the role of Christ—'I compelled him as best I could to take up the task and burden which Christ, whom men know to

---

[37] Fo. 4^r.
[38] Ibid. The parallel with the experience of Amadeus will be obvious, see above, Morisi-Guerra, Ch. 2.
[39] Fo. 4^v.
[40] Ibid.

have been assumed, had previously fulfilled at God's command. When he
had done this, to the rejoicing and applause of all, the holy man blessed
all those things in the name of the Father, and of the Son and of the
Holy Spirit.'[41]

In late medieval papal apocalyptic views the one who will take upon
himself the role of Christ in the last days is none other than the *Pastor
Angelicus*, the final holy Pope.

The description of the 'sanctus presbyter' in the second letter to
Clement VII makes this identification secure. Paulus here recounts how
he went to the 'servum Dei veracem presbiterum vere senem' to show
him the first letter to Clement. The holy one received it with joy, but
he then gave Paulus a treatise he had long worked upon, none other
than the 'Quintus Angelus tuba canens', and ordered Paulus to have it
printed. Then, like Jesus with the Apostles, he asked Paulus who men
say he, 'the man sent from God', is.[42] Like a new Peter, Paulus is
inspired to respond that the holy one is God's true messenger, his
instrument for the renewal of the Church. Finally, to Paulus's question,
'Tu ne es qui venturus es, an alium expectamus?' the holy one responds
with a speech praising the faith of his servant, Paulus Angelus, and
promising 'one eastern and western Church renewed in supreme peace,
with all errors extinguished, because there will be one divine flock and
one Pastor Angelicus.'[43]

Although Paulus Angelus never comes right out and calls the holy old
priest the *Pastor Angelicus*, possibly because he felt that this would not be
politic in materials addressed to a reigning pontiff, it is clear that he
framed his treatise as one communicated to him by the apocalyptic
representative of all that was holy in the papal office. This distinctive
use of the *Pastor Angelicus* myth is the most interesting feature of the
confusing mélange that is the *In Sathan ruinam tyrannidis*.

How are we to evaluate the apocalyptic figure of Paulus Angelus? It is
clear that he impressed some people in his time. Silvestro Meuccio, the
diligent Augustinian who edited Joachim's works, in the preface to his
1525 edition of the Joachite *In Hieremiam*, recounts his initial excitement
in reading Paulus's treatise.[44] Paulus's subsequent association with

---

[41] Fo. 6ᵛ.

[42] Fo. 17ʳ: 'O frater et fili mi. Quem dicunt homines esse hominem missum a Deo' (cf.
Matt. 16: 13).

[43] Fo. 17ʳ⁻ᵛ.

[44] The text has been transcribed in Secret, 'Paulus Angelus', 214–16. Meuccio was so
struck by the vision of the book with the seven seals that he copied out much of it in his
Preface.

Meuccio, who became his confessor and found him entry to present his next apocalyptic work, the *Mirabile interpretatione*, in the highest circles of Venetian society, is well known. In this mingling of prophecy and politics Paulus's activities fit a model of learned apocalyptic advisors to princes and prelates frequently evident in medieval and Renaissance society. It is more difficult to know just how important a role Paulus played in Rome in the early 1520s. If he did have actual contact with Pope Leo X through Cardinal Carvajal, his role in Rome would have been quite similar to that which he later enjoyed in Venice, but the single reference he gives us is rather slim evidence. We also get the impression that his attempts to make direct contact with Hadrian VI and Clement VII met with little success and that he left Rome towards the end of 1524 to concentrate his prophetic energies in northern Italy. For all that, Paulus Angelus remains an interesting minor witness to the strength of belief in the apocalyptic role of Rome and the papacy in the turbulent times of the early sixteenth century.

V

PROPHECY AND
POPULAR CULTURE

Introduction

Prophetic expectations in this period were frequently stirred up to fever pitch by apocalyptic preachers appearing dramatically in the streets and by news of terrible prodigies and portents reported anxiously in flysheets and letters. Although these phenomena were widespread in Italy, the chief focus for preachers and for prophetic signs was Rome. Once more some unexpected aspects emerge. Ottavia Niccoli shows in her essay that there was no clear dividing line between the study of prophecy among the intellectuals and the popular dissemination of prognostications; rather there was a two-way traffic between 'high' and 'low', demonstrating their common concerns. Angus Mackay's short study of the novel by an Andalusian priest, Delicado, plunges us into the Roman underworld where the 'Babylon' denounced by the prophetic preachers comes to life. Thomas Cohen's note on Fra Pelagio shows the continuing attention paid by highly placed clerics and aristocrats to the crazy babblings of an ill-balanced hermit immured near St Peter's.

Cultures mingle here in unexpected ways. Folklore receives attention among scholars; wandering preachers demonstrate their theological learning; biblical and classical signs are interpreted side by side. Of fundamental importance in whipping up this prophetic excitement was the new instrument of the printing press. Whether in the form of the ephemeral flysheet, tract for the times, popular ballad, Latin verse, or substantial treatise, the printed word conveyed the apocalyptic message to readers on a scale never known before.

IO

High and Low Prophetic Culture
in Rome at the Beginning of the
Sixteenth Century

OTTAVIA NICCOLI

In a sonnet composed towards the end of 1831 the poet Giuseppe Gioacchino Belli traced a picture of the Roman plebeian beliefs regarding the end of the times and the Antichrist. The latter was believed to be the child of a nun and a friar with 'the body of a giant and a sad eye'. His preachings were to be opposed by a singular personage who would emerge from a concealed pit near St Paul's Basilica, 'er Nocchilia'.[1]

This bizarre word came from the fusion of two names, Enoch and Elijah, who had been identified, through a widespread pseudo-Joachimist tradition, as the 'two witnesses' spoken of in the Apocalypse ('And I will give power unto my two witnesses, and they shall prophesy a thousand two hundred and threescore days, clothed in sackcloth').[2]

This was a tradition which had an ancient history going back to the Judaic apocalyptic. It is not, however, sufficient to understand the reference. It is necessary also to enquire how and why this learned tradition had penetrated to the lowest levels of the Roman populace from which Belli had gathered it, deformed and distorted, but still recognizable, rather as an antique object will emerge from archaeological excavation, damaged and dented, but still capable of being repaired. The task of the archaeologist is not merely to restore and date the object, but also to place it in its historical context. If we find an Oriental object in Roman subsoil, we must not only ask ourselves what its origin is, but also how and when it came to be in its actual position.

---

[1] G. Belli, *I sonetti* (Milan, 1965), i. 300 (no. 274, *La fin der monno*).
[2] *Apoc.* 11: 3. For the significance of Enoch and Elijah, see O. Niccoli, *Profeti e popolo nell'Italia del Rinascimento* (Rome and Bari, 1987), 128.

Returning to Enoch and Elijah, transformed into 'er Nocchilia', it is probable—though there is no definite proof—that their entry into Roman lower-class culture can be dated between the end of the fifteenth and the beginning of the sixteenth centuries, when prophetism in Italy underwent an exceptional expansion, linked with the traumas of war which were throwing the peninsula into confusion. These years also saw an exceptional fluidity in the relationship between high and low culture, and a great readiness on the part of socially and culturally 'high' Roman circles to gather and utilize elements of folklore and reconvert them to their own uses. The advent of Enoch and Elijah, transformed into 'er Nocchilia', should not therefore be read as the sign of a simple cultural degradation—the incomprehension on the part of primitive people of two unusual names—but rather as the indicator of an open-ended channel of communication. Of the numerous threads of this variegated canvas, we will follow here only two—those which appear the most significant—the role of preaching and the presumptive science of signs.

I

Through preaching, as we already know from numerous researches, all types of cultural material filtered down to the lowest social levels of society.[3] Using residual material from classical culture, the preachers poured out *exempla* drawn from scholasticism or from the Fathers of the Church. But they also utilized folk traditions which they had gathered during their periods of office and which returned thus, transformed and reconstituted, to the popular circles from which they had originated. Thus preaching is one of the best areas on which to focus our attention when considering the problem of the interaction of cultural levels.

More precisely, we know that in the decades bridging the fifteenth and sixteenth centuries prophetic inspiration found one of its principal outlets of transmission in preaching.[4] The announcement of the imminent Antichrist and the end of the world were insistently recurring themes, both in the official homiletics and in the extemporary preaching of itinerant hermits—at least until 1516, when it was abruptly prohibited by the Fifth Lateran Council.[5] After that date prophetic preachers pre-

---

[3] C. Delcorno, *Giordano da Pisa e l'antica predicazione volgare* (Florence, 1975); J.-C. Schmitt, *Le Saint-Lévrier, Guinefort, guérisseur d'enfants depuis le XIIIᵉ siècle* (Paris, 1979).
[4] Niccoli, *Profeti e popolo*, 123–60.
[5] *Conciliorum oecumenicorum decreta* (Bologna, 1973), 635–7. See above, Minnich, Ch. 4.

ferred to avoid these themes, and above all to eliminate allusions to precise dates, although they continued to refer to imminent catastrophes, such as the Turks, the earthquake, as divine retribution for men's sins. We have numerous testimonies on the subject, many of which concern Rome itself:

Do you remember bearded William who preached throughout Rome with the cross the first year of Sixtus in every place predicting everything bad in a loud voice? His cross was taken away and he was beaten but woe to them who harmed and hurt him; the blessed William of Morano (he) is called in the world today by every Christian.

Another who looked like St Paul and preached in the Campo de' Fiori had a long beard reaching to his chest and always went dressed in blue, and told you all the bad things which were dogging you.[6]

We know from contemporary chronicles that these were two hermits who wandered about Italy during the winter of 1472–3,[7] and who are described to us in the act of preaching in the Campo de' Fiori. The author of these verses was the Roman canon, Giuliano Dati. They may describe his personal memories, not yet completely dimmed by the more than twenty years which had elapsed (the text is from the beginning of 1496).

These were not the only prophets who preached in the squares of Rome at the end of the fifteenth century. On 26 April 1485, the Roman chronicler Antonio di Vascho wrote in his diary: 'how in this day . . . came to Rome a hermit of the order of St Francis who rode astride a steer with saddle and bridle and carried an image of the crucifix on his breast, who . . . was much admired . . . above all because of the small amount of justice which was administered in Rome'.[8] We do not know the tenor of this personage's preaching. We note, however, that the chronicler relates his person and aspect to the 'small amount of justice' in Rome at that time. His image, which signified more than words, is sufficient to make him a warning and sign. Some years later another

---

[6] *Del diluvio di Roma del MCCCCLXXXXV adi iiii di dicembre et daltre cose di gran meraviglia* (n.p., n.d.), sig. aiii^v.

[7] A. Volpato, 'La predicazione penitenziale-apocalittica nell'attività di due predicatori del 1473', in *Boll. ISI*, 82 (1970 [1974]), 113–28.

[8] A. di Vascho, *Il diario della città di Roma dall'anno 1480 all'anno 1492*, ed. G. Chiesa, *RIS* 23, 3, 523. See also B. Nobile, ' "Romiti" e vita religiosa nella cronachistica italiana fra '400 e '500,' *CS* 5 (1984), 322, to which may be added the mention of the *Notationes* by the Roman grammarian Paolo Pompilio: A. Mercati, 'Paolo Pompilio e la scoperta del cadavere intatto sull'Appia nel 1485', *Opere minori*, iv (Vatican City, 1937), 276.

similar figure arrives in the city. He is a roughly dressed man, 'like a beggar', and holds a small, wooden cross in his hand. He roams the Roman squares, mounts 'in some high place'—we may imagine a stone, a broken column, a balcony—and calls the people to him, promising to talk of the Gospel. 'He created a harmony between the Old and New Testaments and spoke well and acutely, showing himself to be a person of great culture and eloquence . . . and he said: "I say to you, oh Romans, that many will weep in this year of 1491; and there will be tribulations, killings and blood upon you, because many of you will be killed this year, and there will be a reason for the tribulation which awaits you. Powerful citizens will take the wheat away from the poor and the hungry will rise up and riot against you, and there will be a great tribulation. The following year this tribulation will begin to spread throughout all Italy, and in that year Florence, Milan, and the other cities will be deprived of their liberty and will live under the domination of others. The Venetians will be stripped of all that they possess on the mainland. In the third year, 1493, will come a cleric with no temporal power, who will be, therefore, an Angelic Pope who will care only for the life of the soul and spiritual causes." This said, he departed.'[9]

We note that this hermit also put an emphasis on existing social injustices. This was clearly an element which attracted public attention; witness the hermit on the steer cited above. We also note that the reference to the harmony between the Old and New Testaments leads us inevitably to Joachim of Fiore, and more precisely to the *Liber de Concordia Novi et Veteris Testamenti*. Furthermore, it is significant that the chronicler draws attention to the doctrine and eloquence of the unknown preacher. The reference to the Angelic Pope and the punishments of God completes the picture and affords us a full insight into the way in which the current political situation, the expectation of change, and the receptiveness of the Roman populace to the message of tribulation and renewal were integrated into a general prophetic mood. 'Well meaning men welcomed these things, though not so vehemently as the plebs', writes Antonio di Vascho on the subject of the hermit on the steer.[10]

This predisposition in Rome, as in the rest of Italy, to listen to itinerant preachers and prophets, emerges also in later evidence. 'You

---

[9] S. Infessura, *Diario della città di Roma*, ed. O. Tommasini (Rome, 1890), 264–5. The original is in Latin.
[10] Di Vascho, *Il diario*, 523.

have already seen many prophets and scribes and sages preaching', admonished the Roman Giuliano Dati at the beginning of 1496, and an unknown ballad singer in 1511 evoked a miraculous apparition:

> An ancient hermit
> with long beard and hair
> goes crying through Rome
> "peace, peace";
> then whenever he wishes
> he goes away invisible
> and many believe firmly
> that he is Elijah.[11]

We already know that the personage of Elijah has a precise eschatological significance. This 'ancient hermit' presented himself, therefore, as a prophet, and it is possible that he was that Bonaventura who, in August of the same year, 1511, was imprisoned on one of the Colonnas' estates because, as Count Girolamo da Porcia wrote in a letter, while his preaching was 'admirable' and he had a 'great following', he 'talked a lot of nonsense'—that is, he spoke too much and too imprudently.[12] We know the content of that preaching, which probably underwent a certain evolution during the course of the following years, even if only in a fragmentary form. Its main theme was the announcement of imminent tribulations at the hands of the Turks which would befall Italy and the Church. The Turks would, however, finally be baptized. This would be the beginning of the third age because, as Bonaventura said, 'in the Old Testament there was the rule of the Father, then came that of the Son, therefore, of necessity, the rule of the Holy Ghost must follow, and will last until the end of time.'[13]

There is an evident Joachimist inspiration behind these words. The unknown writer, who sent his letter to Paris from Rome on 8 April 1514, affirms moreover that Bonaventura 'seems to know the whole of the Bible, Augustine and Jerome by heart'.[14] With these sources others, of quite a different kind, could also be interwoven. These themes were recurrent in the vernacular prophecies which were circulating in Italy during those years, both in prose and in verse:

---

[11] *Memoria delli novi segni e spaventevoli prodigii comparsi in più loci de Italia: et in varie parti del mondo: lanno mille cinquecento undese*, n.p., n.d., fo. 1ᵛ.

[12] Sanuto, *Diarii*, xii, col. 323.

[13] 'Admirabilis epistola noviter ex urbe Roma Parrhisius delata . . .' *Mirabilis Liber qui prophetias Revelationesque necnon res mirandas preteritas presentes et futuras aperte demonstrat* (Rome *recte* Lyons, 1524), fo. 88.

[14] Ibid.

The Turk will pursue his victory as far as Rome, with
much cruelty, such as has never been seen in this world,
most of all towards the priests. He will ruin citadels
and churches, friars and priests. Many cardinals and
prelates will be hacked to pieces, and the whole church
will be orphaned. . . . Then our Lord, Jesus Christ,
will visibly summon a holy man. . . . The Turk,
because of the great miracles which the preachings of
that holy man will reveal, will become a true and
perfect Christian. . . . Then the whole world will be
content for the Turk to be Emperor and this holy man to
be Pope, and they will be obeyed throughout the world.

You will see with that holy one
the angel of God will come . . .
he will make the Turk so pious
that, thanking Jesus Christ,
for all his sad sins
he will soon be confessed.
The Turk will be baptized
through the miracle of a holy man.

When the Turk sees
such a miracle sent by God
he will quickly be baptized
and repent of his errors.
He will become an Emperor.[15]

These texts must have been well known in Rome. In the spring of 1519 a
possessed woman, when conducted to St Peter's to be exorcized, emitted
incoherent predictions on the current Imperial elections, quoting from
one of the rhymed prophecies above, which must, therefore, have been
known to her.[16]

To return to Bonaventura, we know that he was born somewhere
between Bologna and Ferrara, but the true fulcrum of his activity during
the brief period that it is known, between 1511 and 1516, was centred in

---

[15] The first quotation is taken from an untitled prophecy, inserted in Ravenna, Bibl.
Classense, MS Class. 287, *Poesie italiane e latine*, fos. 9ᵛ–11ʳ. The other two quotations are
taken from two prophecies in verse published in numerous editions in the early sixteenth
century: *Prophetia trovata in Roma. Intagliata in marmoro in versi latini. Tratta in vulgar
sentimento* (Rome c.1510), fo. 2ᵛ; *Questa è la vera prophetia prophetizata dal glorioso Santo
Anselmo: la quale declara la venuta de uno Imperatore: el qual mettera pace tra li christiani: et
conquistara li infideli trovata in Roma* (? Ferrara, c.1510?), fo. 2ᵛ.
[16] Sanuto, *Diarii*, xxvii, col. 224.

and on Rome. 'God sent him to Rome, to the centre of Rome, so that the people should repent. Tribulation would befall the Christians, beginning in the city of Rome.'[17] So wrote the anonymous author of the letter quoted above. He also informs us that Bonaventura had been imprisoned in the Castel Sant'Angelo during the last eleven months of Julius II's life for having predicted Julius' imminent death and for having spoken against him: 'he who sows swords, will reap swords', he had said amongst other things.[18] It is possible—the dates would substantially coincide—that these words were intended as a reproach to the Pope for the disastrous outcome of his anti-French politics which culminated in the defeat at Ravenna on 11 April 1512.

Bonaventura's pro-French position emerges explicitly in another, later document. This, too, is a letter from Rome, dated 12 May 1516, which was sent to Prince Charles of Gurk by an Imperial official close to the Holy Seat.[19] Bonaventura, he recounts, has once again been imprisoned in the Castel Sant'Angelo for having declared himself to be the Angelic Pope. Furthermore, he had sent the Doge of Venice a book entitled *Liber Venturati de apostatrice abiecta et a Deo maledicta meretrice ecclesia romana* in which he claimed that all Christians should show obedience to himself, excommunicated Pope Leo, and affirmed that he would baptize the Emperor and transport the Holy Seat to Zion with the backing and support of the King of France. Many threads are here joined together. In the prophecies quoted above the holy man, presented as a hermit, similar to Bonaventura, also had the tasks of converting and baptizing the Turk who would become Emperor, and then revealing himself as the Angelic Pope. We can deduce that this *Liber Venturati*, or extracts from it, was circulated in some way from the fact that in a small work dated 1524, the *De mirabili temporis mutatione ac terrene potestatis a loco in locum translatione*, a priest from Vallese, Jean Albertin, also declares, in words similar to those of the mysterious book, that the power of Peter will be removed from Rome because of the abuses committed by the Roman clergy, and transferred to the church of Zion (not Zion/Jerusalem, but Zion in Vallese).[20] The same hypothesis appears to be supported in

---

[17] 'Admirabilis epistola', fo. 88ʳ. See also G. Tognetti, *DBI* 11 (1969), 611–12.

[18] 'Admirabilis epistola', fo. 87ᵛ.

[19] C. Höfler, 'Exemplum literarum domini Stephani Rosin Caesareae Majestatis apud S. Sedem Sollicitatoris ad Reverendum principem D. Carolum Gurcensem', *Analecten zur Geschichte Deutschlands und Italien, Abhandlungen der hist. Cl. del K. Bayer. Akademie der Wissenschaften*, 4 (1846), 56.

[20] *De mirabili temporis mutatione ac terrene potestatis a loco in locum translatione. Iohannis Albertini presbyteri Vallesiensis Declaratio* (Geneva, 1524), fo. 2ʳ⁻ᵛ. On Albertin see H. Naef,

Giovanni Battista Nazari's *Discorso della futura et sperata vittoria contro il Turco*, published in 1570, in which he expressly quotes from a prophecy made around 1512 by 'Santo Bonaventura' (certainly our Bonaventura), in which he foretold the Turkish invasion of Italy and the Turk's final conversion and baptism by the Pope (thus with counter-reformatory prudence, does Nazari adapt the prophecy).[21] The last information we have on Bonaventura dates back to 19 August 1516, and shows him still imprisoned in the Castel Sant'Angelo, angrily intent on prophesying the imminent death of Leo X, as he had, with greater success, already prophesied the death of the elephant Hanno which had been given to the Pope by Emanuel of Portugal in March 1514 and which actually died on 8 June 1516.[22]

The varied types of prophetic preaching experienced in Rome during these decades had increasingly presented a negative image of Rome and the Papacy, while emphasizing the expectation of a radical purification through the work of an Angelic Pope and a cruel enemy, transformed into the Emperor of Peace, who would inaugurate the third age. These were themes which the preachers had gathered from the texts of Joachimism—just beginning to be published in 1516—or from transcribed prophecies which passed from hand to hand or were spread by ballad singers or the people's press and redistributed among a vast and varying public. But it is noteworthy that the theme of the Turk turned Good Emperor was not part of the Joachimist tradition and appears to be new.

This tradition of anti-Roman preaching became progressively more explosive in the weeks preceding the Sack, when Brandano da Petroio arrived in the city and scaled a statue of St Paul in St Peter's Square to turn directly to Clement VII who was blessing the crowd and call him a 'sodomite bastard'. 'Rome will be destroyed because of your sins. Confess. Convert. If you don't believe me you will see within fifteen days.'[23] He continued his prophetic preaching in the Campo de' Fiori (like the hermits who had passed through Rome in the winter of 1472),

---

Les origines de la Réforme à Genève (Paris, 1936), i. 427–35; Niccoli, *Profeti e popolo*, 196–8.

[21] *Discorso della futura et sperata vittoria contra il Turco estratto da i sacri Profeti et da altre Profetie, prodigij et Pronostici et di nuovo dato in luce per Gio. Battista Nazari Bresciano* (Venice, 1570), fo. Cl[v].

[22] Sanuto, *Diarii*, xxii. cols. 474–5. On the elephant Hanno, see D. Gnoli, *La Roma di Léon X* (Milan, 1938), 111–19 and Silvio A. Bedini, 'The Papal Pachyderms', *Proceedings of the American Philosophical Society*, 125 (1981), 75–90.

[23] G. Tognetti, 'Sul "romito" e profeta Brandano da Petroio', *Rivista storica italiana*, 72 (1960), 32; A. Chastel, *The Sack of Rome*, trans. B. Archer (Princeton, NJ, 1983), 87.

surrounded by 'charlatans, tooth-drawers and those who cure hernias', as a contemporary witness noted.[24] We are dealing, therefore, with a pattern of preaching which had a great and constant success in Rome (and, in a slightly different form, in the whole of Italy).

The traces of its success can be discerned in the sparse pieces of news from the *Diarii* of Sanudo which discuss the preaching of a radically different personage from the hermits we have so far encountered, Egidio of Viterbo. It has been observed that we know Egidio best today for those writings of his which were not available in his time, while we have very little knowledge of his preaching, to which, more than anything else, he owed his fame among his contemporaries.[25] A few manuscripts have been identified, and we have some printed texts. What emerges strikingly from Sanudo's evidence is the levelling out between the themes of Egidio's preaching and those of popular apocalyptic preaching. We know—he said so himself—that Egidio's homiletic style was tumultuous and torrential, mixed with complex, intricate arguments from Hebrew learning and tending to offer an unknown revelation ('verbum . . . novum, inauditum . . . et numquam intellectum').[26] Quotations from Rabbinic and cabbalist sources were mixed with those from Platonic texts, with biblical amplification and astrological allusions. Very little of this flood of erudition—which was to be found in a similar form but with much greater breadth in his *Historia* and *Scechina*—probably remained in the understanding of the listeners. The citizens of Rome tended to adapt arguments and quotations mechanically, within the restrictive frame of the prophetic preaching they listened to so often. Egidio's physical aspect favoured this adaptation. In July 1508, Tommaso di Silvestro, a canon from Orvieto, tells of having heard the sermon of 'a master Egidio from the region of Viterbo. An old man (in reality Egidio was not yet 40), with a long black beard in *the style of a hermit*',[27] similar, that is, to those hermits who with their habits, long, straggling beards and bare feet voluntarily imitated the current iconography of St John the Baptist and Elijah,[28] and thus presented themselves to their listeners as prophets. This prophetic image can be linked, moreover, to

---

[24] F. Delicado, *La Lozana Andalusa*, ed. L. Orioli (Milan, 1970), 62. On this source, see below, MacKay, Ch. 11.

[25] J. Monfasani, 'Sermons of Giles of Viterbo as Bishop', in *Egidio da Viterbo, O.S.A., e il suo tempo* (Atti del V Convegno dell' Istituto Storico Agostiniano; Rome, 1983), 138–9. On Egidio, see above, Reeves, Ch. 5, and for his preaching, see Bibliography.

[26] Monfasani, 'Sermons of Giles of Viterbo', 168.

[27] Tommaso di Silvestro, *Diario*, ed. L. Fumi, *RIS* 15, 5.2, 370.

[28] See Niccoli, *Profeti e popolo*, 129, and the bibliography in n. 20.

various constant themes in Egidio's preaching. One was that of a secret
wisdom, miraculously revealed ('Listen to the most marvellous mystery.
It is written in secret books. . . . In this book they tell a stupendous
secret . . .')—thus from a sermon preached at Viterbo on Easter Monday,
1527.[29] But the themes to which Egidio returned with greatest insistence
were those of the crusade and of the Islamic domination of holy places
and of the Mediterranean Basin. These featured, for example, in the
famous sermon at the opening of the Fifth Lateran Council on 3 May
1512 and in the sermon to the faithful of Bologna on 22 October 1525.[30]
On the occasion of the treaty between the Pope and Emperor Maximilian
on 25 November 1512, he exhorted his listeners:

Therefore bring together the spirits of your princes, settle their dissensions,
recall the people to peace and harmony. . . . For two plagues remain for you to
subdue, one at home and one abroad; at home the licentiousness of making light
of religion, and abroad the audacity of invading and oppressing the Faith. . . .
The one lies in ambush against faith and morality; the other threatens massacre
and bloodshed. Therefore . . . amend your morals, luxury and license . . . and
come forth to destroy the impending danger from Mahomet, and the terrible
plague.[31]

What remains to us of Egidio's preaching is generally in Latin. We
should remember, however, that he had the appalling habit (or so it was
judged by the Master of Ceremonies, Paride de' Grassi)[32] of repeating
his own sermons in the vernacular so that everyone could understand
them. We may imagine how the exhortations to a crusade, the frequent
references to the sins of men and of the Church, and the quotations from
'secret' texts would have produced a reaction in the public that was not
very dissimilar from that excited by the itinerant preachers. A letter
from Rome dated 10 January 1517, addressed to the Venetian Girolamo

---

[29] Monfasani, 'Sermons of Giles of Viterbo', 144, which cites Naples, Bibl. Naz., MS
V.F. 14.

[30] *Oratio Prima Synodi Lateranensis habita per Egidium Viterbiensem Augustiniani Ordinis
Generalem* (n.p., n.d. [Rome, 1512]), ed. C. O'Reilly, '"Without Councils we cannot be
saved . . .": Giles of Viterbo Addresses the Fifth Lateran Council', *Augustiniana*, 27 (1977),
166–204; Monfasani, 'Sermons of Giles of Viterbo', 185; J. O'Malley, 'Fulfillment of the
Christian Golden Age under Pope Julius II: Text of a Discourse of Giles of Viterbo, 1507',
Traditio, 25 (1969), 320–7 (repr. in *Rome and the Renaissance: Studies in Culture and
Religion* (London, 1981)).

[31] C. O'Reilly, '"Maximus Caesar et Pontifex Maximus": Giles of Viterbo Proclaims
the Alliance between Emperor Maximilian I and Pope Julius II', *Augustiniana*, 22 (1972),
110–12.

[32] 'Habuit sermonem primo latinum, deinde contra bonas cerimonias vulgarem', cited
in O'Malley, 'Fulfillment', 269.

Lippomano, and referred to by Sanudo, describes Egidio speaking of 'Turkish matters' for four hours in a Roman church, making gloomy allusions to previous Popes and identifying Pope Leo X with 'that most serene lion mentioned by St. John in the Apocalypse: "veniet Leo de tribu Juda"; this is that lion which will conquer our enemy in Christ's faith, that is the Turk.' On hearing 'such a terrifying sermon', Lippomano's anonymous correspondent adds, 'all the women . . . were crying as if the Turks were already at the gates of Rome'.[33] The following year at Saragozza he spoke again of 'Turkish matters and made prophecies'[34]—a further illustration of the fact that these were recurrent themes in his preaching.

II

To summarize, therefore, at the beginning of the sixteenth century Rome found itself at the centre of a flood of prophetic interpretations of contemporary history, originating from the heights of the most refined humanist culture but descending to those who spoke in the squares surrounded by charlatans and tooth drawers. This confluence of cultural levels appears evident not only in preaching circles, but also in those of divination, and more specifically, those which paid attention to the so-called signs. We may define these signs as natural phenomena which were observed with attention and interpreted as foretelling imminent and terrible events. The signs are not, however, simply extraordinary facts, but represent a sort of scientific orchestration which was transmitted orally (mostly, but not exclusively, at a popular level), and which claimed to foretell the future on the basis of mutations in what was usually considered the natural course of events. Humanist culture, with the attention it paid to classical divination, certainly contributed to the spread of interest in the signs and in popular divination, but it was not so much a case of a univocal relationship as one of reciprocal influence. In 1517, when Giovanni Antonio Flaminio, the humanist from Imola, wanted to write an epistle exhorting Pope Leo X to a crusade, and affirming that the ever more frequent signs and portents represented in themselves an urgent warning to take action against the enemy of Christianity, he was not able to think of better evidence than to enclose a good part—twenty-six quatrains, faithfully translated into Latin—of

[33] Sanuto, *Diarii*, xxiii, cols. 486–8.
[34] Ibid. xxv, col. 600.

a minstrel's tale of 1511, the work of an anonymous ballad-singer whom we have already quoted in connection with Bonaventura's preaching.[35]

During these years much attention was paid to signs and portents throughout all Italy,[36] but Rome was definitely a hot spot in this sense. The attention given to the hermit who wandered round Rome astride a steer has already been recorded. Chroniclers were constantly associating his figure with other portents such as a baby friar who preached, the uncorrupted corpse of a young Roman girl recovered at Casal Rotondo, and the theft of the Papal mitre.[37] An eclipse of the sun was also expected at that period, 'over which the people remained in suspense and doubt', as Antonio di Vascho wrote (we recall that he also said: 'well meaning men welcomed these things, though not so vehemently as the plebs').[38] The portents came, however, to be linked with ever increasing frequency to high ranking Roman society and the Papal Court itself. In November 1497 the Venetian Domenico Malipiero reported a series of portentous events which had occurred recently in Rome in his *Annali* and said that they were implicitly related to the papacy of Alexander VI: 'the arrow has struck in Castel Sant'Angelo and consumed all the munitions and a great part of the battlements. Great portents followed in the time of Pope Alexander. The arrow was in his antechamber. The flooding of the Tiber was in his time. His son was killed in that way and there is much ruin in the Castello.'[39]

The same pieces of news were reported by Sanudo with the comment 'concerning the great portents . . . now in Rome they say that some prodigious event will follow'.[40] The fear of the portents seems to have mounted progressively upwards from the lowest levels of society. This ascending movement becomes most evident in December. Sanudo, who was always well informed on Roman affairs, is again our witness:

From Rome. By letter of the 17th instant (the month of December) how the note of a most terrible voice was heard in the Castel Sant'Angelo, terrifying every inhabitant of the Papal Palace. They did not know what voice it was. Now the whole court is afraid. The voice was heard several times, and they said in

---

[35] G. Flaminio, *Epistolae familiares* (Bologna, 1744), 60. Cf. *Memoria delli novi segni*, fos. 1ʳ–2ʳ.

[36] See Niccoli, *Profeti e popolo*, 47–9.

[37] A. Mercati, 'Paolo Pompilio e la scoperta del cadavere intatto sull'Appia nel 1485', *Opera minori*, iv (Vatican City, 1937), 276–80; G. Pontani, *Diario romano*, ed. D. Toni, *RIS* vol. iii, pt. 2, 47; di Vascho, *Il diario*, 522–3.

[38] Di Vascho, *Il diario*, 520, 523.

[39] D. Malipiero, 'Annali veneti dall'anno 1457 ad 1500', *ASI* vii. (1843) 497.

[40] Sanuto, *Diarii*, i, cols. 814–15.

Rome that it was a spirit. . . . For this reason the Pope and others were in great tribulation and very frightened.[41]

Several years later, in August 1506: 'From Rome. A comet appeared in the night, there on the instant, and it was big. Its tail was towards Castel Sant'Angelo, and they said it was menacing the Pope.'[42] There is no doubt that in this case the comet that seemed to menace Julius II was an erudite second reading of that 'star with flowing locks' which, according to Suetonius, accompanied the death and funeral of Julius Caesar.[43] But the examples given so far will have shown how reductive it would be to place the concern with the portents wholly within the framework of the renewed classical culture in Renaissance Rome. The memory of pagan divination gave authoritative confirmation to the prophetic forms which, in Italy between the fifteenth and sixteenth centuries, originated with the common people, while also enjoying great success in intellectual circles. The intensity with which the signs were regarded becomes evident, for example, in February 1527, when, on the slopes of the Subasio, a shepherd boy, helped by his companions, captured an eagle which was carrying off a sheep. This was interpreted in Rome 'as an auspicious sign, and a certain indication of a victory in favour of our Lord'.[44] Obviously the eagle was seen to represent the Emperor and the shepherd the Pope, and it was with such consoling interpretations that Rome attempted to ward off the prospect of its own imminent ruin on the eve of the Sack.

We should probably read the attention focused on monstrous births, which can be verified as being close to the Roman court, in a not dissimilar sense. Of all the signs, the monsters were probably the ones most loaded with significance. 'These signs signify great tribulation for the city from which they come', wrote Luca Landucci in 1489, speaking in fact of some Venetian monsters.[45] The first deformed creature to have real notoriety in Renaissance Italy was the so-called Donkey Pope, who was found in Rome on the bank of the Tiber between Tor di Nona and Castel Sant'Angelo at the receding of the river after the disastrous flood of 4 December 1495.[46] The notoriety of this monster did not,

---

[41] Ibid., col. 842.

[42] Ibid., vi, col. 394.

[43] Suetonius, *Iulii Caesaris vita*, 88.

[44] Sanuto, *Diarii*, xliv, col. 34.

[45] L. Landucci, *Diario Fiorentino dal 1450 al 1516*, ed. I. Del Badia (Florence, 1883), 57. Cf. Niccoli, *Profeti e popolo*, 49–52.

[46] For the first description of the monster, see Malipiero, *Annali veneti*, 422; for the place where it was found, see A. Coniger, *Cronache, Raccolta di varie croniche, diarii ed altri*

however, depend so much on Roman circles (in fact there are no well-known contemporary Italian interpretations), but rather on the fact that a copy of an Italian drawing or print (now lost) which depicts it probably got into the hands of two Bohemian brothers who were passing through Rome. They took it back to their country and identified the monster with the Roman papacy. The subsequent fortunes of the image, which led to the enormous notoriety given to it by Lucas Cranach and by Melancthon, were described by K. Lange at the end of the last century.[47]

In 1496, however, Alexander VI's Rome did not pay much attention to the monster. Only the Modenese Francesco Rocociolo saw it as a forewarning of Italian political and military disasters, though destined to show off the greatness of Ercole d'Este, who would be called to 'overcome such a monster'.[48] This lack of attention close to the Roman court could perhaps be explained by fear of compromising interpretations which the strange creature did in fact receive subsequently. Attention to monsters appears already to be active in Roman circles by the time of Julius II. In July 1506 a Florentine woman of low social condition gave birth to a deformed creature (Sanudo described it as 'a most horrifying thing'),[49] which died after a few days because nutrition and care were deliberately withheld from it, according to a not unusual practice. This solution turned out to be economically advantageous to the mother, who 'took the dead monster to be embalmed and went to Rome'—considered, therefore, as the place which offered the greatest opportunity for the commercial exhibition of such creatures. This type of exhibition was backed up by the distribution of coloured drawings, and perhaps by printed leaflets, but these did not offer any interpretation of the creature. One of these drawings was despatched to Sanudo (we know that Rome provided him with one of his main sources of news), who stuck it in his diaries with the comment: 'what is to come, God alone knows'. Another

opuscoli così italiani come latini appartenenti alla storia del regno di Napoli, ed. A. Pelliccia (Naples, 1782), v. 37.

[47] K. Lange, *Der Papstesel* (Göttingen, 1891).

[48] F. Rocociolo, *Ad illustrissimum ac excelentissimum principem divum Herculem Estensem Francisci Rococioli mutinensis libellus de monstro Romae in Tyberi reperto anno Domini MCCCCLXXXXVI* (n.p., n.d.) fo. 4ᵛ. On Rocociolo, see G. Bertoni, 'Intorno a tre letterati cinquecentisti modenesi. Francesco Roccocciolo', *Giornale storico della letteratura italiana*, 85 (1925), 376–7; A. Rotondò, 'Pellegrino Prisciani', *Rinascimento*, 11 (1960), 106–7.

[49] Sanuto, *Diarii*, vi. col. 390.

identical drawing—or perhaps a leaflet—arrived in Germany, where an engraving was made of it with a caption in German.[50]

Already in 1506, therefore, there existed in Rome a notable impetus for the dissemination of news concerning monstrous births. This obsession perhaps reached its peak in 1512, in the period of the 'monster of Ravenna'.[51] The creature was probably born on 6 March, at a dramatic moment for the Holy See and at the highest point of tension between Julius II and Louis XII. The pro-French cardinals had gathered at the so-called schismatic meeting at Pisa. Gaston de Foix had reconquered Bologna only a month before, snatching it from the Church. Michelangelo's great bronze statue of the Della Rovere Pope had been broken and melted, and the head rolled in the Piazza Maggiore out of disrespect.[52] The French army had then marched on Brescia, sacked it, and crossed the Po Valley again, to turn towards Ravenna. The monster was born during that very period, the moment, as will be remembered, at which Bonaventura had admonished the Pope that 'he who sows swords, will reap swords', and been punished for it by papal anger. The governor of Ravenna, Marco Coccapani, immediately despatched to Julius an announcement of the monstrous birth with a description and drawing of the creature: evidently he linked it with the political situation.[53] The description was particularly alarming because it was said that the monster was the offspring of a nun and a monk, the fruit, therefore, of the corruption of the clergy and a pre-announcement of Antichrist. The clamorous defeat of the League's troops at the Battle of Ravenna— the city in which the monster had been born—on 11 April 1512, confirmed the creature's significance. Egidio of Viterbo opened the Fifth Lateran Council on 3 May with one of his passionate, torrential sermons. After having drawn a catastrophic picture of the moral decadence of the Church, he exclaimed: 'At what other time have there appeared

---

[50] This drawing is still kept in the original in Sanuto's *Diarii*: MS Marc. it. VII. 234 (= 9.221) in the Bibl. Marciana, Venice, fo. 179ᵛ. The German leaflet is in the Bayerische Staatsbibl. in Munich and is catalogued in G. Ecker, *Einblattdrucke von den Anfänge bis 1555* (Göppingen, 1981), no. 189.

[51] Cf. R. Schenda, 'Das monstrum von Ravenna: Eine Studie zur Prodigien-literatur', *Zeitschrift für Volkskunde*, 56 (1960), 209–25; O. Niccoli, 'Il mostro di Ravenna: teratologia e propaganda nei fogli volanti del primo Cinquecento', in D. Bolognesi (ed.), *Ravenna in età veneziana* (Ravenna, 1986), 245–77.

[52] M. Butzek, *Die kommunalen Repräsentationstatuen der Päpste des XVI. Jahrhunderts in Bologna, Perugia und Rom* (Bad Honnef, 1978), 322.

[53] S. di Branca Tedallini, *Diario romano dal 3 maggio 1485 al 6 giugno 1524*, ed. P. Piccolomini, *RIS* vol. xxiii, pt. 3, 327.

with so much frequency and such a horrible aspect monsters, portents, prodigies, signs of celestial menaces and terror on earth? When could there ever be a slaughter or a battle more bloody than those of Brescia or Ravenna? . . . This year the earth has been drenched by blood rather than by rain.'[54] The disastrous results of papal politics were linked, therefore, with ecclesiastic corruption and with monsters—Egidio was certainly referring to the monster of Ravenna—which were a sign of sin and would lead to punishment.

But significantly, right from the beginning, a different interpretation was placed upon the monster of Ravenna. Both Pietro Martire d'Anghiera and the Spanish historian Andrés Bernaldez speak of it in Spain as something noteworthy, but they insist that it was of Roman origin and that the spreading of its image and description were by papal initiative.[55] Sanudo was informed of it by the Venetian orator in Rome, Francesco Foscari.[56] The chief French evidence is a pamphlet which appeared at Valence on 18 September 1513, entitled *Les avertissements es trois estatz du monde selon la signification de ung monstre ne lan mille v. cens et xii.* Its author, François Inoy, was well informed on Roman matters (he was able to quote 'ung proverbe commun a Rome'),[57] and had perhaps stayed for a short time in the city. In these accounts not only does the allusion to a sacrilegious birth of a creature born to a monk and a nun disappear, but the contrary affirmation is made, at least by some authors (Pietro Martire d'Anghiera for example), that the monster was born to a married woman. Moreover, the circle of Latin poets from Giovanni Goritz and Angelo Colocci's Roman "accademia", to which Egidio of Viterbo also belonged,[58] produced a small work which interpreted the monster in a pro-Roman and anti-French sense. This was the *De monstro nato* by the Palermitan Giano Vitale, already published on 16 March 1512.[59] His interpretation of the creature was to identify it substantially with the 'holy religion', transformed into a horrible monster

---

[54] O'Reilly, 'Without Councils', 202–3.

[55] *Opus epistolarum Petri Martiris Anglerii* (Amsterdam, 1670), 256; A. Bernaldez, *Historia de los Reyes Catolicos, don Fernando y dona Isabel* (Seville, 1870), ii. 373.

[56] Sanuto, *Diarii*, 15, col. 200.

[57] Fo. D1ᵛ.

[58] As we understand from F. Ubaldini, *Vita di Mons. Angelo Colucci. Edizione del testo originale italiano (Barb. Lat. 4882)*, ed. V. Fanelli (Vatican City, 1969), 111, 114–15.

[59] We have two editions of this operetta, one which is certainly Roman, but without place or date; the other was printed as *Erfordiae per Mattheum Pictorium anno novi seculi XIƒ Mensis Iulio.* On Vitale, see G. Tumminello, 'Giano Vitale umanista del sec. XVI', *Archivio storico siciliano*, NS 8 (1883), 1–94; F. Ascarelli, *Annali tipografici di Giacomo Mazzocchi* (Florence, 1961), 68–9.

by the French schism. Naturally a laudatory conclusion was not lacking, in which Vitale augured that the stars were favourable to the court of Rome, rendering it sublime in the name of the 'divine Julius' and restoring world government to him, even placing the 'new Thule', America, under his command.[60]

The propagandist capacity of the monsters seemed inexhaustible. When, on 7 March 1513, another deformed creature was born in Rome four days before the election of Pope Leo, it was immediately linked to the event and, almost as a reply to the *De monstro nato* of the previous year, another erudite poet from the same circle, Giovanni Battista Ruberti 'Pegaseo', composed a *Monstrum apud Urbem natum*[61] also in Latin verse. After having dealt, using long mythological references, with the different possible causes of the monstrous births ('our life and limbs goaded by such great dangers so that while we are being formed in our mother's belly . . .'), the poet finally arrives at the point that interests us most. The monsters 'most often call for arms and preach Mars' horrible wars': thus the battle that laid the wretched Ravenna to waste was 'inwardly foreknown by the evil monster'. Ruberti seems to have participated in it, or at least to have seen the 20,000 dead lying on the plain after the massacre ('Ah, how many limbs of great men I saw surrounded with those of other soldiers! The youth of Spain and Italy lies avenged by the blood of the savage enemy.'). But now he wishes to banish these memories: 'may the omens of war be far from Italy, and the bloodthirsty desire for slaughter!' A wholesome breeze has calmed everything down, warlike fury is trembling impotently, its hands and feet are tied, the cruel gates of Janus will not open; 'peace will reign when mighty Leo reigns'.[62] Ruberti's *Monstrum* must therefore be added to the long list of compositions in praise of Leo X, or rather should be placed within that specific laudatory type which, according to a rhetorical scheme followed also by Erasmus, compared the peace-loving Leo's 'century of gold' to Julius's 'century of iron' (and Ruberti's references to a 'bloodthirsty desire for slaughter' and to the 'fury of war' now bound in chains, are certainly intended as an allusion to Julius).

He wished in short that the monster should be the harbinger of peace

---

[60] Vitale, *De monstro nato*, fo. 4ʳ.

[61] *Ioannis Baptiste Ruberti Pegasei Monstrum apud Urbem natum* (n.p., n.d. [Rome, after 11 March 1513]). We note that Ruberti makes the birth of the monster coincide with the election of Leo X, while the German pamphlet mentioned below leads us to understand that it was in fact born on 7 March.

[62] Ibid., fos. 2ᵛ, 4ʳ.

or at least that God and Pope Leo should avert the portents of war from
it. A similar interpretation was also made by Laurent Fries in a leaflet of
German verses, which carried a similar picture to the one in Ruberti's
pamphlet, under the long title: 'Nach dem geburt unsers herren Jesu
Christi tausent funffhundert unnd im xiij jar am vii tag des Merzen ist
dises selczam wunderliches unnd erschrockenlichs Monstrum nit weyt
von Rom von eynem weybs silt geboren wie nach volget.'[63] After describing the frightening appearance of the creature, which shared the joint
natures of man, pig, and monkey (since Ruberti did not speak of this,
the leaflet must have also used some lost Italian print), he explains that
such marvels are certain forerunners of epidemics, wars, and plagues,
and recalls the menace to Christianity from the 'infidel Turk'. But finally
Fries lightens this gloomy picture: God through his grace has granted
the Christians a Pope who does everything for the best, who, through his
authority, has made peace between kings, princes, and cities, and who
has established that all war will cease within a year. At the beginning
of Leo's papacy the monsters appear, therefore, to have been used as
vehicles of propaganda in the peace programme attributed to the Medici
Pope. The negative origin of their message came to be cancelled or
directly inverted by the beneficial effects of papal authority.

There are other examples which share this common ground. On 15
January 1514, Tommaso Giannotti of Ravenna printed a double edition,
Italian and Latin, of a *Prognosticon super Monstri ex Felsinea urbe oriundi
. . . significationibus*, followed by a *Tetrasticon sub enigmate*: 'If now mercy
remains fixed in highest heaven, may the stars change *u* to *o* and add *a* to
a *p* and indeed to another, so that he of these together may officiate with
the holy diadem. Then is the golden age; and all things will be governed
in peace.'[64]

Since the text affirmed that the monster of Bologna foretold 'cruel
lue' (plague) over and above other calamities, the solution to the riddle
will be 'Leo papa'. Thus for Giannotti, too, Leo X was in a position
to transform the malignant elements of the monstrous creatures into
benefits and inaugurate a new age of gold.

During these years Rome was not only a centre of diffusion but also
one which attracted monsters. Thus the mother of the creature born in
Florence in 1506 had it embalmed and brought to Rome. The news

---

[63] Kept in the Staatsbibl., Munich, and reproduced in E. Holländer, *Wunder, Wundergeburt und Wundergestalt in Einblattdrucken des 15.–18. Jahrhunderts* (Stuttgart, 1922),
312.
[64] It deals with a leaflet which carries the date 'Bononie die. XV. Januarii. M.D.Xiiii'.

of the monster of Ravenna in 1512 and that of Bologna in 1514 was transmitted to the Papal Court through official channels by the governors of the respective cities.[65] We should not, therefore, be surprised when we learn from Sebastiano di Branca Tedallini's diary that 'in the month of October 1513, . . . a Spaniard of not yet fourteen years came to Rome. He had a baby sticking out of his body with all its members showing, apart from its head.'[66] The Spanish boy had a Siamese twin attached to his chest, therefore, of much smaller dimensions than his own and without independent life. He undertook a highly successful tour of Italy together with the adults who accompanied him and exhibited him for money,[67] but it was at Rome, the first stop on his tour, that his stay caused particular comment. Giano Vitale, the author of the *De Monstro nato*, tried to repeat the success he had enjoyed two years before by composing another operetta in Latin verse entitled *Teratorizion*.[68] The mythological scenario was more muddled than ever. Italy presents herself at a meeting of the gods with trembling hands and hair hanging loose on her shoulders and laments the epidemics, invasions, and wars, recalling, amongst other things, the rivers of blood and heaped up corpses, 'lying without honour' after the battle of Ravenna, and declaring her aspirations towards peace ('may grim battles be lulled to rest, the wrath of the gods and the noise of the trumpets').[69] The fates prepare to reveal the future to her. They open the mouths of the caverns which enclose the monsters and a frightening creature comes forth: 'a boy comes out who balances an useless trunk under his breast, and a body without a name projects from his breast, which has a neck as its head and little arms instead of hands: its back seems to gaze at the sky.'

But then comes a positive explanation from the oracle: 'Now a new offspring smiles on better years: one faith is to be reverenced by all . . . honour with you . . . rises again, worthy child, which no ancient Roman saw'.[70] The missing members of the twin's body indicate the lost provinces of the Roman Empire. Europe is the head, Propontis and Africa the hands. But now a new age is beginning under the rule of Leo, 'great prince, beloved by the heavens, father of all men'. The Gods order that

---

[65] See Tedallini, *Diario romano*, 327; Sanuto, *Diarii* xvii, col. 516.

[66] Tedallini, *Diario romano*, 351.

[67] His presence is recorded in Florence and Venice; some drawings which represent him arrived in Ferrara. See Landucci, *Diario fiorentino, dal 1450 al 1516*, ed. I. Del Badia (Florence, 1883), 343; Sanuto, *Diarii*, xviii, col. 33; ibid., xx, cols. 179, 183.

[68] Giano Vitali, *Teratorizion* (Rome, 1514).

[69] Ibid., fo. A4^v.

[70] Ibid., fos. B4^v, C1^r.

the Empire should be totally reformed. Leo will know how to curb the barbarian hordes and all those desirous of war with strong bridles; he will be able to reassemble the mutilated body of Christianity: 'Devoutly add to it a head, and place hands on the arms, and let its limbs be joined, with their proper shapes.'[71]

Here we meet yet another illustration of popular culture—in the form of an obsessive concern with monsters and their interpretation— which was taken up by scholarly Roman cultural circles and adapted for propagandist ends. The monster of Ravenna had emerged in one of Renaissance Italy's most terrible moments, yet, alongside the interpretations which made it the precursor of disaster and a sign of divine retribution, were those which turned it into a favourable portent for the papacy. The three monsters which marked the first year of Leo X's papacy could be understood, as they partly were outside Rome, as a truly fatal sign, and yet in Rome they became, perforce, the announcers of prosperity and peace, because Leo's politics seemed to be addressed towards such ends and because they adapted well to the myth of the *aurea saecula* in which they were placed. These techniques had also been used in other contexts (such as Francesco Rocociolo's *Libellus de monstro*), but not so pointedly. The poets close to the papal court were probably putting on a rival show to the prophets. The latter were rushing to define the city as a Rome/Babylon stereotype (a definition which would reach its peak and fulfilment during the Sack, as André Chastel has admirably demonstrated), while the former were presenting Rome as the 'aurea urbs'. These phenomena embody two contrasting urban stereotypes, reflecting different images of Christianity, which were disseminated through leaflets and drawings passing from hand to hand or through other channels of popular culture. The trauma of the Sack and the spread of the Reformation cancelled out or at least weakened these propagandist possibilities.

[71] Ibid., fo. C2^r.

II

The Whores of Babylon

ANGUS MACKAY

In the years leading up to the Sack of Rome, there lived in the capital of Christendom a New Christian Andalusian priest who became intimately acquainted with the low life of the many prostitutes who lived in and around the district of Pozzo Bianco.[1] One result of his experiences was that he wrote a treatise on the treatment of syphilis, a disease from which he himself suffered.[2] But between 1513 and 1524 this priest, Francisco Delicado, also wrote a novel entitled *La Lozana Andaluza*, in which we follow the fortunes of the chief protagonist, Lozana, from her childhood in Andalusia, through her years as a prostitute and procuress in Rome, down to the point where Lozana decides to give up the low life.[3] The novel is famed for its realism. It is almost as if Delicado had equipped himself with a tape recorder and plunged into the streets and houses of Pozzo Bianco, catching all the nuances of the various languages and dialects to be heard in this district of Rome.[4] The book is obscene and funny, but it also has a serious message and is prophetic at two different levels. Since it is written in dialogue form, it is possible to eavesdrop on the conversations without necessarily identifying each speaker—for after all there are over one hundred protagonists, including Delicado who has written himself into his own novel.

From the conversations we gather that Rome is a cesspool of iniquity,

I would like to thank Dr G. Dickson and Mr P. Hersch for helpful advice and suggestions.

[1] On Francisco Delicado in general, see B. Damiani, *Francisco Delicado* (New York, 1974).

[2] Francisco Delicado, *El modo de adoperare el legno de India occidentale* (Venice, 1529).

[3] Francisco Delicado, *La Lozana Andaluza*, ed. B. Damiani (Madrid, 1969). Henceforth this work is cited as *Lozana* and quotations from it are translated into English.

[4] On *Lozana* as a masterpiece of realistic literature, see Damiani, *Francisco Delicado*, 9, 21, 42, 45, 121; A. Foley, *La Lozana Andaluza* (London, 1977); M. Criado de Val, 'Antifrasis y contaminaciones de sentido erótico en *La Lozana Andaluza*', *Homenaje ofrecido a Damaso Alonso* (Madrid, 1960), i. 432.

corruption, and rampant sexual immorality. The whole city is envisaged as a vast brothel and people call it 'Rome the whore'.[5] Others talk about the city in a proverbial manner: 'Look at Rome, the glory of great lords, the paradise of prostitutes, a purgatory for the young, a hell for every-body, a nightmare for beasts of burden, a fraudulent deception for the poor, and an open market for swindlers.'[6] The golden age for the *filles de joie* had been the pontificate of Alexander VI (1492–1503). In those days, according to an expert consulted by Lozana, there were more prostitutes in Rome than there were friars in Venice, philosophers in Greece, physicians in Florence, surgeons in France, moneys of account in Spain, stoves in Germany, tyrants in Italy, and soldiers in Campania.[7] Even so, according to Lozana herself, a leading banker has told the Pope that 100,000 ducats are spent on the prostitutes of Rome.[8] In fact people can do whatever they want in Rome—nobody will stop them because Rome is Babylon, *illa Babylon meretrix magna*, the haven and protector of sinners.[9]

Once inside this world of the sexual low life of Rome, bewildering complexities and subtle distinctions become apparent. For it is not simply a question of observing prostitutes in general but of distinguish-ing between different kinds of prostitutes—prostitutes of different nationalities or from different regions; prostitutes of different religious persuasions; prostitutes who ply their trade in an infinite variety of ways and in different places; prostitutes who cater for men of different back-grounds and occupations; prostitutes who fornicate in different ways and are expert in dealing with different sexual fetishes. Moreover there is a vast army of part-time prostitutes made up of maids, washerwomen, seamstresses, and women of other similar and humble occupations. For twenty-four hours a day, every day, the whores of Babylon ensure that supply is adequate to demand.[10]

Venereal disease is of course rampant. But even more tragic is the nature of the lifespan of these women. A prostitute, according to Lozana, can start her career at twelve years of age, but by the time she is forty she is finished. Thereafter she has to become a procuress, and finally

---

[5] *Lozana*, 64.
[6] Ibid. 81.
[7] Ibid. 131.
[8] Ibid. 146.
[9] Ibid. 120, 252.
[10] For a discussion of the activities and socio-economic problems and status of these prostitutes, see A. MacKay, 'Averroistas y marginadas', in *Actas del III Coloquio de Historia Medieval Andaluza* (Jaén, 1984), 247–61.

she has to resort either to peddling cosmetics, dubious medicines, and religious incantations for the cure of illnesses, or to trying to eke out an existence as a *beata*, hanging round churches and living with other similar women in order to share expenses. As a contemporary proverb put it: 'Prostitute in the spring, procuress in the autumn, and *beata* in the winter.'[11]

As Delicado unfolds the fate of these women, a human tragedy is being enacted. But the wrath of the God of the Old Testament will be inflicted on Rome, and throughout the novel there are specific prophecies to this effect. In Campo dei Fiori Lozana and her pimp–lover, Rampín, come across a fanatical preacher who foretells that Rome will be destroyed in 1527; Rampín comments that the corrupt and proud cardinals of the city will pay for their sins in 1527; and Delicado, as participant, passes on a prophetic jingle of advice to one of his friends: 'In 1527 leave Rome and go home.'[12]

These are of course pseudo-prophecies for, shortly after the Sack of 1527, Delicado fled to Venice, taking the manuscript of his work with him. Then, in Venice, he 'doctored' his novel before publication in 1528 in order to endow it with a moral, both by inserting prophecies into the text and by adding explanatory passages at the end. In fact Delicado states that he added the Epilogue after witnessing the destruction of Rome and the ensuing pestilence 'giving thanks to God for letting him see the punishment that God justly inflicted on such a great city'. Here once again the point is made explicit: Rome is Babylon, and the Sack represented a divine judgement.[13]

The pseudo-prophecies and the additional passages at the end of the novel, however, pose an important problem. Clearly the Sack of Rome and the divine punishment could not have prompted Delicado to write his novel because it was written up between 1513 and 1525. But does this mean that Delicado had originally simply intended to write an obscene and humorous book? In fact there is another more serious side to his work, as well as another prophetic level of a more complicated nature. We need to look at the protagonists and their backgrounds more closely.

Most of those who inhabit Delicado's world are Andalusians of humble wealth and status. New Christian females of Jewish origin—that is, they

[11] *Lozana*, 66, 48, 92–3, 167; L. Martínez Kleiser, *Refranero general ideológico español* (Madrid, 1953), 618.
[12] *Lozana*, 62, 64, 82, 120.
[13] Ibid. 252; Damiani, *Francisco Delicado*, 15, 21.

are *conversas*.[14] Moreover these women are aware of their *converso* status and the problems this involves. When Lozana first arrives in Rome, for example, the women she begins to make friends with are very much on their guard until they subtly establish that she is *de nostris*—that is 'one of us', a *conversa*.[15] They then begin to speak freely, telling Lozana that they have been in Rome ever since the Spanish Inquisition began to operate in Andalusia, and advising her that the Jews of Rome are friendly to *conversos* and will help her.[16] Throughout his work Delicado faithfully records the characteristics of these *conversos*. Their speech pattern constitutes a good example: whereas the Virgin Mary is not mentioned once and Christ receives a mere two incidental references, the God of the Old Testament is invoked well over a hundred times.[17]

But if the women betray no signs of being remotely Christian, they are equally not crypto-Jews except in the sense that they are influenced by the habits of their cultural past—for example in the way that they speak or in the kinds of food they eat.[18] They are at the same time both extremely intelligent or cunning and profoundly ignorant. On Lozana's arrival in Rome another *conversa*, Teresa de Córdoba, accurately sums her up in the following way to her friends: 'Within eight days she'll know everything there is to know about Rome. I can see that she's the type who will be a Christian among the Christians, a Jewess among the Jews, a Turk among Turks, a lady among gentlemen, a Genoese among the Genoese, and French among the French, because she'll outsmart everyone'.[19] But this picaresque-type intelligence and cunning does not reflect any profound religious knowledge. Lozana is theologically ignorant, indeed even illiterate. If in the novel's no man's land of religion

[14] See the excellent article of F. Márquez Villanueva, 'El mundo converso de *La Lozana Andaluza*', *Archivo Hispalense*, 56 (1973), 87–97.

[15] *Lozana*, 51–3. The women, of course, do not question Lozana on the matter. Instead they watch her prepare pastry to see whether she uses water or olive oil—the use of olive oil indicating that she is a *conversa* (Old Christians use water).

[16] Ibid. 54–5.

[17] The complete absence of the Virgin is surely extraordinary. In addition to the two incidental references to Christ there is an indirect Jewish allusion to him. There are some twenty-eight references to saints, but most of these are grotesque remarks about imaginary saints, such as St Nefija 'who used to surrender her body for charity' or 'who let men ride her for charity'. In contrast I calculate that there are some 116 references to God (in the form of *Criador, Deus, Dio, Dios, Dominis, Hacedor*, and *Señor*): *Lozana* 52, 108, 198; Márquez Villanueva, 'El mundo converso', 93.

[18] Márquez Villanueva, 'El mundo converso', 89, 94. The *conversas* partake of dishes that are predominantly Hispano-Muslim; sarcastic references are made to ham, but it transpires that Lozana herself keeps *prosciutto* in her larder: see especially M. Espadas Burgos, 'Aspectos sociorreligiosos de la alimentación española', *Hispania*, 131 (1975), 537–65.

[19] *Lozana*, 56.

Lozana can pass herself off as being a Christian or a Jewess, this is simply due to her sharp wits. Yet Delicado, author and protagonist in his own novel, is a priest and a *converso* who is aware of religious boundaries and obligations. Was he indifferent to the plight of these Andalusian *conversas*?

In fact Delicado took care to indicate that the reader should delve below the surface of the text. He claims that his book is a mixture of truth and entertainment, and he warns: 'I don't want anyone to add or subtract anything, *for if readers examine this work carefully they will find that what is missing at the beginning is to be found at the end.*'[20] Moreover, at the beginning Delicado rather mysteriously cites the fifteenth-century Castilian chronicler Fernando del Pulgar in part explanation of his purpose in writing: 'And as the chronicler Fernando del Pulgar says, "in this way I will forget my grief".'[21] What grief was this? In Delicado's case it could be argued that it was the Sack of Rome and its aftermath, even though he wrote most of the book prior to the Sack, but obviously this could not have been the case with respect to Pulgar's 'grief'. The remarkable fact is that Pulgar, who was also a *converso* and an influential political figure, was the only person who, in a prophetic mode, attempted to defend the young New Christian girls of Andalusia when the Inquisition first began its operations there. When Pulgar defended these girls in 1481, he prophesied that they would flee abroad in order to escape the Inquisition. Years later Delicado reveals their fate in Rome. Lozana asks her new-found friends: 'How long have you been here [in Rome]?' And Beatriz replies: 'My lady, since the year in which the Inquisition began.'[22]

When the inquisitors began their work in Seville in 1481, Pulgar wrote an open letter of protest to Cardinal Pedro González de Mendoza, the archbishop of Seville.[23] But the letter goaded an anonymous Old Christian to engage Pulgar in a bitter and public pamphlet-war. He attacked Pulgar for his widely circulated letter, and he made sure that his counter-attack was well publicized before reaching Pulgar. The latter, of course, did not know who the anonymous writer was, so he had to defend himself by writing an open and public reply.

---

[20] Ibid. 36 (the italics are mine).
[21] Ibid. 34.
[22] Ibid. 54–5.
[23] For what follows, see F. Cantera Burgos, 'Fernando de Pulgar y los conversos', *Sefarad*, 4 (1944), 295–348, which not only gives the texts of the letters but also provides an exhaustive commentary.

What had Pulgar written to provoke such a storm? He argued that the vast majority of crypto-Jews among the *conversos* of Andalusia were not being deliberately wicked but were simply confused and ignorant. Moreover the Church and the Old Christians were not setting them a good example. Addressing the cardinal, he summed up the situation in the following way:

> I believe my lord that there are some
> there (in Andalusia) who sin because
> they are bad, but the others, who are
> the majority, sin because they follow
> the example of those who are bad, whereas
> they would follow the example of good
> Christians if there were any of them there.
> But since the Old Christians there are such
> bad Christians, so the New Christians are
> such good Jews. I am certain, my lord,
> that there are ten thousand young girls
> between ten and twenty years of age in
> Andalusia, who from the time they were
> born have never left their homes or heard
> of or learned any (religious) doctrines
> save that which they have observed
> of their parents indoors. To burn all
> these people would be a very cruel thing.[24]

In effect, then, Pulgar was posing this question to the inquisitors: How can you possibly burn innocent New Christian girls who have never been instructed or given the chance to acquire any knowledge of religious doctrine?

Pulgar's humane and intelligent appeal was immediately subjected to a savage and dangerous attack by his anonymous opponent. Was there not something sinister in the fact that a *converso* was defending *conversos*? Accordingly the anonymous cast doubt on Pulgar himself, cautioning him to 'willingly swallow the great name of Jesus'.[25] Moreover, by criticizing Pulgar for using humour and sarcasm when dealing with such an important issue, his antagonist also implied that he was a political rebel. After all, since the Inquisition had been set up by the Crown, was not Pulgar's attack on its activities also an attack on royal authority as well?

---

[24] Ibid. 308.
[25] Ibid. 314.

Only with great difficulty and considerable ingenuity could Pulgar now extricate himself from the danger he was in. He easily disposed of the criticism that he had been too facetious by citing an impressive list of authors who had imparted profound matters by mixing the salt of humour with the food of truth.[26] However, his alleged attack on royal authority was more problematical and he could only defend himself by drawing a distinction between intentions and actions. Even though the inquisitors' actions may have been bad, he emphasized that the intentions of Queen Isabella the Catholic in setting up the Inquisition had been good.[27] In effect his argument might be summarized by the Latin aphorism 'quidquid agunt homines, intentio salvat omnes'.

But what about the Andalusian *conversos* and the young girls who had received no religious instruction and were innocently ignorant? Pulgar prophesied what would happen in a remarkably accurate way: 'To burn all these people would be a very cruel thing and even difficult to do, because they would flee in desperation to places where there would never be any hope of ever correcting them; and this would be a great sin, as well as a great danger for the responsible officials.'[28] Fleeing not because of religious conscience but because of their fear of the inquisitors, many of the young girls ended up in Rome itself, the Babylon 'where there would never be any hope of ever correcting them.' Delicado's novel, therefore, provides us with episode two in the drama of this lost generation of *conversas*. And there are other remarkable continuities between Pulgar and Delicado.

Immediately after invoking Pulgar's 'grief', Delicado explains that 'my intention has been to mix humour with the truth'.[29] Like Pulgar, therefore, he believed in 'salting' truth with entertainment and, more important, like Pulgar he stresses the importance of intentions as opposed to actions. As a protagonist in his own novel, he has to face up to the problem of Lozana's actions. She has been a prostitute and a procuress, but with the passing of the years she has turned to peddling cosmetics, curative charms, and magical incantations. Delicado visits her and explains that what she is doing is wrong and wicked. Then he says 'So, therefore, I ask you to tell me what your intention is.' Lozana replies 'What you have told me is good and holy, but mark my reply well, and it is that, in order to make a living, I have to say that I

[26] Ibid. 325.
[27] Ibid. 326.
[28] Ibid. 308–9.
[29] *Lozana*, 34.

know much more than I actually do know. . . .' It is at this point that
Delicado absolves her 'And I must tell you a truthful saying which
I have often read, and it is *quidquid agunt homines, intentio salvat omnes.*
Clearly, your intention is to make a living as best you can. . . .'[30] Pulgar
had argued that the inquisitors' intentions were good, although their
actions were bad. Delicado imputes good intentions to the inquisitors'
victims by stressing their socio-economic predicament. Thus when
Lozana arrives in Rome alone and destitute, he entitles the relevant
chapter 'How she learned to make a living, inasmuch as it was necessary
to use boldness instead of *sapientia.*'[31]

But what is *sapientia* and does Lozana possess it? The validity of good
intentions, when combined with bad actions, fundamentally depends on
not knowing that the intentions will have evil results. Pulgar had given
specific examples: When King John II of Castile had entrusted the town
of Toledo to Pedro Sarmiento his intentions had been good; he was
not to know that Sarmiento would subsequently lead the town into a
rebellion against the Crown.[32] It is ignorance or lack of knowledge,
therefore, which validates the intentions or, as the Talmudic precept has
it 'where there is no knowledge at the beginning, but there is knowledge
at the end' the sin is held in suspense.[33] Hence, too, Delicado's admonition to his readers that, if they examine his work carefully 'they will find
that what is missing at the beginning is to be found at the end'.[34] But
what exactly can we find at the end that is missing at the beginning?

As far as the young Andalusian girls were concerned, Pulgar had
argued that it was shameful to persecute them because they had never
had religious knowledge or instruction imparted to them. For Delicado
the word *sapientia* or wisdom also has a religious meaning, and his
verdict on Lozana at first sight appears curious:

Lozana was a very bold woman, and since women know that they are a consolation to men and their customary recreation, they think and do what they
would not do if they possessed the principle of wisdom (*sapiencia*), which is fear
of the Lord, and the woman who acquires this wisdom or intelligence is more
precious than any diamond, and those who do not are on the contrary most
vile.[35]

---

[30] Ibid. 176–8.
[31] Ibid. 45.
[32] F. Cantera Burgos, 'Fernando de Pulgar', 326.
[33] Babylonian Talmud, *Shebu'oth*, 5a.
[34] *Lozana*, 36.
[35] *Lozana*, 247–8.

Logically it would seem to follow that Lozana, who has been a sexual consolation to dozens, if not hundreds, of men must lack *sapientia* and be most vile. Yet Delicado reaches the opposite conclusion and acquits Lozana on two grounds. First, as has been noted, her style of life was a necessary strategy for survival. But secondly, what was missing at the beginning *is* to be found at the end because, as will be seen, Lozana finally does acquire *sapientia* and departs from Babylon. So without knowledge at the beginning were her sins, as well as those of the other girls, held in suspense?

Of necessity then, said Abaye, Rabbi holds that the knowledge gained from a teacher is also called knowledge. But if so, said R. Papa to Abaye, the statement in Mishnah 'where there is no knowledge at the beginning, but there is knowledge at the end' is incomprehensible, for is there anyone who has not even the knowledge gained from a teacher? He replied: Yes! it is possible in a child taken into captivity among heathen.[36]

Delicado's definition of *sapientia* is 'fear of the Lord', and his description of it derives directly from Job 28: 28: 'Et dixit homini: Ecce timor Domini, ipsa est sapientia; et recedere a malo intelligentia.' In the end Lozana acquires *sapientia* and by leaving Rome she departs from evil and shows *intelligentia*. But this wisdom has been acquired without religious instruction and without teachers. Lozana has realized the vanity of all life in Rome and in fact she even has a vision of 'the tree of vanity.'[37] So she departs from Babylon and goes to the island of Lípari.

But perhaps Delicado is also trying to tell us something else which no *converso* would dare say openly. When Lozana leaves Rome, she decides to change her name.[38] Implicit in this change there is, surely, a change of identity. But her new name, Vellida, is not a Christian name—it is a typically Jewish name.[39] Moreover she announces her change of name at the point when she determines to join her 'peers' or 'equals' in Lípari (which is what the name of the island means).[40] Is she then returning to

---

[36] Babylonian Talmud, *Shebu'oth*, 5a.

[37] *Lozana*, 244.

[38] *Lozana*, 245.

[39] This point is emphasized by Márquez Villanueva, 'El mundo converso', 93. In effect, after checking through Inquisition records, it becomes obvious that Vellida is so Jewish a name that even *conversas* would not retain it. See, for example, the lists of names in the indices of *Fontes Iudaeorum Regni Castellae*, ii and iii, ed. C. Carrete Parrondo (Salamanca, 1985–6).

[40] *Lozana*, 245: 'Let us go . . . to the island of Lípari with our peers/equals, and I will change my name, and call myself Vellida.'

Judaism? When she leaves the evil life of Rome, she promises her friend Rampín that if she sees Peace, she will send it back to him tied up in a knot of Solomon, and this mysterious knot is reproduced in the book.[41] Obviously the reference to Solomon once again suggests wisdom or *sapientia*. But the knot is also remarkably like the cross-section of a *tzit tzit*.[42] Now according to Num. 15: 37–41 the children of Israel were commanded to make fringes or knots 'in the borders of their garments throughout their generations' for the following reasons:

And it shall be unto you for a fringe, that ye may look upon it, and remember all the commandments of the Lord, and do them; and that ye seek not after your own heart and your own eyes, after which ye use to go a whoring.

That ye may remember, and do all my commandments, and be holy unto your God.

I am the Lord your God, which brought you out of the land of Egypt, to be your God: I am the Lord your God.

Did Delicado intend the whoring of his book to be both real and meta-phorical? On the one hand the young Andalusian girls, fleeing from inquisitorial persecution, have to survive in Rome by prostituting them-selves. This predicament is bad enough. But they are also theoretically New Christians, even if no one has instructed them. If Delicado's knot is a reminder of a *tzit tzit*, then it emphasizes the evils of both the whoring sought after by the eyes and the whoring after false Gods. What is missing at the beginning of Delicado's novel is there at the end: Lozana adopts a Jewish name, acquires wisdom or fear of the Lord, and uses her intelligence to depart from evil. The evil is Rome and all it stands for. The Sack of Rome is a divine punishment. Pulgar, Delicado, and the whores of Babylon have been vindicated.

---

[41] Ibid.
[42] I am very grateful to Dr G. Dickson for drawing my attention to this similarity.

12

A Note on Fra Pelagio,
a Hermit-Prophet in Rome

Thomas Cohen

On 5 August 1559, just days before the death of Paul IV, there appeared in the Tor di Nona jail before the magistrates of the court of the Governor of Rome, one Fra Pelagio, a hermit, healer, and prophet.[1] Until a few days earlier, Pelagio had dwelt immured in a little house behind St Peter's. The Jesuits may well have had more than a little to do with winkling him out, for their General, Lainez, and two others gave well-concerted depositions against him. Over the past two years, they said, when several of their number had visited him, they had found him ignorant, vituperative, arrogant, and garrulous to an extreme, responding to their questions with torrential obloquy. As Father Riera put it, 'It seemed to us that there was neither juice nor spirit in that man. . . . For his salvation and for the edification of others, he would be better off out of there than walled up.' Furthermore there were rumours that he kept money in the *banchi* and that he refused the sacraments on grounds he did not need them. Two other clerics, a Leo di Barberi and a Bono di Boni di Cortona, the latter an official of the cardinal of Trani, confirmed this picture of an erratic and choleric recluse, as did a druggist who had some years ago befriended the man. Furthermore the hermit was a prophet. As Leo di Barberi recalled:

I began to ask him what he was doing in there. . . . He began to get worked up and to insult me. I don't remember well, but it seems to me that he told me I was presumptuous to interrogate him about these things and he began to shout.

---

[1] Rome, Archivio di Stato, Governatore, Tribunale Criminale, Processi (XVI secolo), busta 48, caso 14. There are 20 fos. This was a preliminary hearing. We have no sentence and so cannot say if a proper trial ever took place. The death of the Pope and the violent Empty See riots that soon erupted could have brought the whole affair to an end. All quotations are from this document.

And I remember that almost in a prophetic spirit he began to turn towards the church and he said, 'This Babylon will destroy itself one day! There will come one who will extirpate it!' and other such words against our holy Church. He kept shouting at me until I arrived at the foot of the hill.

Pelagio had also been foretelling the death of the Pope. The druggist had heard him say that he himself would then become a cardinal; di Barberi had heard gossip that he had reserved the tiara for himself. Such words cannot have helped the hermit's case.

Before the court, the garrulous Fra Pelagio parried the magistrates' questions with a flood of words, punctuating his narration with imprecations, prayers, and forensic demonstrations of the most extreme piety, overwhelming the notary whose usual task it was to record faithfully a witness's every word. Though the judges, to prod him back to their inquiry, sometimes cut him off, the recluse poured out his life's story. In it, with caution, we can trace the outlines of the career of a minor prophet at the end of his tether. In his tale we see how in Rome, even in the harried years of Paul IV, a man of little schooling and the most marginal social origins could use extreme asceticism and a prophetic voice to garner the favours of the élite of both Church and lay society.

Pelagio's tale-telling had its strategy. Though years of privation and solitude may have skewed the hermit's emotional balance and blunted his tact, they did not drive him mad. Thus his narrative, although often reminiscent of the lives of desert fathers, had not only an aesthetic, but a distinct programme. Against the court's suspicion, the hermit wished to prove his holiness. Thus, he related his privations, his moments of ecstatic elevation, and his miraculous cures of the blind, the dumb, the halt, and the burned, whom, by spittle, prayers, and invocations of the Immaculate Conception, he relieved of their afflictions. In these stories, a circle of doubting on-lookers by their change of heart sometimes authenticated the miracle. As he recounted his past gifts of grace, Pelagio re-enacted them in the present, throwing himself on the ground and babbling of his unworthiness and God's generous mercy. Another purpose of the testimony was to point out how many and how potent were his protectors. Thus, though other witnesses confirm some of it, we must question this self-serving picture of his social network. At the same time, Pelagio, sensing danger, rushed to protest over and over his obedience, submission, and loyalty to the Church. He knew well enough that his cures and prophecies could cause him trouble. To cover himself, he claimed at once physical and mental imbecility and prophetic urgency. He sometimes tottered demonstratively before the judges and sometimes excused his prophetic utterances on the ground that he could not keep silent, lest his heart die. Finally, Pelagio, like many a suspect, tried to identify his detractors, such as, he opined, Bonsignore Cacciaguerra, master at San Geronimo, a zealot for frequent communion, and all his 'sect', and to blacken their motives.

Pelagio's life argues that a man with ascetic and prophetic skills could go far. He was born in Cosenza, the bastard son of a married woman named Isabella Megliorese and of a recent convert from Judaism. Shortly after his mother's death, barely bearded, he left home. He had horses and servants, but he soon gave his property to his kinfolk. What happened next is obscure. He attached himself to several lords in Naples, in Sicily, at Polignano, and then went to Vicenza, where he took the habit of a hermit, resolving to live without Holy Orders. He then changed his name to Pelagio, 'from Pellago, spurner of the world, as if it were a sea [*pelago*] of troubles'. He then wandered, going first, briefly, to Provence, to the Magdalen's cave at St Maximin, and then to Barcelona, where first the archdeacon Jacob Cassador, future bishop of the city, housed him, and then his mother, Sor Maria, whom Pelagio sought out as holy and, 'because she had predicted many things'. At her monastery, Pelagio fasted strenuously, begged in vain to be immured, and experienced ecstacies. Then, after 1542, he went to Valencia, where the saintly duke, Francisco Borja, supported him but again forbad him to be immured. Before the death of Paul III he returned briefly to Rome, bearing letters of reference from the Duke of Savoy. There he negotiated with cardinal Cesarini about his living arrangements and then, spurning offers of support, retreated to a hermitage he built on Monte Cucuzzo, above his native Cosenza, where he worked cures and lived off the charity of his neighbours. He dwelt there a year and a half, taking on a fellow hermit, whom he renamed Fra Serva Maria. On a cold winter's night, he said, a dove of fire entered their hut, circled it, and warmed them both. Pelagio then tried to go to the Holy Land, but, failing to cross Slavonia, returned to his mountain to find the hermitage a ruin. He went to Naples, failed once more to be immured and, in 1553 or so, came back for good to Rome.

Rome, symbol of power, seems to have brought out the prophet in Pelagio. The hermit soon made himself a little dwelling and a garden at Porta Portese, 'in an uninhabited place where they throw garbage and dead horses'. Pelagio must have moved to St Peter's after the death of Julius III, for that pontiff's demise found him dwelling in the palace of Cardinal du Bellay. He built his house, with several patrons' help, shortly afterwards and, with du Bellay's leave, in June 1556, walled himself in.

Pelagio's network of supporters cross-cut the coalitions and parties in the city. He had backers both pro-French and imperialist. Thus, he said, both Cardinals du Bellay and Pacheco helped build the house and both sent chaplains to visit him. Likewise Pelagio frequented the house of the francophile Ippolito d'Este but sought access to masses through the imperialist Pio di Carpi. Baronial quarrels seem to have made little difference. The women of the Colonna baked his bread and their staunch ally, the baron Giuliano Cesarini, helped pay for his hut, but the women of the Caraffa, who had usurped the Colonna lands at Paliano, paid calls to his cell. Pelagio's testimony is crammed with the names of the friendly great. One finds the women of the nephews of Julius III, who sought prayers and prophetic counsel, the ambassador of Ragusa who relayed prophecies, the bishop of Verona, Lippomani, who gave a Latin book, the count of Policastro who visited and gave alms, the prince of Bisignano, who sent letters. There are also various underlings who carried messages to their betters. We see the regulars as well. Pelagio cultivated the Jesuit Babadilla, showed prophecies to friars at the Aracoeli, left his money to the Theatines, sent a Dominican to buy his new cloak. Some of these patrons may have fallen away as Pelagio grew less stable or, as in the case of the Caraffa, as they lost power. By mid-1559, despite the friends, his detractors could prevail.

As a prophet, like St Bridget, who figured on one of his crucifixes, Pelagio had mixed political with apocalyptic prophecy. After he was immured, many would come to his cell, 'more than seven battles' worth a day', seeking not only prayers, cures, his spit, his blood (which he refused), but also advice on practical affairs. Pelagio claimed to have refused many, asking, 'Do you want to know the secrets of messer Dominiddio?' But, before and after the immuring, he had in fact not begrudged his insights. He had obliged the women of the papal captain, Ascanio della Corgna, who asked when the king of France would release him. He had told the people of the prince of Salerno that Don Pedro de Toledo, viceroy in Naples, would prosper because he backed the Inquisition and had informed the ambassador of Ragusa that the Theatine cardinal would be the next Pope. To a servant of Cardinal Morone, he had predicted his master's arrest, he claimed, a year before the event. He had also foretold the 'troubles of England' and mixed imperial fortunes in North Africa. The papal succession figured repeatedly in his prophecies. In court, Pelagio described an apocalyptic vision, devoid of such plainly political content, which mixed traditional and novel elements.

Before I was walled in, when I was in a trance in my room in the house of the
Cardinal of Paris (du Bellay), in the middle of the Borgo, there was shown to me
a vision of an antique wall with an ancient woman who seemed to be rich. And
she was crying because many people were shooting at her with arquebuses to
do her harm. They were men of war, the woman was crying, 'Ah! Ah!' And I
answered her, 'Oh lady, do not be troubled, for God will help you! The more
you resist and cry out, the more they will hurt you!' I worked this all out. That
woman was the Church and the wall was Rome. So, on a piece of paper I had
written out—by someone called Antonio, who lives in the Borgo—this vision,
making him paint a lady on that paper. And I painted a lady on that paper and
I wrote there three 'V's', which I interpreted as 'Ve! Ve! Ve!', and a 'P' with a
line through it, that said 'per' and an 'M' with a period that said 'molte' and
another 'P' that said 'parti' and a 'D' that meant 'della Christianità'. So these
letters mean, 'Woe, woe, woe for many parts of Christendom'.

Despite its military imagery, Pelagio's vision fell between wars. In it we
see not the high prophecy of the Joachimist tradition, but, among the
guns, some of the threads of visionary eschatology that had long been
available to the Roman imagination. Access to such images, even in a
man so ill balanced and, at times, so disagreeable as Pelagio, continued
in mid-century to open doors to a kind of power.

VI

PROPHECY AND
POLITICAL THEMES

Introduction

Writing on Gattinara, John Headley describes his *Oratio* of 1517 as
feeding 'the expectancy of a historical moment at a time and to an
audience caught up in possibly the most unique public intoxication ever
experienced in the Western development: namely, a charged atmosphere
prepared by generations of prophecy, driven to a new pitch of expectation by the dual eschatological signs of the Turk in the East and the
uncovering of a new world and unconverted people in the West . . .' The
dual prophetic message of tribulation and *renovatio*, inherited from the
Middle Ages and communicated so widely through the new medium of
print, could not fail to become entwined with political issues. Once
more we see that pessimism and optimism go hand in hand. Years ago
André Chastel showed that the expectation of Antichrist and Judgement
was as characteristic of this period as the themes of the Dignity of Man
and the Golden Age.[1] The emotional build-up to the Sack of Rome
highlights the fatalism of a brilliant society which was haunted by the
spectre of a Babylonish Judgement on Rome. Anxiety and hope mingle
in an extraordinary manner with the motivations of power politics and
diplomatic manœuvring. It is one of the ironies of the sixteenth century that the period which has (albeit unfairly) so often been stamped
as 'Machiavellian' in the secularity and divisiveness of its politics was
characterized, at least in its first half, by the most widespread dream of
a unified world. The prophetic signs of its approaching fulfilment were

---

[1] A. Chastel, 'L'Antéchrist à la Renaissance', in E. Castelli (ed.), *Cristianesimo e ragion
i di stato, Atti del II Congresso Internazionale di Studi Umanistici* (Rome, 1952), 177–86.

discerned in the empire of Charles V and the expansion of horizons to a new world. Even a seasoned politician such as Charles's Chancellor, Gattinara, was influenced by this vision. In the age when Europe was falling apart, the ideal concepts of law, concord, and harmony, as summed up in the haunting *leitmotiv* 'There shall be one shepherd and one sheepfold', still acted powerfully on the imagination.

13

Rhetoric and Reality: Messianic, Humanist, and Civilian Themes in the Imperial Ethos of Gattinara

John M. Headley

The extraordinary fascination exercised by Charles V's Grand Chancellor upon the historian lies in the coalescence of mind and will, of an imperial ideology and the political capacity to attempt its realization. A man of uncompromising integrity, of unquestionable loyalty to the house of Habsburg, and of commanding intellectual endowment, Mercurino di Gattinara possessed immense organizational abilities for the political and administrative realization of his ideals. The danger of the doctrinaire, the ideologist, in the field of politics would with time become partly tempered by his training as a civilian in Roman law and by his increasing exposure to Erasmian humanism. But his imperial messianism was also constrained by the intractable complex of the emerging nations of Europe as well as by the limited support that the emperor gave his great minister. For although he respected his chancellor, Charles never took too seriously the inflated imperial rhetoric among whose expositors Gattinara loomed significantly; here Habsburg diffidence probably afforded a greater political wisdom. Nevertheless under the dual apocalyptic pressures of America's discovery and the revival of Ottoman aggression, coupled with the internal threats to Christendom's unity occasioned by the Reformation and the emerging territorial states, Braudel's extended sixteenth century sees a recrudescence of empire both in theory and in fact. Among the outstanding exponents of this

I wish to thank my friends Ronald G. Witt (Duke University) and Peter Iver Kaufman (University of North Carolina) for their reading and valuable criticism of this essay which draws heavily in places on my previous publications, listed in the bibliography at the end of this book. Here I give a summation of my views on the specifically prophetic element in Gattinara's imperial programme.

imperialist revival, impelled by an apparently providential, if not apocalyptic expectation, must be listed Gattinara. Ultimately distinguishing him from those of his contemporaries singing imperialist paeans is the bald reality of the exalted bureaucrat, dynastic servant of long standing, and eminent counsellor who would now seek to seize the imperial *occasio* and realize its potential import.

As a disciple of Justinian Gattinara shared the view prevailing among civilians that the reality of universal empire existed in the body of Roman law itself irrespective of existing political incongruities. As a child of Dante he advocated an Italy that would forever be the seat of empire and an emperor who might reassert justice in a universal secular jurisdiction paralleling the church. The Dantesque vision of a jurist–emperor, who is the guardian and expositor of Roman law and who as *dominus mundi* champions justice and the law by a pre-eminent moral and juridical authority but does not impose them by force, had received wide dissemination during the fourteenth century in the teaching of Bartolus and Baldus. This vision of the sovereignty of law would now be reasserted. As one born and brought up in the last decades of the fifteenth century, Gattinara participated in the heady atmosphere of imperial messianism and eschatological expectation driven to a new level of excitement by the possible realization of a world emperor in the person of Charles of Habsburg.[1]

To appreciate the excitement and the foreboding that the young Habsburg prince awakened, we need to remind ourselves of the mental outlook that allowed men to experience themselves as participants in a prophetic scheme of history. It is well known that medieval religious and political prophecy had received its ultimate stamp under the influence imparted by the twelfth-century Calabrian abbot, Joachim of Fiore. Here was to be found the idea of that progressive trinitarian elaboration of world history, culminating in the Age of the Spirit with its profound sense of *renovatio*, renewal. As developed in the later Middle Ages, the Joachimist pattern looked to an outstanding ruler, a monarch of the whole world, a second Charlemagne, repeatedly identified either with a current French *Rex Christianissimus* or with a German *Rex Romanorum* who would renew the church, chastise its ministers, conquer the Turk, and gather all sheep into one fold. Some expositors of the Joachimist tradition believed that an Angelic Pope, a *Pastor Angelicus*, shunning

---

[1] J. Headley, 'Germany, the Empire and *Monarchia* in the Thought and Policy of Gattinara', in H. Lutz (ed.), *Das römischdeutsche Reich im politischen System Karls V* (Munich and Vienna, 1982), 15.

temporal goods and collaborating with the Saviour–Emperor, would rule the Holy See. The astrologer–prophet Johann Lichtenberger represented one of the most recent and well-known expressions of this cast of thinking. In his *Prognosticatio* of 1488 he heralded the appearance of a Burgundian world emperor who would arise as a Second Charlemagne, the prince and monarch of all Europe, in order to reform the churches and the clergy. Throughout this extensive literature that captured the historical imagination of successive generations, associating grave foreboding with great hope, there moved as a continual refrain a single text in which profound anxiety gave way to joyous release, the text of John 10: 16 'et fiet unum ovile et unus pastor'.[2]

When on 8 September 1517 Charles of Burgundy set sail from Flushing, Zeeland, in order to claim his Spanish kingdoms, he who embodied Habsburg destiny and represented for many the Saviour–Emperor had left behind in the Charterhouse of Scheut, forgotten and apparently removed from further political assignments, the man who would soon orchestrate the various themes of imperial renovation, reform, and world order adumbrated by poets and humanists into a vast scheme of Ghibelline realization. But the events of the autumn of 1517—whether the departure of Charles for Spain or the posting of theses by a Saxon monk—seemed to have left Mercurino Arborio di Gattinara irrevocably on the sidelines of the developing imperial drama. Scion of local nobility in Piedmont, product of a stiff juristic training and of almost a decade of successful practice as a lawyer in Turin, champion of a most exalted appreciation of Roman law, Gattinara had distinguished himself by his zeal in the service of the Habsburg dynasty. As chief architect of the notorious League of Cambrai against Venice, as a negotiator for the Archduchess Margaret and the Emperor Maximilian in a number of capacities, and as president of Burgundy since 1508, Gattinara had sacrificed every waking moment to the advancement of the dynasty. His rigorous and inflexible application of Roman law against a fractious nobility in the province of Franche-Comté had produced such turmoil there that this Habsburg praetorian prefect of Burgundy, as he was inclined to envisage himself, had witnessed in these past months the erosion of both his personal

---

[2] J. Headley, 'The Hapsburg World Empire and the Revival of Ghibellinism', *Medieval and Renaissance Studies*, 7, ed. S. Wenzel (1978), 94–5. For Joachimist developments, see above, Reeves, Ch. 1. See also M. Reeves, *Influence of Prophecy in the Later Middle Ages: A Study in Joachimism* (Oxford, 1969), *passim*, esp. 350, 365, 386–7, 431, 447, 507. Scriptural references are to the Vulgate, *Biblia Sacra* (Stuttgart, 1969) and the translations have been guided by Douai.

finances and his political support. His eight years of presidency of the
Parlement of Dole in Franche-Comté had been increasingly dogged by
personal litigation involving the purchase of a country estate. The case
had wound its way up through the judicial hierarchy, collecting in its
wake the recriminations and accusations of Gattinara's enemies, com-
pounded by the resistance that his high-handed measures had provoked
among the local nobility. The case had been taken to the supreme court
of the Burgundian inheritance, the Grand Council of Malines, which
had handed down an adverse judgement. To his enemies in Franche-
Comté as well as to those in the archduchess's government at Malines—
and probably to himself—Gattinara's career as diplomat, administrator,
and principal magistrate for the Habsburgs appeared to be over.[3]

His relentless and brilliant services to the Habsburgs had prevented
his fulfilling a long-standing vow, in lieu of pilgrimage, taken in 1513
and commuted by the pope. This obligation now brought him to the
Charterhouse of Scheut near Brussels in August 1516, where he was to
remain for the next nine months.[4] In his autobiography Gattinara later
mentions how in the months of his withdrawal at Scheut he composed
a small book, dedicated to Charles and designed to be presented to him
before his departure.[5] Nothing more effectively manifests the extent to
which Gattinara entered into the prevailing mood of apocalyptic exuberance and imperial aspiration than the recently unearthed *libellus*,

---

[3] Headley, 'Ghibellinism', 95.

[4] For this period in Gattinara's career, see J. Headley, 'The Conflict between Nobles
and Magistrates in Franche-Comté', *Journal of Medieval and Renaissance Studies*, 9 (1979),
60–76.

[5] I am indebted to Paul Kristeller who kindly alerted me to this manuscript, Brit. Lib.,
Add. MS. 18008 (See *Catalogue of Additions to the Manuscripts in the British Museum*,
1848–1853). The MS, described as being in the original stamped leather binding, is a
scribal copy in a beautiful humanist hand of the same *libellus* that Bornate believed to be
lost and that Gattinara refers to in his autobiography: 'Nec ob hec claustrum illud exire
voluit, donec tempus commutationis dicti voti impletum fuisset. Interim tamen tribula-
tionum tempore, ut spiritus aliqualiter recrearetur, conscripsit Mercurinus libellum ad
divum Carolum regem tunc catolicum, orationem super hiis supplicatoriam continentem,
somnium interserens de futura orbis monarchia ac futuro christianorum triumpho in per-
sonam ipsius divi Caroli, quem et Cesarem et maximum et omnium monarcham variis
rationibus futurum predixit, ipsumque libellum eidem Divo Carolo, prius quam ex gallia
belgica solveret in hyspaniamque navigaret, presentari fecit velut apertum futuri sui suc-
cessus presagium. Voto igitur perfecto, anno illo 1517, Mercurinus ad Margaretam pre-
dictam accedens de gestis conqueritur.' See C. Bornate (ed.), 'Historia, vite et gestorum per
dominum magnum cancellarium . . . con note, aggiunte e documenti', *Miscellanea di storia
Italiana*, 48 (1915), 233–568; 'Vita', 266. The date given for the completion of the *libellus*
is 8 Dec. 1516, at Brussels (fo. 107ᵛ). If it was written at the Carthusian monastery at
Scheut, Brussels—and the language of the prefatory letter to Marlianus suggests a place of
withdrawal—then the preferred dates for Gattinara's stay at Scheut, Aug. 1517 to May
1518, must be revised backwards. It is agreed that he entered the monastery in late August

composed during this period of withdrawal, following upon personal anxiety and mental distress. For, at the very time that he withdrew to cool off, the Netherlands was heating up: Charles of Burgundy's accumulation of crowns and titles promoted an acute expectancy. During his monastic retreat in the peace afforded by the cloister the fallen dynastic servant composed this newly discovered work, entitled *Oratio supplicatoria somnium interserens de novissima orbis monarchia ac futuro Christianorum triumpho*. It might be read as an impassioned petition, humanistically couched, wrapped in a dream, a vision, or, as he confessed to his closest friend Marlianus, even as a joke.[6]

Having delivered himself of his petition in the first twenty-one folios, amounting to a fifth of the total composition, Gattinara shifts to ponder the reason for the *vitiorum exuberantiam*; his dream reveals the cause: *principatuum pluralitas*. The prince of philosophers has offered the antidote: it is fit that among bees—as well as cranes—there be but one king. Universal monarchy recommends itself by necessity, for without the supreme prince, all ills pullulate. Sprinkling his route with appropriate supporting biblical texts as he proceeds—especially psalm 84: 11, *justitia* and *pax* embracing—Gattinara comes to identify the seventh vial of the Apocalypse with the ultimate plague of the Muhammadan sect and the occupation of the Eastern Roman Empire. Among the stock pieces of eschatological furniture is the idea of *discessio* advanced by II Thess. 2: 3. Gattinara interestingly understands it as the falling away of the other four patriarchates from the Roman obedience—namely those of Alexandria, Antioch, Jerusalem, and somehow Ephesus, instead of Constantinople.[7]

With only the Latin empire persevering in the faith of the Lamb, the fourth and ultimate state of the Church will be fulfilled in the universal

and departed the following May; disagreement persists as to the years: 1517–18 or 1516–17. P. De Wael, *Collectanea rerum gestarum et eventuum Cartusiae Bruxellensis*, i (1625), (Bibl. Royale Albert Ier, MS 7043), fo. 157ᵛ, confidently states the years to be 1517 to 1518, but given the uncertainty in the late medieval period on year-dates, an error may have been made here. Although neither Gattinara's letter to Marlianus nor his later autobiography conclusively assigns the composition of the work to the Carthusian monastery between 1516 and 1517, the weight of the evidence encourages such a conclusion. F. Ferretti, the most knowledgeable scholar of the evidence for dating events in Gattinara's life, gives the autumn 1516, for his entry into the monastery, see F. Ferretti, *Un Maestro di Politica; L'umana vicenda di Mercurino del nobile Arborio di Gattinara gran cancelliere di Carlo V re di Spagna e imperatore* (pub. by the Commune of Gattinara, 1980), 40.

[6] *Oratio supplicatoria somnium interserens de novissima orbis monarchia ac futuro Christianorum triumpho*, fo. 2ʳ⁻ᵛ. On Marlianus, Bishop of Tuy, see E. Rosenthal, 'The Invention of the Columnar Device of Emperor Charles V at the Court of Burgundy in Flanders in 1516', *JWCI* 36 (1973), 198–230.

[7] *Oratio*, fo. 2ʳ⁻ᵛ.

triumph of the Christian people over the Saracen whore. In this significant departure from the traditional Joachimist triadic structuring of time, by defining a fourth state or period for the Church, Gattinara had been preceded by no less a heretic than Fra Dolcino in 1300. Since this notable heretic had gone down to defeat and death in the very area of northwestern Italy where Gattinara had grown up, his influence here is not too improbable. The president of Burgundy's adoption of the idea of the fourth *status* has the effect of accenting the institutional Church's decay since the time of Constantine and the need for a reform that would recover the apostolic ideal of life. The result will be essentially a politicization of the Dolcinist vision wherein the effective force is the secular arm and a *Papa Angelicus* is absent.[8] Concomitantly the fourth state of the Church coincides with the fifth monarchy of Daniel.

Gattinara here allows himself to be guided through the apocalyptic of Revelation by no other than Giovanni Annio of Viterbo, humanist, notorious forger, and later Master of the Sacred Palace. His *Tractatus de futuris christianorum triumphis in Saracenis*, published in 1481, had enjoyed considerable circulation and even provided Gattinara with part of the title to his own work.[9] Yet he decisively transforms Annio's vision. For where the Dominican identifies the fulfilled third *status* of the Church as well as the fifth monarchy with the Papacy Triumphant, Gattinara deliberately excises all power from the ecclesiastical side of the apocalyptic ledger. What for Annio pertains to the development and consummation of the Church, for Gattinara belongs to the *monarchia* of the empire. Christ will rule its emperor, who successfully fights the Saracens and realizes universal monarchy. The president emphasizes that the subsequent millennium of peace, when there shall be one sheepfold and one shepherd, is the *monarchia christianorum* and not the *monarchia christi* which will occur in the ultimate resurrection. The prince of this future victory over the Saracens is called the Word of God or King of Kings. By distinguishing between the two *monarchiae* Gattinara tells us that he

[8] On Fra Dolcino, see Reeves, *Influence*, 243–8.

[9] *Oratio*, fos. 26ᵛ–27ᵛ. The edition of Annio's work used here was published in Nuremberg (n.d.). An earlier version with the title *Glosa sive Expositio super Apocalypsim . . . De statu Ecclesie ab anno salutis MCCCCLXXXI usque ad finem mundi . . .* was published, according to a later colophon, in 1507 at Cologne. Another undated edition was published at Louvain. See above, Reeves, Ch. 5, for the question of whether Annio influenced Egidio of Viterbo. Dependant on Annio for the general framework and for details, Gattinara also obtained the following from the Dominican: *Pluralitas principatuum* (sig. fviiiᵛ) when he cites XI *Metaphysics* of Aristotle; Ephesus as the fourth Eastern patriarchate (sig. av); the fourth *status* (sig. aviii); a brief quotation from the pseudo-Joachimist *Super Hieremiam*, 49 (sig. ciii).

wishes to free the Church militant from all arms and armies, all soldiers and their ministers, for, as Christ said to Pilate 'If my kingdom were of this world, my ministers would certainly contend with arms according to the habit of the world.'[10] (John 18: 36). Gattinara then proceeds to reject Annio's defence of papal universal monarchy.

After this important deviation from his source, Gattinara resumes his pursuit of Annio's argument by adopting the friar's association of the apocalyptic number 666 with the Antichrist in the form of the *secta mahumetica*. Dissension will now plague the Turkish order. The recovery of the *Imperium Constantinopolitanum* marks the first victory of the Christians, followed by the universal Church's regaining of the four patriarchates and the twenty-four archbishoprics represented by the four animals and twenty-four elders. As the Catholic King is enticed into accepting his role in the last world monarchy, Gattinara urges a concordance among scriptural texts, astrology, and Abbot Joachim for the impending universal monarchy, the Catholic faith and the Roman Church. Indeed, given the exuberance of the moment, he extends the Pauline unities, while claiming the support of Paul's authority: 'unus princeps, unus monarcha, unus dominus, una fides, unum baptisma, una omnibus vivendi lex.'[11]

Gattinara's departure from Annio culminates in that part of his vision that treats the salutary persecution of the clergy and the papacy itself. The president begins by musing that, whether true or not, the Donation of Constantine, according to the common opinion of the lawyers, is the poison that has come to be disseminated in the Church:

Certainly it would have been holier, if in accordance with the primitive church, tithes and first fruits [*primicias*] and other spiritual rights, which are of God, had been honestly obtained and they had been content with these, not aspiring to temporal powers and goods, which by the judgement of God himself are Caesar's. . . . In order that the realization of the last monarchy might by all means be fulfilled for men, the persecution of the clergy proceeds, and thus the saying of the gospel is realized: 'You are the salt of the earth, but if the salt shall lose its savour in which it is salted, it is no longer worth anything unless it be scattered abroad and contemned by men' [Matt. 5: 13]. In the first instance prelates and priests are meant; in the second instance 'in which it is salted'

---

[10] *Oratio*, fos. 27ᵛ–29ʳ; cf. Annio, *Tractatus*, sig. cvᵛ. This section (sigs. cᵛ–cviiᵛ), in which Annio argues for papal universal monarchy, Luther later assailed in a tract of 1537, reprinting the offending passage and denouncing its author in his own inimitable way; cf. *Luthers Werke* (Weimarer Ausgabe), 50, 96–105.

[11] *Oratio*, fos. 29ᵛ–34ʳ; cf. Eph. 4: 5.

the doctrine of the ecclesiastics and their lives are to be understood, and if not worthy they are to be 'scattered abroad', that is, deprived of their benefices and despised by men.[12]

In accordance with the prophecies of Malachi and Zechariah and the revelations of St Bridget, the Church will be devastated, but Gattinara generously allows that some clergy will have abstained from the infection of the greater part. This reformation of the Church will occur before the destruction of the Turk.[13] Once again Gattinara has gone beyond his Viterbian source, quietly changing all the markers, while appearing to follow Annio's lead.

With his customary pertinacity the President of Burgundy does not limit his reformation to the clergy in general but shortly comes to direct it upon Rome and the papacy itself. In shifting to a new aspect of his dream, he asks what prince will realize this fifth monarchy. The reader is hardly kept in an agony of suspense regarding Gattinara's answer that lights upon Charles, the Catholic King. It is interesting to note in passing that in his examination of the tangible resources as well as the concordance of authorities pointing to Charles, Gattinara refers to the recently discovered new world, the antipodes of the ancient cosmographers, 'with a new map having shifted from the arctic to the antarctic pole where a naked, uncivilized almost bestial people may be led to the cult of religion and the civil and rational customs of living that from thence may be yearly rendered a great supply of gold for his Catholic Majesty'.[14] Having considered the evidence—prophecies, scriptures, stars, the destiny of the House of Austria, and *reliquas inferiores qualitates*, that is his many lands and territories, the superior soldiery and armaments, as well as more numerous foundries, even the recent gains along the African coast as a precursor of that *monarchia*—Gattinara turns to that subject for which he apparently considered himself eminently qualified, the moral education of the prince, an indulgence which Charles would come to suffer with mounting restlessness.[15]

---

[12] *Oratio*, fo. 35. In connection with the expected disciplining of the clergy by the laity and the secular power—an idea based on late medieval prophecy and current among contemporaries, for example Jean Lemaire de Belges and Johann Lichtenberger—Gattinara claimed in his autobiography to have refused a cardinal's hat in 1519 because he felt that a reformation of the clergy was imminent and implicitly that he could accomplish more in this regard as a layman (see Bornate, 'Vita', 277–8).

[13] *Oratio*, fo. 36.

[14] Ibid., fos. 38ᵛ–39ʳ.

[15] Ibid., fos. 45ʳ–51ʳ.

Beginning his pedagogical ascent to the higher realms of imperial morality and deportment, Gattinara leads off with laziness, neglect, or torpor as a disastrous quality in the ruler:

For the empire had great growth, but subsequently the neglect of princes converted that splendour of the Roman name into obscure darkness. For then especially with the relaxation of virtue they gave themselves up to pleasures, to the baths and to that which makes the bodies of the young effeminate. Likewise even it is evident that the same has happened to the pontifical dignity. For with sanctity and good teaching [*doctrina*], which were not achieved without great labour and expenditure of virtue, the pontifical dignity, at first quite lacking in political powers, had grown amidst so many enemies and those opposed to the Christian name. But soon with political power the Church of God began to become wanton, having shifted its spiritual adornment from severity to lasciviousness; such license for sinning spawned for us these monstrosities and extravagancies whereby with office seeking and trafficking the most holy see of Peter was occupied rather than possessed. Lest therefore, O Catholic king . . .[16]

Having delivered himself of a latent but profound antipapalism which nevertheless fell short of rejecting the Holy See at Rome as the spiritual centre of pastoral leadership, Gattinara will soar off on his moral trajectory. Nevertheless this critical posture towards Rome persists and what makes the big difference is that it will persist not in a poet or literateur but in one who will come shortly to political influence and power.

In the subsequent deployment of the four cardinal virtues recommended for Charles's appropriation and exercise, Gattinara's resort to the classical authors appears more extensive and profound than in either of his earlier remonstrances—that to the Emperor Maximilian in 1514 or the most recent one to the Archduchess Margaret of Austria.[17] But apart from the resort to Xenophon's *Cyropaedia*[18] the same emphases are present: recourse to Aristotle, Plato, and Claudian, but with a prevailing preference for Seneca. In apparent contrast to his contemporary Machiavelli, Gattinara tells his would-be disciple that nothing is more appropriate in the governance of empires and kingdoms than to be loved and nothing more alien than to be feared. Despite having argued three years earlier for the precise reverse, Machiavelli will also cite with respect the same Senecan topos as does Gattinara here: 'unum est inexpugnabile

---

[16] Ibid., fos. 51^v–52^r.
[17] Headley, 'The Conflict', 54–5, 60–73.
[18] *Oratio*, fo. 53.

munimentum amor civium.'[19] The extent, depth, and familiarity of classical citation would suggest a humanism that goes well beyond the medieval recourse to *florilegia*. Nevertheless citation from the psalms predominates and comes to inform Gattinara's thinking and work. His fondness for the psalms remains a constant in his life. This special affinity is evident in his political *consultas* and breathes through his autobiography. However, Johannine texts also figure significantly: besides John 10: 16, another will play an important role in his dealing with successive emperors, when he instructs Charles: 'May you say openly "I and the Father are one", (John 10: 30) not only professing it verbally but manifesting it by your deeds and works.'[20]

So, to be sure, Gattinara will need all his humanist skill and spiritual, prophetic insight for the *peroratio* where he seeks to call Charles to the role of the Last World Emperor;

Perhaps the fear of death will terrify your Catholic Majesty in what is written regarding the future that with the task having been completed and the monarchy restored, the empire will be restored to Christ. This monarchy will continuously send forth the spirit, as the blessed Methodius is said to have predicted, when he says in his revelations: 'Then the King of the Romans will ascend to Golgotha . . . remove the crown from his head and place it upon the cross and extending his arms to heaven, bear the kingdom of the Christians to God the Father, and substituting himself, the King of the Romans entrusts repeatedly his own spirit [to Him].'[21]

With sharp doses of Cicero, Horace, and Baptista Mantuanus plus appeals to the Constantinopolitan hermit and King Valentinus, not to mention the marvellous concordance of all the evidence, Gattinara exhorts the possibly unconvinced young monarch:[22]

May the fear of death not deter you, O Catholic King, for here is a glorious end—a deed that will be remembered through the ages. [94ᵛ] . . . For what more glorious to you or any other is able to happen than to be claimed, chosen, called, for conserving this tossing bark of Peter in high seas, for congregating the sheep of Christ that had been dispersed when their pastor was struck so that now they are reduced to one flock and there shall be one sheepfold and one shepherd, for the repulsion and extinction of the enemies of the Catholic religion and the

---

[19] Seneca, *De clem.* 1.19.6; cf. Machiavelli, *The Prince*, IX, XIX, and XX; *Oratio*, fo. 75ᵛ.

[20] *Oratio*, fo. 56ᵛ. Like Gattinara, Egidio of Viterbo draws on the psalms extensively, see Reeves, Ch. 5, for his *Historia*, based on the interpretation of the first twenty psalms.

[21] *Oratio*, fos. 93ᵛ–94ʳ.

[22] Ibid., fos. 94ʳ–95ʳ.

exaltation and restoration of its orthodox faith so that all the rites of superstition may everywhere cease [97^v–98^r] . . . for reducing everywhere the entire world unto the true monarchy of the Christians and this monarchy may be restored to Christ. . . . What more glorious, blessed, praiseworthy death . . . you the restorer of the Christian name, . . . restoring the empire to Christ and to the Lamb its Vicar . . . Happy death . . . glorious end . . . *Exurge igitur Rex Catholice.* . . .[23]

At this point, part stunned, part bewildered, part intoxicated the modern reader may naïvely ask: Really? What are we to understand by this tract from the pen of a highly placed dynastic servant? What were contemporary readers supposed to believe? For apparently, according to the prefatory letter that Gattinara wrote to his friend Marlianus, the president did aspire to have his work published and to engage a diverse readership.[24] Following the above exhortation he gradually takes his leave, reminding us that it is a dream, a vision, but also very much a petition and supplication and here he once more returns to the charge.[25] Exhortation, prophecy, but above all supplication—in the agony of his own deprivation and the injustice he has suffered, he seeks to justify himself. He concludes the *libellus* by mobilizing extracts from the psalms in such a way as to transfer the divine lordship to that of the newly proclaimed Catholic King: ' "Hearken to my words, O Lord, understand my cry . . . I will enter into your house, in fear I will worship in the holy temple" [Ps. 5: 1–13] . . . "My enemies thrive and mount against me; those who with iniquity hate me are multiplied" [Ps. 37: 20–2]. "From the depths I have called to thee, O Lord" [Ps. 129: 1–4], . . . I seem to hear you with the same Psalmist responding to me: "Await the Lord, be strong, let your heart be comforted, sustain yourself in the Lord" [Ps. 26: 14]. I however have reposed all my hope in your Catholic Majesty.'[26]

Oratio supplicatoria somnium interserens: wrapping his supplication in

---

[23] Ibid., fos. 94^r–98^v.

[24] Ibid., fos. 4^v–5^r. The celebratory prophecies associated with Charles's advent would seem to make it marketable, but the facts that it is from a discredited and shortly to be deposed dynastic servant, that the personal, supplicatory element might appear to queer the rest of the work, and that Gattinara had sold his plate and was virtually destitute, would all militate against the expense of publication. In fact his friend, the Burgundian *maître des requêtes*, Claude de Chassey, wrote urging him to publish what appears to be the present work; see Bornate, 'Vita', 402. But, given the prejudices prevailing in the early years of printing, a splendid scribal copy rather than a printed version might well have been thought more appropriate as a presentation copy. Possibly more than one copy was made, for the existing one has occasional marginalia, largely from Gattinara's hand but some from an unidentified hand.

[25] *Oratio*, fos. 103^v–104^r.

[26] Ibid., fos. 106^r–107^r.

a prophetic vision, he feeds the expectancy of a historical moment at a time and to an audience caught up in possibly the most unique public intoxication ever experienced in the Western development: namely a charged atmosphere prepared by generations of prophecy, driven to a new pitch of expectation by the dual eschatological signs of the Turk in the East and the uncovering of a new world and unconverted peoples in the West, accented further by the attractive young figure of Charles of Burgundy piling crown upon crown, title upon land, and all finally mobilized by the new currency and community afforded by the printing press. It was a context capable of sustaining the creation of More's *Utopia* and pregnant with much else.

Did Gattinara mean it? Was his vision a true depiction of reality? Of course it was real; of course he meant it, although as time would reveal, not all aspects with the same intensity: certainly his self-justification, his imperial messianism, and his antipapalism would persist; the eschatological emphasis on the Emperor of the Last Days would, however, not be pressed upon his master in later years. Yet in 1517 all these aspects shared in a moment, both charmed and tense. As his friend Marlianus replied: 'Would that [the foolish] might see themselves in your book as if in the mirror of virtue and of truth, about which you write, and those keeping watch and awake dream what you dream and take hold lest such great assignments of God, nature and fortune slip away as in a dream.'[27] For an irrecoverable moment in the past such was the reality. At the same time we cannot disallow a certain playfulness that Philibert Naturelle, Chancellor of the Golden Fleece, had advised not only for his friend's own mental balance but also to provide Gattinara's reader relief from 'the long and prolix narration of injustices suffered'.[28] In advancing the prophetic, Gattinara was using the current language of political power and was satisfying the expectations of his presumed readership.

After emerging from his meditation and penance at the Charterhouse of Brussels, Gattinara suffered the long awaited deposition from office. He withdrew from Flanders and first visited his old master, the Emperor Maximilian at Augsburg. On his way homeward he apparently entertained notions of entering the service of his original master, Duke Charles of Savoy. But he was never to reach his destination. For in Spain Charles's grand chancellor, Jean Le Sauvage, had died in June; this second most important office at court and the one that was at the bureaucratic centre

[27] Ibid., fo. 6.
[28] Ibid., fos. 2^r–3^r. On Philibert Naturelle, see Erasmus, *Opus epistolarum*, ed. P. S. Allen, i (Oxford, 1906), 394–5.

of government needed a skilled jurist and experienced administrator. The gravity of the problem did not end here. In Germany, the ageing emperor and in the Netherlands his gifted daughter Margaret, who embodied the imperial aspirations of the dynasty, must both now attend to the future election of a Habsburg to the Holy Roman Empire. The narrow Flemish nationalism of Chièvres, who as grand chamberlain was the most prominent person in the entourage of Charles, could not provide the necessary vision of and justification for a world empire. The archduchess and the emperor knew their man. Despite his recent fall from office, the ex-president of Burgundy's pre-eminent legal knowledge, his awesome capacity for work, and his absolute devotion to the dynasty were never in doubt and made him the only logical candidate. On 15 October 1518 Mercurino di Gattinara took the oath of office between the hands of Charles in Saragossa.[29]

Gattinara arrived in the peninsula to head the royal secretariat at a time when his master was collecting titles and crowns that suggested to many contemporaries an astounding concentration of power and a recrudescence of the apparently moribund empire. Through his office the new chancellor would seek to orchestrate these imperial themes in support of Charles V. For on 29 November 1519 the delegation from the electors and the estates of the Holy Roman Empire represented by Frederick Count Palatine reached Charles at Molins de Rey outside Barcelona and presented formal recognition of his election in a speech by the jurisconsult Bernard of Worms.

Gattinara's responding oration is interesting both for what it emphasizes and for what it omits. He begins by addressing Charles, for what would appear to be the first time in the Habsburg's long reign, by that title reduced later to the simple SCCM that would appear on correspondence and *consultas* to the Emperor—*Sacratissima Caesarea & Catholica Majestas*. Having established this precedent of correct address, he develops the image of the imperial double-headed eagle. Eschewing the geographical possibilities of this figure—the eastern gaze including the Byzantine inheritance and the western focused on Rome and possibly extending to America—he individualizes its significance to meet the necessity of paying respect to the recently deceased emperor Maximilian: the darkening western sun is associated with grief, while the eastern sun by its orient rays greets the new reign. In the association of Christian with Roman imperial images, so reminiscent of Dante, the scriptural

---

[29] Headley, 'Ghibellinism', 96.

comes to predominate and inform the work at hand with a special imperial resonance. For the theme of the oration is the expected recovery of the long faltering imperial authority and jurisdiction. Thus verbs of restoration, renewal, revival abound. The realization of John 10: 16's one sheepfold, one shepherd approaches; the unplumed eagle of the prophet Ezekiel (17: 3–7) recovers; the embrace of *justitia* and *pax* sung by David (Ps. 84: 11) and meditated upon by Augustine strains towards its fulfilment. The divinely inspired election of Charles, we are told, signifies the restoration and renewal of the empire hitherto diminished and almost effaced. With the renewal of *sacrum imperium* the Christian Commonwealth may receive necessary care, the Christian religion be increased, the Apostolic See stabilized, and the enemies of Christians exterminated, so that the promise of the Saviour that there will be one sheepfold and one shepherd may be fulfilled. The oration reaches its climax in the confident belief that the empire, instituted by God alone according to the attestation of all laws, declared by the prophets, approved by Christ, preached by the apostles and sanctified by the holy canons, although divided under Charlemagne, may now be restored and led back to the obedience of a true and living pastor, *Carolus Maximus . . . felicior Augusto & Trajano melior.*[30]

We need to examine this speech more closely, with its rhetorical flights, its development of messianic and imperial themes, together with its context in Barcelona at the beginning of Charles V's reign. Certainly the event of Charles's election as Holy Roman Emperor, ostensibly the supreme legal magistracy of Latin Christendom, signalled a moment of imperial gratification and presented the opportunity for some florid prose. Moreover, the context afforded by Barcelona at this time provides

---

[30] Headley, 'Gattinara', 66–7. For Gattinara's speech (quoted n. 42 below), see P. Hane, *Historia sacrorum* (Kiel, 1728), 57–60. Its earliest printing is in *Legatio ad sacratissimum Caesarem Carolum ab principibus electoribus* (Antwerp, 1520), cf. W. Nijhoff and M. Kronenberg, *Nederlandsche Bibliographie van 1500 tot 1540* ('s-Gravenhage, 1923–61), no. 3369. J. Beneyto, *España en la gestación historica de Europa* (Madrid, 1975), 234–5, giving a date one year later than Gattinara's speech at Molins de Rey, attributes the newly devised 'SCCM' title incorrectly to Bishop Mota. On the words *felicior Augusto & Trajano melior*, see Eutropius, *Breviarium historiae romanae* (Oxford, 1703), viii. 2, who identifies it as a formula used in the fourth century in the Roman Senate's acclamation of the candidate for the imperial office. Gattinara uses this formula twice earlier in his *Oratio* of 1517 (fos. 74ᵛ, 102ᵛ). While the kings of England and France would appropriate in the 1530s the term *majestas*, which heretofore had properly belonged to the Holy Roman Emperor alone, Gattinara may be the first European statesman to steal it for his master when he repeatedly uses Catholic King and Catholic Majesty interchangeably in the earlier *Oratio*, before Charles began his campaign for the imperial office. On Dante's specific association of the Roman Imperial and the Christian, see C. T. Davis, *Dante's Italy and Other Essays* (Philadelphia, 1984), 23–41.

us with a peculiarly charged atmosphere supportive of these imperial claims. Thus, although we shall find occasional reiteration of this rhetoric in Gattinara's correspondence and memoranda during the next decade, never again will there be such an exuberant concentration of imperialist motifs and images.

At Molins de Rey the Grand Chancellor in his responding oration placed before his audience the image of the double-headed eagle with orient and tenebrous aspect and later emphasized the division of the Roman Empire caused by Charlemagne's coronation. He apparently sought thereby to free the idea of empire from papal dependence and instead of the conventional notion of *translatio* resorted to an earlier notion of parity between eastern and western emperors which in fact had considerable relevance from the time of Charlemagne to the conquest of Constantinople in 1204. In the halcyon years of Charles's coming to power Gattinara experienced the elation of prophetic, providential mythology associated with the Emperor of the Last Days and its post-Byzantine reference. The currency of such thinking is strikingly evident in high places at this time. Indeed Charles's apparently calculating grandfather Ferdinand the Catholic had eagerly obtained the rightful claims to the Byzantine empire from the last of the Paleologi, and even on his deathbed the king had been led to believe that he would still be spared in order to fulfill the prophecy and conquer Jerusalem.

In a similar vein the magistrates of Barcelona, spiritual heirs of Ramon Lull, writing now to Charles to congratulate him on his coronation at Aachen, hail him as the direct successor of Charlemagne the original founder of the western empire, and proclaim that he, Charles, will be the uniter of the western and the eastern empires, of western and eastern Christianity. From Barcelona as *capital de l'imperi* he will go forth to recover the Holy Land. In addition, at his coronation at Aachen Charles would be exhorted to expel the infidel from Christian lands and assume the Empire of the East together with Byzantium, the Holy Sepulchre, Egypt, and Araby. Moreover at Charles's coronation as king of the Lombards immediately preceding the coronation at Bologna, Paolo Giovio, historian and master of ceremonies, reported that in accompanying the emperor back to his rooms after the service he (Giovio) stood at the doorway and in a resonant voice proclaimed: 'Rex invictissime hodie vocaris ad coronam Constantinopolis'; at which Charles merely smiled and waved such an aspiration aside. Gattinara partook of this expectant mood, this charmed mentality but with a difference. Unlike the poets, philosophers, and mystics, as an exalted bureaucrat, dynastic servant of long standing, and imperial counsellor who after Chièvres' death in May 1521 apparently counted for more during the next decade than anyone else around the emperor, the 'Gran Canciller de todos los reinos y dominos del Monarca' would seek to seize the *occasio* and realize the dream.[31]

Three major themes—interlocking and pervasive—inspire this oration of imperial acceptance: namely the universal, the pastoral, and the providential character of empire. In studying each theme in turn, we shall find that the examination of one inevitably involves the other two. For example, beyond the fact that no other dignity *in universo orbe* may appear greater than that conferred by the Electors, this imperial dignity enjoys a unanimous consent, providentially informed by the Holy Spirit. If we attempt to isolate the universal for further exposition, it can be noted that prior to this oration Gattinara had drawn up his first, often quoted, *consulta* or memorandum with Roman law as the pre-eminent vehicle for both the announcement and the realization of universal empire:

Sire: God the creator has given you this grace of raising you in dignity above all Christian kings and princes by constituting you the greatest emperor and king who has been since the division of the empire, which was realized in the person of Charlemagne your predecessor, and by drawing you to the right path of monarchy in order to lead back the entire world to a single shepherd.

In pursuing his argument Gattinara warns Charles that the exaltation of the Christian faith, the growth of the Christian commonwealth, and the preservation of the Holy See, all for the attainment of universal peace, will be impossible without monarchy. After correlating peace and monarchy, he then raises the theme that would be the continuing preoccupation of a lifetime of dynastic service—justice. He calls upon the new Charlemagne to overhaul and codify imperial laws:

since God has given you the title of emperor and legislator and since it belongs to you to declare, interpret, correct, emend, and renew the imperial laws by which to order the entire world, it is most reasonable that in conformity with the good emperor Justinian, your Caesaric Majesty should early select the most outstand-

---

[31] Headley, 'Germany', 17–18. On Ferdinand, see J. Doussinague, *La politica internacional de Fernando el Catolico* (Madrid, 1944), 490; J. Sanchez Montes, *Franceses, Protestantes, Turcos. Los españoles ante la política internacional de Carlos V* (Pamplona, 1951), 98–9; on Barcelona, see J. Reglà Campistol, *Introducció a la historia de la Corona d'Aragó. Dels origens a la Nova Planta* (Palma de Mallorca, 1969), 102–4; on coronations, see K. Morrison, 'History malgré lui: A Neglected Bolognese Account of Charles V's Coronation in Aachen 1520', *Studia Gratiana, XV Post Scripta* (Rome, 1972), 681–4.

ing jurists [for] the reformation of the imperial laws . . . that the entire world may be inclined to make use of [these laws] and that one may say in effect that there is but a single emperor and a single universal law.[32]

The timeless quality of a justice inherent in Roman law and in the Roman empire again comes to the fore seven years later when the chancellor responds to the *Apologia* of Francis I for the peace of Madrid. In denying the legality of Charles's having to do homage for Artois and Flanders, he directly challenges that Romano-canonical tradition, starting with Innocent III's *Per Venerabilem*, whereby what had heretofore *de jure* pertained only to the emperor, began to be diffused among the kings and rulers of Europe:

As this very imperial dignity acknowledges no one to be superior in temporal matters, the emperor was not bound, nor by any law whatever could be distrained, so that to the king of the French, especially for those [lands] which [the emperor] previously possessed, he would perform the vow of homage or recognize [the king] as superior. Furthermore all this pretended superiority is considered restrained under the supreme imperial power itself whence it seems earlier to have departed [*digressa*]. For from the empire as if from a fountain, as the laws [of Justinian] attest, all jurisdictions come forth and thence they flow and flow back, for certainly anything will return to its original nature. The sovereignty of Roman law attests to the overriding supremacy of the imperial jurisdiction from which all other jurisdictions derive. And the superiority of the emperor endures irrespective of existing political contingencies even if it may not be recognized for over a thousand years.[33]

Italy and its role with the *monarchia* constitute an integral part of our inquiry into Gattinara's universalism. To Dante Italy was the garden of the Empire; to Petrarch, a land most holy, destined to be the mistress of all the world. Even to Paolo Giovio, the contemporary historian of Charles V's reign, she was still that infallible ladder of true monarchy. Nor did one have to be a Ghibelline to admire Italy's centrality within Christendom. A century after Giovio, Richelieu would allow in his *Political Testament* that Italy was deemed the heart of the world and the preeminent part of the Spanish empire. By the beginning of the sixteenth century Italy had become the decisive arena for the clashing ambitions and rival claims of Valois and Habsburg, yet the twin challenges to an imperialist interpretation from French legal humanists and Spanish

[32] Headley, 'Ghibellinism', 98–9; cf. Bornate, 'Vita', 405–6, 408.
[33] Headley, *The Emperor and his Chancellor; A Study of the Imperial Chancellery under Gattinara* (Cambridge, 1983), 103–4.

international lawyers were yet to emerge, each promoting a devastatingly pluralist view of political reality. The first third of the sixteenth century augured a different configuration, a Habsburg hegemony, if not world order.[34]

On first coming to the office of chancellor and continuing throughout the succeeding decade Gattinara in all his *consultas* drummed into his master's ear that Italy was the principal foundation of his empire, and lacking it, his honour was void and the growth of his empire jeopardized. He who would counsel ignoring Italy, counselled the emperor's shame and ruin. Repeatedly over the years Gattinara urged Charles V to come to Italy, secure his justice and order there, and complete the pacification of the land. When Gattinara speaks of Italy, he can mean the entire peninsula well known to the Roman jurist that he was, but he can also suggest the traditional notion of the *regnum Italicum*, which included only the north and central portions of the peninsula. He frequently distinguishes Italy from Naples and Sicily and continually reverts to the problems of the comity of independent states north of the Neapolitan kingdom.

Another distinctive feature of his attitude towards Italy is that this Burgundo-Piedmontese statesman has no feeling for Rome and its *Mystik*, which he seems to transfer to Italy as a whole. Rather than Rome, the cities of Milan and Genoa occupy his constant attention, and he never tires of insisting that they are together the gateway to Italy: the two duchies are the keys and bastions for keeping all Italy subject to the emperor and are recognized as the seat and sceptre for dominating the world. Here he defined a strategic truth that made the Genoa–Milan axis the veritable hinge of the entire Habsburg position in Europe, an axiom that would be affirmed long after the chancellor had passed from the scene.

How did Gattinara approach the problem of Italy's liberty, namely the independence of her city–states, which for his contemporary, Francesco Guicciardini, provided the social and political basis for Renaissance culture? The solution to the Italian riddle lay through Milan. He who controlled that city would be master of Italy. Gattinara steadfastly opposed the generals—Lannoy, De Leyva, Moncada, Pescara—who would seize Milan outright, impose a military solution upon the Italian problem, and reach an accord with France to the detriment of Italy. For this rea-

---

[34] See Headley, 'Ghibellinism', 107–12 and the accompanying annotation for the last paragraph.

son he brooked the emperor's wrath and refused to apply the seals of office to the treaty of Madrid on 14 January 1526. Instead he urged that the present duke, Francesco Sforza, be invested with the imperial fief of Milan and that the imperial army be reduced to a small effective force whose maintenance would not ruin the Milanese.

Since the end of 1523 and with increasing urgency during and after the negotiations for the treaty of Madrid Gattinara advocated an Imperial union with the Italian states. Italy was in his mind always the centre and basis of the Habsburg empire. 'Italy is to be preserved for him more by love than by force and with their love he will be able to dominate all the world and without it His Majesty will thrust his kingdoms and affairs into peril and never will he be able to recover himself without necessity and work.' Once having secured Italy, the emperor might extend the pacification of his land to a general pacification within Christendom, which would bring the kings of England and France into an alliance against Turk and heretic. In adjusting these two essentially opposing conditions—the independence of the Italian states and Italy's imperial role—Gattinara invoked the relationship between ancient Rome and her client states within the Roman Empire. He could warn that by using force rather than love and humanity, Rome had taken longer to master the rest of Italy than to conquer the world.

Gattinara's appeal to the Roman Empire seems to have been fundamental to his solution to the Italian problem. Gasparo Contarini, after five years as Venetian ambassador to the imperial court, emphasized this feature in his *relazione* to the Senate. He distinguished the two rival parties within the imperial council, one led by Gattinara, the other by the Viceroy Lannoy. In characterizing the former, he says that the Romans, Cyrus, and others who have produced something like universal monarchies have nevertheless not ruled all directly but have had other kings and other friendly republics that have favoured them, enjoying their fraternity. This was the way the chancellor guided His Imperial Majesty. The contemporary sources for the basis of his appeal to the Roman Empire lie beyond the scope of our present inquiry, but it can be observed that as a jurist trained in the Roman law Gattinara was doubtless aware of the *jus Italicum* that, developing out of the Republic and constituting a basic feature of the pre-Diocletian empire, extended the legal status of Italian cities to non-Italian provincial cities and communities; it comprised various rights of a public and private character such as self-government and exemption from supervision by the governor of the province. Thus, particularly for Italy and to some extent elsewhere, *monarchia* connoted ideally not a uniformly organized empire but a looser Habsburg hegemony that would leave room for local privileges, provincial customs, native institutions.

In the construction of Charles's universal monarchy the special care that Gattinara directed towards Italy and her imperial role supports his claim that he was an Italian, seeking the liberty of Italy. *Libertà d'Italia*—the words appear repeatedly upon the lips and from the quills of the leading Italian political thinkers and actors during the very years that saw the death agony of that independence which had made the Italian Renaissance possible. Both Gattinara's Florentine contemporaries, Machiavelli and Guicciardini, could agree with him in perceiving a vital connection between the destiny of Milan and the liberty of Italy. Yet both could now strenuously oppose the Habsburg *monarchia d'Italia* as they had earlier opposed French domination. However marginal an Italian Gattinara might be and whatever his tangled motives, he appeared to some Italians as one trying to moderate the impact of foreign domination upon the peninsula. The papal nuncio to Castile, Baldassare Castiglione, writing to the archbishop of Capua, could observe that no one in Spain had such a good mind for Italian affairs as Gattinara. Contarini, who often had caustic comments to make about Gattinara, could remark that he was a second Joseph in that both had the opportunity to benefit their people—Joseph the Hebrews, and Gattinara the Italians. He did not disguise his relief when in July 1525 Gattinara's resignation was refused, and he considered the chancellor's return to the Council of State a reason for all Italians to rejoice. For his own part Gattinara could afford to represent himself to Contarini as ready to dare all for the liberty and welfare of Italy. In 1522 he could prematurely claim the merit of having freed his country from the barbarians. And ever on the prowl for funds to support the insatiable war machine thus bringing its numbers and its pillaging under effective control Gattinara could approach Contarini as a fellow Italian and in his efforts to obtain a contribution from Venice, express his deep desire to remove the ruinous Spanish soldiery from the country.

In the supreme crisis of the Italian Renaissance that culminated in the sack of Rome, *libertà d'Italia* had a variety of meanings to the minds of the leading protagonists. In so far as he could realize and maintain it during the few years remaining to him, Gattinara's accommodation was certainly the most realistic; apart from Italy's role in his Dantesque theory of empire, Gattinara squarely confronted the problem of reconciling the peninsula to the facts of Spanish might and French menace.

In this respect Castiglione was more perceptive of current political realities than were his two Florentine contemporaries. The respective solutions of Machiavelli and Guicciardini either failed to materialize or were shattered by events. Gattinara, the child of both Dante and Bartolus, strove to construct a broad enough conception of monarchy to allow for those two poles around which Italy's life continued to move—universalism and particularism. By the treaty of Barcelona 29 June 1529, the emperor capitalized on the recent destruction of the French army before Naples and his alliance with Genoa and the fleet of Andrea Doria—an alliance for which Gattinara had provided the groundwork and impetus. The treaty, described by contemporaries as Gattinara's masterpiece, provided for the decisive accommodation between emperor and Pope. And while it sounded the death knell of the Florentine republic, it led to the pacification of Italy, the chancellor's last and greatest service rendered to his master. Certainly the treaty of Barcelona could not help but soften Gattinara's pronounced anti-papalism at the end of his life.

Before turning to the pastoral role of the emperor in Gattinara's imperial ethos and its antipapal current, we need to emphasize the justification for this present consideration of Italy as a function of universalism and to appreciate its legal component. The notion, supported by the Post-Glossators' tradition of Roman law and reinforced by Dante's understanding of Italy as the *giardin dell' impero*, encourages and promotes in Gattinara the belief that the essential *locus* constituting universal empire is not the city of Rome but rather Italy and specifically northern Italy. As direct heirs of Cino da Pistoia, Dante's jurist friend, Bartolus and Baldus presumed the continuing authority of the emperor in the sovereignty and universality of Roman law. Both jurists claimed it to be a sacrilege to deny that the emperor was *dominus mundi*. Within the tradition of Roman law itself the twin challenges to this imperialist interpretation—that of the French legal humanists and the Spanish school of international law—were yet to emerge, each promoting a devastatingly pluralist view of political reality. But for the present, the first third of the sixteenth century, political realities appeared to define a different configuration, auguring a Habsburg hegemony, if not world order.[35]

[35] Headley, 'Germany', 29; cf. Headley, *The Emperor*, 11. On Bartolus's appreciation of the emperor's power rather than the jurist's usual representation as expositor of particularistic sovereignty, see J. Baszkiewicz, 'Quelques remarques sur la conception de Dominium mundi dans l'œuvre de Bartolus', *Convegno commemorativo del VI centenario di*

In addressing the pastoral aspect of Gattinara's imperial ethos, we are again reminded of how integrally each of these three aspects involves the other two. To ask who is the true pastor of Christendom raises immense spiritual not to mention political issues linked both to an imperial messianism and to a tough even truculent antipapalism. A consideration of the trajectory of the pastoral motif in Gattinara's imperial ethos must begin with the two statements of late 1519. In the oration Gattinara has recourse to John 10: 16 that there will be one sheepfold and one shepherd in a way that implicitly identifies the emperor with this pastor, yet remains ambiguous. Such ambiguity, however, momentarily dissolves with the further plea that God might lead back this reconstituted *imperium* to the obedience of its own true and living pastor. In the earlier, the first of the great *consultas*, he calls upon the new Charlemagne to pursue 'the right path of monarchy in order to lead back the entire world to a single shepherd'. The import remains ambiguous, if evocative of Ghibelline memories.

In representing the emperor to that monarch himself and to his Spanish subjects Gattinara tempers the messianic note with a moral concern, feeding on classical models and examples that show the influence of humanism. Yet just as Dante in the letters hailing the advent of the Emperor Henry VII in Italy applied to the ostensible Saviour–Emperor scriptural texts properly pertaining to the Messiah, likewise Gattinara manifests the same practice. In 1514 he had seen the Emperor Maximilian and the protection he must afford his daughter and grandson symbolized in the eagle of Deuteronomy 32: 11 tending her young, and the statement of John 10: 30: 'I and the Father are one'. This reference would shortly be extended to Charles, as we have seen. In accordance with

Bartolo (Perugia, 1959), 9–25. In a notable passage of the *Oratio* (fo. 34) Gattinara presents an idealization of the emperor's power which conforms to later representations in its indebtedness to Bartolus but claims a more explicitly military role for the emperor, presumably for crusade, while manifesting Gattinara's persisting sensitivity to the papal decretal *Per Venerabilem*, 1202: 'Non dico Caesari tradendum omnium rerum particulare dominium, nec quod omnia regna, et domina ad eius manus sint particulariter reponenda, nec reges et principes spoliandi, aut privandi eorum regnis, et dominiis; sed id duntaxat curandum censeo, ut omnes reges, et principes superioritatem imperialem recognoscant, ut de iure tenentur, fateantur monarcham, ac lites eorum, quae tot bellis causam dedere, iudicio monarchae dirimant, eidem parcant, et assistant, eo quoque imperante in communes Christianae religionis hostes arma communia dirigantur, ipsaque divina impleatur sententia. Non enim quovis iure, quamvis secus iam de facto attentatum discernamus, potest quispiam se in temporalibus ab ipsa Imperiali superioritate exemptum contendere, nec validum exemptionis privilegium in medium adducere, quum etiam ipsemet Imperator sua sponte quempiam, in praeiudicium successorum ab ipsa superioritate eximere non posset: nec quamvis alienationem facere, per quam omnis superioritas ab Imperio abdicaretur.'

earlier Ghibelline practice reminiscent of the Hohenstaufen chancellery, the sacralization of the imperial office perilously approaches the blasphemous. Indeed there lurked in any imperial chancellery, early modern as well as medieval, this potentially blasphemous spectre which would soon be conjured by Gattinara's most loyal secretary, Alfonso de Valdés. Yet in the *consultas*, where the chancellor directly addresses the emperor, the moral strain is uppermost.[36]

A second aspect of Gattinara's Ghibellinism is his attitude towards the Pope. The Pope is considered chiefly a political and administrative figure—the ruler of the papal states, the grantor of *cruzadas* and taxes upon the clergy, so valuable to late medieval secular authorities. Like many notables in pre-Tridentine Europe before Luther began to draw attention to the papacy, Gattinara participated in a broad current of belief that saw the Pope pre-eminently in political terms and sought to reduce him to his originally pastoral office.[37] Nevertheless it would be a major misconstruction of Gattinara's personality and career as chancellor to view him as religiously indifferent or as making religious issues the mere instruments of political ends. He was the first of Charles's counsellors to assert the need for a council. The papal nuncio at Worms, Aleander, himself the epitome of a politicized ecclesiastic, saw Gattinara and Chièvres as using the issue of Luther to gain political ends. Yet while Chièvres thought that the whole Luther disturbance could be handled, Gattinara was impressed by its popular dimensions and saw the necessity of a council. His frequent recourse to monastic retreats, his enthusiastic support of Erasmus as the pre-eminent teacher of the orthodox faith and of a middle way, his continuing desire to see the life of the clergy reformed—all argue for a seriously experienced Catholicism that was neither Roman nor papal.[38]

---

[36] Headley, 'Ghibellinism', 101. Gattinara's first imperialist application of John 10: 30 occurs in his remonstrance to the Emperor Maximilian I, Sept. 1514, entitled 'Remonstrances de Messire Mercurin de Gatinare [*sic*], President de Bourgogne faictes à Maximilian I Empereur sur les traversés causés à sa personne et au Parlement par le Marschal de Bourgogne', Bibl. municipale de Besançon (Collection Chifflet, 187, fo. 128ᵛ). Following the Sack of Rome, Gattinara's secretary, Valdés, composed in 1528 the *Diálogo de las cosas occurridas en Roma* which concludes with the declaration 'que Jesu Cristo formó la Iglesia y el Emperador Carlos Quinto la restaura' (Madrid, 1969), 155.

[37] On Gattinara's incitement of the Council of Castile in the late summer 1526 to war against the papacy and the resulting imperial propaganda campaign, see Headley, *The Emperor*, 88–113. What is here so striking is that in spite of the papacy's spiritual authority, Gattinara tends to treat it like any other territorial state. On the dual character of pontifical sovereignty and its secular dimension, see the splendid study by P. Prodi, *Il sovrano pontifice* (Bologna, 1982).

[38] Headley, 'Ghibellinism', 102–3.

Gattinara was ready enough to withstand any effort on the part of the Pope to intervene to secular matters while advancing the claims of his imperial master to intrude upon the realm of the spiritual. In 1522 he continued to support Juan Manuel, the imperial ambassador at Rome, despite the determined campaign of Pope Adrian to get rid of him. Reading over the diplomatic correspondence, Gattinara grumbles, 'The pope will content himself with what is reasonable and leave it to us to ask advice of whom we like.' When he drew up the instructions for Miguel de Herrera as special envoy to Italy and to the Pope in November 1525 his tone becomes more menacing: '[Tell] His Holiness that if he does not want to use his office of common pastor for the tranquillity of Italy and of Christendom, then we will be forced to use our office as emperor, and His Holiness ought to take note that we still have in our hands the King of France and that he is in our power to leave when we wish it.'[39]

The imperial victory at Pavia in February 1525 encouraged Gattinara to urge the emperor, shortly after the receipt of the news, to make the Pope call a council to extirpate the errors of the Lutheran sect, reform the affairs of Christendom, and mobilize effective action against the Turk. If the Pope sought to excuse himself, then the emperor as *advocat et protecteur* of the church should undertake to convoke a council. As the diplomatic situation darkened and the conniving of Pope Clement with France and sundry Italian states threatened to remove the imperial grip upon the peninsula, Gattinara's insistence that Charles seize the initiative became more strident. In July 1526 he composed one of his longest *consultas*. Therein he urged the emperor to realize the goal of one sheepfold and one shepherd by going to Italy and by convoking a council for the reform of the Church and the extirpation of heresies. He would subject the Lutherans to the truth of evangelical doctrine, in which he believed the sect to be for the most part grounded, win them over as much as possible by amnesty, clemency, and pardons, and thus turn with renewed strength against the Turk.

Amidst the deepening diplomatic crisis the grand chancellor found an outstanding spokesman for his ideas in the person of Alfonso de Valdés, who had been a permanent scribe in the imperial chancellery since 1521 and became one of its Latin secretaries in 1526. Under the direction and

---

[39] Ibid. 103. The instructions to Herrera, which are not available in the Calendar of the State Papers, Spanish, are to be found in the Bibl. Reale, Turin, Stato d'Italia, 75, fos. 37^r–40^v. The present quotation is at fo. 39^v. See Headley, 'Ghibellinism', 122 n. 38 for the original text.

guidance of the Council of State and the grand chancellor, Valdés had been entrusted with the task of composing the official government report on Pavia. Therein Valdés identified the Spaniards as the elect people of God and presented the imperial victory as releasing Charles to attack the Turks and the Moors, recover the empire of Constantinople, and retake the Holy Sepulchre in Jerusalem, thus fulfilling the words of the Redeemer, 'Fiet unum ovile et unus pastor.' Now with the slipping of Pope Clement VII back into the French orbit and the materialization of the League of Cognac, Gattinara was able to look to Valdés's stalwart assistance in leading the diplomatic offensive against Pope Clement during the summer and autumn of 1526.

In those forceful replies that were delivered over to the papal nuncio, Baldassare Castiglione, at Granada on 17 and 18 September 1526, if the hand proclaimed the work of Valdés, the voice was clearly that of Gattinara. In a conscious effort to obtain the understanding and approval of secular and ecclesiastical princes, magistrates, and citizens throughout the Habsburg empire, the grand chancellor arranged to have the correspondence between Pope and emperor together with related materials published at Alcalá, Antwerp, Cologne, and Mainz. The first imperial reply to Clement and the letter to the Sacred College of Cardinals interest us here. The former letter, while raking up all past wrongs inflicted by Rome upon Charles, repeatedly finds the Pope neglectful of his pastoral duties, which is another way of saying that he is deeply involved in preparations for war directed against the very one who is the most obedient prince in all Europe and seeks only the good of Italy. Rather than a shepherd and mediator, the Pope has become a wolf, partisan, begetter of war. The letter reaffirms that those *duo luminaria*, both instituted by God, should rule co-operatively, bring peace to Christendom and war against the Turk. It concludes on the menacing note that if the Pope refuses to exercise his responsibilities as a father and pastor, the emperor must have recourse to a general council, which he now begs the Pope to convoke in order to heal the wounds of Christendom. Pressing the point still further, the imperial chancellery on 6 October dispatched a letter to the College of Cardinals, asking it to call a council if the Pope demurred. The consternation that these letters created in Rome was surpassed only by the impact made by the sack of the city itself.[40]

In the light of Gattinara's predilection for the psalms of David, as well

---

[40] Headley, 'Ghibellinism', 103–5. For a full treatment of these issues, see Headley, *The Emperor*, Ch. 5.

as for the greatest of Ghibellines, Dante Alighieri, we must examine
more closely some critical passages of that polemical confection, which
the imperial chancellery launched under the title of *Pro Divo Carolo* as
a propagandist blast against Pope Clement VII and Francis I. In the
most important of these assembled materials, the first response to the
Pope, the passage that mobilizes a text from the psalms (84: 11) to associate peace and justice and from the *Aeneid* (vi. 853) to promote clemency
is strongly reminiscent of the earlier usage of these passages made by
Gattinara, suggesting his direct intervention or Valdés's recourse to
materials the chancellor made available to him. For although no author
possesses a monopoly over the use of well-known texts, the use of the
psalms, so strongly evident in Gattinara's other writings, intimates that
he had a direct hand at certain points in the composition of the imperial
reply. Following psalm 2: 1–2 'Why have the Gentiles raged and the
people devised vain things? The kings of the earth stood up and the
princes met together against the Lord and against his Christ', we read, 'I
do not say against the Lord and Christ but against the minister and lamb
divinely established by Christ himself from whom all our authority and
power depend.' The text, which was ordinarily understood to pertain
first to David, then to Christ, is now directed to the emperor as the
Lord's anointed. Dante's use of these verses at the end of the first book
of *Monarchia* and the beginning of the second may well have prompted
their more explicit application here. Likewise both Dante and Gattinara
resort to the stock Virgilian imperial text that concludes with 'parcere
subjectis et debellare superbos' (vi. 853), but whereas Dante gladly
includes the preceding lines pertaining to the Romans and their providential role, Gattinara, understandably this time, sees no reason to
include the reference. The messianic note as it applies to the emperor
reverberates in other writings of the chancellor. Its import becomes the
more impressive when we realize that the force of the text is directed
against that pastor or common father who, rather than affirming the
established lines of obedience, by his leagues and conspiracies threatens
to violate the imperial state and dignity. The thought as well as the
specific psalm (2) is completed in Gattinara's 'Refutation' of King Francis's
'Apology'. The 'Refutation' presents the psalm's injunctions to embrace
discipline and good faith, lest they perish from the right way.[41]

Admittedly, Gattinara failed to enlist the services of Erasmus for
bringing out the *editio princeps* of Dante's *Monarchia*, which would

---

[41] Headley, 'Ghibellinism', 94–5; *Pro divo Carolo* (sigs. Hiiii, Kii^v, Lv.).

have given added weight to this propaganda campaign. Nevertheless the attempt itself is not something incidental but would appear to be profoundly indicative of the chancellor's mentality and purposes. He expresses thereby a basic political conformity with Dante in a common Ghibellinism or imperial messianism that is more providential than apocalyptic or eschatological. At the heart of Gattinara's Ghibellinism lingers the strong possibility that the Pope has abandoned whatever pastoral role he might claim and that it is the proper task of the emperor to exercise and realize this responsibility for the world. Moreover, in turning to Erasmus for the definitive publication of Dante's work, Gattinara registers the other intellectual force that would significantly influence his thought and policy in the last year of his life.

Returning once again to the *consulta* and the *oratio* of 1519, we find in both compositions the attestation of divine providence and of realized prophecy in the fact of the young world emperor whose presence works to reconstitute the hitherto divided and scattered empire as *monarchia*. The apparent fulfilment of John 10: 16 and of the Davidicum is less the product of a sudden miraculous intervention bringing a new aeon than of the fact of empire itself, 'instituted by God alone . . . announced by the prophets, approved by the events of the Saviour Christ's birth and death, preached by the Apostles and sanctified by the sacred canons themselves', now faced with the opportunity for its realization in Carolus Maximus.[42] The inherence of the providential motif in the universal and the pastoral makes unnecessary its further consideration, except to observe that whatever the messianic overtones (reaching their climax from 1525 to 1527), the apocalyptic and eschatological are muted and levelled down in a divinely providential secular time. The very effort to reconstitute the empire from its vestiges would serve to mute the eschatological. The chancellor's legal training, the heavy burdens of statesmanship, his growing exposure to Erasmian humanism worked to

---

[42] Hane, *Historia*, 59–60: '. . . ut divino satisfiat obsequio, Reipublicae consulatur, sacrum Imperium restauretur, Christianae religioni incrementum accedat, Apostolica sedes stabiliatur, ipsa Petri navicula diu fluctuans, in salutis portum de[d]ucatur: perfidorum quoque Christiani nominis hostium exterminatio sequatur, hincque Salvatoris sententia impleatur, ut fiat unum ovile, & unus pastor . . . Quid praeterea laudabilius iis adscribi posset, quam quod ex ipsa praeteritarum, praesentium & futurarum rerum animadversione eum Imperatorem futurum decernerent, qui diminutum ac fere exhaustum Imperium restaurare posset, qui implumem Aquilam refoveret, renovaret, ac ad propriam naturam deduceret . . . Faxit itaque Deus optimus maximus, ut hujusmodi Imperium sub Carolo magno divisum, & ut plurimum a Christianae religionis hostibus occupatum, sub Carolo Maximo valeat instaurari, ad ipsiusque vivi & veri pastoris obedientiam reduci: sitque electus ipse Caesar noster felicior Augusto, & Trajano melior.'

promote a political and moral cultivation of Charles that would make it impossible for Gattinara to cast his master in the role of the Last World Emperor.

Yet Gattinara is no modern nor does he experience time as a modern. For whatever his own energies and motivations, the context of his age, saturated in prophecy and the apocalyptic, obtrudes upon his consciousness. Consider again a time that labours under the dual apocalyptic impact of America's discovery and the revival of the Turkish menace. Consider also the charged atmosphere of Barcelona in 1519 during the emperor-elect's presence there. And, as a curious irony, consider the fact that when the Dresden printer Wolfgang Stoeckel decided to bring out a German vernacular edition of book 1 of the *Pro divo Carolo* directed against the pope, he chose to append two short prophecies, one of which pertains to the Emperor of the Last Days in the form of a Second Charlemagne prophecy.[43] For his part Gattinara avoids the extra-scriptural and follows Dante, who directs him back to the psalms. Perhaps the most convenient yardstick for determining the difference between the chancellor's mentality and that of a modern is to consider the probable reasons that compelled Erasmus to refuse the venture of publishing Dante: entertaining a healthy distrust of any one total secular order, the humanist did not share the chancellor's belief in the divine basis to world empire, nor the legal claim, nor the providential motif.

In the last years of his chancellorship both the turn of events and the development of his own mind and policy moved towards accommodation and compromise rather than crisis or confrontation; the language of reasoned argument came to prevail in a new way over that of prophetic challenge and imminent catastrophe. Beginning in July 1526 Gattinara broached the possibility of an accommodation with the Lutherans in central Europe, and in the process he drew more closely to Erasmus and to the humanists in his own *familia*. The political opportunities of 1528–9 permitted Gattinara to shelve his Ghibellinism in order to achieve a long-

---

[43] The first one (sigs. Nii^v–Niii) purports to derive from a learned doctor in 1440, although it betrays all the marks of a contemporary re-presentation of the pseudo-Joachimist schema of prophecy. It predicts that Charles, the son of Philip, will be crowned at seventeen and by the twenty-fourth year of his empire he will have extended his power to all peoples from the English and Lombards to the Arabians, Palestinians, and Georgians and even across the seas to the heathen, subjecting all to his empire 'und ein gemein Gesetze machen und in aller werlt [*sic*] aussruffen lassen'. In the thirty-sixth year of his empire he will come to the holy city of Jerusalem, be anointed, and his soul will be assumed into heaven, crowned and hailed by the holy angels 'ein Vater des Vaterlandes', being the first crowned emperor since Frederick III. I am here using the Brit. Lib. copy of the *Prognosticon*, cf. Reeves, *Influence*, 328, 389–91.

sought accommodation between Pope and emperor through the treaty of Barcelona 29 June 1529. As he proceeded with the imperial entourage from the pacification of Italy and the emperor's coronation at Bologna to the Diet of Augsburg, even the Turkish menace, which harboured apocalyptic motifs and would release a spate of apocalyptic exhortation, began to recede from the walls of Vienna. Although the chancellor never reached Augsburg but died at Innsbruck, everything suggests that he was decisively drawing towards an Erasmian solution to conjuring the Lutheran spectre. In the letter of his friend and subordinate Cornelius Scepperus, written to Erasmus three weeks after Gattinara's death, we learn that the last works read by the dying chancellor were Erasmus's *Deliberation on How the War with the Turks is to be Conducted* plus his *Exposition of Psalm 22* and other items freshly arrived from the press. While in the end humanist motifs had clearly come to prevail, there still remained the inevitable context of the prophetic and potentially apocalyptic.[44]

In conclusion, whose rhetoric? Whose reality? It has not been our intention here to oppose the two. Rather rhetoric and reality at least for Gattinara are indissociable. Rhetoric conveys the perceived reality. His most apocalyptic of statements derived from a moment in his life torn with profound dismay, yet buoyed up by the groundswell of expectation marking the inception of young Charles of Habsburg's rule. The monastery provided the appropriate context for nurturing these darkly exuberant thoughts. But with his promotion to the chancellorship the exigencies of high political office, accentuated by the Revolt of the Comuneros, reinforced and brought to the fore the legal and humanist aspects of his training. Except for outstanding occasions such as that marked by the *Oratio* of November 1519 and the antipapal stance that generated the Ghibellinism of the imperial chancellery's propaganda campaign in 1526–7, the practicalities of office tended towards avoiding the apocalyptic and prophetic. In the final years the association with Erasmus and Erasmian humanism would lend its definitive mark to Gattinara's thought, expression, and policy. Thus on balance, according to the needs of the occasion, an otherwise flamboyant apocalyptic rhetoric comes to be politicized, commensurate with the practicalities of the Habsburg empire in the 1520s.

---

[44] Headley, 'Gattinara', 95–6.

14

A Note on Prophecy and the Sack of Rome (1527)

MARJORIE REEVES

In the 1520s the two prophetic images of Rome[1] confronted each other in a conflict which seemed to contemporaries to reach a cosmic climax in 1527. The splendour of Rome, adorned by architects and artists, which dazzled the eyes of pious pilgrims, made the figure of the New Jerusalem ('as a bride adorned for her husband'[2]) a reality. It was both venerably ancient and magnificently new. Contemporary guidebooks show how 'pious tourism blended the *mirabilia* of Christianity and antiquity into a joint celebration of this city above all cities. This constant amalgam, these mingled histories, made Rome irresistibly appealing to intellectuals, poets and artists, as well as to the throngs of pilgrims.'[3] This glory of Rome was, of course, focused in the papal office. Here we may see the decoration of the four *stanze* of the Vatican as embodying 'an exalted defence' of the papacy and its politics.[4] Begun under Julius II, the programme was carried forward by Leo X and completed under Clement VII. In the third *stanza*, where the deeds of Leo III and IV are evoked, the pre-eminence of the papacy is reaffirmed. 'The very existence of the Church is portrayed as a continuing miracle, but with precise consequences within the political order.' The fourth *stanza*, the Hall of

---

[1] See above, Reeves, Ch. 1. Sadly, Professor André Chastel was unable to write this chapter as promised. With his kind permission I have drawn very heavily on his essential work, *The Sack of Rome*, in the translation by B. Archer (Princeton, NJ, 1983). Many of his evocative phrases which I have used are acknowledged in the text by quotation marks.

[2] Apoc. 21: 2.

[3] Chastel, *Sack of Rome*, 6. On the four *stanze*, see D. Redig de Campos, *Le stanze di Raffaello* (Rome, 1950); J. Shearman, 'The Vatican Stanze: Function and Decoration', *Proceedings of the British Academy*, 57 (1971), 363–424; S. Freedberg, *Painting of the High Renaissance in Rome and Florence* (Cambridge, Mass., 1961), 112–31, 151–67, 293–312, 568–75.

[4] Chastel, *Sack of Rome*, 50–65.

Constantine, culminates in the Donation to Pope Sylvester, forming 'the most authoritarian assertion ever made of the legitimacy of pontifical claims'. The frescoes would seem to be a deliberate reply to Lorenzo Valla's pamphlet showing the Donation to be a forgery. 'In the eyes of the Vatican the presumed relationship between Church and State was a symbolic illustration of a divine and incontrovertible mandate that confirmed the supremacy of Rome's vicar over the Empire.'

The myth of *Roma aeterna* and the image of the New Jerusalem were blended in one. 'As that monumental cycle in the Vatican ceremoniously reaffirmed', the city in the eyes of the pious occupied a unique historical position. Far from declining, this notion was nourished by all kinds of scholarly proofs. Egidio of Viterbo, who 'synthesizes one whole aspect of theological thinking under Julius II and Leo X', represents this line of thought. As we have seen[5] he brought together the pagan and Hebrew/Christian heritages: Janus, the god of Rome, carries keys like St Peter. Such an image entailed 'monumentality and grandeur'. It is significant that Egidio, who passionately desired simplicity and purity in his Order, supported the sale of indulgences for the glorification of Rome. But at this point the dark shadow of that other image inherited from the Middle Ages begins to edge towards the shining spectacle of the New Jerusalem. It was not solely that the antipapal forces were gathering. Rome as New Babylon haunted the imaginations of the faithful as well and the anguished denunciations of wickedness and corruption were heard in the innermost ecclesiastical circles. The 'Whores of Babylon' were in full evidence and Rome was seen as a vast brothel.[6] The fervent reforming groups in Rome, typified in the Observant movements, contributed by their criticisms of clergy and Curia to the gathering storm which in the long run made people receptive of alarmist prophecies. The *Vaticinia de summis pontificibus* were circulating in early Italian printed editions. Although these end with an Angelic series, a large number of them are concerned with bitter denunciations of successive Popes. The 1525 edition was dedicated to Clement VII. One picture which with hindsight must have seemed pointedly prophetic shows the Pope riding away from Rome.[7]

[5] See above, Reeves, Ch. 5.
[6] See above, MacKay, Ch. 11.
[7] Chastel, *Sack of Rome*, 66. This picture is usually number eight in the printed editions. It belongs to the later series of Pope Prophecies, produced before 1356, and appears to be linked to one of the Leo Oracle pictures in which a king with a hawk on his wrist rides away watched by a queen standing in a doorway (Oxford, Bodl. Lib., MS. Barozzi 179, fo. 19ᵛ). In the *Vaticinia* this becomes a mitred Pope riding away with his hand raised in

But it was, of course, among the antipapalists north of the Alps that
the image of Rome as the New Babylon grew to such alarming proportions. The *Vaticinia* themselves were seized upon by German reformers.
Andreas Osiander, the Lutheran minister of St Lawrence, Nuremberg,
found copies of the work in two libraries and published an edition with
a German commentary and a verse for each picture by Hans Sachs.[8]
Anti-Roman polemic restated the primary thesis of Roman doctrine in
reverse: the dogma of the City's providential role was inverted as that
of Babylon and papal authority as that of Antichrist. 'The desanctifica-
tion of Rome could only be achieved by a diabolization, a sanctification
in reverse.'[9] The very forces that urged Roman humanists to glorify
the providential nature of Catholicism led those who rejected it to be
scandalized by the sumptuous manifestations, the ceremony, and the
inclination to the pagan.

The great eschatological myth of approaching catastrophe was shared
by intellectuals and people alike. On the eve of the Sack Italy was rife
with superstitions, calculations, and obsessions. Elsewhere in this book
Ottavia Niccoli describes some of the rumours of portents and monstrosities flying around Rome in the first two decades of the sixteenth
century.[10] The atmosphere of anxiety and general confusion was intensified in the 1520s, fed by 'the magnitude and diversity of the many
predictions'. These were 'so numerous, so intertwined, so contradictory,
that one can only speak of a state of collective madness.'[11] J. Carion's
Prognosticatio of 1521 has a drawing which has prompted the comment:
'Without the text one might think that the Sack of Rome was already
being depicted.'[12] In 1526 the *Prognosticatio* of Lichtenberger, first
published in 1488, was reprinted. In chapter 35,[13] following the extraordinarily apt prophecy of Telesphorus of Cosenza, Lichtenberger writes
that before the *renovatio* God will permit the great eagle, with an army

benediction while a grieving widow stands in front of a church.

[8] See Reeves, 'Some Popular Prophecies from the Fourteenth to the Seventeenth Centuries', in G. Cuming and D. Baker (eds.), *Popular Belief and Practice* (Cambridge, 1972),
122.

[9] Chastel, *Sack of Rome*, 67, 76.

[10] See above, Niccoli, Ch. 10. See also id., *Profeti e popolo nell' Italia del Rinascimento*
(Rome and Bari, 1987).

[11] Chastel, *Sack of Rome*, 81.

[12] Cited ibid. 82. In 1532 Carion returned to prophetic speculation at the end of his
Chronica (lat. ed. Paris, 1551), 577–9, where he puts together many of the well-known
prophecies on the expected great emperor and declares himself to be standing on the eve of
the *renovatio*.

[13] J. Lichtenberger, *Prognosticatio* (Strasbourg, 1488), ii, cap. 35.

not only of Germans but of all the worst elements of the people to enter Rome, kill many clergy, and put the rest to flight. The accompanying picture has the caption 'Hic Imperator ingreditur Romam cum sevitia et eius timore fugunt Romani clerici et laici ad petras et silvas et multi detruncabuntur'. The late medieval myth of the great King–Chastiser seemed about to be—and was—dramatically fulfilled. It could be understood in its two opposite senses.[14] The Italian faithful could see in the rise of the great heresiarch Luther, with his 'locusts' from the bottomless pit, and the approaching Imperial army, signs of the long-expected crisis of Antichrist, in which the German King and nation were the traditional agents. But the Catholic German interpreters could see the Emperor as the just chastiser, the Second Charlemagne, the *rex pudicus facie*.

It was in 1526, too, that Polydore Virgil published his *De prodigiis*. It has been calculated that between 1520 and 1530 fifty-six known authors were concerned with prophecy and astrology and 133 pamphlets of this type were produced.[15] A Spanish witness notes alarming 'signs' on three successive Holy Thursdays in St Peter's: in 1525 the altar cloth was burnt; in 1526 the tabernacle of the Holy Sacrament fell down; in 1527 a madman insulted the Pope.[16] A succession of prophets cried Woe! Woe! in the streets of Rome, of whom Brandano da Petrojo was the most considerable.[17] The Florentine historian Varchi summed up the prevailing mood: 'It had reached such a point that not only the monks from pulpits but even ordinary Romans went around the public squares proclaiming in loud threatening tones not only the ruination of Italy but the end of the world. And there were people who, convinced that the present situation could not get worse, said that Pope Clement was the Antichrist.'[18]

The political events leading up to the Sack of Rome display a strange mixture of chance, miscalculation, and fatalism. The fate of Rome hangs in the balance. A Venetian pamphlet published early in 1527 illustrates this very aptly.[19] Entitled *Triompho di Fortuna*, its title-page shows the globe of the world vulnerable to the contradictory pulls of good and evil,

---

[14] See above, pp. 15–16.

[15] Chastel, *Sack of Rome*, 82.

[16] Ibid., 86–7, citing A. Rodriguez-Villa, *Memorias para la historia del asalto y saco de Roma en 1527* (Madrid, 1875), 140 ff.

[17] See above, p. 210. See also Niccoli, *Profeti e Popolo*, 151–2; G. Pecci, *Notizie storico-critiche sulla vita di Bartolomeo da Petrojo chiamato Brandano* (Lucca, 1763).

[18] Chastel, *Sack of Rome*, 87, citing B. Varchi, *Storia fiorentina* (Florence, 1858), x, cap. 8.

[19] Described by Chastel, *Sack of Rome*, 89–90.

symbolized by an angel and a devil who turn the great wheel of the globe in opposite directions, while the Roman pontiff balances precariously above the globe. One is instantly struck by the contrast which this cynical picture offers to those glorious, dominating pontiffs depicted three years earlier in the *stanza* of Constantine who had seemed so inviolable. 'The documentation reveals a series of so many accidents and mistakes that it was hard to think of so great an accumulation of mishaps as anything but an act of fate. Some kind of underlying determinism seemed to control this sequence of fortuitous events.' 'There are cases in which the modalities of the imagination become the stuff of historical moments.'[20] Chance interacts with the force of symbols. 'Images intrude on the action while the event continues to be projected on the imagination.'[21] Thus everything conspired to make the humiliation of the Holy See and the destruction of the Eternal City a necessary disaster. The collective sub-conscious in Italy as well as in Germany was stirred up by the popular belief in portents and celestial omens: it viewed the attack on Rome as symptomatic of Christianity's ultimate crisis.[22]

The first reaction to the actual event is symbolized in the terrible silence that hung over Rome in the immediate months following the Sack. The horrors of destruction and death, the flight of the Pope and Curia, as well as many intellectuals, the vacuum in leadership, all combined to produce a paralysis of fear. But soon the great question had to be faced: Was this God's judgement? What would then follow? The uncertainty of contemporary historians, seeking to assess the event, is well expressed by Guicciardini: 'The councillors of princes would be put to too hard a test if they were obliged to take into account, not only human arguments and considerations, but also the opinions of astrologers, the prognostications of spirits or the prophecies of monks.'[23] But Spanish historians did not hesitate to declare that the Sack of Rome had more than material causes.[24] Churchmen were irresistibly drawn to see it as an expression of the wrath of God. In the midst of the events themselves Cardinal Gonzaga wrote (in Rome on 7 May): 'One can safely say now that our Father in heaven wants to scourge Christianity' and on

---

[20] Ibid. 16–17.

[21] Ibid. 19.

[22] Ibid. 83–4, citing F. Guicciardini, 'Consolatoria fatta di settembre 1527 . . .', in *Scritti autobiografici e rari di Francesco Guicciardini*, ed. R. Palmarocchi (Bari, 1936).

[23] Cited Chastel, *Sack of Rome*, 86.

[24] This was the conclusion of the Count de la Roca, *Epitome de la vida y echos del emperados Carlo* (Madrid, 1646); see Chastel, *Sack of Rome*, n. 70.

16 May in Ostia: 'There is every reason to suspect that from day to day
new suffering and devastation will occur and that the whole world will
go to ruin and be annihilated. One can be assured that God has drawn
the sword of justice and spilled the cup of his ire on human kind.'[25] And
again, 'All this did not happen by chance but through divine justice. For
there were more than ample warnings.'[26]

Soon the prodigy-mongers were busy with retroactive prophecies.
'An event of this magnitude had to have been explicitly inscribed in
the annals of celestial history.' Pamphlets appearing in 1527–8 include
phenomena such as comets, a celestial hand holding a dagger in the sky,
a shower of weapons and severed heads. Later it was said that the same
comet which appeared in 1527 had been seen before the destruction of
Jerusalem in AD 72 and the Sack of Rome by Alaric in 412.[27] Prophetic
pamphlets reappear in large numbers in 1528–30, often reinterpreting
earlier texts or pretending to have been written earlier. Two contrasting
attitudes appear: on the one hand, fate is invoked; on the other, providence calls for repentance. Sanuto, writing from Orvieto in 1528, where
the Curia was still in residence, says that on Palm Sunday Pope Clement
exhorted cardinals and prelates to do penance for their sins, for that was
the cause of the scourge which had befallen Rome.[28] Bishop Stafileo
gave an address in the same year in which he declared: 'These calamities
fell upon us because of our sins. We were no longer citizens and inhabitants of the Holy City of Rome but of the harlot-city of Babylon.'[29]
He argued that the fall of Babylon described in the Apocalypse did not
apply to biblical times but was a prophecy of modern Rome. Thus his
summons to contrition assumed apocalyptic dimensions.

A cataclysmic event, however, has the power to resurrect old images
of authority and security. On the imperial side old Ghibelline ideas
of world rule by the Emperor were renewed in favour of Charles V.[30]
Charles was seen as the just instrument of chastisement. In October 1528
Clement returned to Rome and on 24 February 1530, he solemnized the
coronation of Charles V at Bologna. Engravings of the elaborate procession show Clement and Charles raised to the same level of venerability.
This event represents 'the valorization of hierarchies and symbols' in the

---

[25] Cited Chastel, *Sack of Rome*, 87.
[26] Ibid. 108, citing A. Rodriguez-Villa, *Memorias*, 134.
[27] Chastel, *Sack of Rome*, 115.
[28] Ibid. 184–5, citing M. Sanuto, *Diarii* (Venice, 1879–1903), xlvii, col. 349. See also
S. Schard, *Historicum opus*, 2, *Rerum Germanicarum scriptores* (Basle, 1574), 1858 ff.
[29] Chastel, *Sack of Rome*, 185.
[30] Ibid. 86. See above, Headley, Ch. 13, for the Ghibellinism of Gattinara.

reorganization of a world emerging from confusion.[31] Universal symbols were again back in currency. The hard realities of the political power struggle were once more simplified and masked by a return to the medieval models of empire and papacy. This resurrection of images marks the transition from the negative to the positive aspect of prophecy. Thus in the early 1530s a revived Ghibellinism finds expression in a new cluster of prophetic works in which all the old oracles (Methodius, Hildegarde, Joachim, Bridget, and others) are wheeled out in support of the expectation that Charles would speedily triumph over the Turks and bring in the *renovatio*.[32] On the other side, the Church soon reasserted its own prophetic role. There had been, it was claimed, a providential element in Clement's escape from Castel Sant'Angelo. This was likened to St Peter's escape from prison and this providential protection was affirmed in the re-erection of the angel's statue on the bridge by which the Pope had escaped.[33] When the Emperor entered Rome in 1536, the Church under Paul III reaffirmed her authority in the many symbolic scenes enacted.[34]

Nowhere is the post-1527 mood better expressed than in Egidio of Viterbo's last work, *Scechina*. In July 1530 Clement sent an urgent letter to Egidio asking him to seek the meaning of these cataclysmic events by interpreting the Scriptures through the great Hebrew esoteric texts. Scriptural occultism might give the key to troubled times. In the *Scechina* the supposed divine voice, addressing itself to Charles V, reveals the working of providence in recent events:

I made Rome the capital: forgetting my goodness it abandoned itself to sin more than any other city. . . . I often tried to frighten it with prophetic voices. . . . I made thy hand intervene . . . I took the city, I pillaged it. . . . Moved by pity I invited thee . . . to spare the city . . . to go into Italy and make peace with thine enemies. . . . Returning to the city, I expected that, having first received the wounds and then the honours, it would understand the action of God's two hands, the left destroying . . . the right embracing, raising up, restoring at Bologna. And I see that nothing has been received . . .[35]

If the warnings are not heeded, God will chastise Rome with the scythe of ferocious enemies, overturning the old order and establishing a new.

---

[31] Chastel, *Sack of Rome*, 179–84.
[32] See Reeves, *Influence of Prophecy in the Later Middle Ages: A Study in Joachimism* (Oxford, 1969), 367–9.
[33] Chastel, *Sack of Rome*, 191.
[34] Ibid. 210.
[35] Egidio of Viterbo, *Scechina e Libellus de litteris hebraicis*, ed. F. Secret (Rome, 1959), 105, using Chastel's translation, *Sack of Rome*, 117–18.

Yet still there is hope that the old agencies will bring in the new order. Charles is the chosen instrument of God's justice and, in alliance with the Pope, the agent of renewal which, by God's grace, may now follow. So the reassuring images are recreated and—in spite of fundamental changes on the realities of politics—they are the old medieval ones.

15

New Heaven and New Earth: Prophecy and Propaganda at the Time of the Discovery and Conquest of the Americas

ADRIANO PROSPERI

I

The first year of the new century was not a happy one for Christopher Columbus. Having been arrested and taken back to Spain in chains to face serious charges, at the same time that his venture on the route to the Indies seemed doomed to failure, he was obliged to appeal to all the help on which he could still count, and all available arguments which might present him in a favourable light before the Spanish sovereigns. This is the situation to be borne in mind when reading the letter he wrote to Prince Don Juan's nurse. Recalling the merits of his service, Columbus presented himself as God's chosen ambassador for that 'new heaven and new earth' of which the prophet Isaiah had spoken and which recurs in the Apocalypse: 'I came with the profoundest love to serve these princes, and I have served them with a devotion that has never before been seen or heard of. I became the message-bearer of the new heaven and earth created by our Lord God, according to the Scripture of St. John in the Apocalypse and told by the word of Isaiah.'[1] A private document from the pen of a man in disgrace, this letter had absolutely no influence on the genesis of the expression 'New World'. It fell to Amerigo Vespucci to give a lasting name to the newly discovered lands in his famous report published under the title of *Mundus Novus*. The 'new world' of which it spoke was purely geographical, with no relation to biblical pronounce-

---

[1] C. de Lollis (ed.), 'Carta al ama del principe Don Juan', *Scritti di Cristoforo Colombo. Raccolta di documenti e studi pubblicati dalla Reale Commissione Colombiana pel quarto centenario dalla scoperta dell'America* (Rome, 1892–4), ii. 66–7.

ments nor to apocalyptical tradition. Thus, Latin and the printing press, those international vehicles of Humanist culture, had a greater success than Columbus's Castilian and the mixture of prophecies and images of crusade which the Genoese navigator compiled in support of his venture. Of course the Apocalypse was a book too much in evidence and too much meditated upon at the time not to occur in more or less explicit quotations and allusions, even in texts and circumstances apparently as far as can be from what we might expect under the heading of apocalyptic. A case in point is offered by the publication of *Mundus Novus* itself, which dramatizes and colours Vespucci's description of the large number of inhabitants found in the discovered lands by a reference to the Apocalypse 7.[2] But it is precisely because knowledge of the Apocalypse was widespread and because apocalyptic themes and millenarian prophesying were common elements of that period's culture and sensibility that it is necessary to make careful distinctions: we must understand what meaning those subjects and quotations had in their context.

In the case of Christopher Columbus, this has been rendered particularly difficult by his venture's extraordinary and unforeseeable success. His use of prophetic images and metaphors comes with the enormous number of problems thrown up by the discovery of the Americas and its relative effects on European history. The result has been to isolate and magnify his references to Joachim of Fiore and the minor prophetic literature of the Middle Ages. One wonders if his suggestive reference to the prophecy of Isaiah ('I create new heavens and a new earth . . . For as the new heavens and the new earth, which I will make, shall remain before me, saith the Lord, so shall your seed and your name remain', Isa. 65: 17, 66: 22) conceals an image of himself as Messiah, and could even be proof of his secret adherence to Hebraic tradition.[3] In reality, it is precisely the obscurity and ambiguity of prophetic language, its capacity to be adapted to a variety of uses, which, when associated with a person of such historic importance as Christopher Columbus, enables aspects totally different in themselves to cohere under his name in apparent unity.

Agostino Giustiniani, another Genoese, alludes to Columbus's prophetic pronouncements in his commentary to a polyglot edition of the

---

[2] See anastatic edn. of the *Mundus Novus*, ed. L. Firpo, *Prime relazioni di navigatori italiani sulla scoperta dell'America: Colombo, Vespucci, Verazzano* (Turin, 1966), 87.

[3] See C. Kappler, 'La vocation messianique de Christophe Colomb, Voyage, quête, pélerinage dans la littérature et la civilisation médiévales', *Senéfiance* (Cahiers du CUERMA, Aix-en-Provence, 1976), 255–71.

psalms in 1516. But here the question that preoccupied theologians, faced with the populations of the Americas who had no record or word of Christianity, was whether the Apostolic word had ever reached them or not.[4] Exactly a century later, still another Genoese imitated Columbus's use of Isaiah in order to strengthen his citizens' patriotism. Odoardo Ganducio, in his *Ragionamento della conversione de' Gentili e particolarmente de' Genovesi predetta da Esaia profeta*,[5] dwelt at length on Columbus's letter to Prince Don Juan's nurse stating that it foretold the Genoese territories' future greatness, for which one only had to consult millenaristic prophecies, ancient and modern. Columbus's letter therefore reflects periods distant in time and quite different levels of Italian and European political culture, each claiming to embody the image of Columbus as prophet.

The reality was probably much simpler. However much the grandiose consequences of Columbus's discovery might lead to an emphasis on awareness of a mission received from God, the Genoese navigator's language does not differ greatly from that which was currently in use in the circle of the Catholic kings. The circumstances in Spain at the close of the fifteenth century were characterized much more by the events of the *Reconquista* and by phenomena such as the expulsion of the Jews, than by Columbus's discoveries. In the translation of the *Libellus de Antichristo*, published in Saragossa in 1496 by the Aragonese humanist, Martin Martinez de Ampies,[6] obstinate Jews ('judios obstinados') and Muslims ('la secta mahomética' of the 'moros en Affrica') were held to be followers of Antichrist preparing to attack the Christians of Spain. In those years statements such as these were obligatory for anyone who had any ties with the Catholic kings' politics. Columbus's *Prologue* to his logbook sets the caravels' venture against the background of the *Reconquista*, seeing it as an act destined to save souls beset by idolatry and 'sects of perdition': as the Pope would not listen to the requests sent him by the populations of the Indies and by the 'Great Khan', the Spanish sovereigns had assumed his place and taken up the mission.[7]

---

[4] See A. Iustiniani, *Psalterium Hebraeum, Graecum, Arabicum et Chaldaeum, cum tribus latinis interpretationibus et glossis* (Genoa, 1516), fo. CVIIʳ ff.

[5] O. Ganducio, *Ragionamento . . . della conversione de' Gentili a particolarmente de' Genovesi predetta da Esaia Profeta* (Genoa, [1615]). Cf. the recent facsimile edn., ed. M. Cipolloni (Rome, 1988). For a copy of Colombo's letter 'in the archive of the Most Serene Republic of Genoa', see Ganducio, *Ragionamento*, 49.

[6] On de Ampies, see A. Milhou, *Colón y su mentalidad mesiánica en el ambiente franciscanista español* (Valladolid, 1983), 13 ff.

[7] De Lollis, *Raccolta*, ii. 81.

But these analogies and appearance of family ties which surround Columbus's writing and that of other texts circulating in the Spanish court at the time should not conceal or diminish the original features of Columbus's position. On the one hand, he stubbornly sought to set his venture within the frame of a providential design, and on the other, its success re-invigorated the cultural strand of visions and prophecy in an unforeseeable way. Although he was not widely read, he was able to establish a relationship between books and reality, in the sense of finding in the written page the confirmation of what it was his destiny to realize in practice. It is of course no surprise that Columbus had read Marco Polo or the fabulous voyages that go under the name of Sir John Mandeville, but it is worth noting that he also 'extracted' the prophetic passage about 'ultima Thule' from Seneca's *Medea*. This adaptation and amassing of literary references in order to throw light on and give significance to the venture underway (and also perhaps to solicit help and protection from the sovereigns by adulating them in the language which would most stir them) is typical of a man of small learning, who brought to his reading the search for 'information' on the hidden meanings of Scripture. It is because of this, more than his few explicit references to Joachim of Fiore, that Columbus may be placed in that large current which flowed from the Calabrian abbot's success in the Middle Ages, which Marjorie Reeves's studies have documented. It is a superficial Joachimism, as Alain Milhou in his vast researches on the subject[8] has expressly concluded, and, in sum, has very little to do with the much more prophetic strands which in those years were seeking to take nourishment from Scripture for their message of world and Church renewal.

At the same time that Columbus was completing his prophetic book, another one in Rome came to be opened, with much solemn religious ceremony, and its contents promulgated for the consolation of troubled religious spirits who were intent on catching any sign of apocalyptic messages being fulfilled. This was not a book on the Apocalypse, but nothing less than an *Apocalypsis Nova*, attributed to Amadeus Menezes da Sylva, a Franciscan whose geographic and social origins found him at the Castilian court before he renounced his career there for life in an Italian monastery. He was defined by Benigno Salviati as more 'evangelist than prophet', in order to elevate him in the eyes of those who confused him with other prophetic traditions of equally wide success like that of

---

[8] See Milhou, *Colón y su mentalidad mesiánica*, 474: '. . . el joaquinismo de Colón no era mas que superficial.'

St Brigit.[9] Those pages which Salviati turned so reverently, and with such fear, prophesied miraculous events in European Christendom: the reform of the Church, the conversion of the infidel, the advent of an Angelic Pope, and peace. The context is not far from that which surrounds Columbus's book: the international network of convents (in particular, those of the Franciscans), a charismatic figure, and a public waiting for signs of reform, peace, religious unity, and the restoration of a perfect and original form of Christianity.[10] But, in contrast to the *Apocalypsis Nova*, and so many groups and movements of visionaries and prophets that animated Italy and Spain in those years, Columbus's *Libro de las profeccias* limited itself to the not very exacting task of compiling Biblical *auctoritates* relating—as the title says—to the reconquest of Jerusalem and the conversion of peoples (especially those of the *insulae* of the Indies)[11]. So it was not strictly speaking 'prophecy', but 'information' from the Scriptures, an exercise which was only tenuously related to Joachim of Fiore; that this exercise afterwards came to be applied to the pronouncement of quite positive and salutary future events depended also in good measure on the disposition Columbus knew he would find in his royal employers and protectors.

The idea of the *Reconquista* and the conversion of people to Christianity through the agency of the Catholic kings was certainly well accepted in court circles. If one later reflects on the importance of the missionary and evangelizing task as a deed which legitimated the Christian sovereigns' possession of the conquered lands, one can well understand how Columbus found listeners well-disposed to his pronouncement of a renewal of Apostolic preaching and of a rapid movement towards that ancient objective of world unification under a single pastor. The accusation in his *Libro* that the Popes had been deaf to the requests of the peoples of the Indies for preachers of the Gospel was made to please Isabella and Ferdinand, because one is forced to note the absence in his writings of any more explicit and harsher criticism of the Roman ecclesiastical world. It is this line which differentiates his religiosity and adaptation of Holy Scripture to problems of the present from the prevalent

---

[9] See Morisi, *Apocalypsis Nova: Richerche sull'origine e la formazione del testo dello pseudo-Amadeo* (Studi Storici, 77; Rome, 1970), 30 n. 56.

[10] See Vasoli, *Filosofia e religione nella cultura del Rinascimento* (Naples, 1988), 211–29.

[11] 'Incipit liber sive manipulus de auctoritatibus, dictis ac sententiis et prophetiis circa materiam recuperare sancte civitatis et montis Dei Syon ac inventionis et conversionis insularam Indie et omnium gentium atque nationum, ad Ferdinandum et Helysabeth...' (De Lollis, *Raccolta*, pt. 1, vol. 2, 76).

methods and tones of the age. Thus, in contrast to the appeals for European Christianity's internal reform that were being woven together from various quarters in those years, and which were often based upon prophetic and visionary signs, Columbus's prophecies directed attention beyond Europe's confines: while Savonarola saw visions of fire and blood over a Babylonian Rome, Columbus pronounced the coming triumphal completion of world evangelization. Such a pronouncement was sure to find a suitable audience in the Spain of the *Reconquista*. But in Columbus we see a quite different and more lasting success developing from the discovery he made without being aware of it. Columbus's case remains isolated and exceptional: only in him do we find the venture of discovery united with its interpretation in terms of prophecy and missionary conquest. After him, the paths return to their various ways; only the members of the religious orders who were engaged more or less directly in the task of propagating the faith outside Europe continued to foster interpretations of that kind, while the actual conquerors and those who followed in order to comment on their ventures displayed little curiosity. As to the sources which fed Columbus's interpretation, they were both very familiar at the time and after him remained so for whoever desired a key to interpreting the comprehensive scheme of history, with the exception that those sources were applied only to European events and served but to feed the recurrent fears of catastrophic events there. This is the case with Pierre d'Ailly's treatise *De concordantia astronomie cum hystorica narratione* from which Columbus took information on the fate of the Muslim religion which must have pleased the Catholic kings. In the age which followed, that scheme of the naturalistic theory of history would enjoy little success among writers of history, while the study of astral conjunctions maintained an undoubted importance, but was no longer applied to events outside Europe.[12]

II

The idea of world unity that was emerging with the Atlantic voyages and Portuguese ventures strengthened the concept of religious unity in Europe: a single flock and a single shepherd, 'unum ovile at unum pastor'. It was an ancient Christian dream, which had become stronger

---

[12] 'It was a dead-end. It does not seem to have exerted the slightest influence on the practice of studying history . . .' (K. Pomian, 'Astrology as a Naturalistic Theology of History', in P. Zambelli (ed.), '*Astrologi hallucinati': Stars and the End of the World in Luther's Time* (Berlin and New York, 1986), 29–43, quoting p. 43.

during the Crusades and was current again in the Spain of the *Reconquista*. Columbus duly noted the passage from John 10: 16 in his *Libro*, but we have to wait until the years around 1510 to register a widespread application of the image. In the years of the attempted French-led Council of Pisa–Milan and the opening of the Lateran Council, the discussions about the ecumenical prospects of European Christianity were kindled with renewed vigour, falling on the question of the newly discovered lands on the other side of the Atlantic. The political and religious conflict of those years, characterized by the wars in Italy and the aggressive politics of Julius II, on the one hand fuelled an unusual growth of prophecies and visions and, on the other, zealous propaganda in favour of one or the other forces in the field. Prophecy and propaganda were often closely interconnected and put to the service of one or other of the political protagonists. The context was the struggle for supremacy between France and Spain, which as a corollary embraced the question of Church reform. Precisely on this question we must remember the success of the image of the Angelic Pope, who was destined to bring Christianity's traditional enemies (Jews and Muslims) into the faith and also initiate an era of peace and happiness. The *Apocalypsis Nova* attributed to the Blessed Amadeus had the function of channelling this kind of expectation.[13] Thus news and evaluations relating to the progress of Christianity in the Atlantic 'islands' became intermingled with the expectation of the prophesied Angelic Pope.

For example the Dominican Isidore Isolani speaks of islands in his *De imperio militantis Ecclesiae*, composed between 1513 and 1514.[14] Its author asks the question whether the lands discovered in the middle of the Ocean would ever be placed under the authority of the Church and furthermore declares himself convinced of the advent of world unity under Christ. But, according to Isidore, before this happy fulfilment of human history there would be an age of sorrows and decadence dominated by Antichrist. This was the era, according to him, which current signs indicated was imminent: wars, the division of Christians, the spread of witch cults worshipping the devil, were further proof of the coming of

---

[13] See above, Morisi-Guerra, Ch. 2.

[14] Isidoro Isolani, *De imperio militantis Ecclesiae libri quattuor* (colophon: Finis... Veronae in coenobio S. Anastasiae in die Sancti Jacobi Apostoli idest XXV Julii MDXV. Impressi vero Mediolani ... die XIV Octobris MDXVI). The long title indicates for the reader the book's subject: 'de praedicatione Evangelii apud insulas magni maris Occeani, de conversione infidelium, de hereticis, de scismaticis, de connexione omnium scientiarum, de cognitione futurorum secundum divinas litteras'. Isolani's quotation of psalm 18 is echoed by Egidio of Viterbo, see above, Reeves, Ch. 5.

Antichrist. Antichrist's defeat would then see the advent of the Angelic Pope and the beginning of the blessed millennium, during which the Jews would be converted and the Atlantic islands would be visited by the Gospel. It is not difficult to explain why Isolani insisted on talking of islands; the controversial passage of psalm 18 ('In omnem terram exivit sonus eorum') which had always been understood as a prophecy of apostolic teaching throughout the world, had been the cause of 'some anxiety'—as Francesco Guicciardini would ironically say—in the face of the West Indies' inhabitants' evident ignorance of the Christian God. It was not only a case of biblical exegesis; on the one hand people wanted to know if the apostolic age was truly forever over or still not yet begun and, on the other, what they should think of the relationship between the population of the Americas and religion. On both counts the question necessarily generated thought about European culture. Isolani cleverly thought he would avoid the problem with a little artful interpretation: he attributed to the Apostles the merit of having preached over all the 'lands' of the three continents of Asia, Africa, and Europe, saying that it was the 'islands' placed on the ends of the earth which had been left to his own time. This suggestive hypothesis went with a hypothesis of geographical facts which, at best, revealed little information. However, it is evident that Isolani too accepted that the task of apostolic preaching in the new lands was reserved for his age and on this point he found himself in agreement with Columbus. But in Isolani the image of world unification under the same faith found a much stronger conviction on which to anchor itself, that of the Joachimist vision of a felicitous *status*, which would begin with Antichrist's defeat.

It has already been indicated that this conviction was at the centre of the *Apocalypsis Nova*. Isolani's treatise was dedicated to Denis Briçonnet, who was a key figure in the small Milanese group dominated by the visionary nun Arcangela Panigarola. Isolani shows himself well informed about the expectations which Giorgio Benigno Salviati and Sister Arcangela Panigarola had with regard to the illustrious French prelate: in the work's dedication he explicitly affirms that because of Briçonnet's saintliness, his theological doctrine, and his assiduous eucharistic devotion, he shows himself worthy to be the *Angelicus Pastor* of whom there was need in that period of wars and serious dislocations in European Christianity.[15] As biographer of the Blessed Veronica of Binasco, Isolani

---

[15] 'Hos libros ad te direxi, cuius vota precesque assiduae, divinae Eucharistiae quottidiana pene oblatio, mores omni sanctitudine pollentes, litterarum divinarum studia, imperii militantis Ecclesiae maiestatem obsecrant, venerantur, adorant . . . Catholicum patrem et

was particularly sensitive to prophecies; he was also strongly convinced that the present was teeming with signs of a profound crisis in the Church and in Christian life, which could only be explained by the imminence of Antichrist. Once Briçonnet's meteoric brightness was ended, Isolani continued to search for other possible candidates for the role of the Angelic Pope. He proposed one in the figure of Adrian VI when, discussing the astrological prediction of a great flood, he maintained that they should rather be looking for a sign of divine ire and a new premonition of Antichrist.[16]

The question of Church reform and the gradual darkening of the political-ecclesiastical horizon in Europe had by now become interwoven with the anticipation of Antichrist, which would be the next necessary stage before the happy millennium and peaceful unity of the Christian flock could come about on earth. But before Luther's Reform drew attention to the European dislocations, the dream of 'unum ovile et unus pastor' had taken on a rosier hue. In presenting his edition of Plato's works to the new pontiff, Leo X, Aldus Manutius had shown how his age had taken giant steps in the progress of knowledge and world unification, greatly outdistancing the ancient world, and he invited the Pope to urge forward the preaching of the Gospel outside Europe.[17] In that same period, that is immediately after Leo's election, and concomitantly with the re-assembly of the Lateran Council, another urgent appeal for missionary work reached the Pope from Vincenzo Querini and Tommaso Giustiniani, who had taken the names of Peter and Paul on becoming Camaldolensians. In their *Libellus ad Leonem X* the two monks did not follow prophetic lines, but took a decisive path of pragmatically calculating the things to be done in order to overcome religious differences in the Christian body and to extend it over the rest of the world.

D. D. Guilielmum Meddensem Episcopum.' Cf. G. Zarri, 'Le sante vive. Per una tipologia della santità femminile nel primo Cinquecento', *Annali dell'Istituto storico italo-germanico in Trento*, 6 (1980), 371–445, esp. 384 ff. and for the same argument, M. Veissière, 'Guillaume Briçonnet et les courants spirituels italiens au début du XVIe siècle', in M. Maccarrone and A. Vauchez (eds.), *Échanges religieux entre la France et l'Italie du Moyen Age à l'époque moderne* (Geneva, 1987), 215–28. On prophecy in the Briçonnet circle, see above, Vasoli, Ch. 7.

[16] Isidoro Isolani, *Ex humana divinaque sapientia tractatus de futura nova mundi mutatione* (Bologna, 1523), fos. 19ᵛ ff. On the conjunction 'in Piscibus' of 1524, see P. Zambelli, 'Many Ends for the World: Luca Gaurico, Instigator of the Debate in Italy and in Germany', in P. Zambelli (ed.), *Astrologi hallucinati*, 239–63.

[17] Manutius was primarily thinking of the Portuguese ventures in India, but also of the peoples 'quos in Oceano occidentali Hispani superioribus annis invenere' (*Aldi Pii Manutii ad Leonem X Pontificem Max. pro Republica Christiana, proque re literaria supplicatio, Omnia Platonis Opera* (Venice, 1513).

They fixed their attention on the cases of the Jews and the populations of the Americas: the former, who were very familiar with Christianity and resisted it only through their perfidy, were to be propelled to conversion by every possible means of persuasion or else expelled from Christian countries; the latter, who had probably never known apostolic teaching in their history and consequently had no record of it, were to be urgently evangelized. In this the Pope was to follow the model of his great predecessor Gregory, who was responsible for sending the missionaries to Britain.[18] It was the kind of argument that was usually presented in a visionary and prophetic context, but neither the Venetian publisher nor the two monks—who came from Venetian territory—decided on that kind of presentation.

In contrast to the two Venetians, Egidio of Viterbo admitted having long searched the *Apocalypse* and other prophetic books in order to decipher future events: his speech at the Lateran Council (which was taken as proof of the Church's aspiration to reform even by Gerdes[19]) and his other writings of those years have the same characteristic of envisaging a future of rapid and positive transformation, rich in potential and promise. The confrontation between ancient and modern was re-solved in favour of the moderns as a direct result of the discovery of America and the new prospects that were opened up there: it was a subject which in those years began specifically to reveal the various implications to which it would give rise. The extension of both the known world and commercial traffic was proof of how much the age was superior to that preceding it, and even to mythical antiquity. Discoursing on the spice trade, the Venetian Gian Battista Ramusio was among the first to formulate an image of what we now call the Middle Ages as being wrapped in the 'shadows of a long night'.[20] On the subject which

---

[18] See H. Jedin, 'Ein Vorschlag für die Amerika-Mission aus dem Jahre 1513', *Neue Zeitschrift für Missionenwissenschaft*, 2 (1946), 81–4. The text of the *Libellus* is in G. Mittarelli and A. Costadoni, *Annales Camaldulenses*, 9 (Venice, 1763), cols. 613–719. The importance of a text by Giustiniani against the *Apocalypsis Nova* has been noted by J. Leclercq, *Un humaniste ermite: le bienheureux Paul Giustiniani (1476–1528)* (Rome, 1951), 97. Unfortunately the studies on this documentation which have been promised for decades by Prof. Eugenio Massa have still not appeared and the manuscripts are still not able to be consulted.

[19] See D. Gerdes, *Introductio in historiam Evangelii seculo XVI passim per Europam renovati doctrinaeque reformatae* (Groningen, 1744), i. 46 ff. and above, Minnich, Ch. 4.

[20] 'It is a truly marvellous thing to think of the huge change and alteration that the coming of the Goths and other barbarians in Italy made throughout the whole of the Roman Empire, although these peoples wiped out all the arts, the sciences and all the traffic and commerce that occurred in the various parts of the world: they [those years] lasted for more than 400 years, almost like the shadows of a dark night...' (in G.

interests us here, church history, the confrontation of ancient and modern brought with it a comparison of the present with the days of the 'Primitive Church' of the Apostles, which meant not only placing under discussion the model of a higher and purer definition of the form of the Church, but also the conception of history which was then commonly accepted. If Aldus Manutius remarked on the superiority of papal compared with Ancient Rome, with regard to the extent of the world to which its missionaries (whom Manutius specifically called 'Apostolos')[21] were to be sent, Egidios of Viterbo went further: the apostles—one reads in the *Scechina*—had only conquered the smallest part of the world, whereas, in his day, they were preparing to conquer it in its entirety.[22] These were the first hints of a different conception of history which was emerging among the rhetorical amplifications of humanist language and the reappearance of Joachimist ideas and terms; and if they turned to Joachim of Fiore in a moment in which the conquests of the present appeared to be such that they gave people the prospect of great future upheavals, this is only understandable. Thus, in these years, one meets his name widely every time the Apocalypse comes to be used for deciphering the signs of a future rich in promise: the Minorite Pietro Galatino, for example, praised his acute penetration of hidden meanings in his own interpretation of the Apocalypse where he announced the coming definitive defeat of the Muslims, whom he interpreted as the Seventh Head of the Beast of Revelation—a promise he made to Charles V.[23] Italian and Mediterranean worries about the Turkish threat saw to it that, in his commentary, the progress of religious conquest in the newly discovered lands remained in the background. But it is to be noted that the authoritative mediator of Galatino's homage to the Emperor was the Minister-General of the Franciscan Order, Brother

---

Ramusio, *Navigazioni e viaggi*, ed. M. Milanesi (Turin, 1979), ii. 967. On the subject of the 'Dark Ages', see C. Ginzburg, 'Vom finstern Mittelalter bis zum Blackout von New York—und zurück', *Freibeuter*, 18 (1983), 25–34.

[21] See G. Ladner, *Images and Ideas in the Middle Ages. Selected Studies in History and Art* (Rome, 1983), vol. ii, pt. 5, 517–763.

[22] 'Per illos (sc. Apostolos) orbem notum per te (Charles V) incognitum acquisivere, per illos partem per te totum circuire, per illos ante oculos omnium positas terras subegere, per te tuamque setatem Hispanam familiam alium ac novum orbem venatae sunt' (Egidio of Viterbo, *Scechina e Libellus de litteris hebraicis*, ed. F. Secret (Rome, 1959), 160. See above, Reeves, Ch. 5, for this quotation but also for Egidio's ambivalence on the question of 'progress' in history.

[23] 'Mahumeticam sectam, septimo bestiae capite, in quo tot annos regnavit abscisso, penitus abolitum iri, ut tandem, orbe universo ad Christum redacto, fiat ubique unus pastor, unumque ovile' (P. Galatino, *Commentaria in Apocalypsim*, Ms. Vat. Lat. 5567, fo. 1ᵛ). On Galatino, see above, Rusconi, Ch. 8; also Vasoli, *Filosofia e religione*, 183 ff.

Francisco Quiñones de los Angeles, for whom the question was very close to the heart.

III

In Franciscan tradition, the epic of the New World's missionary work has a solemn beginning—30 October 1523—when Francisco de Quiñones signed the letter of *missio* for the twelve Franciscans who were to be sent to the Americas to assist in Cortez's military conquest. Quiñones text had an important place in the compilation of the *Annales Minorum* by Luke Wadding, a monk who was Irish but whose religious life was shaped by the Spanish Franciscans.[24] In an age when the more important religious orders were committing themselves to the historical and scholarly order-ing of their own recent, and less recent, past, Franciscan historiography could boast a continuity of spiritual inspiration and appeal to an apostolic model which neither the Dominicans nor the Augustinians nor even the Jesuits were able to record. It was a model coterminous with the order's actual foundation, but the way in which Francisco de Quiñones shaped it in his text of 1523 clearly shows that apocalyptic and spiritual interpreta-tion took precedence over the original rule dictated by St Francis. The world was rapidly on the road to its end—Quiñones wrote—and it was against this backcloth that the missionaries were moving, like workers at the eleventh hour in the Lord's vineyard (Matt. 20). Their number was twelve because this was the number of apostles Christ chose for convert-ing the world.[25] His idea was to complete the Gospel's conquest of the world, which had only been half finished by the Apostles, and which now, with the imminent fulfilment of time, had found an instrument in the Order which more than any other felt the call to be the 'figura' of the Redemption.[26] These kinds of echoes and references to the apostolic model were to become familiar in the Order's missionary work. Even the Society of Jesus—which was generally contrasted with the Franciscans for its (presumed) greater capacity of self-control with regard to apoca-lyptic and prophetic urges—when, in 1590, it drew up instructions for missionaries working in Europe, thought to organize them in groups of

---

[24] L. Wadding, *Annales Minorum* (Rome, 1731–6), 182–9.

[25] 'Hic fuit numerus discipulorum Christi pro mundi conversione, et numerus sociorum sanctissimi Patris nostri Francisci pro evangelicae vitae publicatione' (ibid. 188).

[26] On Spanish Franciscans, see Milhou, *Colón y su mentalidad mesiánica*; see also J. Phelan, *The Millennial Kingdom of the Franciscans in the New World: A Study of the Writings of Gironimo de Mendieta (1525–1604)* (Berkeley, Calif., 1956).

twelve.[27] But in the Franciscan missionary tradition this model was deeply rooted and it was in that Order that its implications were manifested both in concept and in practice.

An Apostolic Church also meant an exemplary Church, in contrast to the current European one over which the Lutheran Reformation was starting to cast its long shadow: this was the 'reformatio' that the Franciscans had to perform against the blaze of Lutheranism, as Brother Francesco da Castrocaro wrote in his speech to the general chapter at Carpi in 1521 (at which, among others, Niccolò Machiavelli was present).[28] Attentive to signs which could help explain the plans of the Divine, the Franciscans saw a providential design in the fact that, like Luther himself, the head of the twelve missionaries was called Martin and also that, like his companions, Brother Martin da Valencia came from the area of S. Gabriele of Estremadura, an environment particularly receptive to spiritual ideas, and that in his youth, he had meditated at great length on verse 15 of psalm 58: 'convertentur ad vesperam et famem patientur ut canes'.[29] This was traditionally interpreted as referring to the conversion of the Jews at the end of time and the hunger which that 'wicked' people would then have for being evangelized, but Brother Martin instead applied it to the peoples of the New World. Spanish Franciscan convents had traditionally had a notable familiarity with evangelization: the Jews and Muslims had been traditional targets of missionary and preaching fervour, which was fuelled by the expectation of divine intervention. The discovery of the Americas now offered an unexpected testing ground; the ease and vastness of the enterprise seemed to fulfill their rosiest dreams, while at the same time it created a distance from and diminished the problems encountered with the traditional targets of Jew and Muslim. Only a few years later, driven by the desire for a more marked political and religious subjugation, the theory was put forward of the origin of the peoples of the Americas as the ten

[27] See the document cited by A. Prosperi, ' "Otras Indias": missionari della Controriforma tra contadini e selvaggi', in G. Garfagnini (ed.), *Scienze, credenze occulte, livelli di cultura* (Florence, 1982), 233 ff. For eschatological overtones in some Jesuit writings, see also Reeves, *Influence*, 274 ff.

[28] *Oratio Venerandi P. Fratris Francisci de Castrocaro . . . praesertim adversus Martinum Luterum* (Bologna, 1521). On Francesco da Castrocaro's insistence on the themes of missionary expansion outside Europe, see A. Prosperi, 'America e Apocalisse. Note sulla "conquista spirituale" del Nuovo Mondo', *Critica storica*, 13 (1976), 19 ff.

[29] See Toribio Motolinia, OFM, *Memoriales e Historia de los Indios de la Nueva Espana*, ed. F. de Lejarza (Madrid, 1970), 54.

lost tribes of Israel deported to Assyria, of which II Kings 17 speaks.[30] The negative stereotype of the Jew would then be used to justify many aspects of the violence of that conquest. But in the years of the twelve's mission, a different model for classifying the peoples of the Americas and for defining the missionaries' relationship with them was used: that of pagan idolatry. The reports Cortez sent to Emperor Charles V described the Aztec religion as an idolatrous cult identical with those of pagan antiquity. This way of giving a conceptual frame to the religion of the peoples to be converted was closely linked to the idea of the ease of conversion: idolatry seemed a lesser obstacle than the difficulties met with in monotheistic religions.

With the conversion of the 'pagans' Christian history appeared to be retracing its steps back to its origin in a literal return to its apostolic form. It was then generally held that history did not move in a direct line from past to future. In 1516 Gaspare Contarini wrote concerning Savonarola's prophecies and Church reform that 'human affairs do not run along an infinite straight line but along a circular one; although they do not all make a perfect circle but, when they come to rise to completion, they then fall away'.[31] In contrast to Contarini, Brother Martin's companions had a greater disposition to prophecy, but if we are to believe what remains about their missionary activity, they were just as equally convinced that history was composed of cyclical returns to a point of origin. Considered as new apostles of a religion that was finally returning to the Gospels (Quiñones gave the title Santo Evangelio to their province), they had a central place in sixteenth-century ecclesiastical history. Their mission was the object of close study by their contemporaries who drew from it a narrative, which was to serve as a doctrinal text for the whole of the religious conquest. This remained in manuscript, and was rediscovered in part and later published during this century. Its author was another Spanish Franciscan, Bernardino de Sahagun, famous for his long studies of Mexican religious rites and the immense ethnological work he put together. Arriving in Mexico only in 1529, he did not take part in the twelve's missionary activity but, through personal experience in Inquisition trials—like the notorious one against Don Carlos, the *cacique* of Texcoco (1539)—he became convinced that the

---

[30] See L. Huddleston, *Origins of the American Indians. European Concepts, 1492–1729* (Austin, Tx., 1967); G. Gliozzi, *Adamo e il nuovo mondo. La nascita dell'antropelogia come ideologia coloniale: dalle genealogie bibliche alle teorie razziale (1500–1700)* (Florence, 1976).
[31] F. Gilbert, 'Contarini on Savonarola: An Unknown Document of 1516', *AR* 59 (1969), 145–9, esp. 149. See above, Reeves, Ch. 1, for cyclical and lineal views of history.

Indians' conversion was only superficial. The first Franciscans in Mexico had been 'harmless as doves', as the Gospel decreed, but they had forgotten to be 'wise as serpents' and had not realized that the Indians had renounced their idols only externally.[32] However, the exemplary character of that first initiation into missionary work in Mexico spurred him to compose an elementary catechism devised from reports of conversations the twelve had with the Indians. Thus the *Libro de los Coloquios* was born.

The nature of this text and its reliability on meetings with the Indians remain open questions. On one side there are those who note Sahagun's accuracy in documentation and the discrepancies between his version and those of other historians and witnesses, like Bernal Diaz; on the other, the purpose of the text in its *nahuatl* version for the work of conversion has caused it to be considered as a compilation of doctrine and rules with no relation to what friars and Indians actually said. In reality the historiographic nature of the text and its missionary purpose are not mutually exclusive. The author's monastic culture, nourished by the works of Eusebius of Cesarea and St Augustine, gave him exact historiographic models in the successes of the first Christians against the idolatrous cults of the pagans. But the very nature of this historiography was apologetic, documenting Christianity's superiority and demonstrating the fulfilment of God's plan for the history of mankind. The geographical discoveries which occurred between the fifteenth and sixteenth centuries did not put God's providential plan in doubt; on the contrary, they extended its range into an even greater plan. Thus, Sahagun, who in general was cautious concerning the prophetic enthusiasm of his brethren, could not help turning to a prophecy of St Brigit for making the discovery of the Americas fit into the secrets of divine plans.[33] The enterprise of discovery and conquest seemed to Sahagun the greatest that had been seen on earth since the time of the *primitiva Yglesia*.

In the conversation he records, a fundamental passage is that in which the indigenous gods come to be defined by the missionaries as lying idols. At this point, he says, the priests of the local cults entered the

[32] Matt. 10: 16. Cf. Poú y Marti, PFM, 'El libro perdido de las platicas o coloquios de los doce primeros misioneros de México', in *Miscellanea Francesco Ehrle* (Rome, 1924), iii. 282–333.

[33] 'Nuestro Señor Dios tenia esta tierra . . . ocultada por sus secretissimos juizios hasta estos nuestros tiempos (noticias avia muchos tiempos antes en la yglesia que avia gente y poblacion . . . como parece en las revelaciones de sancta Brigida)' (ibid. 296). On this document cf. J. Klor de Alva, 'The Aztec–Spanish Dialogues of 1524', *Alcheringa: Ethnopoetics*, 4 (1980), 52–193.

discussion, refusing to forsake their gods for the Christian divinity. They actually asked the conquerors to be satisfied with military victory and not demand of them such a painful renunciation: 'We were all of one accord: it was enough to have been defeated, enough that they should have deprived us of power and royal jurisdiction, but regarding our gods, we would rather die than cease to serve and worship them.'[34]

Awareness of defeat was thus linked to a comparison between the victors' gods and those of the losers. In *De civitate Dei* St Augustine had already spoken ironically on the impotence of the pagan gods, which were incapable of preserving the ancient cities from sack and ruin. But in Sahagun's pages on the subject of the defeat of the idol worshippers, in addition, comes the defeat of their gods: 'What are we to do, we humble and mortal men? If we must die, let us die; if we are to perish, let us perish; for in truth the gods also died'.[35]

The gods, therefore, had to die; and, as one learns a little further on, those gods were in reality demons. This gives us enough to recognize in the American satraps' plea an echo of another story of pagan gods who died following the advent of Christ: Plutarch's death of the 'great Pan' as interpreted by Eusebius[36]. Certainly Sahagun's typical method of arguing is clear and precise; his exposition of the principal points of the Christian religion relies upon patient indoctrination, but that does not prevent the model of ecclesiastical history underlying his narrative from being that of a renewed apostolic era on earth. Other minor echoes of prophecy, which Christopher Columbus had already compiled and related to the discovery of the New World, prove that not even the second generation of Franciscan missionaries—more cautious when compared with the enthusiasm of the 'twelve'—was immune from that kind of hope and expectation.[37]

With the epic of the 'conquistadores', the importance of the discovery had become so enormous that it obscured the political and religious

---

[34] Poú y Marti, 'El libro perdido', 313.

[35] Ibid. 312. Virgil, of course, has the theme of the 'victos deos' in *Aeneid*, ii. 320, which, along with other Virgilian passages, was remembered by St Augustine in *De Civitate Dei*.

[36] See P. Borgeaud, 'La mort du Grand Pan. Problèmes d'interprétation', *Revue de l'histoire des religions*, 200 (1983), 3–39. On the subject of the death of the gods, N. Wachtel (*La visione dei vinti: Gli indios del Perù di fronte alla conquista spagnola*, Italian trans. (Turin, 1977), 39) emphasizes the analogies with a Mayan source, the *Chilam Balam*; but he also notes that the source has been 'shaped by strong Christian influences'.

[37] For example, this was the case with Isaiah 60: 8–9, twice alluded to in the *Coloquios*, when it says that the missionaries had arrived 'dentre las nieblas y nubes del cielo' (Poú y Marti, 'El libro perdido', 310–11).

models offered by antiquity. Ecclesiastical historiographers had to modify the scheme offered by ancient Christian literature to adapt it to the extraordinary novelty of those multitudes of potential new Christians (where the adjective new, as J. A. Maravall has noted, assumed a positive meaning against the usage loaded with suspicion then current in Spain[38]). The hypothesis of a renewal of the most important event in Christian history, Christ's return to earth—a *medius adventus*, intermediate between the Incarnation and the Last Judgement—was sometimes a presence hovering in the texts of those years, but it took the dramatic crisis of the struggle between Luther and Rome to push expectations and fantasies in that direction.

IV

The crisis in European religious unity was preceded, accompanied, and followed by a huge flowering of prophecies; the Apocalypse was combed from the most diverse points of view for arguments and evidence in support of one or another opposing force. In that literature it would be impossible not to find an echo of the unheard of expansion of the known world's limits. But for the most part they are distant and summary echoes, where the concrete questions of the discovery and conquest of the Americas remain in the background and what mattered was the sign that the New World's appearance created on the horizon of European conflict. In general it was a sign that was emphasized in order to exalt the Hapsburg Empire's function as bringer of unity and peace: the conquistadores' venture and the first missionaries' preaching brought salutary images to a Europe threatened by the Turks and prey to a violent religious struggle. It is clear that the ventures of Magellan, Cortez, and Pizarro, used as propaganda for the glory of Spain, led to the eclipse of Columbus's importance. It was Venetian culture which duly recorded his name and merits.[39] On the other hand, the concept of the single flock under a single shepherd, so dear to the Hapsburg line, was obviously seen in quite a different light by the Lutheran reformers. This helps to

---

[38] J. Maravall, 'La utopia politico-religiosa de los franciscanos en la Nueva España', *Estudios Americanos*, 1 (1948–9), 199–227 (now in his *Utopia y reformismo en la España de los Austrias* (Madrid, 1982), 78–110). On the concept of a 'Middle Advent' of Christ in late medieval and Renaissance thought, see Reeves, *Influence of Prophecy in the Later Middle Ages: A Study in Joachimism* (Oxford, 1969), index under *Advents*; see also below, Jungić, Ch. 17.

[39] Gliozzi notes 'The defense of Columbus's rights is a recurrent characteristic in the Venetian Republic' (*Adamo e il nuovo mondo*, 279 n.). Cf. F. Ambrosini, *Paesi e mari ignoti: America e colonialismo europeo nella cultura veneziana (secoli XVI–XVII)* (Venice, 1982), 89.

explain the notable absence of appeals to the conquest of the Americas in the equally rich prophetic and apocalyptic literature which they generated on their side. It fell to those who personally experienced the rigour of Spanish religious uniformity and the spirit of conquest to pose such disturbing problems as the eternal salvation of the peoples of the Americas who had been untouched by the Apostolic word. On this matter Juan Luís Vives, *cristiano nuevo*, advanced the theory that natural morality was sufficient to guarantee salvation in the absence of knowledge of the Gospel. It would be taken up by others, with the consequent effect of weakening the ties of the Church—for example in the case of the Italian heretic, Celio Secondo Curione.[40] In the mean time, however, the developing debate over the nature of the native religions in the Americas put traditional categories of European thought to the test.[41]

Among a public troubled by wars between Christian principalities and by the Turkish threat, the prophecies spread by press and pulpit seem to have interpreted signs of immediate interest in Europe. In fact, what stands out in this literature are the pronouncements of wars and misfortunes beside the news of obscure marvels in the Muslim world and its future conversion or defeat. These were the geographical horizons which interested European readers.[42] Italian presses, which for the first half of the century had an importance limited not only to Italy, offer a good example of these tendencies.[43] For instance the Venetians who published writings of Joachim of Fiore, or at least those attributed to him, do not seem to have been particularly susceptible to the subject of the Americas. On the other hand, echoes and news of the New World's missionary successes were circulating along the channels of the religious orders, the Franciscans in particular, and it is precisely from the New World that the most revealing interpretative uses came. The new church of the

[40] On the opinion expressed by Vives in his commentary on St Augustine's *De Civitate Dei*, pub. 1522, see G. Gliozzi, 'The Apostles in the New World: Monotheism and Idolatry between Revelation and Fetishism', *History and Anthropology*, 3 (1987), 123–48, esp. 124 ff. The question is addressed by Celio Secondo Curione in *De amplitudine beati regni Dei* (printed in 1554); cf. D. Cantimori, *Eretici italiani del Cinquecento: Ricerche storiche* (Florence, 1967), 192.

[41] *History and Anthropology*, 3 (1987), 'The Inconceivable Polytheism. Studies in Religious Historiography', is devoted to the question.

[42] The conclusions of G. Atkinson (*Les nouveaux horizons de la Renaissance française* (Paris, 1935), are in this sense still valid.

[43] See Ottavia Niccoli, *Profeti e popolo*. The majority of the prophecies catalogued by A. Schutte, *Printed Italian Vernacular Religious Books, 1465–1550. A Finding List* (Geneva, 1983), 305 ff., deal with 'Muslim and other Turkish matters'. Cf. for example, A. Pegolotto, *Una littera . . . de una prophetia di Santa Brigide* (n.p., 1542). On the Venetian publications of Joachimist works, see above, Reeves, Ch. 5.

Americas offered itself as a possible positive model against the European church's negative one. Thus the idea of transferring the Church's seat from Rome to the soil of the Americas was born and grew (in the same way that Jerusalem had to cede to Rome the function of religious capital because it had not welcomed Christ). The other use to which the evangelization of the Americas was put was that already hinted at by Christopher Columbus: he took it upon himself to announce, along with other disturbing signs of the present, the completion of time and the coming end of the world.

The search for signs and the troubled deciphering of prophecies was a widespread social phenomenon in the Italy of the early sixteenth century. The legislation of the Fifth Lateran Council and the provincial council of Florence of 1517 were not enough to halt a tendency fuelled by uncertainties over the present. According to a preacher who enjoyed great success, it was precisely these uncertainties that signalled the coming of Antichrist: 'in this time a multitude of the spiritual, or rather the possessed, delight in dreams and visions and prophecies and miracles.'[44] This did not prevent him participating in such a tendency, entering into competition himself, and thus proving how popular the deciphering of 'signs' was:

so that we may recognize that the hour [of Antichrist] is approaching, several years ago God permitted the discovery of the new world, which according to the faithful report of many who write, is not less than the whole of Europe in geographical size nor in its population, and where they now sail from Spain every year and where the Gospel, whose laws were not there known, is now preached. And already many have been converted, so that where the Jews' blindness was the occasion for our salvation, so our blindness constrains Christ to go to other lands, and if at the present time what the Apostles prophesied is coming true, that their words would penetrate every country, there being four shores or areas of the world, as Scripture often says, and the Bible having already penetrated Asia, Africa and Europe, that left only the fourth area, recently discovered, called America, where it had to be taken.[45]

According to Serafino da Fermo, therefore, that 'new church' the missionaries' reports spoke of was destined to take over from European Christianity, torn apart by its internal problems and in particular by

---

[44] Serafino da Fermo, *Breve dichiaratione sopra l'apocalisse de Giovanni, dove si prova esser venuto il precursor de Antichristo et avicinarsi la percossa da lui predetta nel sesto sigillo* (Venice, 1541), fo. 52^v. The first edn. appeared in Milan, 1538.
[45] Ibid., fos. 35^r–36^r.

Luther. For this writer, as well, the guides for unravelling the dramatic knot of external problems (the Turks) and internal (Luther and his followers) were those of the Joachimist and spiritual tradition: 'Read what Joachim, Ubertino and many others have written in this respect, and the whole of it will become clear, because all that they prophesied for this time we now see to have been fulfilled.'[46]

A few years earlier, the authoritative Franciscan Niklaus Herborn had published a letter in Cologne to the ministers of the ultramontane provinces of his Order with texts by Cortez, Martin de Valencia, and Zumárraga: he contrasted the spectacle of German churches, despoiled of their ornaments, and become the theatre of heretical rites, with the orthodox fervour which seemed to animate the Indians of Spanish America. His conclusion was the same as that later taken up by Serafino da Fermo: they were approaching the realization of the threat contained in Matt. 21: 43: 'The Kingdom of God shall be taken from you and given to a nation bringing forth the fruits thereof.'[47]

These were dark prophecies, with an evident political and religious significance; the rebellion of the German 'heretics' was placed under accusation, together with all those forces which were opposed to the Imperial myth of 'unum ovile et unus pastor'. Along the channels offered by the religious Orders most committed in the anti-heretical struggle, on the one hand, and in the religious conquest of the New World, on the other, passed political and social messages in the shape of prophecies. Their objective was the consolidation and expansion of the Hapsburg Empire, the guarantor of Church unity, and also of its internal reform. The spread of visionary pronouncements of this kind follows the rise and fall of Hapsburg power throughout the whole century with the aim of directing fears and expectations to a single end. The attempts of the Fifth Lateran Council to restrain the use of apocalyptic and prophetic themes were evidently useless or were let fall tacitly in abeyance in the face of that kind of prophetic preaching, which, more often than not, was instigated by the principal Orders of the Counter-Reformation. But things were obviously different when the prophecies were directed against the institutions of the Church and the authorities in Rome, and also when their political orientation was ambiguous. This was probably the case with the reading of the Apocalypse proposed by Giovanni Antonio

---

[46] Ibid., fo. 31ᵛ.

[47] N. Herborn, OM, *De insulis nuper inventis Ferdinandi Cortesii ad Carolum V Rom. Imperatorem narrationes . . . Item epitome de inventis nuper Indiae populis idolatris ad fidem Christi, atque adeo ad Ecclesiam Catholicam convertendis* (Cologne, 1532).

Pantera, Vicar-General of Parenzo, in his *Monarchia del nostro Signor Iesu Christo*: between his two editions of 1545 and 1552—the first dedicated to Francis I and the second to Henry II—there was an intervention to 'purge [the work] of many errors'.[48] The work's pro-French orientation was in agreement with the prophecy of a blessed millennium which would see the end of wars, of the idea of domination, and of religious intolerance. The religion of which Pantera spoke was not that founded 'upon ceremonies, laws and other external good works' but that justified through the merits of Christ. Furthermore, according to Pantera, one was not to expect an imminent millennium resulting from the abolition of error and heresy; instead one should train oneself to follow the example of the 'Apostles and martyrs and holy men' in dealing lovingly with those who committed errors. Pantera had in mind Reformist propaganda committed 'to denouncing, in biting words, present Antichrist and the collapse of the whole ecclesiastical order' and wished to engage polemically with it.[49] But it was not because of this that his religious and political sympathies were tainted with suspicion. The interpretation of the Apocalypse in those years was battlefield over the whole of Europe, where matters of the Americas were invoked, if at all, for show. Even lay people put forward their analysis of the Apocalypse in order to provide an orthodox direction for the 'lowest and most ignorant', as Anton Francesco Doni put it.[50] If Serafino da Fermo tried to utilize the evangelization of the New World as an argument for supporting his view that Luther was the precursor of Antichrist, Protestant propaganda was equally—if differently—fuelled, attributing the same role to the papacy.

V

It appears that the New World was given small space among the prophetic impulses and visionary expectations that animated groups of

---

[48] *Monarchia del Nostro Signor Iesu Christo di Messer Gioan' Antonio Panthera parentino al Christianissimo re Francesco* (Venice, 1545); the second edn. 'newly amended with the addition of many things' appeared from the same press in 1552; a third and last edn. was published by Comin da Trino in 1573. The dedication to Henry II, which gives corrections to the second edn., is dated 15 April 1548.

[49] Pantera, *Monarchia* (1552), fo. 48ᵛ.

[50] Cf. *Dichiaratione del Doni sopra il XIII cap. dell'Apocalisse contro agli eretici, con modi non mai più intesi de huomo vivente: che cosa sieno la nave di San Pietro, la Chiesa Romana, il Concilio di Trento, la destra della nave, la sinistra, la rete et i 153 pesci dell'Evangelio de S. Giovanni secondo i Cabalisti* (Venice, 1562), 9. (The work was also published in Padua, 1562.)

dissenters and heretics in sixteenth-century Italy. The single exception is perhaps the group of twelve nuns in Reggio Emilia gathered around the physician Basilio Albrisio, like apostles around a reincarnated Christ. They meditated on the writings of Joachim of Fiore and awaited the coming conversion to Christ of all 'those throughout the whole world who are not Christians, or who are Jews or Pagans in one or the other hemisphere'.[51]

As would be obvious, the Americas found a much greater space in the religious preoccupations of Spain. The fact is that the success of the conquest and evangelization of the Americas directly affected the military and religious experience of Spain: as soon as its imperial ambitions came to be frustrated in Europe, those successes offered compensation and an alternative. The visions which populated the dreams of Lucrezia de León in Madrid at the end of the sixteenth century are an extraordinary confirmation of this.[52] If Lucrezia dreamed of transferring the Holy See from Rome to Toledo as the culmination of a victorious crusade, similar dreams fed the heretical preachings of the Dominican Francisco de la Cruz, who was tried and sent to the stake in Lima in 1578. Among the accusations brought against him was that of having presented himself as a new world redeemer, whose destiny it was to transfer the renewed Church to Peru.[53] His was certainly not the first case of its kind: very similar aspirations were those which had supported and given a wide reputation to figures such as Giorgio Siculo. The character of the Spanish crisis should be seen in a millenarian scenario, set in the background of the Americas' colonies. This scenario was enacted on both sides of the Atlantic along the frontiers of the Spanish Empire, but its significance changed, depending on whether that frontier was one of conquest or defeat. However, the propounders were always members of the great religious Orders, more directly committed in the work of religious conquest of the Americas and in theological controversy against the Reformation in Europe.

The frontier where the military and religious power of Spain suffered the clearest defeat was in the Low Countries, whence arose the voice of Johann Frederick Lummen, or Lumnius, the author of a work eloquently

---

[51] See A. Biondi and A. Prosperi, 'Il processo al medico Basilio Albrisio, Reggio 1559', *Contributi*, 4 (1976), 70–1.

[52] See R. Kagan, 'Lucrezia de León: per una valutazione dei sogni e delle visioni nella Spagna del Cinquecento', *Quaderni storici*, NS 68 (1988), 595–607.

[53] See M. Bataillon, 'La herejía de fray Francisco de la Cruz y la reacción antilascasiana', *Études sur Bartolomé de las Casas* (Paris, 1965), 309–24.

entitled *De extremo Dei iudicio et Indorum vocatione.*[54] Fuelled by enthusiastic reading of the religious conquests overseas, such as those given by the Jesuits, Lumnius proposed setting them against the reality of their background circumstances: churches abandoned by their priests and bishops, the ferocious religious struggles and the blaze of heresy which spared no one and placed even the few remaining faithful in danger. For Lumnius, all this meant that the moment of final judgement was drawing near, the surest sign being given by the universalization of Gospel preaching. Spanish weaponry, which had forced the latecomers into the Church like the last guests in the parable of the great supper ('compel them to come in', Luke 14: 23), had a providential function, that of accelerating the fulfilment of time. Lumnius's interpretation took up that already proposed by Columbus and circulating in the pro-Spanish literature of the *Reconquista* and the crusade. For the Jesuits who were endeavouring to propagate the Gospel, Lumnius adapted the prophecy of Isaiah (60: 8–9), which spoke of divine messengers flying like doves towards remote islands.

The Jesuits also found themselves implicated in the case of Francesco de la Cruz and they only managed to emerge unscathed thanks to the Inquisition's favour.[55] But it was precisely a Jesuit—José de Acosta—who took the opportunity of that occasion to put forward a harsh critique of the whole millenarian tradition fuelled by the discovery and, above all, by the conquest of the New World.[56] Behind his critique lay the notable distrust of apocalyptic preaching, both for the elements of aggression towards the ecclesiastical institutions for which it was always a potential vehicle and because the experience of the century had shown the risks of a genre of which the most astute churchmen had been forced to take notice. As has been observed, it was not fortuitous that the only book of the Bible which Calvin did not comment upon was that of the Apocalypse.[57] At the same time, there was a desire to differentiate between the missionary methods associated with the apocalyptic tradition

---

[54] Printed in Antwerp 'apud Antonium Tilenium Brechtanum' in 1567; reprinted in Venice 'apud Dominicum de Farris' in 1569; republished by its author in Antwerp in 1594 'apud Ioannem Keerbergium' with the title *De vicinitate extremi iudicii Dei et consummationis saeculi libri duo.*

[55] Affirmed by the Superior-General C. Aquaviva in a letter of 21 Nov. 1583 (*Monumenta Peruana*, iii. 1581–95, ed. A. de Egaña, SJ (Rome, 1961), 299).

[56] J. Acosta, *De temporibus novissimis* (Rome, 1590).

[57] The observation is by Marguerite Soulié, 'Prophétisme et visions d'Apocalypse dans *Les Tragiques* d'Agrippa d'Aubigné', *Réforme, Humanisme, Renaissance*, 12 (1986), n. 22, 5–10, see esp. 9.

and what they implied. If one began with the conviction of the coming fulfilment of time, the consequent missionary work was summary and hurried, and relied on Spanish troops for its efficacy. 'Those who do not want to listen willingly to the Holy Gospel of Jesus Christ, must do so by force; because in this matter the following proverb holds sway: better to have good by means of force than evil for pleasure'; this was the opinion of one of the most famous Spanish missionaries in the Americas, Brother Toribio de Benavente, called 'Motolinia', on the proposals of Bartolomé de las Casas to preach the Gospel to the Indians without using force.[58] Precisely in those years Geronimo de Mendieta was compiling the *Historia eclesiastica indiana*,[59] the historiographical *summa* of the whole of that phase of the religious conquest. The desire to rewrite the history of the Church had taken shape with the venture of the Indies: the great Christian conquest of the world, begun with the Apostles and later blocked by the stubborn and strong enemies of the Jews and Muslims, appeared to be setting off again with an overwhelming momentum, giving a meaning to the whole course of history. As in the apostolic age, the things of this world gained meaning again from their imminent end. The providential function of the Spanish monarchy was associated with that other equally providential function of the missionary orders. Together they had to compel the last guest to enter so that the feast of which the evangelical word had spoken could begin. We can find widespread traces of this reworking of the providential view of history in the literary production of the religious orders of those years. On more precisely missionary grounds that visionary and apocalyptic interpretation could also be adapted for other political authorities: the Capuchin Claude d'Abbéville, for example, spoke of a renaissance of the Church— 'nouvelle naissance de l'Eglise romaine en ce monde nouveau'—brought into being 'sur le déclin du monde', thanks to the King of France.[60] But in fact during the sixteenth century this interpretation was closely associated with the idea of the religious mission of the Habsburg monarchy. On behalf of the Society of Jesus, José de Acosta began a cautious but determined move away from all this: the preaching of the Gospel and

---

[58] Toribio Motolinia, OFM, *Historia de los indios de la Nueva España*, ed. E. O'Gorman (Mexico, 1973), 211.

[59] The order to compose its history had been given to Mendieta by the general of the order in 1571; the work was finished in 1596. See Geronimo de Mendieta, OM, *Historia eclésiastica indiana*, ed. J. Garcia (Icazbalceta, Mexico, 1945).

[60] Claude d'Abbéville, *Historie de la mission des pères capucins en l'isle de Maragnan et terres circonvoisins* (Paris, 1614); anastatic edn., ed. A. Métraux and J. Lafaye (Graz, 1963).

the conversion of non-European peoples was to be a long and difficult task. It would be necessary to perform it without haste, starting from the conviction that the history of the world had not yet all been written and that the future before them was vast and unknown.

VII

PROPHECY AND ICONOGRAPHY

Introduction

Prophecy expressed through visual images has a long history, going back at least as far as the magnificent series of concrete images given by Old Testament prophets.[1] In the Middle Ages Joachim of Fiore discovered that his prophetic vision of history was often more adequately expressed in *figurae* than in words.[2] The popularity of this method in the later Middle Ages is demonstrated first by the distribution of *figurae* in Joachimist manuscripts[3] and secondly—though in quite a different style—by the numerous pictorial versions of the *Vaticinia de summis pontificibus*.[4]

In the Renaissance period we move into a new phase of what may perhaps be termed prophetic art. Botticelli's great mystical paintings, as we have seen, must be 'read' within the prophetic frame of the Last Age, from which springs the inspiration of their message. André Chastel has interpreted the prophetic message for the Papacy in the *stanze* adorned by Raphael.

In his essay on the Sistine Chapel ceiling, partially reprinted below,[5] Malcolm Bull takes a different approach. He argues that iconographical study is concerned with relationships, both paradigmatic and syntagmatic. That is to say, the reader of the work of art is concerned not only 'with relationships which seek to establish a correspondence between one pictorial element and another extra-pictorial element', but also 'with relation-

---

[1] See, for example, the visual images in the opening chapters of Jeremiah and Ezekiel.
[2] See references given in M. Reeves and B. Hirsch-Reich, *The* Figurae *of Joachim of Fiore* (Oxford, 1972), 20–1.
[3] See ibid. 31 ff., 99–116, 297–9.
[4] For the *Vaticinia*, see above, Reeves, Ch. 1.
[5] This essay originally appeared in *Burlington Magazine*, 130 (August 1988), 597–605.

ships between the elements within a picture or between a group of pictures'. In language syntagmatic relationships realize the potential of words to generate new meanings. It is surely not absurd to argue that the recombination of images may have a similar effect. 'If both syntagmatic and paradigmatic relationships are considered, the range of reference is governed by the existence of sources and the range of connotation by the composition of the images.' Thus iconography involves the search for factors which will help account for the specific configurations of form within a picture.

The three iconographical studies which follow seek to relate their subjects to a possible contextual framework by such a method. They reveal, first, the response of artists and their programmers to the general prophetic mood of Renaissance Rome and, secondly, the strong influence of certain prophetic sources and events. Malcolm Bull's essay suggests that Joachim of Fiore's prophetic typology of Old Testament figures and stories could still form an evocative pattern of images for visual representation. Josephine Jungić's first study demonstrates the impact made by the discovery of the *Apocalypsis Nova* as a prophetic event and her second suggests the all-pervasive influence exercised by the chimera of the Angelic Pope.

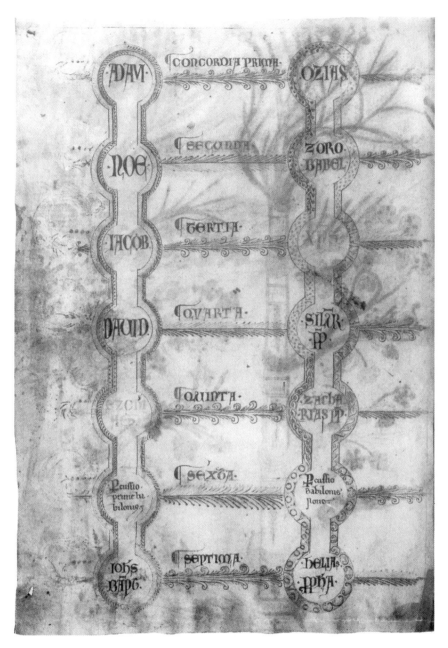

1. The concords of persons. *Liber Figurarum,*
MS. Oxford, C.C.C. 255A, fo. 13ᵛ.

2. The *pavimentum*. *Liber de Concordia*, MS. Corsini 41, fo. 26ᵛ.

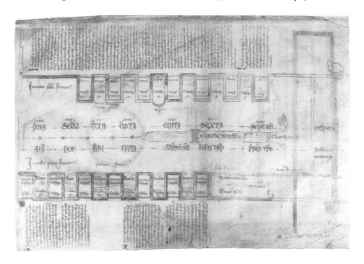

3. The seven *etates*. *Liber Figurarum*, MS. Oxford, C.C.C. 255A, fo. 8ᵛ.

4. Circles. *Liber de Concordia*, MS. Corsini 41, fo. 21ᵛ.

5. Borgherini Chapel. Rome, San Pietro in Montorio.

6. Sebastiano del Piombo, *Flagellation of Christ*. Borgherini Chapel.
Photo: Anderson/Art Resource

7. Sebastiano del Piombo, *Tranfiguration of Christ*. Borgherini Chapel.

8. Sebastiano del Piombo, Two Prophets. Borgherini Chapel.
7 & 8 Photo: Rizzoli

9. Botticelli, *Transfiguration of Christ*. Vatican, Pinacoteca.

Photo: Anderson/Art Resource

10. Writing on the book below the altar. Borgherini Chapel.

Photo: P. Higginson

11. Raphael, *Tranfiguration of Christ*. Vatican, Pinacoteca.
Photo: Anderson/Art Resource

12. Sebastiano del Piombo, *Cardinal Bandinello Sauli and Three Companions 1516.*
National Gallery of Art, Washington, DC, Samuel H. Kress Collection.

13. Raphael, *Cardinal Bernardo Dovizzi da Bibbiena*. Pitti Gallery, Florence.
Photo: Alinari/Art Resource.

14. Raphael, *Pope Julius II*. National Gallery, London.

15. Sebastiano del Piombo, *Ferry Carondelet and Two Companions*. Thyssen-Bornemisza Collection, Lugano, Switzerland.

16. Raphael, *Pope Leo X with Cardinals Giulio de' Medici and Luigi de' Rossi*. Uffizi
Gallery, Florence. Photo: Alinari/Art Resource.

16

The Iconography of the Sistine Chapel Ceiling

Malcolm Bull

The decoration of the Sistine Chapel ceiling comprises about 175 picture units. There are nine scenes from Genesis along the longitudinal axis, four Old Testament histories in the corner spandrels, seven prophets and five sibyls on thrones, forty ancestors of Christ in the lunettes and spandrels above, ten medallions depicting events from the Old Testament, and a multitude of nude figures adorning a fictive architectural frame-work.[1] In most cases the subject of each picture unit is easily recogniz-able. The ancestors, prophets, and sibyls are identified by name, and most of the histories are straightforward representations of well-known biblical stories. Even the obscure apocryphal stories depicted in some of the medallions have now been identified.[2] Few scholars would dispute that the basic programme is that illustrated here as Fig. 1.[3]

The successful identification of the subject matter usually implies the completion rather than the beginning of iconographical work. But recent

I am indebted to Michael Hirst and Marjorie Reeves who commented on an earlier draft of this article, John Dunnett who assisted with translations from the Latin, Giles Darkes who drew the diagram for Fig. 1, and to the President and Fellows of Corpus Christi College, Oxford, for permission to reproduce photographs of the *Liber Figurarum*.

[1] See C. de Tolnay, *Michelangelo II, The Sistine Ceiling* (Princeton, NJ, 1945).

[2] See E. Wind, 'Maccabean Histories in the Sistine Ceiling', in E. Jacob (ed.), *Italian Renaissance Studies* (London, 1960), 312–27.

[3] Some, however, would disagree with the identification of the third history from the altar, in which God is shown hovering with outstretched arms between sea and sky. For various interpretations of this scene, see M. Bull, 'The Iconography of the Sistine Chapel Ceiling', *Burlington Magazine*, 130 (August 1988) 597 n. 3. The most sensible interpre-tation is probably that of De Tolnay, *Michelangelo II*, 138, who argues that this is the second day of creation, the separation of the firmament from the waters (Gen. 1: 6–8), an event which could be represented adequately by sea and sky alone. The chronological order is thus 132/456/879, an arrangement which implies that the histories should be viewed, not in a continuous sequence, but in three groups of three.

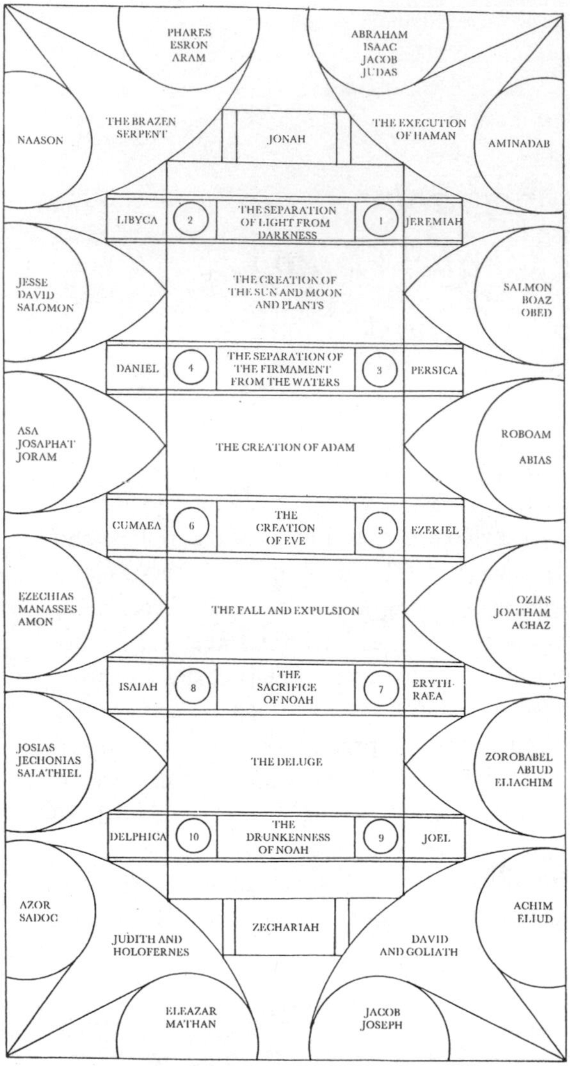
PHARES
ESRON
ARAM
ABRAHAM
ISAAC
JACOB
JUDAS
NAASON
THE BRAZEN
SERPENT
JONAH
THE EXECUTION
OF HAMAN
AMINADAB
LIBYCA
2
THE SEPARATION
OF LIGHT FROM
DARKNESS
1
JEREMIAH
JESSE
DAVID
SALOMON
THE CREATION OF
THE SUN AND MOON
AND PLANTS
SALMON
BOAZ
OBED
DANIEL
4
THE SEPARATION OF
THE FIRMAMENT
FROM THE WATERS
3
PERSICA
ASA
JOSAPHAT
JORAM
THE CREATION OF ADAM
ROBOAM
ABIAS
CUMAEA
6
THE
CREATION
OF EVE
5
EZEKIEL
EZECHIAS
MANASSES
AMON
THE FALL AND EXPULSION
OZIAS
JOATHAM
ACHAZ
ISAIAH
8
THE
SACRIFICE
OF NOAH
7
ERYTH-
RAEA
JOSIAS
JECHONIAS
SALATHIEL
THE DELUGE
ZOROBABEL
ABIUD
ELIACHIM
DELPHICA
10
THE
DRUNKENNESS
OF NOAH
9
JOEL
AZOR
SADOC
ZECHARIAH
ACHIM
ELIUD
JUDITH AND
HOLOFERNES
DAVID
AND GOLIATH
ELEAZAR
MATHAN
JACOB
JOSEPH

1. The Sacrifice of Abraham
2. The Ascension of Elijah
3. The Death of Absalom
4. blank
5. Alexander kneels before the high priest
6. The Punishment of Nicanor
7. The Chastisement of Heliodorus
8. The Destruction of the idol
9. Joab murders Abner
10. Joram thrown from Naboth's vineyard

1. Diagram of the Sistine Chapel ceiling

studies of the ceiling have been concerned with secondary questions: what criteria governed the selection of the subject matter? Who was responsible for making that selection? What is the significance of this complex combination of religious images?[4] The diversity and complexity

[4] For various contributions to the solution of these questions, see Bull, 'Iconography', 597 n. 4.

of this literature means that any further work needs to follow clearly stated practical guidelines. The following are suggested: the histories should be read in their general chronological sequence, and thus in the same direction as every other cycle in the chapel; interpretation should be based on the manifest identity of the figures, and not on any supplementary identity perceived in the pose, expression, or handling of the figure; and a clear distinction should be made between an interpretation based upon a retrospective reading of the ceiling, and a programme which could have been devised in the early sixteenth century before the ceiling existed.

What is represented in the ceiling is, then, with the exception of a few minor ambiguities, quite clear. Still uncertain, however, is the process by which particular scenes and figures were brought together and placed in a specific arrangement. What follows is concerned not with the 'meaning' of the ceiling, or even of parts of the ceiling, but with the factors involved in producing a particular visual configuration. It is argued that one significant factor in producing this arrangement may have been the influence of the ideas of Joachim of Fiore.

The first four books of Joachim's *Liber de Concordia* are concerned to establish concords between the generations from Adam to Christ, and the generations from Christ to the end of the world. Joachim used two complementary frameworks to structure his interpretation: the *prima* and *seconda diffinitiones*. The *prima diffinitio* was represented by the Greek letter *alpha* and formed history into three overlapping *status*, one for each person of the Trinity. The *seconda diffinitio*, symbolized by the letter *omega*, divided history into two *tempora*, one before and one after Christ.[5] This framework is supplemented by subsidiary patterns: seven *sigilla* in the Old Testament and their openings in the Christian era, and seven *etates*, symbolized by the seven days of creation, which divide all human history into seven temporal units.[6] Some of these patterns are illustrated by diagrams, similar to those of the *Liber figurarum*, which show the generations as trees and the three *status* as interconnected circles.[7] The fifth book of the *Liber de Concordia* is a commentary on the Old Testament which focuses on the seven days of creation, the lives of

[5] See M. Reeves, *Influence of Prophecy in the Later Middle Ages: A Study in Joachimism* (Oxford, 1969), 19–20; Reeves and B. Hirsch-Reich, *The Figurae of Joachim of Fiore* (Oxford, 1972), 7–8; Joachim of Fiore, *Liber de Concordia Novi et Veteris Testamenti*, ed. E. R. Daniel (Philadelphia, 1983), pp. xxii–xxvii, xxxv–xlii.

[6] Reeves and Hirsch-Reich, *Figurae*, 5–10.

[7] Ibid. 29–38, 44–5.

David and Solomon and of Elijah and Elisha, and the books of Job, Tobit, Esther, and Judith.

In the Sistine Chapel the ancestors of Christ are painted above the figures of the first thirty sainted Popes. Like the Popes, the ancestors are depicted in chronological order from altar to entrance, the sequence moving from one wall to the other in every bay. For Joachim the Popes marked out the Christian era, just as the generations of Christ had calibrated the progress of history in the Old Testament. It was a link that found expression in one of the *figurae* of the *Liber figurarum*. In one half the generations from Adam to Christ are listed in full, in the other there is a list of the Popes.[8]

In another *figura* Joachim illustrates the concords of persons, pairing seven ancestors with seven biblical characters and Popes, each forty-two generations apart (Pl. 1).[9] The Popes mentioned are later than those depicted in the Sistine Chapel, but the first two concords, those of Adam and Ozias, and Noah and Zorobabel, do refer to persons depicted on the ceiling. In Joachim's diagram Adam is followed by Noah in one column, Ozias by Zorobabel in the other. On the ceiling the lunette and spandrel of Ozias is adjacent to the scene of Adam's expulsion, while Zorobabel, in the next lunette and spandrel of the north wall, adjoins the Flood panel. Thus on the ceiling Ozias and Zorobabel are parallel to Adam and Noah just as they are in Joachim's *figura*.

The concord between Adam and Ozias was of particular importance to Joachim as Adam had initiated the first *status* and Ozias the second.[10] In the *Liber de Concordia* Joachim noted the particular event that linked the two men: 'That Adam was driven out of paradise or Ozias out of the temple for his sins . . . signified that the Jewish people should be driven out of the holy place for their offences'.[11] Not only does Joachim establish a concord between Adam and Ozias, but he points to the expulsion from paradise as a specific concord between them.

Joachim considered that Ozias was the forty-third generation from Adam, and that Zorobabel was the forty-third generation from Shem, concluding that: 'Around him [Zorobabel] should be sought the concords of those things which happened in the days of Shem'.[12] The viewer does

---

[8] Joachim of Fiore, *Il Libro delle Figure dell'Abate Gioacchino da Fiore*, ii, ed. L. Tondelli, M. Reeves, and B. Hirsch-Reich, 2nd edn. (Turin, 1954), tav. 9, 10.

[9] Ibid., tav. 7.

[10] Reeves and Hirsch-Reich, *Figurae*, 7, 37, 142–3; *Lib. de Conc.*, 101.

[11] *Lib. de Conc.*, 196; *Liber Concordie Novi et Veteris Testamenti* (Venice, 1519), fo. 134[v].

[12] *Lib. de Conc.* 106.

not have far to look for events from the life of Shem. On the ceiling, Zorobabel is placed opposite the Flood, with the *Sacrifice* and *Drunkenness of Noah* on either side. Both are scenes in which Shem figures prominently.

In the Sistine programme Ozias and Zorobabel are placed in positions which fit exactly with their significance in Joachim's thought. It is an occurrence hard to dismiss as fortuitous, for it does not result from the pattern employed for the rest of the ancestors, according to which Zorobabel should not be in the lunette and spandrel following Ozias, but in the one diagonally opposite. It may not, then, be improbable that the decision to place Ozias and Zorobabel in these particular lunettes and spandrels was influenced by the awareness of how appropriate this juxtaposition would be to the adjacent histories.[13]

In Joachim's commentary on the creation narratives in book five of the *Liber de Concordia* he expounds the significance of each day of creation according to the species of *intelligentiae*.[14] In the *intelligentia allegorica* Joachim sees the separation of the waters on the second day as prefiguring the division 'inter domum iusti et domos peccatorum' to be effected by Noah's ark and the Christian church. He describes how: 'One of Noah's sons, although he had been saved physically in the ark yet remained outside both mentally and spiritually, namely he who was shown to have no share in his father's blessing. And indeed the eternal blessing itself was a solid firmament from which Ham and all false people have been separated until the present day.'[15] Thus he was able to link the events of the second day of creation with the sin of Ham.

Joachim connects the third day not with the judgement of evil but with the salvation of the just. The separation of dry land is linked to the salvation found in the ark and the formation of a chosen race from the seed of Abraham. The chosen people are also symbolized by the creation of the plants.[16] For Joachim then, the days of creation pointed forward to the events of the Flood. In particular he saw analogies between the second day and the sin of Ham, and between the third day and the salvation in the ark. On the Sistine ceiling the second scene depicts the third day of creation while the eighth shows the ark. The third scene represents the second day of creation and the ninth the sin of Ham. Joachim's commentary thus makes explicit connections between the

---

[13] For another explanation of this interruption, see De Tolnay, *Michelangelo II*, 174.
[14] *Lib. de Conc.* fos. 60ᵛ–73ʳ.
[15] Ibid., fo. 72ᵛ.
[16] Ibid.

events represented in the second and third panels of the creation sequence, and the corresponding panels of the Noachian sequence.

There is little in the *Liber de Concordia* about either the Brazen Serpent or David and Goliath. But there is a very detailed exposition of the stories of Judith and Esther.[17] Joachim found great significance in these two books, considering them to be two of the four 'special histories', and often using them as two elements in any symbolic group of four—for example the four gospels.[18] The spandrel of Judith and Holofernes is located on the entrance wall of the chapel, at the end of the Noachian histories. In the *Liber de Concordia* Joachim linked the two stories through the symbol of the dove:

Then a dove was sent out of the ark and brought in an olive branch as a sign of peace. Here purest Judith went out of Bathulia, truly like a dove beautiful in looks and character, and brought back a word of peace to the children of Israel— when they were weakened by the swelling waves of Assyrians—by the decapitation of Holofernes, who was the head of their bravery and courage.[19]

From a Joachimist perspective, the Judith spandrel would not only have been a natural choice as one of a group of four histories, but its position relative to the fresco of the Flood, which shows the dove flying from the ark, could not have been more appropriate.[20]

Joachim pairs the story of Judith with that of Esther, an aspect of which, the crucifixion of Haman, is represented in the spandrel diagonally opposite. The subject is unusual, for, according to the Vulgate, Haman was hanged not crucified: 'Suspensus est itaque Aman in patibulo quod paraverat Mardochaeo' (Esther 7: 10). The Sistine spandrel, it has been argued, follows Dante, who in the *Purgatorio* described Haman as 'crucifisso, dispettoso e fiero'.[21] However, Joachim's account of the execution also suggests that Haman was crucified: 'On the orders of the king, Haman was finally hung on the high cross which he himself had prepared for Mordecai.'[22] In Joachim's mind this event had additional significance, for with it was accomplished Mordecai's rise to power,

---

[17] Ibid., fos. 117ᵛ–122ᵛ.
[18] Reeves and Hirsch-Reich, *Figurae*, 226–8.
[19] *Lib. de Conc.* 187.
[20] Ibid. 196.
[21] Dante, *Purgatorio*, xvii. 27. See E. Wind, 'The Crucifixion of Haman', *JWCI* 1 (1937–8), 245–8.
[22] *Lib. Conc.*, fo. 122ᵛ: 'Suspensus est tandem Aman iubente rege in excelsa cruce quam ipse paraverat Mardochero.' Joachim has transferred the word 'crux' from Esther 5: 14. The use of the verb 'suspendo' would not rule out the idea of crucifixion. The Vulgate refers to Jesus 'quem occiderant suspendentes in ligno' (Acts 10: 39).

which prefigured the elevation of the great Pope of the third *status*. As Joachim noted: 'Thereafter Mordecai was exalted beyond measure, and the power and glory of Haman was given to him by the king, because the one who will be Peter's successor at that time will be raised on high, as if the most faithful deputy of Christ Jesus.'[23]

Although the sibyls figured prominently in the Joachite tradition—Joachim had been summoned by Pope Lucius to interpret a sibylline text, and there is pseudo-Joachimist work relating to the sibyls, including the *Vaticinium Sibillae Erithreae*[24]—they are not mentioned in the *Liber de Concordia*. There are brief expositions of the prophetic books of the Old Testament, but only one casts any light on the position of the prophets in the ceiling. Jonah was traditionally linked with the entombment and resurrection of Christ, but on the Sistine ceiling, situated between the *Brazen Serpent* and the *Crucifixion of Haman*, some link with the crucifixion is implied. In Joachim's account this connection is made explicit: 'Therefore the sea rose up against Jonah. The uproar of the Jews rose up against Christ when they kept on saying, "Away with him, Away with him, crucify him!" And the Son of Man was in the heart of the earth, just like Jonah, for three days and three nights.'[25]

Most obviously in keeping with Joachimist thought is the division of the twelve seers into seven prophets and five sibyls. According to Joachim's symbolism, which in this respect appears to be unique in medieval numerology, twelve divides naturally into a five and a seven—the seven making manifest what is concealed in the five.[26] It is a pattern which could easily be applied to the sibyls who represented the partial revelation given to the Gentile world, and the prophets who were responsible for the fuller revelation given to the Jews.

On the ceiling each history is flanked either by a prophet or sibyl on a throne or by a pair of ancestors in the spandrels. The practice of associating pairs of Old Testament characters with particular events is well illustrated by the *pavimentum figura* in the *Liber de Concordia*, in which pairs of names are associated with the seven[27] seals (Pl. 2). Five of the names in Joachim's diagram are also used for ancestors or prophets adjacent to the histories in the ceiling. But the similarity in the conception is more striking than that of the execution. However, the figure of

---

[23] *Lib. Conc.* fo. 122ᵛ; see also, fos. 121ʳ, 132ᵛ.
[24] Reeves, *Influence*, 307–8.
[25] *Lib. Conc.* fo. 123ʳ.
[26] Reeves and Hirsch-Reich, *Figurae*, 13–19.
[27] *Lib. de Conc.* 215–17; Reeves and Hirsch-Reich, *Figurae*, 65–6.

the seven *etates* in the *Liber figurarum* (Pl. 3) is a closer parallel. The seven *etates*, which for Joachim were symbolized by the seven days of creation, run down the centre with the names of Old Testament figures, including those of Ozias and Zorobabel, on either side.[28] It is of particular interest that the fourth *etas*, associated with David, has Boaz as the parallel figure of the first *status*. On the ceiling the creation of the sun and moon on the fourth day is shown with Boaz and David on either side.

There are thus several features in the iconography of the ceiling which correspond, sometimes precisely and often generally, to the themes of the *Liber de Concordia*. Most of them are details, which point only to the possibility that the overall programme may owe something to Joachimist ideas. But the sequence of the histories reveals a pattern which accords well with Joachim's central theme.

The histories divide naturally into three groups of three. In the first God the Father is shown alone with the elements; in the second Adam and Eve are the chief protagonists, while the final group is centred around the Flood, with Noah and his family as the main figures. It was a theological commonplace that Adam was a type of Christ, and that the sending of the dove from the ark prefigured the procession of the Holy Spirit. Thus the first group of histories depicts God the Father, the second shows a type of Christ, and the third includes the type of the sending of the Spirit. It is a progression strongly reminiscent of Joachim's division of history into three *status*, one for each member of the Trinity.

When examined in more detail, the ceiling yields further evidence for a Trinitarian division of the histories similar to that of Joachim's three *status*. According to Joachim: 'If Adam and Eve are taken alone, it signifies Christ and the Church. Therefore the Church is the mother of Christ, and the Church is nonetheless the bride of Christ.'[29] For Joachim, as for other interpreters, the creation of Eve represented the formation of the Church, and the formation of the Church was the central event of the second *status*, just as the creation of Eve is the central panel in the second group of the ceiling.[30] In Joachim's thought Ozias inaugurated the second *status*, and on the ceiling he is represented alongside the second group, next to the particularly appropriate scene of the expulsion.

---

[28] *Lib. Fig.* tav. 18.

[29] *Lib. de Conc.* 241.

[30] The centrality of the Eve panel has also been noted by S. Sinding-Larsen, 'A Re-Reading of the Sistine Ceiling', *Acta Instituti Romani Norvegiae*, 4 (1969), 153, and E. Leach, 'Michelangelo's *Genesis*: Structuralist Comments on the Paintings on the Sistine Chapel Ceiling', *Times Literary Supplement* (18 March 1977), 313.

Zorobabel, however, was understood by Joachim to portend the sending of the Holy Spirit, and was reckoned by Joachim's followers to be the type of the *novus dux* who would lead the Church into the third *status* of the Spirit.[31] On the ceiling Zorobabel is placed next to the panel which depicts the sending forth of the dove. Joachim makes explicit the connection between the events of the Flood and the situation of the *viri spirituales* who were to inaugurate the third *status*:

The fact that the flood and the armies of the nations spread over the sinners of the earth while a remnant was saved in the ark and the city of Bathulia indicates that around the end of the ages—to which we ourselves have come—the multitude of the nations should flood over the earth on account of sins, while a remnant should be saved by the device of the spiritual church; until a dove goes forth from her, another Judith as it were, a certain spiritual group of righteous men given to preaching the word of God, who will return and proclaim that peace is to come swiftly over the earth.[32]

Joachim illustrated his conception of the three *status* in several diagrams, sometimes using overlapping circles, and sometimes foliated tree-circles.[33] In the *figura* of the three *status* in the *Liber de Concordia*, Joachim divides each *status* into three sub-sections, creating, as in the ceiling, three groups of three (Pl. 4).[34] The events of the first group relate to the Exodus, the second to the formation of the Church, and the third to the activities of the *viri spirituales* (two monastic orders symbolized by the dove and raven sent from the ark),[35] whom Joachim thought would bring the *status* of the Spirit to its fruition.

In Joachim's *figura* each circle is linked to an event analogous to the events of the circles in corresponding position in the other *status*. Thus

[31] *Lib. de Conc.* 401–2. See Reeves and Hirsch-Reich, *Figurae*, 295. For the possibility that Dante adapted Joachim's Zorobabel prophecy for his own purposes, see Reeves, 'The Third Age: Dante's Debt to Gioacchino da Fiore', in A. Crocco, *L'Eta Dello Spirito e La Fine dei Tempi in Gioacchino da Fiore, Atti del II Congresso Internazionale di Studi Gioachimiti*, (S. Giovanni in Fiore, 1986), 134–9.

[32] *Lib. de Conc.* 196.

[33] *Lib. Fig.* tav. 11, 22. See Reeves and Hirsch-Reich, *Figurae*, 38–40, 43–6, 170–3, 192–8.

[34] *Lib. de Conc.* 163–72: (i) The children of Israel enter into Egypt with Jacob their father; (ii) the children of Israel leave Egypt led by Moses and Aaron; (iii) the children of Israel enter the promised land with Joshua—their leader; (iv) the Apostles preach in the synagogue with Christ their master; (v) the Disciples of Christ go over to (the Gentiles) to preach the kingdom of God, led by Paul and Barnabas; (vi) the Disciples of Christ receive the legacy of the nations while John the Evangelist is still alive; (vii) the spiritual men preach in the world that they may win some; (viii) the spiritual men go over to the stricter life lest they perish with the children of the age; (ix) whoever believes in the spiritual men enters into that rest of which the holy prophets spoke.

[35] Joachim of Fiore, *Expositio in Apocalypsim* (Venice, 1527), fo. 81ᵛ.

the first circles describe a movement towards a situation (*The Israelites in Egypt*, *Christ in the synagogue*, the *viri spirituales* in the world), while the second circles describe a move away from that situation (the *Exodus*, *The Gentile mission*, etc.), and the third circles refer to an entry into a new situation (for example the Promised Land).

Sinding-Larsen has noted a similar set of relationships in the Sistine histories.[36] And Joachim's interpretation of the second and third days of creation suggests links between the second and third Sistine histories and the events of the eighth and ninth. How far does the correspondence between the Sistine histories and Joachim's *figura* extend? As Fig. 2 indicates, although there are not precise correlations in every instance, the majority of the events portrayed in the histories are interpreted by the *Liber de Concordia* in a fashion that links them to the events in the *figura* of the three *status*.

It should be noted, however, that the relationship of type and anti-type is not reciprocal.[37] No one who wished to replicate Joachim's diagram of the three *status* would have arrived at the Sistine programme from reading the *Liber de Concordia*. But someone who wished to give significance and shape to the events of Genesis could easily have been led from their reading to examine the *figura*, and then have structured the programme accordingly.

Within the *Liber de Concordia* there are ideas which could explain some aspects of the ceiling's compositional structure. But it is not permissible to invoke these ideas solely on the basis of their explicatory value. A monk writing in twelfth-century Calabria is unlikely to have had any direct influence on artistic projects in sixteenth-century Rome. Unless mediating factors can be discovered, the correlations must either be dismissed as fortuitous, or else ascribed to some independent cause.

As several essays in this book, however, have already shown, a continuous Joachimist tradition was active throughout the later Middle Ages, fuelled by a considerable body of Joachimist literature widely disseminated in manuscript and, by the sixteenth century, in printed form. In the Renaissance period, as we have seen, the three prophetic agencies of Joachimism—'new spiritual men', Angelic Pope, and World Emperor—were still awaited with expectant hope, to bring about the *renovatio mundi* of the third *status*. This prophetic hope was highlighted by the Savonarolan movement, by the 'discovery' of the *Apocalypsis*

---

[36] Sinding-Larsen, 'A Re-Reading', 155.

[37] The interpretation of the days of creation points forward to the events of the Flood but the account of the Flood does not look back to the creation.

SISTINE HISTORY	The separation of Light from Darkness	Creation of Plants Sun and Moon	Separation of the Firmament from the Waters
INTERPRE-TATION IN *LIB. CONC.*		Plants = Chosen People	Firmament = blessing that excludes sinners (e.g. Ham)
EVENTS OF *FIGURA*	Entry into Egypt	Exodus from Egypt	Entry into Promised land
SISTINE HISTORY	Creation of Adam	Creation of Eve	Fall and Expulsion
INTERPRE-TATION IN *LIB. CONC.*	Adam = Christ	Eve = Church	Expulsion = Rejection of the Jews
EVENTS OF *FIGURA*	Apostles preach with Christ in synagogue	Paul and Barnabas preach to Gentiles	Gentiles accepted into Church
SISTINE HISTORY	Sacrifice of Noah	The Flood	Drunkenness of Noah
INTERPRE-TATION IN *LIB. CONC.*		Flood = Judgement on world Ark = Church Dove = Spiritual Men	
EVENTS OF *FIGURA*	Spiritual men preach in the world	Spiritual men adopt stricter life so they do not perish with the world	Those who believe in the spiritual men enter into rest

2. Diagram showing correlations between the *Liber de Concordia* and the histories of the Sistine Chapel ceiling

Nova and the influence of the Amadeite circle, by expectations among Augustinian hermits, especially the Venetian group who started publishing Joachimist works, and by the apocalyptic preachers who erupted dramatically in the cities from time to time.

In Rome the circulation and discussion of prophecies was actively pursued in various groups. In the person of Egidio of Viterbo prophetic expectations penetrated into the Papal Curia of Julius II, as witness his oration of 1507, *De ecclesiae incremento*.[38] Later he would construct his

[38] First published by J. O'Malley, *Traditio*, 25 (1969), 265–338, repr. in J. O'Malley, *Rome and the Renaissance: Studies in Culture and Religion* (London, 1981), same pag. For further reference, see above, Reeves, Ch. 5.

own prophetic pattern of history, culminating in the tenth (or twentieth) age of renewal. This pattern was not a Joachimist one, but, as we have already seen, he was in touch with the Joachimist enthusiasts of his own order and may well have been drawn to the prophetic image of the *viri spirituales* as a model for the coming role of the Augustinians. It is therefore by no means unlikely, that, at an earlier date, he had studied Joachim's *Liber de Concordia*. Numerous late medieval Italian manuscripts of this work were available. Nine of the extant copies are in Rome.[39] Given Egidio's method of seeking the providential purposes of God in history through type and concord, Joachim's great work of concords would have been attractive to him.

Egidio, of course, is the man suggested by Edgar Wind and adopted by Dotson as the possible programmer of the Sistine ceiling.[40] Was he likely to have been consulted about the ceiling? And if he was, is it plausible to suggest that his ideas on this were influenced by Joachim? The 1507 oration expects the Golden Age to be realized under Julius II, whom he sees prophetically typified in various figures throughout the Old Testament. For instance, interpreting Isaiah 6: 1, which refers to the death of Uzziah (Ozias), he says: ' "I saw the Lord seated." He meant to say, I saw Julius II, the Pope, both succeeding the dead Ozias and seated on the throne of religious increase.'[41]

For Joachim, Ozias had been succeeded by Zorobabel, the symbol of the *novus dux*. Julius was, as Egidio himself pointed out, a follower of Zorobabel, another rebuilder of God's temple.[42] And just as Joachim saw Zorobabel as portending the third *status* following the second initiated by Ozias, so Egidio perceived Julius as the successor of Ozias, and the herald of a new Age. On the ceiling Zorobabel succeeds Ozias, and is situated next to the Joachimist symbol of the third *status*, the sending forth of the dove from the ark.

Joachim is probably unique for his visual representations of the patterns of history. It is unlikely that Egidio had seen his *Liber figurarum*, but the diagrammatic figures of history in the *Liber de Concordia* were easily available. It would seem that Egidio's thinking during the Julian period was becoming focused on a study of the clues for reading providential meanings in history and from such a point of view Joachim's

---

[39] *Lib. de Conc.* pp. xliii–lix.
[40] Wind, 'The Crucifixion of Haman', 82–4; E. Dotson, 'An Augustinian Interpretation of Michelangelo's Sistine Ceiling', *Art Bulletin*, 61 (1979), 250–5.
[41] O'Malley, *Traditio*, and *Rome and Renaissance*, 320.
[42] Ibid. 322.

methodology was most apt. The fact that later his sources, hermetic, Neoplatonic, cabbalist, led him away from Joachim's Trinitarian structure of history does not invalidate the suggestion that he found in the interconnections of Joachim's Old Testament typology, ideas which awoke his imagination to see them in visual terms.[43]

There is no indication that Egidio was trying to impose a Joachimist framework on his interpretation of current events. On the contrary, his methods were always eclectic, and it would have been uncharacteristic if he had attempted to follow any one source. But what these interconnections do suggest is that there are parallels between the way Joachim conceived the future, the way Egidio conceived the pontificate of Julius II, and the way the iconography of the Sistine ceiling is structured.

17

Joachimist Prophecies in Sebastiano del Piombo's Borgherini Chapel and Raphael's Transfiguration

JOSEPHINE JUNGIĆ

Michael Hirst's recent monograph on Sebastiano del Piombo draws attention to the curious circumstance whereby two monumental paintings of the Transfiguration 'never a common subject in Renaissance art' were commissioned in Rome at almost the same time and both ended up in the same church.[1] Pierfrancesco Borgherini, the Florentine banker, commissioned Sebastiano to paint a *Transfiguration* in the summer of 1516 as part of the decoration of his chapel in the church of San Pietro in Montorio. A few months later, Cardinal Giulio de' Medici commissioned two large altarpieces, a *Transfiguration* and a *Raising of Lazarus* for his See, the Cathedral of Narbonne. Sebastiano painted the *Raising of Lazarus*, which was then sent to Narbonne, but the *Transfiguration* painted by Raphael remained in Rome. In March 1524, when Sebastiano's *Transfiguration* painting in the Borgherini Chapel was publicly unveiled, Raphael's *Transfiguration* was already in place on the high altar of the same church.[2]

Why did both patrons choose this unusual theme within months of each other and why was Raphael's work given to the same church? Could the Transfiguration theme have special significance for this particular Roman church? Should a connection be sought between these two works? Very few attempts have been made to interpret the meaning of the Borgherini Chapel programme, whereas much has been written about

I should like to thank Dr Debra Pincus of the University of British Columbia for reading this article and suggesting improvements.

[1] M. Hirst, *Sebastiano del Piombo* (Oxford, 1981), 55.
[2] Hirst, *Sebastiano*, 54–5.

Raphael's last painting.[3] No one, I believe, has ever linked them. I hope to demonstrate that both works do arise from a common source and that the interpretation proposed here for the Borgherini Chapel helps to shed new light on Raphael's *Transfiguration*.

The Borgherini Chapel

The Borgherini Chapel (Pl. 5) is the first on the right of the nave of San Pietro in Montorio and is usually referred to as the *Capella Flagellazione* because of Sebastiano's monumental painting of the *Flagellation of Christ* (Pl. 6) painted in oil directly on the wall above the altar. Christ is tied to a column in the centre of a columned hall and four men, two on either side, raise whips in the air. Standing to the left and right of the *Flagellation* painting, but quite separate from it, are the monumental figures of Sts Peter and Francis. Above, in the half-dome, there is a *Transfiguration* (Pl. 7) showing Christ in the same scale as the Christ below in the *Flagellation*. Although the *Transfiguration* is separated from the *Flagellation* by a painted cornice, both figures of Christ are aligned on the same axis so that a clear relationship exists between the two scenes. Just above the cornice, in the lower left of the *Transfiguration* are the figures of Sts Peter and John with the Apostle James on the opposite side. All three look to the transfigured Christ and above them float the half-figures of Moses and Elijah. In the uppermost tier, painted directly in the nave spandrels above the arch, are two Prophets (Pl. 8).

Most of the literature on the chapel concentrates on the issue of Michelangelo's possible collaboration in the designs of these paintings, which will not be here discussed.[4] Kokot accounted for all the scenes in the chapel, explaining the presence of the saints, Peter and Francis, as a reference to the patron, Pierfrancesco.[5] The combined images of the *Flagellation* and *Transfiguration* he saw as a reference to the revelation that glory will follow the passion and he gave as an example Michelangelo's contemporary figure of *The Risen Christ* in the church of

[3] The most recent: E. Gombrich, 'The Ecclesiastical Significance of Raphael's *Transfiguration:* "Ars auro prior"', in *Essays in Honour of Jan Bialostocki* (Warsaw, 1981), 241–3; C. King, 'The Liturgical and Commemorative Allusions in Raphael's *Transfiguration* and *Failure to Heal*', *JWCI* 45 (1982), 148–59.

[4] See: L. Dussler, *Sebastiano del Piombo* (Basle, 1942); R. Pallucchini, *Sebastian Viniziano* (Milan, 1944); C. Volpe and M. Lucco, *L'opera completa di Sebastiano del Piombo* (Milan, 1980); M. Hirst, *Sebastiano*.

[5] I. Kokot, 'La fonte ispiratrice nei capolavori delle aquile (II): La cappella Borgherini a S. Pietro in Montorio di Sebastiano del Piombo', *Fede e Arte, Rivista internazionale di Arte Sacra*, 2, 4 (1954), 100.

Santa Maria sopra Minerva where the dual themes of Crucifixion and Resurrection are presented simultaneously in one image.

On another level Kokot saw the *Transfiguration* as a reference to the Bull of Callixtus III which instituted the Feast of the Transfiguration for 6 August in celebration of Christian Europe's victory over the Turks at the battle of Belgrade in 1456. During the pontificate of Leo X this took on special significance since the Turks were once again threatening the borders of Christian Europe, and in a letter sent by Leo X to the King of Hungary, Leo referred to the 'mahometanorum arma' as 'Dei flagellum'. Finally Kokot pointed to the speech given in 1515 at the Lateran Council by the Archbishop of Patras, in which the Archbishop expressed grave concern about the Turkish threat, 'Flagellum nostri temporis'.[6]

Volpe and Lucco rejected this hypothesis because San Pietro in Montorio functioned as a college for Franciscan missionaries who were sent to the Orient to evangelize the Turks. They agreed that the physical suffering of Christ in the *Flagellation* was the first step towards his glorification as represented in the *Transfiguration*. The two prophets in the nave spandrels were identified as St Matthew on the left and Isaiah on the right because both prophesied that suffering must come before glorification.[7]

Hirst agreed that the two flanking saints, Peter and Francis, were included because they were the titular saints of the patron and that Borgherini had favoured San Pietro in Montorio with his patronage because the church was dedicated to St Peter and in the charge of the Friars Minor. The choice of the *Flagellation* was appropriate to this Franciscan church because the theme is emphasized in much Franciscan mystical literature. Hirst also suggested that the *Transfiguration* made allusion to the new Turkish threat of 1516 and mentioned the Spanish Cardinal, Bernardino Carvajal, who was intimately associated with this church. Carvajal was the nephew of Juan Carvajal, legate of Hungary during the pontificate of Pius II, who had played an important role in the 1455–6 victory over the Turks. Hirst noted that Bernardino shared his uncle's concerns since in 1517 he served on a papal commission to study ways of attacking the Turks. Lastly, Hirst referred to the speech made by the Archbishop of Patras in 1515 where the words 'per flagellum

---

[6] Kokot, 'La fonte ispiratrice', 100: 'quanta sit latitudo, submersio et periculum imminentis tribulationis Turcarum. . . . Ad nostri temporis flagellum potentissimos fidei nostrae hostes excitavit, et negligentissimos principes christianos pro illius defensione adhuc multis flagellis a somno turpissimo excitare non valuit.'

[7] Volpe and Lucco, *L'opera completa*, 108–10.

Turcarum' may explain the presence of the *Flagellation* beneath the *Transfiguration*.[8]

I propose that the Borgherini Chapel programme becomes meaningful if it is viewed within the context of early sixteenth-century Franciscan reformist circles, in particular those Friars Minor of San Pietro in Montorio who were preoccupied with the final programme of history, believing that they would have a special role in this last age. Responding to the intense atmosphere of apocalyptic expectancy that prevailed in Rome at the beginning of the sixteenth century, these Franciscans placed their hopes in prophecies that promised an imminent renovation of the Church and a new age of peace and joy to be ushered in by a divinely appointed 'Angelic Pastor'. I hope to prove that the paintings in the Borgherini Chapel express these prophetic beliefs and that the Transfiguration theme becomes a focus for the idea of Church renewal. Cardinal Bernardino Carvajal was at the centre of this reformist circle in San Pietro in Montorio and was almost certainly responsible for devising the programme.

From contemporary sources we know that Carvajal was an important patron of the arts in Rome, overseeing the rebuilding of his titular church, S. Croce in Gerusalemme,[9] and the enlargement of S. Giacomo degli Spagnoli in the Piazza Navona.[10] His main interest, however, was in supervising the work of enlarging and embellishing the church and monastery of San Pietro in Montorio on behalf of its patrons, Ferdinand and Isabella of Spain. It is quite probable that it was Carvajal who commissioned Bramante to design the *tempietto* for the courtyard, since Bramante had worked previously for the Cardinal in the church of S. Giacomo degli Spagnoli and on designs for a fountain in Trastevere.[11] Carvajal was a very active patron of art and one writer even suggests that the whole concept for a monument on this site to commemorate the first St Peter was Cardinal Carvajal's.[12]

[8] Hirst, *Sebastiano*, 54–7.

[9] Sanuto, *Diarii*, xxxiv, col. 218: 'Poi a Santa Croce in Hierusalem, titolo dil reverendissimo Santa Croce, fabricha nova, fabricata per sua signoria, et tutavia si fabricha et si lavorava alcune cornise et volti di alcune porte di una preda racolta da le antigaglie, di tanta extrema belleza, che certo un pizol pezeto de essa seria degno di legare in oro et portareo per bellissimo anello.'

[10] A. Bruschi, *Bramante architetto* (Paris, 1969), 859.

[11] A. Bruschi, *Bramante*, 861 n. 11.

[12] E. Tormo, *Monumentos de Españoles en Roma y de Portugueses e Hispano-Americanos* (Rome, 1940), i. 195, 'Si el templo de Montorio no inicia novedad, en cambio, sí, da nota propia inconfundible de arquitecto creador, el templete adjunto, al centro del claustro, también pagado por los Reyes Católicos, y que también en mi hipótesis, debió de ser de la idea y mecenazgo de Don Bernardino.'

This idea seems very plausible, since a special veneration for St Peter is consistent with what we know of Carvajal. Throughout his life Carvajal was obsessed with the papacy and the need as he saw it to reform the Church from the head down. He believed a reform could only come about if a true successor to the first St Peter was elected, a second St Peter who would restore to the universal church its original sanctity. This was the theme of an oration he gave to the College of Cardinals in 1492 when he was still a bishop. The cardinals were to enter the conclave to elect a successor to Innocent VIII and Carvajal reminded them of their sacred duty, which was to choose a Pope who would be a true vicar of Christ and carry out the necessary reform.[13]

This oration, given on the Feast Day of the Transfiguration, 6 August, uses the Transfiguration story as a metaphor for Church renewal. 'And so transform, transfigurate yourselves, fathers, unto different men, transform the Church, the columns of the Church unto a different splendour, so that its appearance shines like the sun and its vestments are white like the snow.'[14] Here is an early example of how the Transfiguration theme is linked in the mind of Carvajal to the idea of Church renewal.

It was this same concern to correct abuses in the Church hierarchy itself that led Carvajal to challenge the authority of Julius II and instigate the schismatic Council of Pisa in 1511. Declaring Julius remiss in his duty because he had failed to call for a council to address the issue of Church reform, Carvajal, together with a small group of dissident cardinals,[15] began making plans for a council in May. Carvajal, head of this council, had the support of King Louis XII of France and Emperor Maximilian I, but this support rapidly dwindled when, in July, Julius confused matters by announcing that his own council would be convened at the Lateran the following year. In October 1511 Carvajal and the other

[13] P. Paschini, 'Una predica inefficace (proposito di riforma ecclesiastica alla fine del sec. XV)', *Studi romani*, 1 (1953), 31–8. On Carvajal, see above, Minnich, Ch. 6, who also refers to the oration of 1492.

[14] *Bernardini Carvajal Episcopi Pacensis Oratio de Eligendo Summo Pontifice* (Rome, c.1492), cited in E. Martène and V. Durand, *Thesaurus novus anecdotorum* (Paris, 1717), ii, col. 1785 'Eia igitur, transformamini, transfiguramini, Patres, in viros alios, transformate Ecclesiam, columnae Ecclesiae in splendorem alium, ita ut resplendeat facies ejus sicut sol, & vestimenta ejus albescant sicut nix.'

[15] There were seven: Bernardino Carvajal, Guillaume Briçonnet, René de Prie, Amanieu d'Albret, Philippe de Luxembourg, Francesco Borgia, and Federico Sanseverino: A. Renaudet, *Préréforme et Humanisme à Paris pendant les premières Guerres d'Italie (1494–1517)* (Paris, 1953), 540 n. 4. See above, Landi, Ch. 3, for a discussion of Carvajal's motives in promoting the Council of Pisa.

opposition cardinals were stripped of their titles and excommunic-ated.[16]

Carvajal's disobedience to the supreme head of the Church was later excused by Mariano da Firenze, who wrote a history of the Minorite Order in 1517, on the grounds that he had been unduly influenced by the famous *Apocalypsis Nova* which he had discovered in the church of S. Pietro in Montorio in 1502.[17] There may have been some truth in this since the book appeared to have a profound effect on him, promising the very thing he desired so much, the imminent appearance of an Angelic Pope who would lead the Church to renewal. It is this book that provides the key to the programme of the Borgherini Chapel since its central message is expressed in the paintings.

The pseudo-Amadeite *Apocalypsis Nova* is described elsewhere in this book. Amadeus, the supposed author, had been called to Rome by Sixtus IV who gave him and his followers the church and monastery of S. Pietro in Montorio. According to one account, when the *Apocalypsis Nova* was found in a 'cavern'[18] in S. Pietro it caused great consternation. The *frati* felt both awe and fear since they believed that to touch the book or open it would bring certain death. Details about the discovery have survived in copies of letters bound together with two of the manuscripts. The letters were sent in 1502 by the Franciscan, Giorgio Benigno Salviati, to his friend in Florence, Ubertino Risaliti. Salviati recounts how, during the time of Sixtus IV, certain *frati* on several occasions tried to open the

---

[16] H. Jedin, *The History of the Council of Trent*, trans. D. Graf (London, 1957), 107–12. The Council opened in Pisa on 1 Nov. 1511 under the presidency of Carvajal but was forced to move to Milan on 12 Nov. After the death of Julius, Carvajal's titles were restored by Leo X in 1513. See L. von Pastor, *History of the Popes*, vols. vii–xxiv, trans. and ed. R. Kerr (London, 1908–33), vii. 44–5; 54–9.

[17] This history of the Order is lost but it was used by Wadding, *Annales*, xiv, *ad a.* 1482, no. XXXIX, 322–3: 'Ferunt, Amadeum librum clausum, sigillisque signatum, Consoldalium commedasse custodiae. Et quidem Marianus narrat, a Dominico Grimano, tituli sancti Marci, Ordinis Protectore, et Bernardino Carvajalio, tituli sanctae Crucis Cardinalibus, post plures annos primo apertum, lectum et exscriptum, et Bernardinum sibi ipsi nimis complacentem ac blandientem, facile ex quibusdam in libro contentis credidisse, se proximum fore Pontificem, ac propterea temere atque insipienter concilia-bulum Pisanum, insanientium audacia celeberriumum, contra Julium II coacervasse, sed a Julio e gradu dejectum omnium risui et contemptui expositum, derisorie ab omnibus Papam Andream vocitatum, etsi postea a Leone X fuerit pristino honori restitutus. Jam tunc coepit publicari, et insanis additamentis foedari liber . . .'

[18] Bruschi, *Bramante*, 990, quotes Fra Mariano's description of the little cave 'subque crucifixione Petri in quadam cavernula orans et ieiunans, ut dicitur, angelo revelante plurima scripsit'. See above, Morisi-Guerra, Ch. 2, for a full study of the *Apocalypsis Nova* based on her book, A. Morisi, *Apocalypsis Nova: Nova Ricerche sull' origine e la formazione del testo dello pseudo-Amadeo* (Studi Storici, 77; Rome, 1970).

book but a little later they fell ill and died.[19] After that no one had dared
to try until the occasion of 1502, which Salviati describes:

Now the Cardinal of Santa Croce, a man well versed in all good behaviour, being
protector of the aforesaid Congregation, at the request of several people, decreed
together with our General, a man of excellent and saintly character, that this
book should be opened and that, as I was told, there was no one better than me
to whom they could entrust this task. I in truth was no little afraid; at length a
bishop of our Order began to comfort me and promised that he wished to be the
first to open it. And so on the established day we went to San Pietro in Montorio
and the Bishop sang Mass; after the mysteries[?] the book on the altar before the
General was opened by us at his command and handed to the two of us to read.[20]

Salviati describes to Ubertino the contents of the book, saying that it
predicted a reformation of the Church, the conversion of all non-believers
and the miraculous election of a new pastor. It foretold that in this time
there would be a pacified, pure, and bright kingdom.[21] Carvajal appears
to have kept the discovery of the book secret and Salviati tells Ubertino
that he cannot send the requested copy because the book is jealously
guarded by Carvajal who keeps it locked in a case and carries the key
with him.[22] At the end of his letter Salviati says: 'Speak of this to no
one.'

The prophecies of the *Apocalypsis Nova* are not unique to the pseudo-
Amadeus but form part of a long prophetic tradition originating with
Joachim of Fiore.[23] As part of his prophetic programme, Joachim prophesied the advent of two orders of spiritual men, *viri spirituales*, who
will lead the Church from the second into the third *status*.[24] By the late
thirteenth century it was a group of Franciscans who had most fully
embraced this prophecy and the fortunes of Joachimism became closely

[19] Morisi, *Apocalypsis Nova*, 29 'certi frati circha a tre volte tentorino daprirlo, infra
pochi di subito amalati moririno. Costui mori al tempo de sixto. Questo libro lo tene
Sixto, tenelo poi et Innocentio, ma nesuno di loro fu ardito ad aprirlo per casi intervenuti
a quelli frati.'
[20] Ibid.
[21] Ibid., 'In quo reformatio ecclesie et conversio infidelium omnium et electio mirabilis novi pastoris, regni [sic] christi pacatum purumque ac nitidum diebus his fore
predixit . . . Nemini dixeritis.'
[22] Ibid. 31 n. 56.
[23] See above, Reeves, Ch. 1, for a brief account of this tradition. See also M. Reeves,
Influence of Prophecy in the Later Middle Ages: A Study in Joachimism (Oxford, 1969),
passim.
[24] Reeves, *Influence*, 140–4.

interwoven with the stricter section of the Franciscan Order.[25] Within this Joachimist circle of the Minorite Order there was a full appropriation of the role of the new spiritual men. These Franciscans saw themselves in this prophetic role and the Testament of St Francis as having the stamp of the Holy Ghost.[26] Petrus Johannis Olivi, a leader in this group, combined Joachim's Trinitarian pattern with a Christocentric view of history, developing the concept of three Advents of Christ. Christ came first in the flesh; His Second, or Middle, Advent would be in the spirit of evangelical reform when there would appear 'quoddam novum seculum seu nova ecclesia'; His Third Advent would be in final judgement.[27] The Middle Coming of Christ, not in the flesh but in a new outpouring of the Holy Spirit, would usher in the third *status*.[28] St Francis and his Order would be assigned the final eschatological task of converting the whole world.[29]

The *Apocalypsis Nova* continues this Joachimist expectation of a third *status* when Christ's appearance would be manifested in the inauguration of a renewal of the Church under the guidance of the Angelic Pope. Indeed the last lines of the book are a moving prayer for the imminent fulfilment of prophecy: 'Veni domini Jesu, et mitte ad nos pastorem promissum. Gratia domini nostri Jesu Christi sit omnibus ipsum expectantibus.'[30] Sebastiano's painting of the Transfiguration in S. Pietro in Montorio expresses this idea of a Second or Middle Coming, where the Christ transfigured symbolizes His return as an outpouring of the Holy Spirit. In devising the pictorial scheme for the chapel, the Transfiguration was chosen because of its allusions to the Second Coming of Christ. This theme had long been associated with Christ's return in glory at the End-Time. Thus John Chrysostom, in his commentary on St Matthew's Gospel, interprets the Transfiguration as a prefiguration of the Second Coming, understood in the usual sense.[31] In this Joachimist context, however, it symbolizes Christ's Middle Advent to usher in the reformation of the Church. Carvajal was familiar with the writings of Chrysostom

[25] E. Daniel, 'A Re-examination of the Origins of Franciscan Joachitism', *Speculum*, 43 (1968), 676. See also Reeves, *Influence*, 191–212.

[26] M. Reeves, *Joachim of Fiore and the Prophetic Future* (London, 1976), 32.

[27] Reeves, *Influence*, 198; D. Burr, 'Bonaventure, Olivi and Franciscan Eschatology', *Collectanea franciscana*, 53 (1983), 28.

[28] Reeves, *Prophetic Future*, 143.

[29] Reeves, *Influence*, 199.

[30] *Apocalypsis Nova*, MS. Vat. Lat. 3825, fo. 217ᵛ.

[31] Migne, *PG* 58 (Paris, 1857–66), 554. Chrysostom sees the Transfiguration as a fulfilment of the second coming described in Matt. 16: 27: 'For the Son of man shall

whom he quotes in his 1492 Oration and, as already mentioned, Carvajal very specifically associates the idea of Church renewal with the Transfiguration story in this early speech. In fact he almost seems to anticipate some of the ideas concerning the Angelic Pastor and reform of the Church, to be found later in the *Apocalypsis Nova*[32] when he describes the three Tabernacles of Christ, Moses, and Elijah as follows: 'The tabernacle of Christ so that there may be a faithful Catholic pontifex and an imitator of Christ as his own successor; a tabernacle of Moses as he was the greatest law maker and most gentle man on earth; a tabernacle of Elijah as he was the most vehement zealot of divine love, of church reformation and seeker of glory.'[33] In the Fourth Rapture of the *Apocalypsis Nova* the mission of the *Pastor Angelicus* is discussed at length. The Angel Gabriel provides many details for Amadeus about this 'true vicar of Christ who will come to sanctify, purge and reform the Church' and, in response to a question about the significance of the mountain, Gabriel replies:

the great mountain is called a 'fiery heaven' by you on which good men follow where the lamb goes. On a mountain the Lord was transfigured, on a mountain he fed a huge crowd from five loaves, on a mountain he was conceived, on a mountain he prayed, on a mountain he ate, on a mountain he was captured, on a mountain he was crucified, on a mountain he was taken to Heaven. When it comes to the Judgement his feet will stand above a mountain. On a mountain your father Franciscus was born. On a mountain he was marked with nail holes in his hands and feet and he revived in a remarkable way the passion of our Lord the Saviour, an act that was a special singular and unusual gift. On a mountain the future shepherd, the man the Lord chose for himself was born, he who will raise Mount Syon, that is, the Church of Christ to the summit of mountains who will snatch it from the clouds and cover of darkness and light it up and its splendour will from that time on last forever. Oh happy are you who will see

come in the glory of His Father with His angels and then He shall reward every man according to his works' which comes just before the Transfiguration story (Matt. 17: 1).

[32] This had led some scholars to believe that, on the instructions of Carvajal, Salviati incorporated into the text of the *Apocalypsis Nova* some of the Cardinal's own political and ecclesiastical concerns, see esp. C. Vasoli, 'Sul probabile autore di una "profezia" Cinquecentesca', *Il Pensiero Politico*, 2 (1969), 464–72, and id., 'Ancora su Giorgio Benigno Salviati (Juraj Dragišić) e la "profezia" dello pseudo-Amadeo', *Il Pensiero Politico*, 3 (1970), 417–21. On Salviati and Carvajal, see also above, Vasoli, Ch. 7.

[33] Carvajal, *Oratio* (n. 14), p. 1785: 'Tabernaculum Christi ut fidelis & catholicus sit futurus pontifex & Christi imitator cujus vicarius. Moisi, ut legislator optimus & mansuetissimus sit super terram. Heliae, ut divini amoris zelator & Ecclesiae reformationis & gloriae aemulator vehementissimus fit.'

this. You indeed see it now in a dream, others will see it for real. . . . And the angel said 'Behold I have given this image to you, pay respect and keep it.'[34]

In Sebastiano's painting, Christ transfigured on Mount Tabor becomes a powerful visual metaphor for the vision Gabriel describes to Amadeus of the Church of Christ renewed in all its splendour.

Within the late fifteenth-century Florentine iconographical tradition of the *Transfiguration*, there is a small painting (Pl. 9), the central panel of a devotional triptych, which appears to be significant. The painting is attributed to Botticelli and is dated about 1500 because of stylistic affinities with the *Mystical Nativity* painted in *c.*1500.[35] An unusual element in Botticelli's *Transfiguration* is the figure of Elijah, who, departing from tradition, is placed on the left. He is wearing a hair shirt and has the long dishevelled hair and beard that identify him as a desert ascetic. This iconographical type contrasts sharply with the type in Sebastiano's work where Elijah is shown wearing the more commonly represented garments of an Old Testament prophet.[36] Elijah in Botticelli's painting resembles the iconographical type usually reserved for St John the Baptist. This relationship is here intentional and significant. St John was often characterized as a second Elijah. This assimilation was based on the eschatological prophecies of Mal. 4: 5 who announced that the Messiah would be preceded by the Prophet Elijah, who will descend from heaven to prepare the way, just as St John prepared the way for Christ.[37] Botticelli's panel, departing from the convention by representing Elijah

---

[34] Morisi, *Apocalypsis Nova*, 17–18; *Apocalypsis Nova* (MS. Vat. Lat. 3825), fo. 31: 'et mons magnus est caelum empyreum a vobis vocatum, in quo boni sequentur *agnum quocunque ierit*. In monte dominus transfiguratus est, in monte satiavit de quinque panibus tantam multitudinem, in monte conceptus est, in monte oravit, in monte cenam fecit, in monte captus, in monte crucifixus, in monte ad caelum assumptus; super montem, cum ad iudicium venerit, stabunt pedes eius. In monte vester pater Franciscus natus est, in monte stigmatibus cum clavis in manibus ac pedibus insignitus, admirabili modo domini salvatoris passionem renovavit, quod fuit praecipuum et singulare atque inauditum donum. In monte et futurus pastor, quem sibi dominus elegit, natus est, qui montem Syon, videlicet ecclesiam Christi, in verticem montium eriget, et de nebulis atque caligine tenebrarum ereptam illustrabit, cuius splendor ex tunc perpetuus erit. O beatos vos qui haec videbitis. Tu quidem vides nunc in imagine, alii videbunt in specie. . . . Dixit vero angelus: Ecce dedi tibi faciem eius, observa et custodi eam.'
[35] R. Lightbown, *Botticelli* (Berkeley, Calif., 1978), ii. 99; the painting is in the Galleria Pallavicini, Rome. Cited in Lightbown, vol. i, pl. 51. See above, Reeves, Ch. 1, for Botticelli's eschatological paintings.
[36] In the paintings of the Transfiguration by Bellini, Perugino, and Raphael, Elijah is also dressed as an Old Testament prophet. I have been unable to find a pictorial precedent for Botticelli's Elijah.
[37] L. Réau, *Iconographie de l'art chrétien* (Paris, 1955), ii. 349.

as an ascetic, clearly depicts Elijah as a type of St John the Baptist who awaits the coming of the Lord. In the Joachimist interpretation of history, the concord between Elijah and St John the Baptist takes on special significance. Elijah is to be the herald of the third *status* or age, as St John was the herald for the second, and according to the doctrine of Joachim, Elijah becomes a symbol of the Holy Spirit, as well as the inaugurator of the third *status*.[38]

Given that Elijah in Botticelli's painting wears a hair shirt which refers to his role as precursor of the Messiah, it is quite possible that Botticelli's *Transfiguration* symbolizes the Joachimist Second or Intermediate Advent of Christ who comes to usher in the new age of Church renewal and that Elijah, standing now on the left to indicate that he has prepared the way for the Lord, is the herald of this event. If this interpretation is correct, it means that a second late painting by Botticelli was influenced by Joachimist ideas. The *Mystical Nativity* of 1500 with its apocalyptic inscription giving the length of time the tribulation must be endured and showing the third *status* of peace and joy, when angels and men embrace and the devil is beaten down, has long been described as one of the greatest documents of Joachimist thought.[39] Lightbown in his recent discussion of the *Mystical Nativity*, concludes: 'The picture tells us then that Sandro believed Italy was passing through the two terrible tribulations foretold in the Apocalypse, and that the Church was sunk in its last days of corruption and decay. But he comforted himself with the hope that the Church would soon be renewed in all its peace and beauty.'[40] It would seem that Botticelli's *Transfiguration* reiterates the message of Church renewal found in the *Mystical Nativity* and shows that in those troubled and anxious years following the execution of Savonarola, Botticelli sought reassurance in prophecies that promised a new age of peace and joy. Lightbown does not link the *Mystical Nativity* with the Joachimist prophetic tradition but instead suggests that Botticelli had been influenced by the prophetic sermons of Savonarola.[41] It is clear, however, that Savonarola himself stood in the Joachimist tradition in his prediction of chastisement and tribulation to be followed by renovation brought about by an Angelic Pope. In his sermon, *Predica della rinnovazione della Chiesa* of 13 January 1495 and printed later in the same

---

[38] M. Reeves and B. Hirsch-Reich, *The* Figurae *of Joachim of Fiore* (Oxford, 1972), 196.
[39] F. Saxl, 'A Spiritual Encyclopaedia of the Later Middle Ages', *JWCI* 5 (1942), 84, for a Joachimist interpretation of the *Mystical Nativity*; see also Reeves, *Influence*, 436 ff.
[40] Lightbown, *Botticelli*, i. 138.
[41] Ibid.

year,[42] Savonarola specifically mentions Joachim of Fiore as one of the prophets who predict that in this time there will be a great punishment, a *flagello*.[43] In another sermon (24 March 1496) Savonarola talks about the coming *rinnovazione della Chiesa* which he says will be a supernatural event—'che è cosa sopranaturale'—and describes the coming of a *Papa santo e buono*. This holy Pope will also be *uno instrumento sopranaturale*.[44]

It is quite possible that there existed a special relationship between the Amadeites in the circle of Carvajal and the Florentine followers of Savonarola. Several writers have suggested that a rapport may indeed have existed between the Amadeite community and the post-Savonarolian *piagnoni* but this has not been studied fully.[45] We know that Savonarola was revered among the Amadeites of San Pietro in Montorio. Salviati, the Franciscan theologian who opened the *Apocalypsis Nova* in 1502, was a disciple of Savonarola, whose book, *Propheticae solutiones*, written in 1497, defended Savonarola against his critics.[46] Both Carvajal and Savonarola were passionately concerned with Church reform and there is evidence that Carvajal called for the canonization of Savonarola at the session of the Council of Pisa on 9 November 1511.[47]

[42] *Predica della rinnovazione della Chiesa, fatta a di 13 gennaio 1494/5*, repr. in M. Ferrara, *Savonarola: Prediche* (Florence, 1952), 238–63.

[43] Ferrara, *Savonarola*, 244: 'Vedi ognuno che pare che predichi e aspetti il flagello e le tribulazione; e a ognuno pare che sia giusta cosa che la punizione di tante iniquita debba venire: lo abate Gioacchino e molti altri predicano ed annunziano che in questo tempo ha a venire questo flagello. Queste sono le ragioni per le quali t'ho predicato la rinnovazione della Chiesa.'

[44] Savonarola, *Prediche italiane ai Fiorentini*, ed. R. Palmarocchi (Florence, 1930), vol. iii, pt. 2, 313; D. Weinstein, *Savonarola and Florence: Prophecy and Patriotism in the Renaissance* (Princeton, NJ, 1970), 175 and n. 61.

[45] G. Spini, 'Introduzione al Savonarola', *Belfagor*, 3, 4 (1948), 420; Tognetti, 'Un Episodio', 192. The Amadeite, Antonio da Cremona, was accused of preaching themes of *flagello* and *renovatio* which were similar to those of Savonarola. See M. Binaghi, 'L'immagine sacra in Luini e il circolo di Santa Marta', *Sacro e profano nella pittura di Bernardino Luini* (Milan, 1975), 57; 'le connessioni tra amadeiti e piagnoni post-savonaroliani si fanno molte strette, sia sul piano die rapporti personali tra i vari esponenti, sia per la sostanziale concordanza dei temi trattati dall'Amedeo e dal Savonarola, la penitenza e la minaccia del castigo, in un contesto di fervido profetismo.' For further references to Antonio da Cremona, see Index.

[46] G. Benigni Salviati Ordinis Minorum *Propheticae Solutiones impressae per Ser Laurentium de Morgianis* (Florence, 1497); Dionisotti, 'Umanisti dimenticati?', *Italia medioevale e umanistica*, 3–4 (1960–1), 308. On this work, see above, Vasoli, Ch. 7.

[47] *Apologia . . . in difesa della dottrina del R. P. F. Girolamo Savonarola da Ferrara del medesimo ordine* (Florence, 1564), 40. 'Quest'huomo da bene (Fra Bartolomeo da Faenza) fu con grandissima instanzia, dal predetto Bernardino Carvagiallo, pregato che, con i suoi frati, a quali eran' uolti gli occhi di tutto'l clero di Pisa, per veder quel che e facevano, consentisse a quel Concilio, promettendogli, che determinerebbon l'oppenione della concezzione per la parte nostra, e canonizerebbon fra Girolamo.' A. Renaudet, *Le Concile gallican de Pise-Milan: Documents florentins, 1510–1512* (Paris, 1922), 495 n. 98. See above, Minnich, Ch. 6, on the question of Carvajal's motives.

As we have seen, the Borgherini Chapel painting of the *Transfiguration* is placed directly above the *Flagellation*. If the *Transfiguration* symbolizes the Second Advent of Christ and the coming age of the Holy Spirit, the *Flagellation* is presented as a metaphor of the great *flagellum* that will take place prior to the coming of this age. Flagellation signifying the scourging of the Church and the punishment that society must undergo for its sins was a theme elaborated by Joachim of Fiore in his interpretation of history where the tribulation—'eadem flagella significatio precesserit'—will come before the third age.[48] This theme of punishment, of a *flagello* was taken up by Savonarola and other preachers at the end of the fifteenth and beginning of the sixteenth centuries. An anonymous book with the title *L'Imminente flagello de Italia* appeared in 1510 and very much reflects the mysterious anxiety that prevailed at this time.[49] Savonarola in his sermon on the renovation of the Church used the image of a *flagello di Dio* repeatedly to stress that punishment must come before the Church could be renewed.[50]

In Sebastiano's *Flagellation of Christ* the Joachimist doctrine of the Mystical Body of Christ is expressed, where the history of the Church is set to the pattern of the Saviour. Christ's Body represents the Church so that the suffering that befalls Christ must also be endured by the Church before it can obtain its final restoration. Just as Christ was resurrected to glory, so will the Church revive itself in this third age of the Holy Spirit.[51] Savonarola's warnings of an imminent *flagello* are symbolically alluded to in the painting of Christ's flagellation. Here the *Flagellation* becomes the focal point of the chapel decoration because the suffering of the Body of Christ symbolically alludes to the present sufferings of the Church and gives proof that the prophecy of a *flagello* has been fulfilled as part of this great pattern of history. That the Church had sunk into a state of wretchedness at this particular time is attested to by the Archbishop of Patras in his speech to the tenth session of the Lateran Council in May 1515:

---

[48] *Lib. de Conc.*, (ed. Daniel) 20 (bk 1, Ch. 1, l. 27).

[49] O. Niccoli, 'Profezie in piazza. Note sul profetismo popolare nell'Italia del primo Cinquecento', *Quaderni storici*, 41 (1979), 513.

[50] Ferrara, *Savonarola: Prediche*, 242 'Quando tu vedi che alcuno signore o capo di reggimento non vuole e' buoni e giusti appresso, ma gli scacciano perchè non vogliono che gli sia detta la verità, di'che il flagello di Dio è presso. . . . Quando tu vedi che tutti gli uomini di buona vita desiderano e chiamano il flagello, credi che ha a venire presto. Guarda oggi se ognuno ti pare ch'el chiami il flagello!' See above, Reeves, Ch. 1, for the chastisement of Florence represented in one of Botticelli's paintings.

[51] E. Benz, *Ecclesia spiritualis* (1934; repr. Stuttgart, 1964), 26–7.

At present sin flourishes, church officials from cardinals and bishops to curial officials and rectors have forgotten Christ's glory and His justice, have committed so many crimes and infractions of law, are without charity and the fear of God, transgress the limits set by the holy fathers and feast on the people of Christ. The times indeed, are evil: charity is extinct and virtue corrupted. Depravity, pride, avarice, concern for earthly goods and neglect of the flock—such are the crimes of the clergy.[52]

The very great pessimism expressed in these lines is echoed in the stark portrayal of Christ's flagellation in the Borgherini Chapel.

Standing on either side of the *Flagellation* are the two monumental figures of St Peter on the left and St Francis on the right. Within this Joachimist scheme, St Peter represents the universal Church and symbolizes the Angelic Pastor, the true follower of St Peter who will lead the Church to renewal. St Francis alludes to the new spiritual men who will usher in the age of renovation, those Franciscans who will be entrusted with converting the unbelievers. St Francis also takes on the eschatological role assigned to Elijah in Botticelli's *Transfiguration* because St Francis had very early on been identified as the Sixth Angel of the Apocalypse 'who ascended from the sun and had the sign of the living God'. To the Franciscan Joachimists of the thirteenth century St Francis was identified as the Sixth Angel who would usher in the new age because he embodied those qualities of life which this age would require.[53] St Bonaventura made this identification official in his *Legenda maiora* and it was reaffirmed once more in a Bull promulgated by Leo X in 1517, *Ite vos in vineam meam*.[54]

Above, in the nave spandrels are painted two prophets. The figure on the left is here identified as Isaiah who prophesied a great scourging of the Lord.[55] On the right the prophet is Ezekiel, whose prophecy spoke of a purified temple filled with the glory of God and of the 'one shepherd' sent to rule over the cleansed and united people of God. This shepherd was Christ, as Head of the Church, but the meaning can be carried further to point to the Pope as Shepherd of the Church and therefore to

---

[52] N. Minnich, 'Concepts of Reform Proposed at the Fifth Lateran Council', *AHP* 12 (1974), 200.

[53] Reeves, *Influence*, 176.

[54] For a history of this tradition, see P. Bihel, OFM, 'S. Franciscus Fuitne Angelus Sexti Sigilli? (Apoc. 7: 2)', *Antonianum*, 2, ser. 2 (1927), 59–90; P. Di Fonzo, OFM, 'La Famosa Bolla Di Leone X "Ite vos" non "Ite et vos" (29 Maggio 1517)', *Misc. Franc.* 44 (1944), 164–71.

[55] Isa. 10: 26; 50: 6; 53: 2–10.

the divinely appointed *Pastor Angelicus*.[56] The identification of these two prophets as Isaiah and Ezekiel is made more secure by comparing them to the prophets in the Sistine Ceiling where some important correspondences do exist. For example, the pose of the Isaiah figures are similar, with heads averted to the left, right arms pulled across the twisted upper torso and fingers inserted in closed books to mark the page. As Hirst observed, for the figure of the second prophet on the right, Sebastiano evidently reworked the composition of Michelangelo's *Joel*[57] but to keep his identity as Ezekiel, Sebastiano gave the figure a turban to conform to Ezekiel in the ceiling.

Considering the Borgherini Chapel decoration as a whole, confirmation that the programme was inspired by the prophecies of renewal in the *Apocalypsis Nova* is found below the altar table beneath the *Flagellation* painting. There, set obliquely below the monogram IHS, is a painting of a large book, inscribed on the cover with the words *Aperietur | in | Tempore*—'Let it be opened in time' (Pl. 10). This book represents the *Apocalypsis Nova*, as was first pointed out by Vannicelli.[58] Amadeus was buried in the church of Santa Maria della Pace in Milan, another Amadeite convent founded by him, and according to an oral legend that quickly became established in the sixteenth century, he was laid in his tomb with his book, on the cover of which the words *Aperietur | in | Tempore* were inscribed. All subsequent effigies of the saint show him holding the book with the inscription.[59] In the chapel, to the left and right of the altar, stand the figures of St Peter and St Francis, both shown with open books. This too must refer to the *Apocalypsis Nova* which has now, in the chapel, been opened 'in time'.

Very little information about the patron, Pierfrancesco Borgherini, has come down to us and there is no way of knowing whether this Florentine banker was a sympathetic follower of Savonarola, drawn to this particular Roman church because of the veneration accorded Savonarola by the Amadeites.[60] If the programme for the chapel was inspired by the

---

[56] Ezek. 34: 23.

[57] Hirst, *Sebastiano*, 57. See above, Bull, Ch. 16, for a study of the Sistine Chapel ceiling.

[58] P. Vannicelli, *S. Pietro in Montorio e il tempietto del Bramante* (Rome, 1971), 38.

[59] *AS*, August, ed. Joannes Carnandet (Paris and Rome, 1857), 2, 562.

[60] It is interesting to note, however, that the so-called *Borgherini Holy Family* painted by Andrea del Sarto, *c.*1528, for Giovanni, brother of Pierfrancesco, may have allusions to Savonarola; see J. O'Gorman, 'An Interpretation of Andrea del Sarto's *Borgherini Holy Family*', Art Bulletin, 47 (1965), 502–4.

revelations in the *Apocalypsis Nova*, at the same time it also forcefully expresses the message of woe and exaltation found in the prophetic sermons of Savonarola and seems to suggest that there is a possible connection.

In 1515, a year before the commission, Pierfrancesco Borgherini married into the prominent Florentine family of the Acciaiuoli. His father-in-law, Roberto, was in 1516 Florentine ambassador to Rome and could have introduced his son-in-law to Carvajal whom he knew well, since he had served as Florentine ambassador at the Court of Louis XII during the crisis years of the schismatic council, reporting the events from 1510 to 1512 to the *Dieci di Balìa* in Florence.[61] Carvajal would have met Roberto Acciaiuoli at the Court in Lyons where Carvajal had sought refuge after the fall of Milan in 1512. Both worked on the peace agreement between France and Spain which was founded on the proposed marriage of the French princess Renata to a nephew of the King of Spain.[62]

In the Acciaiuoli family there were two very ardent followers of Savonarola. Roberto Acciaiuoli's brother, Alessandro, fell completely under the Dominican's spell and his cousin, Zanobi Acciaiuoli, the humanist scholar, educated like Roberto at the Medici court, was so affected by the Frate's preaching that he entered the Dominican Order in San Marco in 1495.[63] It was Zanobi Acciaiuoli who had persuaded the Franciscan, Salviati, to give Savonarola's sermons his serious attention.[64] Zanobi moved to Rome when Giovanni de' Medici, his childhood friend and companion, was elected to the papacy in 1513. Later he was appointed the Pope's Librarian.[65] It is known that Zanobi was interested in prophecies concerning the *Pastor Angelicus*. This is recorded by the Venetian General of the Camaldolesi, Pietro Delphino, who wrote to the monk, Hieronymus, reporting a visit by Zanobius Acciaiuoli, who had brought prophecies for Delphino to read, including a 'vaticinium quoddam de Angelico futuro pontifice'.[66] Therefore it could have been Zanobi Acciaiuoli, together with Carvajal, who advised Borgherini on the programme for the chapel in 1516.

---

[61] Renaudet, *Concile de Pise–Milan, passim.*

[62] *DBI* xxi. 32 and i. 90–1.

[63] C. Ugurgieri della Berardenga, *Gli Acciaiuolo di Firenze nella luce dei loro tempi (1160–1834)* (Florence, 1962), ii. 620, 648, 669, 702.

[64] D. Weinstein, *Savonarola and Florence: Prophecy and Patriotism in the Renaissance* (Princeton, NJ, 1970), 243.

[65] *DBI* i. 94.

[66] Reeves, *Influence*, 433–4.

In the first quarter of the sixteenth century there prevailed in Rome an atmosphere of great apocalyptic expectancy. A considerable circle of scholars, humanists, and theologians was deeply concerned with future events and the destiny of the Church, and they collected, read, and exchanged oracles and apocalyptic tracts.[67] But the increasing number of apocalyptic preachers prophesying chastisement and renovation in the tradition of Savonarola was perceived by church leaders to be a threat to ecclesiastical authority. Hence the attempts of the Fifth Lateran Council to curb them and hence the prominence given to the investigation of Savonarola's prophecies during the Pope's visit to Florence between December 1515 and late February 1516. Norman Minnich's analysis of the Council's decree on prophecy and the inconclusive discussions at Florence both reveal the ambivalence of many leading churchmen on the question of what constituted valid prophecy.[68] There were certainly false prophets who must be restrained from preaching dangerous doctrines. But too many prominent churchmen believed in the reality of true prophecy for the Pope to obtain a sweeping condemnation.

Carvajal accompanied Leo X to Florence.[69] He was also in the later days of the Lateran Council a member of the Faith Deputation which drew up the decree on prophecy published in 1517. Thus he must have been close to the centre of the discussions on the distinctions between true and false prophets.

If it is assumed that the programme for the Borgherini Chapel was planned during the late spring and early summer of 1516,[70] then surely it must have arisen from the continuing debate about prophecies of Church renewal. Among the Amadeites of San Pietro in Montorio there would have been no doubt whatsoever that the prophecies promising a *renovatio Ecclesiae* under an Angelic Pastor were divinely inspired, because they could find the necessary proof in the revelations of the Blessed Amadeus contained in the *Apocalypsis Nova*. The Borgherini Chapel decoration was planned to affirm this very belief, for even though it dwells on the present evils that afflict the Church as represented in the *Flagellation*, above in the *Transfiguration* it expresses a faith that the time of renewal is at hand, when the transfigured Christ will be present to usher in this golden age of church renewal and peace and joy.

[67] Ibid. 441. See esp. M. Reeves, 'Roma Profetica', in F. Troncarelli (ed.), *La città del segreti: Magia, astrologia e cultura esoterica a Roma (XV–XVIII)* (Milan, 1985), 277–97.
[68] See above, Minnich, Ch. 4.
[69] *DBI* xxi. 32.
[70] Hirst, *Sebastiano*, 51.

Raphael's Transfiguration

The first recorded mention of Raphael's *Transfiguration* (Pl. 11) is in January 1517,[71] therefore, Cardinal Giulio de' Medici would have commissioned the two paintings for the Cathedral of Narbonne some time during the second half of 1516, after the programme for the Borgherini Chapel had been decided. Previous discussions of Raphael's last painting have always noted the unprecedented combination in one painting of two seemingly unrelated stories, the Transfiguration (Matt. 17: 2–13, Mark 11: 2–13) and the Failure to Heal the Boy Possessed by a Devil (Matt. 17: 14 ff., Mark 11: 14 ff.). Although the stories follow each other in the Gospel accounts, there does not appear to be a clear narrative link between them. The painting itself seems to emphasize this division; the groups of figures only half emerge from the darkness at the foot of the mountain, while above, in brilliant light, we see the floating figure of Christ transfigured.

Recognizing that contrast, rather than unity, was Raphael's intention in this work and also that this double scene was required by the patron, Giulio de' Medici, writers have searched for an encompassing theme. Did the transfigured Christ symbolize the triumph of Christianity over the Moslems who are alluded to in the image of the 'possessed heathen'?[72] Or is the Transfiguration a prefiguration of the resurrection and salvation where the luministic imagery, derived from the liturgical Feast of the Transfiguration, serves as a metaphor for purification and healing, and where the faithful must await Christ's Second Coming before the boy can be healed?[73] Or is the inability of the nine Apostles to heal the possessed boy in the absence of Christ an allusion to the primacy of St Peter, Christ's vicar, who like Christ is absent from the scene?[74] And, taken further, could the concept of Christ the Healer, *Christus medicus*, be a direct reference to the Medici Papacy, since this play on the family name had broad currency during the reign of Leo X, and therefore in this instance would be highly appropriate?[75]

[71] V. Golzio, *Raffaello nei documenti nelle testimonianze dei contemporanei e nella letteratura del suo secolo* (Vatican City, 1936), 53. Leonardo Sellaio in Rome wrote to Michelangelo in Florence 'chome Bastiano aveva tolto a fare quella tauola, avuti danari per fare el legname. Ora mi pare, che Rafaello metta sotosopra el mondo, perche lui non lla faca [*sic*] per non uenire a paraghonj.'

[72] F. Schneider, 'Theologisches in Raffaels Disputa und Transfiguration,' *Katholik*, 3 (1896), 11 ff.

[73] K. Posner, *Leonardo and Central Italian Art: 1515–1550* (New York, 1974), 43–7.

[74] Ibid. 45; Gombrich, *Raphael's Transfiguration*, 242.

[75] Posner, *Leonardo*, 45–6. See above, Reeves, Ch. 5 for Egidio of Viterbo's use of the Medici name as a prophetic symbol in his *Historia*.

That very careful consideration was given to the planning of this altarpiece seems to be supported by the fact that even though the Transfiguration and the story of the possessed boy are synchronous actions, to combine them in one painting defies all traditions of Christian art. Not only is the story of the possessed boy extremely uncommon in art, but in those rare cycles where it does appear, the scene always shows Christ exorcizing the devil. Therefore it would seem that Raphael and his advisers have invented a totally new iconography that lays stress on the Apostles' inability to heal. It has been suggested that Raphael's original commission was for the Transfiguration only and that at a later date the conception was revised to include the second story.[76] It is, however, hard to imagine that Raphael would have introduced this second story into the altarpiece simply for practical reasons, as some have recently proposed, to make the composition more crowded and to reduce the scale of his original figures in the *Transfiguration* so that they are compatible with the scale of the figures in Sebastiano's companion altarpiece, the *Raising of Lazarus*.[77]

If Raphael's *Transfiguration* is considered in relation to the Borgherini Chapel programme, the combining of these two particular scenes becomes meaningful as the double message of the altarpiece becomes apparent. The *Failure to Heal the Possessed Boy* placed directly below the *Transfiguration* corresponds to the *Flagellation* set below the *Transfiguration* in the chapel. There results the familiar juxtaposition of woe and exaltation, the double theme of tribulation and blessedness. That two important commissions for the Transfiguration were given within months of each other and that both are placed above scenes of great distress, strongly suggests this is not a coincidence, but that both spring from the same source. As we have seen, the Borgherini programme, drawn from the prophecies of pseudo-Amadeus, Savonarola and the Joachimist tradition, promises an imminent renewal of the church after it has undergone a period of great trial. Raphael's painting undoubtedly echoes this same theme.

In his unusually dramatic rendering of the Transfiguration itself, where intense light radiates from the glorified Christ, Raphael has

[76] This idea was prompted by the discovery in the Albertina of a workshop drawing of a transfiguration (presumed to be after a lost sketch-model by Raphael), which is presented as Raphael's First Idea of the *Transfiguration*, K. Oberhuber, 'Vorzeichnungen zu Raffaels *Transfiguration*', *Jahrbuch der Berliner Museen*, 4 (1962), 116ff.; K. Oberhuber, 'Style and Meaning', in his *A Masterpiece Close-Up: The Transfiguration by Raphael* (Cambridge, Mass., 1981), 10ff.

[77] J. Gere and N. Turner, *Drawings by Raphael* (London, 1983), 216.

created a visual embodiment of the idea that Christ will return not in the Flesh but as an outpouring of the Holy Spirit. The scene below shows the vain attempt by the possessed boy's family to elicit help from the Apostles, who are in disarray and seem powerless to act. The focal point of this lower scene is a kneeling woman, whose classical profile, proportions, and drapery isolate her from the boy and his family group on the left. Who is she, this woman who emphatically points with both forefingers to the boy? When she is mentioned in the literature, she has most often been described simply as the 'kneeling woman'.[78] But surely there must be greater significance given to this figure so prominently placed in the composition. Could it be possible that she represents the Erythraean Sibyl, prophetess of the Apocalypse, whose oracle tells of the signs that will appear to announce the end of the world and that she was included in this lower scene to evoke the necessary atmosphere of apocalyptic dread?[79] Above the darkness and gloom the brilliant light reveals Christ's Transfiguration, witnessed on the upper left by two kneeling figures. These two figures have been the subject of some debate, identified as Felicissimus and Agapitus, martyrs, who are commemorated in the missal on the Feast of the Transfiguration,[80] or as Justus and Pastor, the patron saints of the city of Narbonne, whose festival is celebrated on 6 August, the Feast day of the Transfiguration.[81] More recently they have been identified as the archdeacons, Stephen and Laurence.[82] Perhaps they are the two child martyrs, Justus and Pastor, included here in his altarpiece not only because they have significance for Narbonne Cathedral but also because as 'martyrs' they become the two witnesses of the eleventh chapter of the Apocalypse.

[78] L. Dussler, *Raphael: A Critical Catalogue* (London, and New York, 1971), 54; A. de Rinaldis, 'Una interpretazione della Trasfiguratione de Raffaello in Vaticano', *L'illustrazione vaticana*, 6 (1935), 295 ff., identified the kneeling woman as 'Mater Ecclesia' and the possessed boy as the Reformation.

[79] The Erythraean Sibyl was a popular prophetess because of her apocalyptic predictions. A pseudo-Joachimist work, *Vaticinium Sibillae Erithreae*, purporting to be a gloss by Joachim on a sibylline oracle, appeared between 1252 and 1254 (see Reeves, *Influence*, 56); for the printed text, see O. Holder-Egger, 'Italienische Prophetien des 13. Jahrhunderts', *Neues Archiv der Gesellschaft für ältere deutsche Geschichtskunde*, 15 (1889), 155–73. It was used extensively by the fourteenth-century Minorite, Jean de Roquetaillade and circulated widely in the fifteenth century. In 1516 it was printed in the compilation published by the Venetian Augustinian (fos. 52ᵛ–54ᵛ) of which the main work was the *libellus* of Telesphorus of Corenza.

[80] Schneider, 'Theologisches in Raffael's Disputa', 16.

[81] H. von Einem, 'Die "Verklarung Christi" und "Die Heilung des Besessenen" von Raffael', *Abhandlungen der Akademie der Wissenschaften und der Literatur in Mainz*, 5 (1966), 323–4.

[82] King, 'Raphael's *Transfiguration*', 156.

According to the pseudo-Joachimist tradition, the two orders of spiritual men, the *viri spirituales* who would appear at the beginning of the third age, were associated in Scripture with a series of pairs and therefore sometimes identified as the two witnesses of the Apocalypse.[83] Perhaps these two witnesses who give to this transfiguration an eschatological reading can be compared to the apocalyptic role of Elijah in the Botticelli *Transfiguration* or to St Francis, Angel of the Sixth Seal in the Borgherini Chapel programme.[84]

Early in 1515 Cardinal Giulio de' Medici was appointed Bishop of Narbonne, following the death in the previous December of the former Bishop, Cardinal Guillaume Briçonnet. It will be remembered that Briçonnet, whose body was interred in the Cathedral, had participated with Carvajal in the Council of Pisa. His two sons, Denis, Bishop of St Malò and Guillaume, Bishop of Lodève, were leaders of the Church reform movement in France and both had played a significant role at the Council of Pisa.[85] Denis Briçonnet was intimately associated with a small group of reformers in Milan at the monastery of Santa Marta, and these reformers were in close contact with the Amadeite community of Santa Maria della Pace. The spiritual director of this small group in Santa Marta was the Franciscan, Benigno Salviati. He had arrived in Milan in 1514 and had introduced the group to the so-called revelations of Amadeus in the *Apocalypsis Nova*. From that moment on, within their small circle, there was an air of constant expectancy that the prophecies would be fulfilled and an Angelic Pastor would soon appear.[86]

Denis Briçonnet arrived in Rome in August 1516, having been appointed ambassador extraordinary to the papal court.[87] Perhaps it was at this time that he discussed these same prophecies with Giulio de' Medici as well as with his father's close collaborator in Church reform, Cardinal

---

[83] Reeves, *Influence*, 147–8.

[84] If this essentially Joachimist reading of Raphael's altarpiece is correct, it raises some obvious questions, the most important being whether Sebastiano's companion altarpiece, the *Raising of Lazarus*, commissioned by Giulio de' Medici and also destined for Narbonne Cathedral, could be similarly interpreted. In his typology, Joachim does link the raising of Lazarus story in the second age with the resurrection of the masses from their state of sinning in the third, see Benz, *Ecclesia Spiritualis*, 27. This point needs investigation in a future study.

[85] *DHGE* 10 (Paris, 1938), 676–80; Guillaume Briçonnet became spiritual adviser to Marguerite of Navarre and was at the centre of a group of reformers at Meaux 1518–25, see H. Heller, 'Marguerite of Navarre and the Reformers of Meaux', *Bibliothèque d'Humanisme et Renaissance*, 33 (1971), 271–310. For further references to the Briçonnet family, see above, Vasoli, Ch. 7.

[86] Binaghi, 'L'immagine sacra', 51–60.

[87] L. Febvre, *Au Cœur Religieux du XVI Siècle* (Paris, 1957), 152.

Carvajal. It may well be at this time that the decision was made to send two altar paintings to Narbonne Cathedral and Giulio de'Medici could have sought Briçonnet's advice concerning suitable subjects. Narbonne had special significance for Franciscan Joachites because the great Franciscan leader, Petrus Johannis Olivi, had spent the last years of his life at Narbonne.[88] As we have seen, Olivi adopted Joachim's pattern of history and it is in Olivi's writings that the relationship between Christ and the third age is made with reference to Christ's three Advents, in the Flesh, in the Spirit, and in Judgement.

It is difficult to determine Cardinal Giulio de' Medici's attitude towards prophecies of church renewal, although we do know that he discussed Savonarola's prophecies at length with Girolamo Benivieni, one of the Frate's most ardent followers.[89] As Archbishop of Florence he would have heard the arguments in the case against Savonarola, including the disinterested opinion of the most respected Venetian reformer, Gasparo Contarini, who ended his statement of 17 September 1516 by essentially giving support to the Dominican's basic prophetic doctrine of the coming *renovatio Ecclesiae*.[90] He could also have been influenced by Egidio of Viterbo who, it has been suggested, helped form the religious and theological thought of the papal court during the pontificate of Leo X.[91] As we have seen, this leading reformer had prophesied as early as 1512, in his opening address to the Lateran Council, an imminent eschatological purification to be followed by a final renovation of the church.[92] Much later, at the end of his life, in the *Scechina*, written at the request of Giulio de'Medici, now Pope Clement VII, Egidio returned to the prophetic theme of purification and *renovatio*, to be carried out by God's chosen agents, Pope and Emperor.[93] Direct evidence of Clement's sympathy with the Amadeite community emerges in the fact that he chose one of their number as his personal confessor, first recorded in a letter dated 1526, addressed to Clement's datary, Gian Matteo Giberti. This confessor, P. Juan Antonio Tomas de Locarno was Spanish, but little else is known about him.[94] It would appear then that Giulio was in

[88] D. Douie, *The Nature and the Effect of the Heresy of the Fraticelli* (Manchester, 1932; repr. 1978), 91.

[89] Weinstein, *Savonarola*, 351.

[90] F. Gilbert, 'Contarini on Savonarola: An Unknown Document of 1516', *AR* 59 (1968), 145–50; Weinstein, *Savonarola*, 361.

[91] J. O'Malley, *Giles of Viterbo on Church and Reform* (Leiden, 1968), 9. On Egidio of Viterbo, see also above, Ch. 5.

[92] Minnich, *Concepts of Reform*, 169, and above, Ch. 4.

[93] See above, Reeves, Ch. 5.

[94] P. Meseguer Fernandez, OFM, 'Breves de Clemente VII en favor de la Provincia de

sympathy with the strict reformist rule of the Amadeites, those Friars Minor who believed in a new age of the Holy Spirit, and this may well have been his attitude earlier, at the time he commissioned Raphael's *Transfiguration*.

Raphael's painting was in place on the high altar of San Pietro in Montorio in 1522.[95] In all likelihood, a decision had been taken not to send the last great painting by Raphael to France but to keep it in Rome. Given its theme, the most appropriate place for it in the city would be the high altar of San Pietro in Montorio, that same altar where many years before, in 1502, Mass was said and the *Apocalypsis Nova* had been opened for the first time. Raphael's altarpiece and Sebastiano's Chapel would serve as constant reminders to the Amadeite congregation that the much awaited *renovatio Ecclesiae* would come soon and that they would be called upon to usher in this new age.

S. Pedro in Montorio y de su confesor Juan Antonio Tomas de Locarno, OFM', *AFH* 44 (1951), 167.
[95] G. Vasari, *Vite de' più eccellenti pittori, scultori e architetti*, ed. G. Bottari (Rome, 1759–62), ii. 18–19 in appendix entitled 'Giunta alle Note del Tomo secondo'.

18

Prophecies of the Angelic Pastor in Sebastiano del Piombo's Portrait of Cardinal Bandinello Sauli and Three Companions

JOSEPHINE JUNGIĆ

When Sebastiano del Piombo painted his group portrait now known as *Cardinal Bandinello Sauli and Three Companions* (Pl. 12) in 1516, it was the largest and most ambitious easel portrait painting ever attempted by an artist of the Roman School[1] but until recently the work has received little attention, most probably because the identity of the figures was still in doubt. In 1951 W. E. Suida discovered an inscription on the bell which led to the identification of the young Genoese cardinal, Bandinello Sauli,[2] and now Hirst and Davis have identified the two figures to the right of the cardinal as Giovanni Maria Cattaneo, the humanist poet, and Paolo Giovio, the historian.[3] The portrait showing Cardinal Sauli in the company of humanist scholars suggested to Hirst that the painting was commissioned to record Sauli's patronage of letters for which he was particularly esteemed, but if this is a portrait of the Cardinal surrounded by *famigliari*, there are also carefully rendered details, such as the open

I am grateful to Dr H. G. Edinger for translating the Latin texts and to Dr Rose Marie San Juan for her comments and criticisms. I should also like to acknowledge the research grant I received from the Social Sciences and Humanities Research Council of Canada without which this article could not have been written.

[1] S. Freedberg, *Painting of the High Renaissance in Rome and Florence* (Cambridge, Mass., 1961), 374. The painting measures 126.6 × 149.8 cm. and is located in the National Gallery of Art, Washington, DC.

[2] F. Shapley, *Paintings from the S. H. Kress Collection, Italian Schools, XV–XVI Centuries* (London, 1968), 166.

[3] M. Hirst, *Sebastiano del Piombo* (Oxford, 1981), 99–100; C. Davis, 'Un appunto per Sebastiano del Piombo Ritrattista', *Mitteilungen des Kunsthistorischen Institutes in Florenz* (1982), 383–8.

book of maps, the small bell, Giovio's emphatic gesture, and the *trompe l'œil* fly painted on the Cardinal's rochet that seem to hint at meanings that go beyond a conventional portrait.

In this essay I hope to demonstrate that the portrait of Bandinello Sauli, commissioned by an intimate member of the cardinal's Roman circle, was inspired by prophecy and shows Sauli as the Joachimist Angelic Pastor, the holy Pope whose role was to bring about a renewal of the Church and initiate a new age of spiritual enlightenment. Within a year of the painting Bandinello was implicated in the so-called Conspiracy of Cardinals and accused with four other cardinals of conspiring to poison Leo X; by March 1518, disgraced and ruined, he was dead. It is proposed here that the cardinal was a victim of political hatred and envy and that rather than being a conspirator in a plot to assassinate the Pope, Cardinal Sauli was condemned because some in the papal court feared the prophecy naming him the Angelic Pastor would be fulfilled and that his election to the Holy See would stand in the way of Medicean dynastic ambitions for the papacy.

Sebastiano's group portrait, which bears the artist's signature 'S . . . Faciebat' and the date 1516 on the unfolded paper at the lower right, shows the cardinal seated in an armchair in the foreground gazing out at the viewer. His left hand rests upon the carpeted table top and his fingers direct the viewer's attention to the small silver and gold bell placed a few inches away. The inscription identifying Sauli is found near the bottom of the bell and reads: 'B. DE. SAVLIS. CAR.'. Bandinello Sauli was consecrated Bishop of Gerace in 1509 and two years later he was one of eight prelates nominated cardinal by Julius II, becoming Cardinal Deacon of S. Adriano and titular Cardinal of Santa Sabina.[4]

Michael Hirst identified the figure on the far right as Paolo Giovio because the distinctive profile looked very similar to the profile of a much older man known to be Giovio, painted by Vasari in his Cancelleria mural, *The Reward of Virtue*.[5] This identification has been confirmed by Charles Davis, who discovered a woodcut portrait of the younger Giovio by Tobias Stimmer, illustrating Giovio's *Elogia virorum literis illustrium*, published in Basle in 1577. The portrait in Stimmer's woodcut, except for the length of hair, is exactly like the portrait by Sebastiano. The woodcut is based on a now lost portrait of Giovio which Stimmer saw in Giovio's portrait gallery at Como in 1570–1. Davis was also able to

[4] C. Eubel, *Hierarchia catholica medü Aevi* (Münster, 1898–), iii. 12; L. von Pastor, *History of the Popes*, vols. i–vi, trans. and ed. E. Antrobus (London, 1891–8), vi. 343–4.
[5] Hirst, *Sebastiano*, 99–100.

identify the figure turned towards Giovio as Giovanni Maria Cattaneo, since his portrait is also reproduced by Stimmer in the *Elogia*, and was similarly based on a lost portrait that once formed part of Giovio's gallery of famous people. Because the two woodcuts are almost identical to the portraits in Sebastiano's painting, Davis reasonably concludes that Sebastiano may have actually painted the lost portraits for Giovio's 'Museo' at Como.[6]

Giovanni Maria Cattaneo of Novara (d. 1531?), shown seated behind the table and dressed in black with closely cropped hair, settled in Rome probably in 1511[7] and, as Giovio tells us in his biography of Cattaneo in the *Elogia*, entered the service of Bandinello Sauli as secretary.[8] Early in 1514 Cattaneo published his long Latin poem *Genua*, which he dedicated to the Cardinal's younger brother, the apostolic protonotary, Stefano Sauli. This poem celebrates the beauty of the city of Genoa and its people and history. Cattaneo concludes the poem with the hope that one day Bandinello Sauli may become Pope:

> O Bendinello, the unsurpassed glory of virtues, may I
> see your brow crowned by the third wreath! What a
> Golden Age, when you, united Genoa, and that leader
> unsurpassed in the world, buckle on your lawful swords
> for purposes of war and both of you strike down the
> madness of the Muhammadans![9]

It is significant that Sauli's role as Pope will be to defeat the Turks and usher in a Golden Age because this echoes the expectations of some leading churchmen of the day, who, influenced by the prophecies of Joachim of Fiore, also predicted the overthrow of the Turks and the coming of a Golden Age.

The Dominican theologian, Giovanni Annio da Viterbo (1432–1502) in his commentary on the Apocalypse, *Tractatus de futuris Christianorum triumphis in Saracenos*, or *Glosa super Apocalypsim de statu ecclesie ab anno MCCCCLXXXI usque ad finem mundi*, combined the prophecy of

---

[6] Davis, 'Un Appunto', 383–4; see also B. Fasola, 'Per un nuovo catalogo della collezione gioviana', *Atti del Convegno Paolo Giovio II: Rinascimento e la Memoria* (Como 1985), 169–80.

[7] *DBI*, xxii. 469.

[8] P. Giovio, *Elogia virorum literis illustrium* (Basle, 1577), 147.

[9] J. Catanaeus, *Io, Mariae Catanaei Genua*, 1514, reprinted in G. Bertolotto, '"Genua" Poemetto di Giovanni Maria Cataneo', *Atti della Società Ligure di Storia Patria*, 24 (1891–2), 721–881: O te, maxime, Serto | BENDinelle decus virtutum, tempora terno | Aspiciam redimitum. O aurea tempora quando | Tu concors Genua, atque in terris Maximus ille | Cingetis gladios ad belli munera iustos, | Et Maumetanos contundet uterque furores (461–6).

Joachim of Fiore with astrological theories and his own interpretation of the Apocalypse to predict the defeat of the Turks, whom he identified as the Beast of the Apocalypse. He also prophesied that a new Jerusalem would be established here on earth and that a Pope would be elected who would lead the Church to renewal in a new, more perfected age and that the Johannine promise (John 10: 16) that there would be 'unum ovile et unus pastor' would be fulfilled.[10] This commentary was well known, especially in Genoese circles, because Annio had written it during his lengthy stay in Genoa, where it was published in 1480.[11] Because of its optimistic vision of the future, the work was popular and many editions were printed, both in Italy and elsewhere, well into the sixteenth century.[12]

The same themes, as we have seen,[13] also occur in the discourse written for Julius II in 1507 by Egidio of Viterbo. At first Egidio believed that the great renewal of the Church would come during the pontificate of Julius II, but later he saw Leo X as the first pontiff of this new age.[14] In the last lines of Cattaneo's poem, hopes for the future are placed in Bandinello Sauli, who is seen as that Pope who strikes down the Muhammadans and presides over this new epoch.

Perhaps the sentiments expressed in the poem were in part inspired by a prophecy that promised the papacy to Bandinello Sauli, a prophecy important enough for Giovio to remember long after the death of this cardinal, when he mentioned it in his biography of Leo X, published in 1548. Giovio writes: 'Molti mathematici anchora inclinati alla adulatione per il guadagno, havevano promesso il papato al Sauli.'[15] It is suggested here that the group portrait, painted two years after Cattaneo's poem,

---

[10] Giovanni Annio da Viterbo, *Tractatus de futuris Christianorum triumphis in Saracenos*, 1497 (British Library 40854); C. Vasoli, 'Profezia e astrologia in un testo di Annio de Viterbo', *Studi sul Medioevo Cristiano*, 2 (1974), 1029–34; D. Weinstein, *Savonarola and Florence: Prophecy and Patriotism in the Renaissance* (Princeton, NJ, 1970), 89–90; M. Reeves, *Influence of Prophecy in the Later Middle Ages: A Study in Joachimism* (Oxford, 1969), 173, 354, 463–4; L. Thorndike, *A History of Magic and Experimental Science* (New York, 1934), iv. 263–7.

[11] R. Weiss, 'Traccia per una biografia di Annio da Viterbo', *Italia medioevale e umanistica*, 5 (1962), 430.

[12] Vasoli, 'Profezia e astrologia', 1035 n. 20. See above, Reeves, Ch. 5 (on Annio's possible influence on Egidio of Viterbo) and Headley, Ch. 13 (on his clear influence on Gattinara).

[13] J. O'Malley, 'Fulfillment of the Christian Golden Age under Pope Julius II: Text of a Discourse of Giles of Viterbo, 1507', *Traditio*, 25 (1969), 274. See also above, Reeves, Ch. 5.

[14] J. O'Malley, *Giles of Viterbo on Church and Reform* (Leiden, 1968), 111.

[15] P. Giovio, *Le Vite di Leon Decimo et d'Adriano Sesto Sommi Pontefici et del Cardinal Pompeo Colonna*, trans. L. Domenichi (Florence, 1549), 269.

was devised as an allegorical portrait, intended to embody these same hopes that Sauli would one day succeed to the papacy and lead the Church in the coming perfected age. This allegorizing of portraiture was not unique to Sebastiano who, in his earlier career in Venice, must have been familiar with the symbolic portraits of Lorenzo Lotto.[16] Sebastiano's group portrait-drawing of *Clement VII and the Emperor Charles V*, now in the British Museum, which dates probably soon after their meeting in Bologna in the winter of 1529–30, represents the Pope and Emperor seated in council, surrounded by attendant figures and is a further example of portraiture functioning as allegory, where the globe being placed on the table and the monstrance set between the tiara and crown symbolize the two figures of the Pope and Emperor as the divinely ordained rulers of the world.[17] Perhaps it was the prophecy that also inspired the commission for the Sauli portrait in addition to the speculation in August 1516, that Rome would soon have a new Pope, for Leo X had become so seriously ill with a fistula that some in the Roman Curia had already begun thinking of the preparations that needed to be made for the conclave.[18]

In format, Sebastiano's portrait of Cardinal Sauli differs substantially from his portrait painted in 1515 of Cardinal Antonio del Monte, because in the del Monte portrait Sebastiano appears to have followed the format established by Raphael for cardinal portraits, as seen in Raphael's *Portrait of a Cardinal* (*c.*1510–11) and the copy of Raphael's lost *Portrait of Cardinal Bibbiena* (*c.*1516–17; Pl. 13). In each portrait the cardinal is presented in half-length, sitting close to the viewer with an arm extending along the base of the picture. In painting the portrait of Cardinal Sauli, however, Sebastiano looked instead to Raphael's papal state *Portrait of Julius II* (1512), a pose adopted thereafter as the standard for papal portraits, and a work well known to the artist as well as to a wide audience in Rome since it was hung on a pillar in the Church of Santa Maria del Popolo on special feast days (Pl. 14).[19] In both portraits we see

[16] For the allegorical portraits of Lorenzo Lotto, see J. Grabski, 'Sul rapporto fra ritratto e simbolo nella ritrattistica del Lotto', in P. Zampetti and V. Sgarbi (eds.) *Lorenzo Lotto: Atti del Convegno internazionale di studi per il V centenario della nascita* (Asolo, 1980), 383–91.

[17] P. Pouncey and J. Gere, *Italian Drawings in the Department of Prints and Drawings in the British Museum* (London, 1962), 167; Hirst, *Sebastiano*, 109.

[18] Pastor, *History of the Popes*, vii. 156 n.

[19] K. Oberhuber, 'Raphael and the State Portrait 1: The Portrait of Julius II', *Burlington Magazine*, 113 (1971), 124 n. 4; C. Gould, 'Raphael's Papal Patrons', *Apollo*, 117 (1983), 360–1. 'This pose became, almost immediately, standard practice for papal portraits thereafter, followed in such masterpieces as Sebastiano del Piombo's *Clement VII* (Naples),

much more of the figure, sitting in an armchair adorned with fringes; the chair is placed on the left in the foreground, canted diagonally into the picture. Both Pope Julius and Cardinal Sauli wear the flowing white rochet—a feature not emphasized in earlier cardinal portraits—under the crimson hooded *mozzetta*. By basing his portrait of Sauli so closely on the Julius portrait, Sebastiano's intention was clearly to establish Sauli, in the minds of the viewer, in this papal role. It is interesting to note that several writers, overlooking the cardinal's biretta, thought the sitter was a Pope, identifying him as Hadrian VI.[20] Roberto Longhi also described this portrait as 'Un Papa con i famigliari e i segretari'.[21]

The Pope's right arm is supported on the armrest and he is shown holding a cloth in his right hand. Sauli's arm is similarly placed, but instead of a cloth he holds a pair of gloves. This difference in detail is significant. The cloth held by Julius alludes to classical antiquity, in which a cloth or *mappa* held in the right hand denotes consular rank and therefore imperial authority. It also signifies renewal and regeneration, since in antiquity the *mappa* was thrown down or lowered at the commencement of the great cycle of games held at the beginning of the year.[22] Perhaps here the cloth is an allusion to the Fifth Lateran Council, instituted by Julius in 1512 to address the question of Church reform. The gloves Sauli holds appear to have been deliberately chosen because, although gloves are often seen in secular portraits in the sixteenth century, their inclusion here in a cardinal portrait appears to be unique. Cardinals are usually portrayed holding in their right hand a folded letter, which could perhaps be a reference to their role in Consistory; gloves, however, suggest another idea.

Gloves were worn by officiating priests when consecrating the Holy Sacrament. In the Middle Ages gloves were regarded as an important part of ecclesiastical vestment as visible symbols of the priests' purity of heart and deed, and for this reason the laity were prohibited from wearing gloves in church.[23] The gloves Cardinal Sauli holds symbolically affirm his purity and righteousness. This idea finds further elaboration in

Titian's *Paul III* (Naples), Velásquez's *Innocent X* (Doria, Rome), and innumerable others . . .'; L. Partridge and R. Starn, *A Renaissance Likeness* (Berkeley, Calif., 1980), 1.

[20] J. Crowe and G. Cavalcaselle, *History of Painting in North Italy*, ed. T. Borenius (London, 1912), iii. 232 and n. 4; G. Vasari, *Le Vite de più eccellenti pittori, scultori e architti*, ed. G. Milanesi (Florence, 1880), v. 573 n. 3.

[21] R. Longhi, 'Cartella tizianesca', *Vita artistica*, ii. 11–12 (1926–7), 224.

[22] Partridge and Starn, *Renaissance Likeness*, 55–6.

[23] W. Smith, *Gloves, Past and Present* (New York, 1918), 18–22; B. Schwinekoper, *Der Handschuh im Recht, Amterwesen, Brauch und Volksglauben* (Berlin, 1938), 33.

the prominence given to the flowing white rochet, a liturgical vestment also worn by priests when celebrating Mass. In the Julius portrait, this wearing of the white rochet is interpreted by Partridge and Starn as an allusion to the Pope's priestly role.[24] This idea is emphasized by Sebastiano, as he heightens the pristine whiteness and amplifies the vestment so that it dominates the foregound of the picture. While the gloves and white rochet stress the sacred character of Cardinal Sauli's office, his hypnotic gaze is also important, becoming a vehicle of empathic communication with the viewer. It is perhaps not surprising that one writer believes Pontormo in his Leningrad *Holy Family* 'makes the face of the Virgin a literal quotation from Sebastiano's *Portrait of Cardinal Bandinello Sauli and Suite*'.[25]

This stress on the Cardinal's purity and holiness, unusual in cardinal portraits of the time, is highly significant because these are the very qualities that the Joachimist Angelic Pastor must possess. He was to be a Pope of shining saintliness who would be revealed and elected to the historic Chair of St Peter by divine inspiration. This concept of an Angelic Pastor, as we have seen, was the focus of apocalyptic speculation at the beginning of the sixteenth century.[26] It can be seen as a reaction to the development of the papacy as a world power calling for the election of Popes skilled in administration and diplomacy. In the eyes of church-men for whom the supreme religious ideal was a life of sanctity and poverty, the worldly Popes were wholly compromised.[27] Thus they turned for salvation to the prophetic concept of an Angelic Pope who would guide the Church through the transition to a new age. In the early sixteenth century, as we have seen, the circulation of prophecies embodying this hope was certainly stimulated by the new medium of print.[28] The most dramatic focus of this concern is found in the 'discovery' of the pseudo-Amadeite *Apocalypsis Nova* in S. Pietro in Montorio in 1502, already discussed in several chapters of this book.[29] Reports of this episode and the later circulation of extracts from the text heightened prophetic expectations. No doubt these were fostered especially in the Amadeite community of S. Pietro and it was here that in 1516, the very year in which he was painting the Sauli portrait,

[24] Partridge and Starn, *Renaissance Likeness*, 59–60.
[25] S. McKillop, *Franciabigio* (Berkeley, Calif., 1974), 87.
[26] See above, Chs. 2, 6, 7, 8.
[27] B. McGinn, *Visions of the End* (New York, 1979), 186.
[28] M. Reeves, 'Roma profetica', in F. Troncarelli (ed.), *La Città dei Segreti, Magi, estrologie e cultura esoterica a Rome (XV-XVIII)* (Milan, 1985), 277–97.
[29] See above, Chs. 2, 3, 6, 7.

Sebastiano was commissioned to paint a chapel in S. Pietro in Montorio for the Florentine, Pierfrancesco Borgherini. The iconography of the Borgherini Chapel, already studied in the preceding chapter, was almost certainly inspired by the prophecies of Amadeus, for beneath the altar table there is a painting of a large book, the *Apocalypsis Nova*, and above the altar there are paintings which express the Joachimist double message of tribulation and blessedness.[30] Again, we remember the wide diffusion of Joachimist ideas in the sermons of popular preachers, with their message of woe and exaltation.[31] The *Piagnoni* disciples of Savonarola were still spreading their prophetic warning of the great *flagello*, to be followed by the advent of the 'papa santo e buono',[32] the 'instrumento sopranaturale' whose role would be to convert the Turks and propagate the faith throughout the world.

These ideas find visual expression in Sebastiano's group portrait. Cardinal Sauli sits before a table upon which we see a small bell. The bell is the Sanctus bell rung during Mass to announce the Advent of Christ and here the bell symbolizes the presence of Christ. Bandinello Sauli's name, as we have seen, is inscribed on the bell, confirming that he has been chosen by God as the 'instrumento sopranaturale' who will bring about the great Christian renewal. The table is covered with a Turkish 'arabesque Ushak' or 'Lotto'[33] carpet, with a prominently displayed Kufic border decorating the overhanging section. Sauli lays the palm of his hand upon the table above this patterned border. The bell is also positioned directly above and its handle is centred on the map illustrating the page of the book. This juxtaposition of Sauli's hand, Kufic border, bell, and map is significant.

The Ottoman carpet with its prominent ornamental border is an important element of the iconography. Sebastiano had used the motif in his earlier *Portrait of Ferry Carondelet and Two Companions* of 1512 (Pl. 15) and here the table rug, a variant of the 'Small Pattern Holbein' carpet, also displays a Kufic border.[34] Sebastiano appears to have es-

---

[30] See above, Jungić, Ch. 17.

[31] See above, Niccoli, Ch. 10 and Index. See also Weinstein, *Savonarola*, 348–9, 342–4, 353–8; C. Vasoli, *Studi Storici in onore di Gabriele Pepe* (Bari, 1969), 217–40; Tognetti, 'Un episodio', 190–9; A. Prosperi, 'Il monaco Teodoro: note su un processo fiorentino del 1515', *Critica storica*, 11 (1975), 71–101; O. Niccoli, 'Profezie in Piazza: Note sul profetismo popolare nell'Italia del primo Cinquecento', *Quaderni Storici*, 41 (1979), 500–39.

[32] *Prediche italiane ai Fiorentini*, ed. R. Palmarocchi, 3² (Florence, 1930), 313; Weinstein, *Savonarola*, 175 and n. 61.

[33] So named because Lorenzo Lotto painted carpets of this design on more than one occasion (J. Mills, '"Lotto" Carpets in Western Paintings', *Hali*, 3, 4 (1981), 281).

[34] J. Mills, '"Small Pattern Holbein" Carpets in Western Paintings', *Hali*, 1, 4 (1978), 329.

tablished the tradition in Italian portraiture, which by the 1520s became very popular, of presenting the sitter standing or sitting next to a table covered with a richly designed Turkish carpet.[35] Ferry Carondelet, Imperial Ambassador to the Papal Curia, is shown dictating a letter to his secretary, while in the shadows on the left a third figure approaches from behind. Clearly it is Carondelet's official diplomatic role that is being emphasized here.[36] The letter held in Carondelet's hand is inscribed with the words 'Bisuntino Consiliario' and the Ottoman table rug seen directly below makes a further allusion to his diplomatic relations with the East. This use of a Kufic bordered table rug to signify the Ottoman Empire is not confined to Sebastiano for it was a motif also used by Pintorricchio in his fresco of *Pius II at the Congress of Mantua* (1506), in the Piccolomini Library of Siena Cathedral, a congress specifically called by the Pope to formulate plans for a massive crusade against the Turks. In the centre of the fresco there is, appropriately, a table covered with a Turkish carpet showing a prominent Kufic border. In the Sauli portrait, the displayed Kufic border also alludes to the East, symbolizing the ever-present menace of the Turk and the apocalyptic prophecy that sees the Turk as one of the divine scourges. According to this prophetic programme, as we have seen, the Turks would be subdued and converted to Christianity. Cardinal Sauli's hand laid flat upon the table becomes a visual counterpart to this idea of subjugation. With divine intervention, symbolized by the bell, the Angelic Pastor will subdue the Turks and they will be converted to Christianity.

The open book of maps represents the entire world and alludes to the recent discoveries of the New World which seemed to confirm that the Joachimist prophecy calling for the propagation of the faith throughout the world and the final conversion of all peoples would soon be realized, heralding the dawn of a more perfect age. Even Christopher Columbus,

[35] Turkish carpets of geometrical design first appear in Italian paintings in the second half of the 15th cent. where they are mainly seen on the steps of the Virgin's throne, on low benches in scenes of the Annunciation or hanging from balconies, e.g. in Antonella da Messina's *St Sebastian* (1476). See Mills, ' "Lotto" Carpets', 278–89 and Mills ' "Small Pattern Holbein" Carpets', 326–34; J. Mills, 'The Coming of the Carpet to the West' in D. King and D. Sylvester (eds.), *The Eastern Carpet in the Western World from the 15th to the 17th Century* (London, 1983), 12–17, 68; Mills, ' "Lotto" Carpets', mentions the Lotto *Portrait of Cardinal Pompeo Colonna*, suggesting that if it is by Lotto it replaces the Sebastiano as the earliest representation, c.1509, of a portrait with an Anatolian carpet. The portrait, which no one doubts represents Pompeo Colonna, could not have been painted before 1517, the year he was created cardinal. The attribution to Lotto is questioned and it is now thought to have been painted by an artist of Venetian origins working in Rome c.1520–5. See E. Safarik (ed.), *Catalogio sommario della Galleria Colonna in Roma* (Rome, 1981), 84 and fig. 106.

[36] Hirst, *Sebastiano*, 98.

nourished on Joachimist prophecy, increasingly saw his voyages in a mystical light, believing that they were made under the inspiration of the Holy Spirit and that his discoveries would precipitate the climax of history.[37] The handle of the bell bisecting the map becomes a visual realization of the idea that the world is permeated by the Spirit of Christ and will be reduced to one sheepfold under this one shepherd.

Aligned on the same axis as the handle of the bell and appearing directly above the book of maps is the enlarged hand of Paolo Giovio with its upward pointing index finger. Paolo Giovio (1483–1552) was born in Como and obtained degrees in the liberal arts and medicine at Pavia before arriving in Rome in 1512.[38] Although we know little about Giovio's life in Rome between 1512 and 1519, it is thought that his first Roman patron was Bandinello Sauli.[39] In 1516, however, Giovio could have already entered the service of Cardinal Giulio de' Medici, later Clement VII, whom we know he was to serve for many years in the capacity of physician and humanist.[40] Giovio's index finger points heavenward, suggesting, in the context of our reading of the painting and following the famous example of St Thomas in Leonardo's *Last Supper*, that he is questioning Cattaneo as to whether the prophecy naming Bandinello Sauli as the Angelic Pastor is divinely inspired. In 1516 this question was being asked about all prophecies that proclaimed an imminent *renovatio Ecclesiae* and the coming of an Angelic Pastor, following the Pope's visit to Florence between December 1515 and February 1516, when the ecclesiastical authorities, concerned that this popular enthusiasm for apocalyptic prophecy could become potentially explosive and subversive of authority, moved to investigate all prophecies, particularly those of Savonarola, which the Medici believed were the source of all this apocalyptic ferment. It was decided that Cardinal Giulio

[37] See above, Prosperi Ch. 15. See also J. Phelan, *The Millennial Kingdom of the Franciscans in the New World: A Study of the writings of Gironimo de Mendieta (1525–1604)* (Berkeley, Calif., 1936), 17–23; Reeves, *Prophetic Future*, 128–9.

[38] P. Gini, 'Paolo Giovio e la vita religiosa del Cinquecento', *Atti del Convegno Paolo Giovio, Il Rinascimento e la Memoria* (Como, 1985), 39.

[39] T. Zimmermann, 'Giovio e la Crisi del Cinquecento', *Atti del Convegno Paolo Giovio*, 10. This assumption is based on two of Giovio's letters. In one dated 1515 and addressed to the Venetian Marino Sanudo, describing the meeting between Leo X and Francis I, he says 'Da poi andò dal Papa compagnato da quatro cardinali, quali avevano disnato seco, cioè Monsignor *nostro* de Sauli . . .' In the second letter dated 1520 to his brother Benedetto he writes 'Io corrò fortuna di maggior cosa se la sorte sarà la mia, e così voi non crederete se non quello vedrete, perche se la disgrazia della buona memoria del Cardinale de Sauli non fosse intervenuta, forse che non arebbe a stentar più; e forse che sarà per il meglio, . . .' (P. Giovio, *Lettere*, ed. G. Ferrero (Rome, 1956), i. 85, 86).

[40] E. Travi, 'Paolo Giovio nel suo Tempo', *Atti del Convegno Paolo Giovio*, 318.

de' Medici, cousin of Leo X and Archbishop of Florence, should hold a synod in Florence to examine the credibility of all prophets, but the case against Savonarola was dropped even before it reached the Synod because so many *Piagnoni* were willing to come forward and defend the Dominican's prophecies.[41]

The painting, animated by this added dimension of narrative, captures the atmosphere of intense debate that prevailed in certain Roman circles in 1516. Responding to Giovio, Cattaneo quietly places his hand on the edge of the book, above Sauli's hand, as though proclaiming his faith in the divinely ordained future events and in his patron. A closer examination of the life of Bandinello Sauli may help us to understand why those near him believed he possessed those special qualities that would enable him to fulfill the role of the Angelic Pope.

There is no independent study of the life of this Genoese Cardinal. Information about him comes from four principal sources: church histories,[42] works dealing specifically with Genoese church history and culture,[43] biographies of Leo X, and studies devoted to the Conspiracy of Cardinals.[44] It is the fourth subject, the conspiracy, that appears to dominate all biographical accounts of Bandinello Sauli. The Sauli family, descended from Genoese nobility, became leaders in trade and finance

[41] Weinstein, *Savonarola*, 358–61. It is interesting to note that Giovio was still in a prophetic frame of mind—albeit imperially focused—at the time of Charles V's coronation in Italy (see above, Headley, Ch. 13).

[42] Such as H. Garimberto, *La Prima parte delle Vite overo fatti memorabili d'alcuni papi, et di tutti i cardinali passati, 1506–1575* (Ferrara, 1575), 395; A. Ciaconius, *Vitae et Gesta Summorum Pontificum ab Innocentio IV usque ad Clementem VIII necnon S.R.E. Cardinalium cum Eorumdem insignibus* (Rome, 1601), ii. 1061, 1069, 1083; id., *Vitae et res gestae pontificum romanorum et S.R.E. Cardinalium* (Rome, 1677), iii. 298, 312, 320; F. Ughelli, *Italia Sacra sive de Episcopis Italiae et Insularum Adjacentium* (Venice, 1719), iv. 921; G. Eggs, *Supplementum novum purpurae doctae seu vitae legationes, res gestae obitus* (1729), 263; L. Cardella, *Memorie storiche de' cardinali della Santa Romana Chiesa* (Rome, 1893), iii. 356; G. Moroni, *Dizionario di erudizione storico-ecclesiastica*, 61 (1840–61), 291–2.

[43] P. Bizaro, *Historia Genuensis* (Antwerp, 1579), 447–8; M. Foglietta, *Gli Eloggi degli Huomini Chiari della Liguria*, trans. L. Conti (Genoa, 1579), fo. 85ʳ⁻ᵛ; G. Semeria, *Storia ecclesiastica di Genova e della Liguria dai tempi Apostolici sino all'anno 1838* (Turin, 1838), 163; id., *Secoli Cristiani della Liguria* (Turin, 1843), ii. 398–400.

[44] P. Giovio, *Le Vite*, bk. 4, 263–75; W. Roscoe, *The Life and Pontificate of Leo the Tenth* (London, 1846), ii. 80–8; M. Armellini, *Il Diario di Leone X di Paride de Grassi* (Rome, 1884), 36 ff.; E. Rodocanachi, *Histoire de Rome: Le Pontificat de Léon X 1513–1521* (Rome, 1931), 113–27; L. von Pastor, *History of the Popes*, vols. vii–xxiv, trans. and ed. R. Kerr (London, 1908–33), vii. 170–96; F. Gregorovius, *History of the City of Rome in the Middle Ages*, trans. A. Hamilton (London, 1902), viii, pt. 1, 226–34; M. Creighton, *A History of the Papacy from the Great Schism to the Sack of Rome* (London, 1901), v. 280–6; A. Ferrajoli, *La congiura dei cardinale contro Leone X* (Rome, 1920), 51–3, 64–5, 98–101; G. Picotti, 'La Congiura dei Cardinali contro Leone X', *Rivista storica italiana*, 40–1 (1923–4), 249–67; F. Winspeare, *La Congiura dei Cardinali contro Leone X* (Florence, 1957).

and in 1493 established their bank in Rome. Later they were favoured by Julius II, himself a native of Genoa, and during his pontificate they became Chief Collectors of papal revenues and engaged in a wide variety of business transactions for the Church.[45]

A Venetian envoy in Rome at the time described Bandinello as 'Sauli Zenoese, practicho di mercanti come e li soi, et e bon merchadante'.[46] Other sources emphasize his great wealth,[47] his candour and modest manners, and his generous patronage of letters,[48] but none of these sources describe his religious life or record that he was one of the most generous benefactors of his titular church, Santa Sabina.[49] He was also, as we shall see, at the centre of the beginnings of the Catholic reform movement in Rome.

Cardinal Sauli's special attachment to his titular church is apparent in his concern to enlarge and embellish the Dominican convent of Santa Sabina. In 1512 he commissioned work to begin on the restoration of the cloister, building new vaults and supporting columns, and providing for additional cells for the monks above. In the following years he commissioned a series of fresco paintings to decorate the north cloister, paintings presenting the most important events and miracles in the life of St Dominic in Rome. The cycle also included a painting of the Virgin Mary with Santa Sabina, a portrait of the cardinal, and a painting of his coat of arms. These works were apparently covered up at the beginning of the eighteenth century.[50]

During those few years in which Sauli was cardinal in Rome, he appears to have spent much of his time at Santa Sabina: 'Le Cardinal Bandinello Sauli, ayant éprouvé assez de déboires pour être degoûté des hommes, s'était retiré dans la solitude de Sainte-Sabine.'[51] Santa Sabina seems to have been one of the centres of reformist thought in Rome.

---

[45] F. Gilbert, *The Pope, his Banker and Venice* (Cambridge, Mass., 1980), 75; B. Hallman, *Italian Cardinals, Reform, and the Church as Property* (Berkeley, Calif., 1985), 138.

[46] Winspeare, *La Congiura*, 116.

[47] Ferrajoli, *La Congiura*, 52.

[48] Ciaconius, *Vitae et Gesta Summorum Pontificum ab Innocentio IV usque ad Clementem VIII*, 1061; G. Moroni, *Dizionario*, 291.

[49] J. Berthier, *L'Église de Sainte-Sabine* (Rome, 1910), 519; only Moroni refers to his restoration work at Santa Sabina, *Dizionario*, 292.

[50] Vatican MS. Lat. 9167, *Notizie storiche della chiese e convento di santa Sabina martire in Roma* (Rome, 1755), cited in Berthier, *Sainte-Sabine*, 536; C. Piazza, *La Gerarchia Cardinalizia* (Rome, 1702), 435; J. Berthier, *Le Couvent de Sainte-Sabine à Rome* (Rome, 1912), 6–7; E. Rodocanachi, *Una Cronica di Santa Sabina sull'Aventino* (Rome, 1898), 14 n. 2; P. Taurisano, OP, *Le Chiese di Roma illustrate*, no. 11 (Rome, n.d.), 81.

[51] Berthier, *Sainte-Sabine*, 519, 536.

The Dominican monks were annexed to the Lombard Congregation, a strict observant group committed to the ideal of religious renewal through a return to the faithful observance of the rule. For a brief time, however, from 1496 to 1498, when Alexander VI formed a new congregation of reformed convents, the monks came under the jurisdiction of San Marco in Florence, during the very time Savonarola was preaching reform there.[52] After Savonarola's downfall, the monks returned to the Lombard Congregation but there may have been some who were swayed by Savonarola's message of a *renovatio Ecclesiae* and it is possible that some Dominicans saw him as their prophet and placed their hopes in his prophecies.

An interest in prophecy may also have existed at this time in Santa Sabina because the convent possessed a history of the Order, an anonymous *Brevis Historia*, dated 1367, in which there was a pseudo-Joachimist prophecy that appears to have been especially created for the Dominican Order.[53] It records how the Abbot Joachim of Fiore had foretold the coming of the Friars Preacher saying that a new order of teaching brethren would soon arise in the Church of God: one great man would be the head, and with him and beneath him would be twelve men ruling the order. Just as the patriarch Jacob went down to Egypt with twelve sons, so he, with those twelve in that Order after him, would go out and enlighten the world. Joachim also showed his brethren the distinctive habit of the Dominicans which was later miraculously revealed by the Virgin Mary in a vision to Reginald of Orleans. The prophecy records that Joachim had a picture of the habit painted in a certain monastery of his order in Calabria.[54]

One wonders how the Dominicans of Santa Sabina interpreted this Joachimist prophecy of the Preachers in this climate of intense prophetic speculation. It could have created a current of expectancy, encouraging some monks to believe that since Joachim's third age was to be dominated by the *vita contemplativa*, in which monastic orders would take

[52] Berthier, *Le Couvent*, 431–3.

[53] Reeves, *Influence*, 163–4.

[54] *Brevis Historia Ordinis Fratrum Praedicatorum auctore anonymo, 1367*, in *Veterum Scriptorum et Monumentorum Historicorum, Dogmaticorum, Moralium, Amplissima Collectio* (Paris, 1729), vi, cols. 347–8. 'Venerabilis etiam abbas Joachim, Florensis ordinis institutor, fratribus suis habitum, quem dictus magister Raynaldus a B. Virgine acceperat, prophetice demonstrans, in quodam monasterio ordinis sui depingi fecit in Calabria, dicens: "Cito surrecturus est in ecclesia Dei ordo novus docentium, cui praeerit unus major et cum eo et sub eo erunt duodecim praefatum ordinem regentes: quia sicut patriarcha Jacob cum duodecim filiis ingressus est Aegyptum, sic ipse cum illis duodecim in illo ordine post ipsum majoribus ingredietur, et illuminabit mundum."'

precedence, they, like the Friars Minor of San Pietro in Montorio, would play a special role in the coming reformation. This 1367 prophecy makes no mention of the *Pastor Angelicus* but there may have been some Dominicans at the convent in the early sixteenth century who, in their desire to bring about an imminent reform, saw their Cardinal, Bandinello Sauli, as the one who would lead them into the new age.

In the quiet seclusion of Santa Sabina on the Aventine Hill, Bandinello preferred the company of a small group of Genoese, who were known for their piety and dedication to Church reform. One of these was the Dominican Cardinal Nicolò Fieschi (1456–1524) whose titular church of Santa Prisca is also situated on the Aventine Hill. We know that there was close collaboration between Santa Prisca and Santa Sabina, for in 1516 Cardinal Fieschi gave a commission to the Dominican brothers of Santa Sabina to celebrate Mass in Santa Prisca every Sunday.[55] Fieschi, who was particularly zealous in matters of reform, instituted a synod especially devoted to reform of Church discipline at his bishopric of Frejus, which he held from 1510 to 1518.[56]

Cardinal Fieschi's spiritual outlook was no doubt formed in part by the example of his sister, Caterina Fieschi Adorno (1447–1510), whom we know as St Catherine of Genoa. This remarkable woman profoundly influenced a group of pious religious in Genoa by her mystical doctrine of divine love and her lifelong commitment to the poor. She experienced the love of Christ through the fervent practice of daily private devotions and frequent Communion, and expressed her love for Him through service to the poor. As Superior of the Hospital Pammatone in Genoa, St Catherine was tireless in her devotion to caring for the sick, especially plague victims and those destitute people suffering from incurable diseases.[57] Her most ardent disciple was the Genoese notary, Ettore Vernazza, who followed her example and worked with the chronically ill, building the first hospital in Genoa to care for poor incurables. In 1497 Vernazza established a religious confraternity in Genoa, the Oratory of Divine Love. Its purpose was to promote the personal moral regeneration of the individual through faithful practice of religious devotions and through the performance of works of charity. The members of the confraternity

[55] E. Rodocanachi, *Una Cronica*, 14.

[56] Moroni, *Dizionario*, 24, 254; Hallman, *Italian Cardinals*, 23. On Fieschi, see also above, Minnich, Ch. 4.

[57] P. Bonzi da Genova, *Teologia mistica di S. Caterina de Genoa* (Genoa, 1960); *Dictionnaire de spiritualité*, ed. C. Baumgartner SJ (Paris, 1953–), ii. 290–326; F. von Hugel, *The Mystical Element of Religion as Studied in Saint Catherine of Genoa and her Friends* (London, 1923).

numbered thirty-six laity and four clergy but their identity and activities were kept strictly anonymous.[58]

Ettore Vernazza arrived in Rome sometime between late 1511 and early 1514 to establish a Roman Oratory of Divine Love and to build a hospital for incurables modelled on the one in Genoa. The primary source of information for Vernazza's stay in Rome is a letter written by his daughter, Battista Vernazza, in 1581, to a Father Gasparo de Piacenza. She reveals that not only was there a close rapport between her father and the Sauli family, but also that Cardinal Sauli and his brothers actively assisted Vernazza in his work. When Vernazza sought help to found his Roman hospital, S. Giacomo of the Incurables, Battista writes 'Moved by great piety he resolved to stay in Rome and seek by prayers and the aid of others to found a hospital for Incurables; he enjoyed the Cardinal Sauli's favour who told him: "At any time when you do not have enough funds, come to me." '[59] While in Rome Vernazza stayed in the house of Bandinello's brother, Sebastiano, who, together with his brothers Giovanni and Agostino, managed the affairs of the bank. Battista writes that her father 'stayed two years in his house, living with his family and wife, without their asking for any payment'.[60] Battista also relates that when Cardinal Sauli was dying in 1518, his mother Mariola wrote to Vernazza who was building another hospital in Naples, urging him to come at once to Rome. Making no reference to the conspiracy or to the sad circumstances of Bandinello, Battista writes that her father did not go to Rome because if he had, the hospital in Naples would have fallen into ruin.[61] However, it is more likely that his reluctance to go to Rome was prompted by fear, for Rome had become a hostile place for the Sauli family in the aftermath of the conspiracy trial.

The Oratories of Divine Love established by Vernazza in Genoa, Rome, and Naples fostered a living personal faith based on the love of God and the principle that personal sanctification must be achieved by

[58] A. Bianconi, *L'Opera delle Compagnie del Divino Amore nella riforma Cattolica* (Città di Castello, 1914); P. Paschini, *La beneficenza in Italia e le Compagnie del Divino Amore nei primi decenni del Cinquecento* (Rome, 1925; repr. in *Tre Ricerche sulla Storia della Chiesa nel Cinquecento* (Rome, 1945), 3–88; C. da Langasco, *Gli Ospedali degli incurabili* (Genoa, 1938); A. Cistellini, *Figure della Riforma Pretridentina* (Brescia, 1948); M. Lumbroso and A. Martini, *Le Confraternite Romane nelle loro Chiese* (Rome, 1963), 122–5; J. Olin, *The Catholic Reformation: Savonarola to Ignatius Loyola* (New York, 1969), 16–26.

[59] B. Vernazza, *Lettera biografica del padre e della madre*, in *Delle Opere Spirituali della reverende et divotiss. Vergine di Christo donna Battista da Genova* (Verona, 1602), cited in Bianconi, *Compagnie del Divino Amore*, 65.

[60] Bianconi, *Compagnie del Divino Amore*, 65.

[61] Ibid. 66.

the performance of good works on behalf of others. Historians have long recognized the Oratory of Divine Love as the beginning of an effective Catholic reform, especially since some of the original members of the Roman Oratory, Giampiero Caraffa, later Paul IV, Gaetano de Thiene, and Marcantonio Flaminio became major figures in the Catholic reform movement of the 1520s and 1530s.[62] We have seen how Cardinal Sauli and his brothers were Vernazza's close personal friends, eager to support and make possible his programme of action. Although the activities of the Oratory were under a veil of secrecy, so that we may not know the full extent of Cardinal Sauli's involvement, enough is known to suggest that he shared Vernazza's desire to create a reform of Christian life.

The Genoese Dominican and renowned humanist, Agostino Giustiniani (1470–1536), Sauli's cousin, was another influential member of Sauli's Roman circle. He had spent many years in Genoa in the ambit of the Lombard Congregation so that, although there is no direct evidence of a personal relationship between Giustiniani and St Catherine, it is highly likely that there were some encounters,[63] especially since the Dominicans are thought to have played a large part in instituting the Oratory of Divine Love.[64] However, Giustiniani's great admiration for St Catherine is revealed much later when he wrote the first account of her life in his *Castigatissimi Annali*, published in 1537; he spoke of her with great reverence, comparing her to St Catherine of Siena, recognizing fully her profound influence on the religious life of the Genoese.[65] Giustiniani was in Rome during the critical years of 1514 to 1517 'per piacere et per servire il Cardinale mio cugino et mio signore'.[66]

Another key member of the cardinal's circle and the one most likely to have commissioned Sebastiano to paint the group portrait in 1516 was his devoted younger brother, the apostolic protonotary, Stefano Sauli, who appears, quite possibly, as the fourth figure in the painting, standing behind the cardinal's chair. This figure has been described as 'the attendant secretary-type',[67] but the way he leans earnestly forward and

---

[62] Pastor, *History of the Popes*, x. 388–92; H. Jedin, *The History of the Council of Trent*, trans. D. Graf (London, 1957), i. 146; Olin, *The Catholic Reformation*, 16–17.

[63] C. da Langasco, 'Il Giustiniani e le tensioni ecclesiali del suo Tempo', *Agostino Giustiniani Annalista Genovese de i Suo Tempi. Atti del Convegno di studi a Genova* (Genoa, 1984), 113.

[64] Paschini, *La Beneficenza in Italia*, 20 n. 2.

[65] A. Giustiniani, *Castigatissimi Annali . . . della Ecclesia et illustrissima Repubblica di Genoa* (Genoa, 1537), 267; da Langasco, 'Il Giustiniani', 113; von Hugel, *The Mystical Element*, i. 382.

[66] Giustiniani, *Castigatissimi Annali*, 224.

[67] Hirst, *Sebastiano*, 99.

the close juxtaposition of the two heads, suggests a more intimate relationship. Stefano, who also chose a clerical career was, like Bandinello, a patron and friend to humanist scholars, most notably Marcantonio Flaminio (1498–1550) and the Belgian, Cristoforo Longolio (1490–1522). It is Longolio's letters, written from Padua to Stefano and Marcantonio in Genoa in 1521, that confirm Stefano's personal friendship with Ettore Vernazza.[68] Stefano, much preoccupied with religion, had lasting ties to the Catholic reform movement. Gregorio Cortese, the Benedictine monk who later, as the abbot of San Giorgio Maggiore in Venice in the 1530s, played a central role in establishing the Oratory of Divine Love in that city,[69] was a close friend. Stefano's visit to the reformed monastery of St Onofrio on the island of Lerins when Cortese was abbot there in 1518 is well documented. Stefano was accompanied by Longolio and may have made the journey to Lerins shortly after Bandinello's death. Stefano's continuing friendship with other important figures in the reform movement include those with Reginald Pole and Benedetto Ramberti. His religious attitude is presumably revealed in the book he wrote, unfortunately lost, entitled *De homine Cristiano*. It is recorded that this work was very highly acclaimed by Reginald Pole.[70]

If Stefano did indeed commission Sebastiano to paint the group portrait as an affirmation of his faith in the Joachimist prophecy and from a passionate belief that his brother, if elected Pope, would inaugurate the much desired reforms, the arrest and imprisonment of the cardinal must have been devastating. Undoubtedly Stefano must have shared Giustiniani's harsh judgement of the Pope and the Roman Curia: 'the conduct of the Pope was attacked and cursed by many as was the conduct of those who governed the Roman court. As for them it is said, that their actions in this affair were calculated, following the example

[68] Bianconi, *Compagnie del Divino Amore*, 35–7.
[69] S. Tramontin, 'I Teatini e l'Oratorio del Divino Amore a Venezia', *Regnum Dei*, 29 (1973), 67.
[70] G. Tiraboschi, *Storia della Letteratura Italiana* (Milan, 1824), vii, pt. 1, 286–8; S. Seidel Menchi, 'Passione Civile e Aneliti Erasmiani di Riforma nel Patriziato Genovese del Primo Cinquecento: Ludovico Spinola', *Rinascimento*, 2nd ser., 18 (1978), 114–15; see esp. A. Pastore, *Marcantonio Flaminio* (Milan, 1981), 40–4; P. Manutius, *Epistolarum libri IV* (Venice, 1560), 5–21. While painting the Sauli portrait, Sebastiano was also working on the Borgherini Chapel commission which, as we have seen, is also influenced by Joachimist ideas. It is perhaps impossible to determine whether there was any communication between the respective patrons although we do know that Cardinal Carvajal had a special regard for the Sauli family and for Bandinello. A few months after his release from prison in July 1517, Cardinal Sauli, gravely ill and dishonoured by the revelations of his alleged complicity in the plot to kill the Pope, went to live in Carvajal's Roman palace, where he died in March 1518 (see M. Armellini, *Un Censimento della città di Roma sotto il pontificato di Leone X, tratto de un codice inedito dell'Archivio Vaticano* (Rome, 1882), 136).

of the Pope, in such a way as to leave them more open to accusations of too much mildness and clemency, rather than to give others the opportunity to point to their greed and avarice.'[71]

In addition to Pastor, the Conspiracy of Cardinals of 1517 has been studied by three modern historians, namely Alessandra Ferrajoli, G. B. Picotti, and Fabrizio Winspeare. According to the traditional account which Ferrajoli, Winspeare, and Pastor believe,[72] Cardinal Alfonso Petrucci plotted to poison Leo X to revenge his brother Borghese, whom the Pope in 1516 had deposed from the *Signoria* in Siena in a move to consolidate Medici control in Tuscany. Since Alfonso's father had actively assisted in the Medicean restoration in Florence and the cardinal had played a considerable part in getting Leo elected, the Pope's ingratitude angered Alfonso even more. It seems certain that he did publicly threaten the Pope and intrigue to restore Siena with Leo's enemies, particularly Francesco della Rovere, who had just regained his possessions in Urbino, after losing his dukedom to the Pope's nephew Lorenzo in 1516. In March 1517 the Pope wrote to warn Petrucci, who had sought refuge in Gennazzano, not to stir up revolution in Siena. The following month, Petrucci's secretary, Marco Antonio Nino, was arrested in Rome and interrogated under torture. During the interrogations Nino alleged that there was a plot to poison the Pope: a Florentine surgeon, Battista da Vercelli, was to go to Rome on the pretext of curing the Pope of his fistula and administer poison.

With the guarantee of safe conduct, and an agreement from the Pope that Borghese would be restored to Siena, an agreement countersigned by Petrucci's intermediaries at the Vatican, Cardinals Sauli and Cornaro, Petrucci came to Rome, but when he appeared in the Vatican on 19 May, accompanied by Sauli, both were arrested and imprisoned in Castel Sant' Angelo.[73] In the next days Cardinals Rafaello Riario, Francesco Soderini, and Adriano Castellesi were also accused of complicity in the plot. All five cardinals, to a greater or lesser degree, were found guilty of plotting or giving ear to Petrucci's schemes. Cardinal Petrucci was executed in the early days of July, following the public executions of Nino and Vercelli. Riario was pardoned after promising to pay the

---

[71] Giustiniani, *Castigatissimi Annali*, 272; Menchi, *Passione Civili*, 114: '... così da molti fu biasimato e vituperato il contegno del Papa, e di coloro, quali governano la corte Romana. E si dice che da parte loro era a diportasi in questo fatto per tal modo, che potessero anzi essere incolpati di troppo mansuetudine e clemenza seguendo le vestigia del sommo pastore, che dar occasione di esser notati di cupidità e di avarizia.'

[72] See above, n. 44; Rodocanachi, *Histoire de Rome*, 113–27 gives a general account of the conspiracy.

[73] Ibid. 115–16.

enormous sum of 150,000 ducats, with a further sum of 150,000 imposed as a guarantee that he would observe all the conditions of the pardon.[74] Cardinals Soderini and Castellesi were also pardoned, each fined 12,500, but when the fine was suddenly increased to 25,000, they both fled Rome fearful of their safety.[75] Cardinal Sauli remained in prison until 31 July when he was brought before Consistory and made to confess publicly.[76] He was fined 25,000 ducats, reluctantly pardoned by Leo, and reinstated as cardinal, with his voting rights withheld.

Important documents relating to the trial and conviction of the cardinals which were never made public at the time, were long ago removed from the Vatican archives[77] and those documents that survive contain many inconsistencies and contradictions. G. B. Picotti, in his critical study of Ferrajoli's book, reviewed all the remaining evidence and came to an entirely different conclusion from the other three historians, believing that Leo X took advantage of the political transgressions and intriguing of Petrucci to fabricate a plot against his own life as a means to rid himself of his enemies while at the same time extorting from them huge sums of money which he desperately needed to continue his war with Urbino. Picotti also believes that the Pope's larger plan becomes evident when we consider how he used the conspiracy as a pretext greatly to enlarge the College of Cardinals,[78] which he did on 1 July 1517, creating an unprecedented thirty-one new cardinals, including three of his nephews and many others who were pro-Medicean, thus greatly increasing the chances that a Medici would succeed him on the papal throne. Since the new cardinals had to pay well for their elevation, Leo raised a further 300,000 for his war effort.[79] As K. M. Setton observes, referring to Picotti's interpretation of events, 'Unfortunately for Leo's memory little that we know of his character makes this grave accusation impossible or even unlikely.'[80]

---

[74] Winspeare, *La Congiura*, 161–2.

[75] Pastor, *History of the Popes*, vii. 180.

[76] Ferrajoli (*La congiura*, 64–5) raises doubts about the value of this confession because, according to Paris de Grassis, the Pope had intimated to Sauli that if he did not confess, he would be left to die in the dungeons of Castel Sant'Angelo (see also Picotti, 'La congiura', 262). There is also the question of whether Sauli had been tortured. Leo X denied the cardinals had been racked but others believe they were (see Pastor, *History of the Popes*, vii. 176 n.); Giovio says: 'Alfonso adunque fu convinto al martirio con gli indici del Vercelli e d'Antonio suo segretario, i quali lungamente tormentati havevano confessato ogni cosa: il Sauli apena puote vedere il martirio.' Giovio, *Le Vite*, 270.

[77] Pastor, *History of the Popes*, vii. 185 n.

[78] Picotti, 'La congiura', 266.

[79] Butters, *Governors and Government*, 295.

[80] K. Setton, 'Pope Leo X and the Turkish Peril', *Proceedings of the American Philosophical Society*, 113 (1969), 395.

In fact there were many at the time who came to exactly the same conclusions as Picotti, saying that the crimes were false, that innocent people had been condemned in order for Leo to find money to continue waging war with Francesco Maria in Umbria, and that the Pope had enlarged the College of Cardinals with unprecedented liberality to serve his own interests, without taking any notice of the old cardinals who were all too frightened to open their mouths.[81]

Those modern historians who are persuaded that the plot was real admit that 'how far each was individually involved cannot be ascertained by the material at our disposal'.[82] But they nevertheless believe that Sauli was 'deeply' involved. None provides a motive for Sauli's alleged participation and indeed they point out that Leo had always favoured him and, because of Sauli's 'candore e la modestia de' suoi costumi', seemed genuinely fond of him, nominating him to the bishopric of Albenga in 1513 and conferring upon him generous pensions and benefices, even three months prior to the alleged conspiracy.[83] Giovio tries to explain the cardinal's actions by asserting that he was angry with Leo because he did not receive the bishopric of Marseilles, which was given instead to Giulio de' Medici,[84] but, as Ferrajoli concedes, this is not sufficient cause to desire the death of the Pope. Ferrajoli concludes, as does Winspeare, that Sauli's motive remains a mystery.[85]

Ferrajoli is persuaded of Sauli's guilt only because Leo X had no political motive himself for persecuting Sauli. He argues that, unlike Petrucci, whose political intriguing against Leo was not in doubt, or Soderini, whom Leo could see as another political enemy, since his brother had been deposed as head of Florence's republican government

[81] Giovio, Le Vite, 273–4: 'alcuni altri dicevano, che quei delitti si gli apponevano al falso; & che a torto erano condannati huomini innocenti, per ritrovare con malvagissimo modo danari da far guerra percioche accendedosi tutta via la guerra; di Francesco Maria nell'Umbria, essendo del tutto consumati i thesori in Roma e in Fiorenza, dicevasi chel papa contra la sua natura era diventato crudele per la paura, & rapace per i bisogno; . . . Per queste cagioni mosso a colera & sdegno, giudico che gli fosse bisogno creare uno altro collegio, volendo portrarsi da principe no punto scempio ne ridicolo per iscorno, ma gradeamente savio & valoroso per interesse suo. La onde con incredible & non piu usata liberalità, riempi il collegio di trenta-uno cardinali, tenendo poco conto de cardinali vecchi, i quali tutti spaventati non ardivano aprir bocca.'

[82] Pastor, History of the Popes, vii. 187.

[83] Moroni, Dizionario, 291; Pastor, History of the Popes, vii. 188.

[84] Giovio, Le Vite, 270. Giovio is not convinced that Riario, Soderini, Castellesi, and Sauli did actually participate in a plot but he says even if they did not, they were so consumed by hatred and ambition that they wished that Petrucci might succeed in removing the Pope (267–8). Giovio's opinions are suspect, however, since the Medici were his patrons and his biography of the Pope was dedicated to Duke Alessandro de' Medici.

[85] Ferrajoli, La congiura, 53; Winspeare, La Congiura, 117.

on the return of the Medici, Bandinello Sauli was a person of little political significance and posed no threat. Ferrajoli even goes on to say: 'Thus one can conclude that his guilt must have been even more serious than it appears in the trial documents known to us; even in these documents he is the most seriously inculpated after Petrucci.'[86]

Significantly, not one of these modern writers considered the prophecy recorded by Giovio, foretelling Sauli's elevation to the papacy, to be even worthy of mention,[87] although Giovio, recalling its significance to the cardinal's circle, included it in his comparatively brief account of the conspiracy in Leo's biography thirty years later. Given the prevailing atmosphere in Rome of heightened apocalyptic expectancy, and the demonstrated moves by the Medici to weight the College of Cardinals in their favour in order to ensure a Medici succession, it is highly conceivable that Bandinello Sauli was named as conspirator because some feared that the prophecy would be fulfilled. Perhaps it was the very existence of the allegorical group portrait that provided the catalyst for his arrest. We know that prophecies were taken very seriously at this time, as was proved during Leo's illness in 1516. Paris de Grassis, the Pope's Master of Ceremonies, described how Leo began to weep openly, convinced that he was going to die because a certain Franciscan hermit, Fra Bonaventura, proclaiming himself to be the Angelic Pope, had prophesied that Leo would die that year, and this same prophet had predicted the death of Julius II. Fra Bonaventura was imprisoned in the Castel Sant'Angelo and Leo ordered that he be interrogated repeatedly. Thereafter we hear no more about him.[88]

Fears of the power of prophecy could also explain why Cardinal Castellesi (1458–1522?) was implicated in the conspiracy, thereby removing him as a potential successor to the papacy. Castellesi of Corneto, created cardinal by Alexander VI in 1503, was a renowned Latin stylist and Hebrew scholar who had worked his way up in the curial bureaucracy on the strength of his scholarly abilities. Giovio records that Castellesi, unlike the others, did not wish the Pope's death out of any feelings of hatred or ill-will, but because he had been unduly influenced by a pro-

---

[86] Ferrajoli, *La congiura*, 98–9: 'Quindi è da credere che la sua colpevolezza dové risultare anche più grave di quanto apparisse dai processi a noi noti, nei quali pure è il più aggravato dopo il Petrucci.'

[87] The only writer to refer to the prophecy in Giovio's biography of Leo X is Gregorovius, *City of Rome*, viii, pt. 1, 227.

[88] Pastor, *History of the Popes*, vii. 156 n.; see also v. 224–5; F. Secret, 'Aspects oubliés des courants prophétiques au debut du XVIᵉ siècle', *Revue de l'histoire des religions*, 173 (1968), 178–9, 197–9; Rodocanachi, *Histoire de Rome*, 108.

phetess who had told him Leo would die young and that an aged man of obscure birth called Adrian, who was famous for his letters, would become Pope.[89]

Bandinello Sauli remained in the Vatican a virtual prisoner of the Pope until September when he was allowed to return to Santa Sabina. In October he went to the Orsini's at Monterotondo, but because of illness he was forced to return almost immediately to Rome where he spent the last months of his life, living in the palace of Cardinal Carvajal, attended by his mother, brothers, and Giustiniani. He died on 29 March 1518.[90] Many believed that he died of poison that had been administered to him in prison,[91] but Ferrajoli is not persuaded and says 'First of all, the pope could not have been actuated by any political or financial motive to procure the death of Sauli since he was neither a dangerous nor a rich adversary.'[92]

Sauli was buried in Santa Sabina and, according to Picotti, in order to 'tranquillasse la sua coscienza', Leo ordered that the cardinal be buried with full honours.[93] The tomb, which has since disappeared, was inscribed with this revealing epitaph, recorded by Ciacconius in 1677:

> When evil fate, enemy of virtue, discerned in You
> Genius and transcendent gifts of spirit,
> It could not bear it and stood in the way of your success,
> Daring to prevent the ascent toward Glory.
> Happy were you if you and your Glory kept on the course
> Where you began to unfurl your sails through the deep.
> For when virtue was decking you out with the purple cloak,
> It then promised you the triple crown.
> But Fate, which has earlier taken away well-deserved rewards,
> Could not keep you out of the Celestial Choir.

[89] Giovio, *Le Vite*, 268–9; on Castellesi, see *DBI* 21 (1978), 665–71; P. Paschini, 'Tre illustri prelati del Rinascimento', *Lateranum*, NS 23 (Rome, 1957), 1–4, 43–130; D'Amico, *Renaissance Humanism*, 16–19. The motives for implicating Cardinal Riario also become apparent. He was not only exceedingly rich but he was also universally respected and much loved by the people of Rome and, narrowly missing being elected Pope in 1513, would have been the one most likely to have succeeded Leo.

[90] Ferrajoli, *La congiura*, 99–100.

[91] Giustiniani, *Castigatissimi Annali*, 273: 'non senza gran sospizione che li fosse stato dato tosico terminato'; Garimberto, *La Prima Parte*, 395 'che appresso ottennero ancora la restitutione della dignita e liberta sua, dopo aver pagato grossa somma di danari, e bevuto il veleno terminato, per quanto su detto e guidicato da gli inditij in vita e dei i segni nella morte, la quale fu poco dopo la sua liberazione.'

[92] Ferrajoli, *La congiura*, 100–1.

[93] Picotti, 'La congiura', 267.

Here you, 'midst the Heavenly Host and the Holy Congregation
Despise wealth, the weight of gold, kingdoms, Glory.[94]

When Leo X died suddenly in December 1521, Stefano Sauli, together
with Marcantonio Flaminio, returned immediately to Rome to seek a
revision of the original trial and in April Cardinal Fieschi was appointed
president of a review commission.[95] A letter sent by Longolio in Padua
to Stefano gives an illuminating description of Rome at this time:

Of course I commend your family feeling since you have decided to pursue at
law and in court the crime against your brother Bendinello, but I strongly dispute the advantage of the timing of it, because, my dear Stefano, I am afraid that
it is not very safe for you to fight without weapons against armed men, both
because of the unrivalled power of your enemies and because of their free use of
swords. Remember that you are at Rome, and, what disturbs me the most, in a
city which is not only without a government itself, but is even now in the grip of
the powerful forces of your adversaries. But since you have taken the affair to
the point that you cannot retreat honourably, be ever more careful how far you
trust any individual, even one of your friends. I do not really doubt that everything is going to be safer there and easier for you once the Pope arrives.[96]

Despite Pope Hadrian's receptiveness to a re-examination of the trial, the
process moved very slowly. When Hadrian died and Giulio de' Medici
was elected Clement VII in 1523, there was no longer any hope of restoring the reputation of Cardinal Sauli.

Returning to Sebastiano's group portrait, the last detail to be considered is the fly seen on the Cardinal's rochet, a fly painted so realistically

[94] A. Ciaconii, *Vitae et res gestae pontificum romanorum et S.R.E. Cardinalium* (Rome, 1677), col. iii, 298; Berthier, *Sainte-Sabine*, 520. 'Invida virtuti mala sors cum cerneret in Te | Ingenium et dotes eximias animi, | Non tulit atque tuis successibus obstitit ausa | Ad decus intentum continuisse gradum. | Felix si quo coepisti dare vela per altum | Hoc ires cursu tuque tuumque decus! | Nam virtus cum purpureo te ornaret amictu, | Promisit triplex tum diadema tibi. | At sors quae meritos olim subtraxit honores | Non potuit superis te prohibere choris. | Hic tu Coelestes inter sanctumque Senatum | Spernis opes, auri pondera, regna, decus.

[95] Ferrajoli, *La congiura*, appendix, 350; Pastore, *Marcantonio*, 44 n. 15.

[96] C. Longolius, *Epistolarum liber quartus* (Basle, 1558), 254. 'Nam quod Bendinelli fratris tui iniuriam legibus & iudicio persequi statuisti, laudo equidem pietatem tuam, sed temporis in eo opportunitatem vehementer desidero, vereor enim mi Stephane charissime, ne cum in summa inimicorum tuorum potentia, tum in ista gladiorum licentia, tibi non satis tutum sit inermem contra armatos contendere. Memento te esse Romae, & quod me maxime conmovet, in urbe, quae non solum sine imperio ipsa hodie sit, sed magnis adver-sariorum tuorum opibus etiam nunc teneatur. Sed quoniam eo res iam e te deducta est, ut honeste pedem referre non queas, etiam etque etiam diligenter videto quantum unicuique vel tuorum committendum sit. Quin adventu Pont. Maximi omnia istia tutiora sint futura, & tibi etiam aequiora, non equidem dubito.'

that printers have sometimes corrected their reproductions to omit it. Although *trompe l'oeil* flies are seen in Italian paintings of the late Quattrocento, where they appear to be included as a naturalistic tour de force, a sign of the artist's virtuosity, flies also appear in devotional works as symbols of death and decay.[97] Such is the example of Giovanni Santi's *Christ in the Tomb* (c.1490) in the Ducal Palace, Urbino, where the fly, placed on the body of Christ, makes explicit reference to His mortality. By a curious coincidence, in Lorenzo Lotto's portrait of *Giovanni Agostino della Torre and his Son Niccolò*, dated 1515, in the National Gallery, London, a fly is painted on the white cloth held in Giovanni's hand. Giovanni died the following year[98] and it is thought that the portrait of his son Niccolò was added at a later date.[99] Given the symbolism of the fly, perhaps it was painted on at this time to indicate that the father had died. In Sebastiano's portrait, the fly likewise could have been painted after Sauli's death to signify to the viewer that death had intervened to prevent Sauli from fulfilling his role as the reforming holy Pope. Its meaning would therefore echo the epitaph of his tomb. Perhaps this poignant detail was painted just before the portrait was transported to Genoa for safekeeping, where it was first recorded in the palace of Giacomo Balbi in 1780.[100]

In this study of Sebastiano's painting, one important question remains. In the winter of 1517–18[101] Raphael painted a group portrait of *Leo X with Cardinals Giulio de' Medici and Luigi de' Rossi* (Pl. 16). The painting is so similar in both form and content to the Sebastiano that some writers believe Raphael used Sebastiano as a direct source for his work.[102] But why would Raphael use the Sauli portrait as his source, knowing full well that the cardinal had been convicted of plotting the murder of his patron? Raphael's painting, also unusually large,[103] shows Leo seated in the foreground in a fringed armchair, before a table covered with a

---

[97] A. Chastel, 'A Fly in the Pigment', *FMR*, ed. Franco Maria Ricci, 4 (1986), 75.

[98] F. Bosco, 'Il Ritratto di Nicolo della Torre Disegnate da Lorenzo Lotto', in P. Zampetti and V. Sgarbi (eds.) *Lorenzo Lotto: Atti del Convegno internationale di studi per il V centenario della nascita* (Asolo, 1980), 317.

[99] C. Gould, *The Sixteenth-Century Venetian School* (National Gallery Catalogues) (London, 1959), 50.

[100] C. Volpe and M. Lucco, *L'opera completa di Sebastiano del Piombo* (Milan, 1980), 106.

[101] B. Davidson, *Raphael's Bible: A Study of the Vatican Logge* (University Park, Pa. and London, 1985), 11; see also R. Sherr, 'A New Document concerning Raphael's Portrait of Leo X', *Burlington Magazine*, 125 (1983), 31–2.

[102] Freedberg, *Painting of the High Renaissance*, 341; Davis, 'Un appunto', 386.

[103] It measures 154 × 119 cm. and is in the Uffizi Gallery, Florence.

heavy crimson table rug, upon which we see an open book and a small bell. On the left is Giulio de' Medici and on the right is Leo's nephew, Luigi Rossi, nominated cardinal in the great creation of 1 July, standing behind the papal chair which he clutches with both hands. We observe the same psychological dislocation of the figures as they emerge from the shadows of the background and the same chromatic scheme of red and white. As others have suggested, Raphael's group portrait is clearly a declaration of Medici dynastic aspirations for the papacy, in the manner of Melozzo da Forli's Vatican fresco showing Sixtus IV with his two Cardinal nephews.[104] However, this does not explain the presence of the book and bell, motifs obviously appropriated from the portrait of Bandinello Sauli. The book is the Hamilton Bible, a fourteenth-century illuminated codex which is reproduced in such detail that the actual page can be recognized as folio 400 verso, where, in addition to the small miniatures, the viewer is able to read the opening lines of the Gospel of St John: 'In principio erat verbū & verbū.' Davidson, in emphasizing the ecclesiastical content of the portrait and the allusions to the Pope's priestly role, suggests that a reference is made specifically to Leo's baptismal name, for on the same page of the Hamilton Bible, not included by Raphael because of limitations of space, there is the verse: 'There was a man sent by God whose name was John.'[105]

Given the context of Sebastiano's group portrait, however, where, as we have seen, Joachimist concepts have been translated into visual imagery, it does seem to be more than a coincidence that there is a reference in Raphael's portrait to the Gospel of St John. In Joachim's interpretation of history, St John the Evangelist typifies the third *status* of the Holy Spirit, the much awaited last age of the Church, which according to Joachim will be dominated by Johannine spirituality, the *vita contemplativa*, and it was for this reason and in anticipation of this new life to come that Joachim called his monastery S. Giovanni in Fiore.[106] Therefore the Gospel of St John is entirely appropriate to a Joachimist reading of the painting. The small silver and gold Sanctus bell symbolizes, as it does in the Sauli portrait, the presence of Christ, because this new age will be ushered in by Christ; its scarlet-tasselled handle points directly to the opening lines of the Gospel, announcing the advent of Christ as the incarnate Word of God.

[104] L. and H. Ettlinger, *Raphael* (Oxford, 1987), 195.
[105] Davidson, *Raphael's Bible*, 12–13.
[106] Reeves, *Influence*, 137, 209; *Prophetic Future*, 3.

Why does Raphael portray Leo looking out to the left, thereby directing the viewer's attention to something beyond the painting, unseen, but anticipated? Could this be an allusion to the expected coming of the new age, and is Leo here presented as the Angelic Pastor, the one called to lead us into this new epoch? Is he the one 'sent by God'? Were the Joachimist allusions appropriated from the Sauli portrait in order to recast Leo in this role? Donald Weinstein believes that for a time Leo may have entertained the idea of promoting himself as the Angelic Pastor to take advantage of the prevailing enthusiasm for millenarianism in Florence and Rome.[107] In the wake of the conspiracy trial and the challenges posed by circulating Joachimist prophecies, Leo may well have felt the need to reassert his own papal supremacy.

Looking at the figures in these two group portraits and considering the context in which they were created, we can sense even more acutely and understand better than before the psychological strain and tension that others have observed in the past. The decoding of Sebastiano's group portrait reveals for the first time the full intentions of the artist and patron. The portrait of Bandinello Sauli as the Angelic Pastor speaks to us differently now, leaving us to wonder whether the painting is the one surviving document that could serve to exonerate this cardinal, whose personality and spiritual temperament were so clearly incompatible with the political machinations of the Medici papal court.

[107] Weinstein, *Savonarola*, 353.

VIII

Marcellus II, Girolamo Seripando, and The Image of the Angelic Pope

WILLIAM HUDON

Marcello Cervini (1501–55), Cardinal of Santa Croce, a scholar, papal diplomat, Inquisitor, and sometime legate to the Council of Trent, became Pope Marcellus II on 9 April 1555, after a five-day conclave.[1] Cervini died suddenly, apparently from a stroke, twenty-two days later, and this demise signalled the beginning of an historical assessment of the man which rapidly became hagiographical. Cervini was, or so maintained writers from Onofrio Panvinio in the sixteenth century to Hubert Jedin in the twentieth, a reform-minded and industrious man who aroused great hope and expectation upon his election and general dismay and consternation upon his death.[2] One of these historians even related an elsewhere unconfirmed and geographically quite impossible story of a vision of the Virgin Mary which Cervini allegedly experienced while making his way to Rome for the conclave in 1555.[3] He was modest

I wish to thank three scholars for their help and encouragement. Constantin Fasolt of the University of Chicago first suggested to me several years ago that Cervini and his connection with the Angelic Pope prophecies might make an interesting study. John M. Headley of the University of North Carolina and Marjorie Reeves both gave a careful reading to an earlier version of this essay and their suggestions improved this work immensely.

[1] On Cervini's life, see P. Pollidori, *De vita gestis et moribus Marcelli II* (Rome, 1744) and A. Cervini, *Vita di Marcelli II*. The latter is a MS in the Bibl. pubblica, Ferrara, published in part in *ASI* 7 (1848), 248–51. Another copy is in the Archivio di Stato, Florence, Carte Cerviniana, Filza 52, fos. 124ʳ–131ᵛ. That MS collection, henceforth cited as *C. Cerv.*, is in 54 vols. and contains correspondence and personal papers of Cervini. See also the biographical sketch in G. Moroni, *Dizionario di erudizione storico-ecclesiastica* (Venice, 1840–79), 42, 238–46.

[2] O. Panvinio, *Historia della vite de i summi pontefici* (Venice, 1594), 283–6; Jedin, *Council of Trent*, ii. 47–9.

[3] The Virgin allegedly assured him, as he was saying Mass on the feast of the Annunciation (25 March), of the future pontificate (Moroni, *Dizionario*, 42, 243). Other narrative sources on Cervini's journey to the conclave cast doubt on this story. He travelled south

and scholarly, many of them insisted, as a result of the continuing humanistic education he received under the tutelage of his father and others in his home town of Montepulciano, and later in Siena and Rome. They added that his education and virtue bore fruit in his public administration of papal policies and concerns during the pontificate of Paul III.[4] A man of diverse interests and abilities, he was, according to these historians, a versatile and creative problem-solver, attending to innumerable particular issues from preaching and parish organization in the three dioceses he oversaw successively as administrator, to theological disputes and petty infighting at the conciliar congregations of Trent, where he frequently employed humanistic methodology to find solutions.[5] Some of these historians maintained that the exhortations to peace he issued through his ambassadors in the few days of his reign were indicative of the hope and expectation he represented for the restoration of peace throughout Europe.[6] Still others, and indeed the greatest nineteenth-century historians of the papacy, considered Cervini to epitomize the reform spirit prevailing in the mid-sixteenth century in Rome.[7] Had he lived long enough as Pope, he might have begun, these writers maintained, the real and effective reform of Church practice which was the hope of so many. The favourable view shared by all these authors was made explicitly hagiographical in 1962 by Stanley Morison, in an article

---

from Gubbio to Rome via Perugia between his receipt of the notice of Julius III's death (which he could not have received before 24 March 1555 and more likely did not receive before 25 March) and 1 April 1555. Loreto is some 85 km. east of Gubbio and it would have been virtually impossible for him to go there before travelling to Rome by the route his companions described, all before the first of April. The journey is described in a letter from Galieno Benci to Cervini's half-brother, Alessandro, on 3 April 1555 (see *C. Cerv.*, lii. 10[r]).

[4] J. Pogianus, *Epistolas et orationes* (Rome, 1762–68), i. 105–6, 116–17; C. Baronius, *Annales ecclesiastici* (Lucca, 1738–57), xxxiii. 550–3.

[5] He was successively administrator of the dioceses of Nicastro (1539–40), of Reggio Emilia (1540–4), and of Gubbio (1544–55). On these administrations, see M. Sartio, *De episcopis eugubinis* (Pesaro, 1755), 222–5; U. Pesci, *I vescovi di Gubbio* (Perugia, 1919), 108–18; C. Saccani, *I vescovi di Reggio Emilia* (Reggio Emilia, 1902), 118–20; G. Alberigo, *I vescovi italiani al Concilio di Trento* (Florence, 1959), 140–4, 150–2; G. Costi, in N. Artioli (ed.), 'L'episcopato a Reggio Emilia (1540–44) del cardinale Marcello Cervini poi papa Marcello II', *In Memoria di Leone Tondelli* (Reggio Emilia, 1980), 203–29. For his humanistic methodology at Trent, see G. Calenzio, *Saggio di storia del concilio generale di Trento sotto Paolo III* (Rome and Turin, 1869), 51, and the records of the debates at Trent and the portions of Cervini's correspondence published in *Concilium Tridentinum* (Freiburg, 1901–30), henceforth cited as *CT*.

[6] G. Piatti, *Storia critico-cronologica de' Romani Pontefici e de' generali e provinciali concili* (Naples, 1765–8), 10, 320–1; G. Brilli, *Interno alla vita e alla azioni di Marcello II pontefice ottimo massimo, orazione* (Montepulciano, 1846), 21.

[7] L. von Ranke, *History of the Popes* (New York, 1966), i. 192; L. von Pastor, *History of the Popes*, vols. vii–xxiv, trans. and ed. R. Kerr (London, 1908–33), xiv. 14–15, 33.

entitled 'Marcello Cervini, Pope Marcellus II: Bibliography's Patron Saint.'[8]

One of the earliest and certainly one of the more interesting, if brief, assessments in this tradition was provided by Girolamo Seripando (c.1493–1563) the cardinal and general of the Augustinians who worked closely with Cervini on theological issues at the early sessions of the Council of Trent. No less a historian than Baronius utilized Seripando's assessment in his own description of the election of Cervini.[9] In a letter to the Bishop of Fiesole, Pietro Camaiani, Seripando explained that he had long urged Cervini to seek the pontificate actively, but that this course of action had been unthinkable to him.[10] Since in fact he later attained the office, Seripando reasoned, then certainly Cervini must have taken some sort of role in the conclave which elected him. 'Either he sought the pontificate, or he did not', Seripando said. If he did seek it, such a move would have been contrary to all his past actions, and in addition, he must have been unaware of the opposition he would face in the curial reform he intended, Seripando maintained. 'If he did not [seek the papacy],' Seripando concluded, 'then undoubtedly he was not a man, but an Angel sent to earth, rising above not only men of his own age, but above those of many past ages as well.'[11] Seripando went on to suggest that in addition to his refusal to lobby for his own election, Cervini was determined to assume the papal office on one condition only: if he were chosen for the qualities he had always represented. He added that Cervini himself was the only one in the conclave to vote for any other candidate, while the other thirty-eight cardinals were in complete agreement.[12] Seripando saw in the elevation of Cervini to the

[8] S. Morison, 'Marcello Cervini, Pope Marcellus II: Bibliography's Patron Saint', *Italia medioevale e umanistica*, 5 (1962), 301–18.

[9] Baronius, *Annales ecclesiastici*, xxxiii. 551. On Baronius as a historian, see C. Pullapilly, *Caesar Baronius, Counter-Reformation Historian* (Notre Dame, Ind., 1975), 33–48, 144–73; *DBI*, s.v. 'Baronio, Caesare' by A. Pincerle; E. Cochrane, *Historians and Historiography in the Italian Renaissance* (Chicago, Ill., 1981), 457–63.

[10] 'Ego nunquam mihi persuadere potui Cardinalem Sanctae Crucis ad Pontificatum pervenire potuisse, immo rem impossibilem existimabam, sicut etiam, cum de ejusmodi rebus loquebamur, aperte afferebam', Baronius, *Annales ecclesiastici*, xxxiii. 551. On Camaiani (Bishop of Fiesole 1552–66), see *DBI*, s.v. 'Camaiani, Pietro' by G. Rill.

[11] 'Ratio satis clara erat, omnesque ejus actiones, quibus ad Pontificatum solemne est pervenire, contrariae mihi videbantur, quare sic in me ipso ratiocinari solebam, vel iste Pontificatum optat, vel non; si eum optat, ignorantior est, et judicio prae omnibus carens, eorum praecipue, qui in curia morantur Romana cum non advertat cuncta omnino agi contra suum consilium; si non optat, certe, homo non est, sed Angelus carne indutus omnes, non solum hujus saeculi homines, sed etiam multos praeteritorum saeculorum superans' (Baronius, *Annales ecclesiastici*, xxxiii. 551).

[12] 'Et quia familiarem ego cum illo jamdiu consuetudinem habens, nunquam ab aliis talem existimatum fuisse percepi, necessario concludebam; vel ipse Pontificatum non curat,

pontificate 'a special dispensation of divine grace, which had directed the votes to the man who would "save Israel" '.[13] In a passage in his diary covering April 1555, Seripando indicated that Cervini was 'given by God in order to guard the Church from every evil'.[14]

This account of the conclave is at best an oversimplification and at worst blatantly misleading, since Gian Pietro Caraffa, the man who was to succeed Cervini as Pope Paul IV, was leading in the early voting, and his shift in favour of Cervini was the turning-point which induced the others to do the same.[15] None the less Seripando's characterization of Cervini as 'an Angel sent to earth' is important for other reasons. By describing him in such terms, Seripando is almost certainly linking Cervini with the expectation of the coming Angelic Pope which, inherited from medieval Joachimism, was still, as we have seen, widespread in sixteenth-century Italy. If this connection was indeed made by Seripando it illuminates the programme followed by Cervini during his brief pontificate and goes a long way towards explaining the hagiographical assessment of Cervini here outlined.

The model here suggested for Cervini, while taking its genesis from Joachim of Fiore's vision of a church-centred 'new age',[16] had been formed under the influence of Franciscan Spirituals, popularized through the series of Pope Prophecies and presented as 'instant prophecy' to sixteenth-century churchmen in the pseudo-Amadean *Apocalypsis Nova*. In the shaping of this angelic model stress had been laid on the renunciation of simony, political intrigue, and worldly pomp in favour of a life of poverty, spiritual devotion, and humility. According to the pseudo-Amadeus the future pastor would observe ancient laws, bring back justice, and establish peace and unity to the astonishment and admiration of all.[17] The popularity of the *Apocalypsis Nova*, together with the

---

vel ad eum pervenire non vult, nisi per suam propriam viam, adeoque constans erat in suis consiliis, et in augustissima et inflexibili semita justitiae, et bonitatis in qua adeo firmum conspiciebam, ... Nunquam credidi fieri posse ut ad Pontificatum efferretur, mirum fuit triginta octo Cardinales in eum qui trigesimi noni numerum complebat summa animorum consensione, nulloque aemulo invento, suffragia aperta contulisse' (ibid.).

[13] Pastor, *History of the Popes*, xiv. 33–4.

[14] G. Seripando, *Commentarii de vita sua*, CT ii. 448.

[15] F. Baumgartner's 'Henry II and the Papal Conclave of 1549', *Sixteenth-Century Journal*, 16 (1985), 301–14 gives some information about the conclave that elected Cervini. See also *CT* ii. 250–3; G. Manucci, 'Il conclave di papa Marcello', *Bullettino senese di storia patria*, 27 (1920), 94–103.

[16] B. McGinn, *The Calabrian Abbot: Joachim of Fiore in the History of Western Thought* (New York and London, 1985), 112.

[17] A. Morisi, *Apocalypsis Nova: Richerche sull'origine e la formazione del testo dello pseudo-Amadeo* (Studi Storici, 77; Rome, 1970), 6–7, 15–24.

famous *Vaticinia* (which went through at least twenty-three editions between 1515 and 1670)[18] kept the expectation of this holy pastor at fever pitch.

More particularly this expectation seems to have been fostered in certain religious circles where serious-minded churchmen looked for an immediate fulfilment of the prophecy, or even possibly put forward a candidate. In Rome, at San Pietro in Montorio, the Franciscan Amadeite congregation kept alive the message of the mysterious book represented in painting beneath the altar in their Borgherini chapel.[19] At the Aracoeli the Franciscan, Pietro Galatino, was assembling all the prophecies of the Angelic Pope for his friends.[20] At Santa Sabina, Professor Jungić suggests, the circle of friends around Cardinal Sauli saw in him the possible fulfilment of the prophecy.[21] Outside Rome, as we have seen, a group of Augustinian hermits in Venice were busy in the second and third decades publishing the genuine and spurious works of Joachim. It is here that we meet the closest link between Seripando and prophetic circles, for Egidio of Viterbo, Prior General of the Augustinians at that time, was in touch with the Venetian group[22] and Seripando was his disciple.[23]

Seripando had taken the habit of the Augustinians in 1507 and had received private tutoring in Greek from Egidio, in addition to accompanying him on a journey to Rome in 1510. He followed Egidio's interest in humanism and Neoplatonism, taking up his mentor's place in the Pontanian Academy in Naples when Egidio went to Rome to reassume direction of the Augustinians in 1523.[24] Seripando followed his master by serving as general of the Order from 1539 and by attempting the reform of the group along the same lines as those begun by Egidio.[25] Thus he centred his efforts on the renewal of the common life in the Order and on the reinvigoration of studies. He hoped that the attainment of these two goals would effectively bring back the Order's most ancient ideals. But besides this idealistic, even ideologically based, re-

---

[18] B. McGinn, 'Angel Pope and Papal Antichrist', *Church History*, 47 (1978), 171.
[19] See above, Jungić, Ch. 17.
[20] See above, Rusconi, Ch. 8.
[21] See above, Jungić, Ch. 18.
[22] See above, Reeves, Ch. 5.
[23] H. Jedin, *Papal Legate at the Council of Trent: Cardinal Seripando*, trans. F. Eckhoff (St. Louis, 1947), 7–10, 56–68. The most recent work on Seripando is R. Abbondanza, *Girolamo Seripando tra evangelismo e riforma cattolica* (Naples, 1981).
[24] Jedin, *Papal Legate*, 61.
[25] Ibid. 118–26.

form effort, Seripando, like his mentor, could be very practical, seeking through a series of inventories and other information-gathering procedures, to end wasteful and irresponsible administration of the Order's property. He saw through, for example, the failure of the Augustinian confraternity in Montepulciano to fulfil its obligation to provide the monastery in that town with food and clothing. He knew it instead to be an attempt on the part of the confraternity to seize possession of property to which it had no right.[26]

Most important, Seripando imitated his master in his work as adviser and theologian in the service of the Holy See. He attended the Council of Trent in his capacity both as theologian and as general of the Augustinians. Although at times his ideas clashed with those of the papal legates at the assembly (and therefore with those of Cervini), he affected intimately the outcome of a number of decrees. He insisted, for example, that the terminology of the various schools be avoided in the discussions of the theological issues the council attempted to solve, and the final form of the decree on original sin is marked by such avoidance.[27] Just as Egidio in his work as adviser to the Holy See urged strong action on the part of contemporary Popes and sought to identify possible leaders in the reform of the church, so did Seripando. He maintained for example, that the nepotism of Paul III in granting the territories of Parma and Piacenza to his son Pierlugi, and in attempting to gain Milan for his grandson Ottavio were actions which would ultimately undermine the assessment of the Pope's desire for real reform.[28] He regarded the pontificate of Paul's successor Julius III as a failure, since he neither spoke about reform nor carried it out, and especially because he failed to effect reform in the papal curia: a strong hand, abolishing abuses and closing loopholes, was necessary in the current situation, he believed.[29]

None of this suggests the vision of the Angelic Pope, yet it was Seripando who urged Cervini to seek the pontificate and it was Seripando, as we have seen, who described him in words which come close to claiming

---

[26] The story is related in a letter of 21 Dec. 1551 from Paulo Mancini da Siena, an Augustinian priest, to Cervini (*C. Cerv.*, xlv. 5^{r-v}). For Seripando's assessment of the situation, see Jedin, *Papal Legate*, 204–5.

[27] 'Non est meo quidem iudicio doctrina haec a scholis petenda, in quibus ardue modum et spinose tractantur quaestiones, utiles ille quidem ad exercenda ingenia adipiscendam sapientiam huius seculi, verum ad cognoscendam sapientiam in mysterium absconditam et erudiendum Christi populum ad iustitiam non satis idoneae.' *CT* xii. 614. See also Jedin, *Papal Legate*, 247–392.

[28] Jedin, *Papal Legate*, 248.

[29] Seripando, *Commentarii de vita sua*, *CT* ii. 448–9; Jedin, *Papal Legate*, 491–2, 497.

him as this divinely sent agent. Although there is no more direct evidence in Seripando's writings of his interest in the explicitly eschatological elements in the image of the Angelic Pope, this view of Cervini can be seen as part of the same tradition as Egidio's expectations with respect to Julius II, Leo X, and later Clement VII. In addition, it might be argued, Seripando had reason to make just that suggestion about Cervini. A number of individuals attempted to predict the outcome of the conclave in which he was elected. Ludovico Bondoni de Branchis Firmani, the papal master of ceremonies, related a story of one such prediction in his diary of the conclave. His unnamed conclavist informed him that 'modern soothsayers' predicted that after the fourth day of the conclave the Cardinal of Santa Croce would be elected, but would live a 'very short time and before long the college would find itself in a conclave once more'.[30] And while that prediction would literally come true, it should be noted that more interesting prophecies, ones involving imminent revolution in Church administration, were still being generated and circulated, such as that of Celio Secondo Curione. In his 1554 tract *De Amplitudine beati regni dei dialecti*, he expressed his belief that reform and the renovation of the Church were beginning in his own time as a diffusion of 'light', which he defined as a combination of humanistic understanding and evangelical truth.[31] A tantalizing, yet ultimately unsatisfying, suggestion of Cervini's eschatological role as Marcellus II serves to underline the importance of Seripando's intimation of the 'angelic' quality of Cervini. In a letter written on 23 April 1555 Ottavio Graccho informed Alessandro Cervini of recent happenings in Rome. On that day, the new Pope regained a bit of his failing health (for the last time, it seems) and Graccho suggested that this gift from God was proof that Galieno Benci, the medical doctor in attendance at the conclave of Julius III and a relative of Cervini, was right—the coming of

---

[30] 'Dum Illustrissimi cardinales essent clausi in scrutinio, cum socius meus et ego essemus custodientes portam scrutinii et simul confabulantes de futuro pontifice, prefatus socius dixit mihi: "Hodie secundum prophetas modernos, qui dixerunt quod post quartum diem introitus conclavis creabitur pontifex cardinal de Santa Cruce, qui parum vivet, et cito redibitur in conclave,—hodie quintus est dies et habebimus pontificem."' Ludovico Firmani, *Diaria*, *CT* ii. 507.

[31] M. Reeves, *Influence of Prophecy in the Later Middle Ages: A Study in Joachimism* (Oxford, 1969), 482. On Curione (1503–69), see also *DBI*, s.v. 'Curione, Celio Secondo' by A. Biondi. Biondi's view of the *De amplitudine* is focused on the idea that the text constitutes a defence of 'nicodemismo', the outward practice of Catholicism with simultaneous assent to the basic tenets of the Protestant reformation, allegedly followed by numerous humanists in this period.

God in an apocalyptic sense was near.[32] But beyond such prophecies, there were other reasons for Seripando's suggestion.

To begin with there was no doubt at all that whoever was elected Pope in the conclave following the death of Julius III on 23 March 1555 would be called upon to provide leadership for the Church in a time of crisis. The papacy was, according to one Italian scholar, in the midst of a fundamental reconstruction, developing for itself a centralized bureaucracy, a foreign policy based on the pursuit of a balance of power, and an image of both spiritual and temporal power which became the prototype of the absolutist, territorial state in Europe.[33] The doctrinal and disciplinary issues dividing Christendom were far from settled, and the council, called to provide the Catholic solution to those issues, was itself divided and currently in a state of suspension which had already lasted nearly three years. Many individuals, moreover, believed that Cervini was just the person to provide such leadership. They delighted in his election and in his statement when asked what he would assume as his papal name. Cervini reportedly said, 'I was Marcellus, I will be Marcellus: the pontifical office will change neither my name nor my ways.'[34] One of those delighted individuals was Reginald Pole, who not only addressed a letter of congratulations to the new Pope on 28 April, but also reiterated his genuine confidence that Cervini would commence serious and proper reform of the Church in a letter to Queen Mary of England.[35] The election had brought him the 'greatest joy', Pole said, because he always felt closely joined to Cervini both in studies and in their general outlook.[36] God, Pole maintained, had always given Cervini the desire to undertake reform and now had given him the necessary power as well, and he expressed his hope that Cervini would be confirmed

---

[32] 'Mi dette la bona nova della sanità di Sua Beatitudine. Entrammo poi in ragionamenti di philosophia speramo come insieme havemo concluso che Dio habia reservato si divino dono in tempo di Sua Santità per il che stamo aspettando la venuta di quella quale secondo mi dice Monsignor Galieno (Benci) sarà presta' (Ottavio Graccho to Alessandro Cervini, 23 Apr. 1555, C. Cerv. lv. 28^{r-v}). Benci is identified as a 'medico' in attendance at the earlier conclave in a letter by Giovanni Battista Cervini to Romolo Cervini, 2 Dec. 1549, C. Cerv. l. 159^r; also printed in CT ii. 530.

[33] P. Prodi, Il sovrano pontefice: Un corpo e due anime: la monarchia papale nella prima età moderna (Bologna, 1982), available in English as The Papal Prince, One Body and Two Souls: The Papal Monarchy in Early Modern Europe, trans. S. Haskins (New York, 1987), 37–42.

[34] 'Marcellus eram: ero Marcellus: nec mores mutabit nec nomen meum pontificatus' (cited by J. Pogianus, Epistolae, i. 125).

[35] Pole's letters are published in H. Lutz, ed., Nuntiaturberichte aus Deutschland, erste Abteilung 1533–1559, 15; Friedenslegation des Reginald Pole zu Kaiser Karl V und König Heinrich VIII (1533–1556) (Tübingen, 1981), 256–8.

[36] Ibid. 258.

with the good health necessary to carry out such work in this office. After all, Pole asserted, Cervini was always very worthy of the position.[37] And indeed, the new Pope rapidly began to undertake ecclesiastical reforms.

Cervini embarked upon a programme after his papal election which was much in line with what was to be expected from the Angelic Pope, that is, a revolutionary change from the policies and pursuits of his predecessors. He immediately set to work to initiate a reform which might produce real results. In his diary covering the days immediately after Cervini's election, Angelo Massarelli, the secretary of the Council of Trent and Cervini's long-time companion and collaborator, wrote of the Pope's desire to eliminate abuses in the Church and to restore it to pristine health and purity.[38] To that end, Massarelli maintained, he provided leadership by example. Marcellus made it clear that he had no desire for a lavish coronation ceremony, and the money that was saved as a result purchased large amounts of food and fuel for the poor of Rome. In the Holy Week ceremonies shortly after his election he demonstrated piety and simplicity, not only through his personal attendance, which was considered unusual in and of itself, but also by going on foot to and from the services. He called Massarelli before him less than forty-eight hours after the election and assigned to him the task of collecting all the materials pertaining to reform drawn up by the commission for reform established under his predecessor.

Most significantly Marcellus demonstrated disdain for the all too-well-established practice of enriching blood relations with benefices, cardinalates, curial administrative offices, and 'reservations' of money for their personal use. Although he did appoint two family members to the posts immediately responsible for insuring his personal safety,[39] Marcellus forbade his half-brother and future biographer Alessandro and his wife even to make the journey to Rome.[40] Similarly, he was incensed when his two nephews, Ricciardo and Herennio, arrived in Rome expecting to benefit from their uncle's new position. He ordered them to go home at once, without so much as granting them an audience.[41] He sought to cut

---

[37] Ibid. 256–7.
[38] *CT* ii. 254–61.
[39] Giovan Battista Cervini, appointed governor of Castel Sant' Angelo, and Biagio Cervini, appointed captain of the Vatican guard (ibid. 258; see also Pastor, *History of the Popes*, xiv. 41).
[40] *CT* ii. 261.
[41] Ibid.; see also Moroni, *Dizionario*, 42, 245.

expenses wherever possible in the curial household, kept a simple table, and demanded virtuous and modest behaviour from all those living there.[42] In this he maintained the approach he had taken in his household as cardinal, as well as the attitude towards clerical demeanour, which he demonstrated in his administration of the dioceses entrusted to him during his career.[43] A frugal, humble, honest, and unworldly pastor, many believed, had finally ascended the papal throne.

Furthermore, Marcellus demonstrated his determination to initiate curial and clerical reforms according to the model he had frequently recommended to Pope Paul III earlier in his career as adviser and legate: direct and effective papal action in the offices and commissions under his control. This, too, squares with the image of the Angelic Pope as one who would advance the papacy beyond its simoniacal and hypocritical past, leading a new age of genuine spirituality. When Holy Week services concluded with Easter Sunday on 14 April, Marcellus immediately turned his attention to practical curial reforms which would eliminate the fundamental sources of traditional abuses. He called before him the auditors of the Rota, as Vittore Soranzo reported in a letter to Cervini's family in Montepulciano, and insisted that they ignore such typically influential factors as familial relations, past service, and all other worldly considerations when cases were argued before them, and that they seek instead simple justice. He also added, according to Soranzo, that 'he does not think auditors should hold bishoprics'.[44] A position in the curia such as that of an auditor, which required regular residence in Rome, was, in his opinion, incompatible with the pastoral responsibilities of a bishop. Likewise, Marcellus summoned the Datarius, ordering 'that no benefices be distributed without his knowledge'.[45] In the future, he would personally ensure that qualified candidates were appointed to

---

[42] *CT* ii. 257, 261–2; Pastor, *History of the Popes*, xiv. 37–40.

[43] Besides the works cited above regarding his episcopal administration, the diary of Antonio Lorenzini, Cervini's vicar, describing the visitation of Reggio in 1543, in the Episcopal Archive of the diocese, and the constitutions resulting from the visitation, were published in part by A. Mercati, *Prescrizioni del culto divino nella diocesi di Reggio Emilia del vescovo Cardinale Marcello Cervini* (Reggio Emilia, 1933). The constitutions regulated, among other things, clerical behaviour, giving precise descriptions, for example, of the kinds and sizes of arms clerics were permitted to carry, and of the fines to be imposed in the future, if such regulations, and others regarding blasphemy, etc. were not followed.

[44] 'Ha sua Santità chiamati li auditori di Rotta et ditoli che avertiscano a la iustitia et non ascoltino né parente né servitor suo, né li habbino un rispetto al mondo et expediscano et che non li par bene li auditori di Rotta habbino vescovati' (Vittore Soranzo to Alessandro Cervini, 20 April 1555, C. *Cerv.*, lii. 58[r–v]).

[45] 'Ha chiamato il Datario et comesso che non dia benefitio senza sua saputo' (ibid. 58[v]).

beneficed positions. He also made known his desire, according to Soranzo, to turn over the Penitentiary to Gian Pietro Caraffa.[46] Hubert Jedin, the great historian of the Council, summed up the situation this way: 'There is no possibility of doubt: Marcellus II had decided finally to act.'[47] He attempted to undertake real reform of the fiscal abuses in the curia which neither his predecessors nor his immediate successors were ever serious about eliminating.[48] It could be argued that the abolition of abuses and administrative loopholes which Seripando hoped for in the pontificate of Julius III had begun. And if we are to believe the testimony of Soranzo and Massarelli, Cervini did stand opposed to the aggrandizement of papal pretensions to secular authority by underplaying the importance of his coronation as Pope, while he sought to concentrate his adminis- tration around religious issues from a pastoral, not an absolutist perspec- tive, by attending to changes in key curial offices.[49]

The importance Cervini attributed to this reform work is underlined by the state of his health, because whatever activity he undertook during those days in 1555, he must have rapidly come to the conclusion that it would be his last. Antonio Lorenzini, Cervini's sometime episcopal vicar, and in 1555 his conclavist, reported frequently to the family on the state of the Pope's health. 'I have been so occupied after the creation of the pope that I have scarcely had time to breathe', he wrote on 13 April. 'The Holy Father similarly', he added, 'is so breathless from the conclave and from the visits that it is a pity to see him.'[50] The first indication of real illness in the new Pope came just ten days after the election. 'Since this morning His Holiness is not feeling too well, and he has had some fever', Lorenzini wrote. He added, 'I believe it will not be anything dangerous.'[51] The fever and *catharro* of the Pope continued

---

[46] 'Sua Santità ha voglia dare la penitentiaria al Reverendissimo Napoli . . .' (ibid.).

[47] Jedin, *Council of Trent*, iv. 25.

[48] B. Hallman, *Italian Cardinals, Reform, and the Church as Property* (Berkeley, Calif., 1985), 162–8.

[49] For Prodi, an 'absolutist' perspective can be seen in such actions as the use of ecclesi- astical record-keeping as a kind of 'police action', in appointments made to ecclesiastical office, and in the use of such institutions as the Datary for the purpose of augmenting specifically secular administration and prestige. Pastoral concerns were downplayed, if not ignored altogether (Prodi, *The Papal Prince*, 53–4, 97–103, 123–56).

[50] 'Sono stato tanto occupato doppo la creatione di Nostro Signore che hauto pur tempo di respirare. Il Papa similmente dalle cappelle, et dalle visite è tanto affannato che è una compassione a vederlo' (Antonio Lorenzini to Alessandro Cervini, 13 April 1555, *C. Cerv.*, lii. 4ʳ).

[51] 'Nostro Signore da stamattina in qua non sta bene et non li manca un poco di febre . . . io credo che non sarà cosa di periculo' (Antonio Lorenzini to Alessandro Cervini, 29 April 1555, ibid. 5ʳ).

to worsen, and on the twenty-second Lorenzini wrote, 'it is causing fear'.[52] Marcellus lived only until the first of May, and died sometime very early that morning, thus spending only twenty-two days as Pope.

These are perhaps some of the reasons why Seripando came to describe Cervini after his death as an 'angel' come to earth, and why he wrote still later to Guglielmo Sirleto, another of Cervini's long-time associates and the man who succeeded him as papal librarian, that 'I never found anyone who had the kind of relationship to me that I had to that blessed man, so that I can say that I served him without seeking any glory for myself, unless it be with God.'[53] It seems clear that the suggestion expressed by Seripando that Cervini could have been the long-awaited Angelic Pope, heightened by his undoubtedly spiritual, pastoral, and reforming emphasis during his few days as Pope, and his sudden death, helped to create the glowing, even hagiographical, assessment of the man which characterized sketches of him over four centuries. That luminous picture is only in the present day being re-examined.

Studies of Italian religious life in the sixteenth century today tend to maintain that prelates in the period were divided into two mutually exclusive and warring camps: the *spirituali*, who were motivated by their reading of the New Testament and seemed conciliatory towards Protestants, and the *intransigenti*, who were traditionalists in the pejorative sense and who by an emphasis on the need for the Inquisition began the work of the 'Counter Reformation'. This ideological dichotomy is considered a useful characterization of the period by a number of scholars on both sides of the Atlantic.[54] It is the identification of Cervini, along with such personages as Michele Ghislieri and Gian Pietro Caraffa as members of the *intransigenti* that has begun to demolish the hagiographical assessment of Cervini which had lasted for so long.[55] Undoubtedly, the

---

[52] 'Io so certo che Vostro Signore hara hauto dispiacere della mia precendenti nella quale gli davo avviso delle indispositione di Nostro Signore quale indispositione per deve il vero da principio ci fece paura' (Antonio Lorenzini to Alessandro Cervini, 22 April 1555, ibid. 7ᵛ).

[53] Quoted in D. Gutiérrez, *The History of the Order of St. Augustine*, trans. J. Kelly (Villanova, Pa., 1979), i. 173.

[54] See M. Firpo and D. Marcatto, *Il processo inquisitoriale del Cardinal Giovanni Morone*, 4 vols. (Rome, 1981–7); P. Simoncelli, *Evangelismo italiano del Cinquecento* (Rome, 1979); A. Schutte, 'The *Lettere Volgare* and the Crisis of Evangelism in Italy', *Renaissance Quarterly*, 28 (1975), 639–88; E. Gleason, 'On the Nature of Sixteenth-Century Italian Evangelism', *Sixteenth-Century Journal*, 9 (1978), 3–25; W. Bangert, *Claude Jay and Alfonso Salmeron, Two Early Jesuits* (Chicago, 1985), 79; D. Fenlon, *Heresy and Obedience in Tridentine Italy: Cardinal Pole and the Counter Reformation* (Cambridge, 1972), pp. ix–x, 137–60, 220–85; C. Stinger, *The Renaissance in Rome* (Bloomington, Ind., 1985), 328–30.

[55] Some historians suggest that Cervini is to be understood as a member of the *intransigenti* by generally equating that group with the list of those who participated in the

earlier assessment of Cervini should be under suspicion as an oversimplification of a complex individual, as should likewise be the characterization which focuses only upon his activity as a member of the Inquisition and brands Cervini as 'intransigent', or indeed any characterization which centres around only one aspect of his busy life. Cervini did actively participate in the restored Roman Inquisition and did frequently recommend financial enticements to motivate prelates to support the papal programme at Trent, yet he also demonstrated genuine pastoral concern for those in his diocese, in line with the so-called Catholic reform identified by scholars from Joseph Kerker and Ludwig von Pastor to Hubert Jedin and beyond, notwithstanding the programme of several leading Italian scholars to demolish that concept today.[56] And while the polite diplomatic correspondence between Cervini and other prelates in this period might reflect conventional ambassadorial language rather than real sentiment, there is other evidence to suggest that personages on either side of this supposed ideological dichotomy were in reality collaborators who supported one another. One of the best indications of this comes from the correspondence of Pole following the conclaves of 1555. To avoid jumping to the conclusion that the words Pole addressed to Queen Mary on the occasion of Cervini's election were merely conventional expressions of praise directed to any new Pope and devoid of real feeling, all that is necessary is to look at the letter Pole wrote about five weeks later, after the election of Gian Pietro Caraffa as Paul IV. It is congratulatory, to be sure, but the words of congratulations come at the end of a long paragraph reciting the evils facing the Church and extolling the reform work Caraffa had done before, in an obvious reference not to Caraffa's recent inquisitorial escapades, but to his earlier sponsorship and co-inspiration of the Theatine order. Pole suggested that the actual undertaking of reform work is the true object of any real praise for the papacy.[57] We may note here that Cervini was even the author of a set

Roman Inquisition after 1542 (for example Fenlon, *Heresy and Obedience*, 73). Firpo places Cervini explicitly in this group in his review of Hallman's book (n. 48), which appears in *Sixteenth-Century Journal*, 18 (1987), 117–18.

[56] P. Simoncelli, in his *Evangelismo italiano* and 'Inquisizione Romana e riforma in Italia', *Rivista storica italiano*, 100 (1988), 1–125, and Firpo, especially in his *Il processo inquisitoriale* are leading this endeavour.

[57] 'Caetera enim, quae summus hic pontificatus honor secum affert (quod nemo melius Sanctite Vostri novit), utcunque ad speciem pulchra atque praeclara, plus quidem habent, hoc praesertim tempore, quo omnia in tot tantisque difficultatibus versantur, cur doleamus, quam gratulemur. Hoc autem reformationis opus, etsi complures habet propter iniquitatem temporum difficultates, tamen eiusmodi est, ut ei, qui hoc ex animo cupiat et in eo laboret, ipsa cura futura sit iucundissima, et piis omnibus eo maiorem allatura causam Sanctiti Vostri gratulandi, quo maiori curae id illi esse viderint' (Reginald Pole to Paul IV, 6 June 1555, pub. in Lutz, *Nuntiaturberichte*, 267).

of instructions to preachers which is in many respects similar to that written by one of the most commonly acknowledged of the *spirituali*, Gasparo Contarini.[58] And in the light of Giuseppe Alberigo's classic study of the Italian episcopacy in the Tridentine era, it seems irrefutable that prelates like Gian Matteo Giberti, Contarini, Cervini, and others did experiment with and undertake practical reforms on the diocesan level which were converted through the decrees of the Council of Trent into a programme for the Church at large, regardless of what term we utilize to refer to that experimentation and conversion.[59] But beyond all this, it is more important, it seems to me, to understand from whence the glowing vision of Cervini, whether or not it reflects 'Catholic reform', might have come.

It was a combination of a number of factors which most likely led to the creation of that 'vision'. The first of these factors was the strong, continuing interest in the sixteenth century in the Joachimist and pseudo-Joachimist prophecies, especially regarding the future Angelic Pope and what actions that figure might take to inaugurate a new spiritual age. The very survival of this strain of Joachimist thought might reflect the frustration felt by those desirous of real reform in this period concerning a papacy which, as Prodi suggests, was reconstructing itself into a territorial and absolutist state. Dissatisfaction over what the papacy was becoming in the sixteenth century could have been just as powerful an influence for the continuation of hope and active expectation of an Angelic Pope as the situation in the medieval Church between the twelfth and fourteenth centuries, which originally led to the identification of this figure. This possibility is extremely interesting, given the position which people like Seripando held—these were members of the papal curia and other persons intimately connected with the ecclesiastical hierarchy, not dagger-brandishing revolutionaries consumed by their interest in astrology, like Benedetto Accolti just a few years later.[60] A

[58] See W. Hudon, 'Two Instructions to Preachers from the Tridentine Reformation', *Sixteenth-Century Journal*, 20 (1989), 457–70.

[59] G. Alberigo, *I vescovi italiani*.

[60] Accolti, the namesake of his cardinal–father, hatched an unsuccessful plot against Pius IV in 1564. He allegedly hoped to liberate Italy and the rest of the world through this deed, since only the removal of Pius from the papal throne would, in his opinion, make room for an Angelic Pope who would bring about the union of the Greek church with Rome, the submission of the Turks, and a perfectly just society. He and two others were executed in 1565 after the plot to kill Pius was revealed by one of the conspirators. Pastor's assessment was that one could only feel 'deep compassion for the deluded man' (see Pastor, *History of the Popes*, xvi. 383–91, 485–95; and *DBI*, s.v. 'Accolti, Benedetto' by E. Massa).

curious detail is relevant here. A note by Seripando in the archetypal copy of Egidio of Viterbo's *Historia* states that it was lost but later found and bought on a stall in Rome by Cardinal Cervini.[61] Without referring to the Angelic Pope, Egidio calls on the Pope to lead the Church into the Tenth Age of *renovatio*. Did Seripando and Cervini respond to that vision?

The second factor was Cervini's embodiment during his brief pontificate of the qualities usually associated with the Angelic Pope: he was a frugal, pious man who began, even though briefly and incompletely, to eliminate the financial practices which debased the medieval and Renaissance papacy, and who, for some, held out the promise of international peace and doctrinal unity. The third was the belief on the part of his contemporaries, and notably Seripando, that Cervini would begin what his predecessors had long avoided: a true and lasting reform of the Church. Seripando and many of Cervini's other colleagues knew that he viewed himself first as a pastor, and this might be the reason they latched on to him as the one who might stand against growing papal centralization and secularization, returning the papacy to its more genuine, apostolic model. Another factor might well have been the death of Cervini itself, untimely by anyone's estimation, which, according to many contemporary observers, snatched away a promising Pope before his time. Still another factor could be the person of Cervini's immediate successor in the pontificate, the fearsome Gian Pietro Caraffa, who ruled as Pope Paul IV from 1555 until 1559. He was more interested in the pursuit of frequently unjustified inquisitorial procedures than in the sort of work Cervini undertook as Pope.[62] These very powerful influences seem to have been active in the contemporary assessments of Pope Marcellus II, a fact which serves to underline the importance of the Joachimist prophecies into the mid-sixteenth century and beyond.

[61] See above, Reeves, Ch. 5 n. 37.
[62] One who suggested precisely this interpretation is none other than Ignatius Loyola. In his 17 June 1555 letter to Manuel Lopez describing the turn of events, a glowing description of the deceased 'holy father Marcellus' is juxtaposed to a short request for prayers that God will make 'the present pope . . . a successful minister' (see W. Young (ed.), *Letters of St. Ignatius Loyola* (Chicago, 1959), 393–5.

CONCLUSION

The studies presented in this book have endeavoured to convey the climate of extraordinary apocalyptic expectation which characterized Julian and Medicean Rome. A climax in history—whether interim or final—was close at hand. The papacy was the prime focus of hope for a supreme Golden Age to follow the greatest tribulation. Although two other prophetic agencies, the World Emperor and the new order of spiritual men, also occupied places on the prophetic stage, in Rome at least the crucial actor to come was the longed-for Angelic Pope. As we have seen, these prophetic hopes invaded the very papal establishment itself, where humanist and classical concepts combined with medieval images in a rhetoric designed to bring these prophetic figures out of the future back into the present. And these intellectual visions chimed with the excitement engendered by apocalyptic preachers and circulating prophecies among the populace. A striking paradox lies in the contradiction between this high apocalyptic mood and the *realpolitik* of the very Popes who listened to the orators and preachers. Certain modern studies have portrayed both the politics and the administrative developments of the papal institution at this time as it steadily moved towards the concept of a modern state. Practical reform, in the sense of cleaning up administration and rooting out abuses, was part of this process. It becomes a central concern in the Council of Trent. By contrast, the programme of reform sketched for an Angelic Pope was always general and visionary. Its striking feature was the chasm it presented between ends and means, for it continually raises the question: how naïve could these humanists and churchmen be in expecting its immediate fulfilment ('iam iam in ianuis tenemus')? Could the two concepts ever meet? In his final study on Cervini William Hudon describes a fleeting moment when there did appear to be a meeting point between the vision and the hard realities tackled at the Council of Trent. The election of Cervini as Pope Marcellus II could still have the effect—for some at least—of confirming a vision brought down to earth. Yet in the short time allowed to him, Pope Marcellus plunged into a series of very practical measures to purify papal administration. The question that remains unanswerable—for him as for earlier Popes—is whether they themselves believed in the angelic role assigned for them. In any case the twenty-two days of Marcellus's rule form probably the last possible occasion in the sixteenth century when rhetoric and reality (to use John Headley's phrase) might have been brought together.

BIBLIOGRAPHY

A. BOOKS

ABBÉVILLE, C. d', *Histoire de la mission des pères capucins en l'isle de Maragnan et terres circonvoisins* (Paris, 1614).

—— anastatic edn., ed. A. Métraux and J. Lafaye (Graz, 1963).

ABBONDANZA, R., *Girolamo Seripando tra evangelismo e riforma cattolica* (Naples, 1981).

ACERBI, A., *'Serra Lignea'. Studi sulla fortuna della 'Ascensione di Isaia'* (Rome, 1984).

ACOSTA, José de, SJ, *De temporibus novissimis* (Rome, 1590). *Acta et decreta sacrosancta secundae generalis Pisani Synodi prout per protonotarios et notarios summarie scripta reperiuntur* (Paris, 1612).

ALBERIGO, G., *I vescovi italiani al Concilio di Trento* (Florence, 1959).

ALBERTIN, J., *De mirabile temporis mutatione ac terrene potestatis a loco in locum translatione* (Geneva, 1524).

ALBERTINI, R. von, *Firenze dalla Republica al Principato. Storia e coscienza politica* (Italian trans.) (Turin, 1970).

AMBROSINI, F., *Paesi e mari ignoti: America e colonialismo europeo nella cultura veneziana (secoli XVI–XVII)* (Venice, 1982).

AMICO, J. d', 'Humanism and Theology at Papal Rome, 1480–1520' (Rochester (NY) University Ph.D. diss., 1977).

—— *Renaissance Humanism in Papal Rome: Humanists and Churchmen on the Eve of the Reformation* (Baltimore, Md., 1983).

ANCONA, P. d', *Le miniature fiorentine* (Florence, 1914).

ANGELUS, PAULUS, *Epistola ad Saracenos cum Libello contra Alcoranum* (1522/3).

—— *In Sathan ruinam tyrannidis* (Venice, 1524).

—— *Mirabile interpretatione di prophetie del fine del mondo* (Venice, 1527).

—— *Expositio novissima . . . supra nonum capitulum Apocalypsis* (after 1526).

—— *Profetie certissime . . . dell' Anticristo* (1530).

—— *Apologia* (Rome, 1537 and 1544).

ANNIO OF VITERBO, *Tractatus de futuris Christianorum triumphis in Saracenos*, alt. *Glosa super Apocalypsim de statu ecclesiae ab anno MCCCCLXXXI usque ad finem mundi* (Cologne, 1507).

—— another edn. (Nuremberg, n.d.).

—— another edn. (Louvain, n.d.).

ANTONIUS DE FANTIS, *Tabula generalis scotice subtilitatis octo sectoribus universam Doctoris Subtilis Peritiam complectens* (Lyons, 1520).

ARMELLINI, M., *Un Censimento della città di Roma sotto il pontificato di Leone X, tratto de un codice inedito dell' Archivio Vaticano* (Rome, 1882).

——*Il Diario di Leone X di Paride de Grassi* (Rome, 1884).

ARNOLD OF SWABIA, OP, *De correctione Ecclesiae*, ed. E. Winkelmann (Berlin, 1965).

ASCARELLI, F., *Annali tipografici di Giacomo Mazzocchi* (Florence, 1961).

ATKINSON, G., *Les nouveaux horizons de la Renaissance française* (Paris, 1935).

BACHELET, X. le, *Auctarium Bellarminianum* (Paris, 1913).

BACON, Roger, OM, *Opus Tertium; Opera inedita; Compendium studi philosophiae*, ed. J. Brewer (Rolls series; London, 1859).

BALIĆ, C., *Joannis Duns Scoti theologiae marianae elementa* (Sibenik, 1933).

——*Testimonia de assumptione beatae virginis Mariae* (Rome, 1948).

BANDINI, A., *Catalogus codicum latinorum Bibliothecae Mediceae Laurentianae* (Florence, 1774–6).

BANGERT, W., *Claude Jay and Alphonso Salmeron, Two Early Jesuits* (Chicago, 1985).

BARONIUS, C., *Annales ecclesiastici* (Lucca, 1738–57).

BASZKIEWICZ, J., 'Quelques remarques sur la conception de Dominium mundi dans l'œuvre de Bartolus', *Convegno commemorativo del VI centenario di Bartolo* (Perugia, 1959), 9–25.

BATAILLON, M., 'La herejìa de fray Francisco de la Cruz y la reacciòn antilascasiana', *Études sur Bartolomé de las Casas* (Paris, 1965).

BAUDOT, G., 'Les missions franciscaines au Mexique au XVIème siècle et les "Douze Premiers"', *Diffusione del Francescanesimo nelle Americhe* (Atti del X Convegno della S.I.S.F.; Assisi, 1984), 123–52.

BAUDRY, L., *La querelle des futurs contingents (Louvain 1465–1475): Textes inédits* (Paris, 1950).

BÄUMER, R. von (ed.), *Lutherprozess und Lutherbann. Vorgeschichte, Ergebnis, Nachwirkung* (Münster, 1972).

BAXTER, C., 'Jean Bodin's Daemon and his Conversion to Judaism', in H. Denzer (ed.), *Verhandlungen der internationale Bodin-Tagung* (Munich, 1973), 1–21.

BAYONNE, E.-C., *Étude sur Jérôme Savonarola* (Paris, 1879).

BELLI, G., *I sonetti* (Milan, 1965).

BENEYTO, J., *España en la gestación historica de Europa* (Madrid, 1975).

BENZ, E., *Ecclesia spiritualis* (1934; repr. Stuttgart, 1964).

BERNALDEZ, A., *Historia de los Reyes Catolicos, don Fernando y dona Isabel* (Seville, 1870).

BERNARDINO OF FELTRE, *Sermoni do Bernardino da Feltre* (Milan, 1966).

BERTHIER, A., *L'Église de Sainte-Sabine* (Rome, 1910).

——*Le Couvent de Sainte-Sabine à Rome* (Rome, 1912).

BIANCONI, A., *L'Opera delle Compagnie del Divino Amore nella riforma Cattolica* (Città di Castello, 1914).

BIGNAMI-ODIER, J., *Études sur Jean de Roquetaillade* (Paris, 1952), rev. edn. *Histoire littéraire de la France*, 41 (Paris, 1981), 75–284.

BINAGHI, M., 'L'immagine sacra in Luini e il circolo di Santa Marta', *Sacro e profano nella pittura di Bernardino Luini* (Milan, 1975), 49–76.

——'Bernardino Luini. Affreschi dalla Cappella di S. Giuseppe in Santa Maria della Pace a Milano', *Pinacoteca di Brera. Scuole lombarda e piemontese 1300–1535* (Milan, 1988), 234–9.

BIZARI, P., *Historia Genuensis* (Antwerp, 1579).

BODIN, J., *De la demonomanie des sorciers* (Paris, 1587).

BONNEFOY, J., *Le vénérable Duns Scot, Docteur de l'Immaculée-Conception; son milieu, sa doctrine, son influence* (Rome, 1960).

BONZI DA GENOVA, P., *Teologia Mistica di S. Caterina de Genoa* (Genoa, 1960).

BOSCO, F., 'Il Ritratto di Nicolo della Torre Disegnate da Lorenzo Lotto', in P. Zampetti and V. Sgarbi (eds.), *Lorenzo Lotto. Atti del Convegno internazionale di studi per il V centenario della nascità* (Asolo, 1980), 313–35.

BRANDI, K., *The Emperor Charles V: The Growth and Destiny of a Man and of a World-Emperor*, trans. C. Wedgewood (London, 1939; repr. 1967).

Brevis Historia Ordinis Fratrum Praedicatorum Auctore Anonymo, 1367, in *Veterum Scriptorum et Monumentorum Historicorum Dogmaticorum, Moralium, Amplissima Collectio* (Paris, 1729).

BREZZI, P., *Storia degli Anni Santi* (Milan, 1975).

BRIDGE, A., *Suleiman the Magnificent: The Scourge of Heaven* (New York, 1983).

BRIDGET, St, *Revelationes S. Brigittae . . . a Consalvo Duranto illustratae* (Antwerp, 1611).

BRILLI, G., *Intorno alla vita e alla azioni di Marcello II pontefice ottimo massimo, orazione* (Montepulciano, 1846).

BRIZENUS, I., *Prima pars celebriorum controversiarum in Primam Sententiarum Joannis Scoti Doctoris subtilis Theologorum facile Principis* (Madrid, 1642).

BRUSCHI, A., *Bramante architetto* (Paris, 1969).

BURDACH, K., and PIUR, P., *Vom Mittelalter zur Reformation* (Berlin, 1912–29).

BUTTERS, H., *Governors and Government in Early Sixteenth-Century Florence 1502–1519* (Oxford, 1985).

BUTZEK, D., *Wie kommunalen Repräsentationstatuen der Päpste des XVI. Jahrhunderts in Bologna, Perugia und Rom* (Bad Honnef, 1978).

BZOWIUS, A., *ΠΡΕΣΒΕΙΑ sive legatio Philippi III. et IV . . . de definienda controversia Immaculatae Conceptionibus B. Virginis Mariae* (Louvain, 1624).

——*Annalium ecclesiasticorum . . . continuatio* (Cologne, 1627).

——*Immaculatae Conceptioni B. Mariae Virginis non adversari eius mortem corporalem opusculum* (Rome, 1655).

CALENZIO, G., *Saggio di storia del concilio generale di Trento sotto Paolo III* (Rome and Turin, 1869).

CAMBI, G., *Delizie degli eruditi toscani*, ii (Florence, 1786).

CANTIMORI, D., *Eretici italiani del Cinquecento: Ricerche storiche* (Florence, 1967).

—— 'Niccolo Machiavelli: il politico e lo storico', in E. Cecchi and N. Sapegno (eds.), *Il Cinquecento; Storia della Letteratura italiana* (Milan, 1966), 7–53.

—— 'Le idee religiose del Cinquecento: La storiografia', in E. Cecchi and N. Sapegno (eds.), *Il Seicento; Storia della Letteratura italiana* (Milan, 1967), 7–87.

CARDELLA, L., *Memorie storiche de' cardinali della Santa Romana Chiesa* (Rome, 1792–7; 2nd edn. 1893).

CARGNONI, C., 'L'immagine di San Francesco nella formazione dell' ordine cappuccino', *L'immagine di San Francesco nella storiografia dall'Umanesimo all'Ottocento* (Atti del IX Convegno internazionale della S.I.S.F.; Assisi, 1983).

CARION, J., *Pronosticatio* (Rome, 1521).

—— *Chronica*, Lat. edn. (Paris, 1551).

CARRETE PARRONDO, C. (ed.), *Fontes Indaearum Regni Castellae* (Salamanca, 1985–6).

CARVAJAL, Bernardino López de, Cardinal, *Oratio de eligendo summo Pontifice* (Rome, *c.*1492; repr. in E. Martène and V. Durand, *Thesaurus*).

—— *Homelia doctissimi Reverendissimi domini Cardinalis sanctae Crucis*, alt. *Omelia habita Mechlinie in collegiata ecclesia sancti Rumoldi Cameracensis diocesis* (? Rome, 1504).

—— [Another edn.] *Homelia doctissima coram maximo Maximiliano Caesare semper augusto* (Rome, 1508).

CASTROCARO, Francesco da, OM, *Oratio Venerandi P. Fratris Francisci de Castrocaro . . . praesertim adversus Martinum Luterum* (Bologna, 1521).

Catalogus manuscriptorum Bibliothecae Regiae (Paris, 1744).

CAVALIERI, G., *Galleria de' sommi pontefici, patriarchi, arcivescovi, e vescovi dell'ordine de' predicatori* (Benevento, 1696).

CERRI, D., *I futuri destini degli stati e delle nazioni ovvero profezie e predizioni riguardanti i rivolgimenti di tutti i regni dell'universo sino alla fine del mondo*, 7th edn. (Turin, 1871).

CHASTEL, A., 'L'Antéchrist à la Renaissance', in E. Castelli (ed.), *Cristianesimo e ragione di stato, Atti del II Congresso Internazionale di Studi umanistici* (Rome, 1952), 177–86.

—— *The Sack of Rome*, trans. B. Archer (Princeton, NJ, 1983).

CIACONIUS, A., *Vitae et Gesta Summorum Pontificum ab Innocentio IV usque ad Clementem VIII necnon S.R.E. Cardinalium cum eorumdem insignibus* (Rome, 1601).

—— *Vitae et res gestae pontificum romanorum et S.R.E. Cardinalium* (Rome, 1677).

CISTELLINI, A., *Figure della Riforma Pretridentina* (Brescia, 1948).

COCHRANE, E., *Historians and Historiography in the Italian Renaissance* (Chicago, 1981).

COLLINS, J. J., *et al*, *Apocalypse; The Morphology of a Genre* (Missoula, Mont., 1979).

Concilium Oecumenicorum Decreta (Bologna, 1973).

Concilium Tridentinum diariorum actorum epistolarum tractatuum nova collectio, ed. Societas Goerresiana (Freiburg im Breisgau, 1901–30).

CONIGER, A., *Cronache, Raccolta di varie croniche, diarii ed altri opuscoli così italiani come latini appartenenti alla storia del regno di Napoli*, ed. A. Pellicia (Naples, 1782).

CONTARINI, G., *De officio episcopi, Opera*, ed. L. Contarini (Paris, 1571).

COSSIO, A., *Il cardinale Gaetano e la Riforma* (Cividale, 1902).

COSTI, G., 'L'episcopato a Reggio Emilia (1540–44) del cardinale Marcello Cervini poi papa Marcello II', in N. Artioli (ed.), *In Memoria di Leone Tondelli* (Reggio Emilia, 1980), 203–29.

CREIGHTON, M., *A History of the Papacy from the Great Schism to the Sack of Rome* (London, 1897; 2nd edn. 1901).

CRIADO DE VAL, M., 'Antífrasis y contaminaciones de sentido erótico en *La Lozana Andaluza*', in *Homenaje ofrecido a Damaso Alonso* (Madrid, 1960), i. 431–57.

CROWE, J., and CAVALCASELLE, G., *History of Painting in North Italy*, ed. T. Borenius (London, 1912).

CURIONE, Celio Secondo, *De amplitudine beati regni Dei* (1554).

DAMIANI, B., *Francisco Delicado* (New York, 1974).

DANTE, *Divina Commedia: Purgatorio*.

DAVIDSON, S., *Raphael's Bible: A Study of the Vatican Logge* (University Park, Pa. and London, 1985).

DAVIS, C. T., *Dante's Italy and Other Essays* (Philadelphia, 1984).

DELCORNO, C., *Giordano da Pisa e l'antica predicazione volgare* (Florence, 1975).

Del diluvio di Roma del MCCCCLXXXXV adi iiii di dicembre et daltre cose di gran meraviglia (n.p., n.d.).

DELICADO, F., *La Lozana Andaluza*, ed. B. Damiani (Madrid, 1969).

——*La Lozana Andalusa*, ed. L. Orioli (Milan, 1970).

——*El modo de adoperare el legno de India occidentale* (Venice, 1529).

DENIS, A., *Charles VIII et les Italiens: Histoire et Mythe* (Geneva, 1979).

DENZINGER, H., and SCHÖNMETZER, A., *Enchiridion Symbolorum* (Freiburg im Breisgau, 1963).

Diario d'Anonimo Fiorentino dall'anno 1358 al 1389, ed. A. Gherardi (Documenti di storia italiana, 6; Florence, 1876).

DIONYSIUS AREOPAGITA, pseudo, *De caelesti hierarchia*, trans. eds. of the Shrine of Wisdom (London, 1935).

DONI, A., *Dichiaratione del Doni sopra il XIII cap. dell'Apocalisse contro agli eretici, con modi non mai più intesi de huomo vivente: che cosa sono la nave di San Pietro, la Chiesa Romana, il Concilio di Trento, la destra della nave, la sinistra, la rete et i 153 pesci dell'Evangelio de S. Giovanni secondo i Cabalisti* (Venice, 1562).

DOUIE, D., *The Nature and the Effect of the Heresy of the Fraticelli* (Manchester,

1932; repr. 1978).

DOUSSINAGUE, J., *La politica internacional de Fernando el Católico* (Madrid, 1944).

——*Fernando el Católico y el cisma de Pisa* (Madrid, 1946).

DUSSLER, L., *Sebastiano del Piombo* (Basel, 1942).

——*Raphael: A Critical Catalogue* (London and New York, 1971).

ECKER, G., *Einblattdrucke von den Anfänge bis 1555* (Göppingen, 1981).

EGGS, G., *Supplementum Novum Purpure Doctae seu Vitae Legationes, Res Gestae Obitus* (1729).

EGIDIO OF VITERBO, Cardinal, *Scechina e Libellus de litteris hebraicis*, ed. F. Secret (Rome, 1959).

EMDEN, A., *A Biographical Register of the University of Oxford* (Oxford, 1958).

ERASMUS, *Opus epistolarum*, ed. P. S. Allen (Oxford, 1906–58).

——*Opuscula*, ed. W. Ferguson (The Hague, 1933).

ETTLINGER, L., and ETTLINGER, H., *Raphael* (Oxford, 1987).

EUBEL, C., *Hierarchia Catholica Medii Aevi* (Münster, 1898–).

——*Hierarchia Catholica Medii et Recentioris Aevi* (Münster, 1923).

EUTROPIUS, *Breviarium historiae romanae* (Oxford, 1703).

FABRONI, A., *Laurentii Magnifici vita* (Pisa, 1784).

FASOLA, B., 'Per un nuovo catalogo della collezione gioviana', *Atti del Convegno Paolo Giovio II: Rinascimento e la Memoria* (Como, 1985), 169–80.

FEBVRE, L., *Au cœur religieux du XVI siècle* (Paris, 1957).

——*Amour sacré et amour prophane. Autour de l'Heptameron* (Paris, 1944; repr. 1971).

FENLON, D., *Heresy and Obedience in Tridentine Italy: Cardinal Pole and the Counter Reformation* (Cambridge, 1972).

FERRAJOLI, A., *La congiura dei cardinali contro Leone X* (Rome, 1920).

——*Il ruolo della corte di Leone X (1515–1516)*, ed. V. de Caprio (Rome, 1984), 531–44.

FERRARA, M., *Savonarola: Prediche* (Florence, 1952).

FERRERI, Z., *Promotiones et progressus sacrosancti pisani concilii moderni indicti et incohati anno domini M.D.XI* (n.p., n.d.).

——*Lugudunense somnium . . . Sylva centesimadecima* (n.p., n.d.).

FERRETTI, F., *Un maestro di politica; L'umana vicenda di Mercurino del nobile Arborio di Gattinara gran cancelliere di Carlo V re di Spagna e imperatore* (pub. by the Comune of Gattinara, 1980).

FICINO, MARSILIO, *Opera* (Basle, 1576).

FIRPO, L. (ed.), *Prime relazioni di navigatori italiani sulla scoperta dell'America: Colombo, Vespucci, Verazzano* (Turin, 1966).

FIRPO, M., and MARCATTO, D., *Il processo inquisitoriale del Cardinal Giovanni Morone*, 4 vols. (Rome, 1981–7).

FLAMINIO, G., *Epistolae familiares* (Bologna, 1744).

FOCILLON, H., *L'an Mil* (Paris, 1938).

FOGLIETTA, M., *Gli Eloggi degli huomini chiari della Liguria*, trans. L. Conti (Genoa, 1579).

FOLEY, A. (ed.), *La Lozana andaluza* (London, 1977).

FOSSI, F., *Novelle letterarie* (Florence, 1790), xx. 609–18.

FREEDBERG, S., *Painting of the High Renaissance in Rome and Florence* (Cambridge, Mass., 1961).

FRIEDLÄNDER, G., *Beiträge zur Reformationsgeschichte* (Berlin, 1837).

GALATINO, P., OFM, *Opus toti christianae Reipublice maxime utile, de arcanis catholicae veritatis, contra obstinatissimam Iudaeorum nostrae tempestatis perfidiam: ex Talmud genere eleganter congestum* . . . (Ortona, 1518).

GANDUCIO, O., *Ragionamento* . . . *della conversione de' Gentili a particolarmente de' Genovesi predetta da Esaia Profeta* (Genoa, n.d. [1615]); ed. M. Cipolloni (Rome, 1988).

GARCIA-VILLOSLADA, R., *Martino Lutero* (Milan, 1985).

GARIMBERTO, H., *La prima parte delle Vite overo fatti memorabili d'alcuni papi, et di tutti i cardinali Passati, 1506–1575* (Ferrara, 1575).

GATTINARA, Mercurino di, *Legatio ad sacratissimum Caesarem carolum ab principibus electoribus* (Antwerp, 1520).

GEIGER, L. (ed.), *Johann Reuchlin: Briefwechsel* (Tübingen, 1875).

GERE, J., and TURNER, N., *Drawings by Raphael* (London, 1983).

GERGES, D., *Introductio in historiam Evangelii seculo XVI passim per Europam renovati doctrinaeque reformatae* (Groningen, 1744).

GERSON, J., *Iosephina*, in *Œuvres complètes*, ed. P. Glorieux (Tournai, 1962), iv. 31–100.

Gesamtkatalog der Wiegendrucke, iii (Leipzig, 1928).

GEYER, B., *Die patristische und scholastische Philosophie* (Berlin, 1928).

GILBERT, F., *The Pope, his Banker and Venice* (Cambridge, Mass., 1980).

GINI, P., 'Paolo Giovio e la vita religiosa del Cinquecento', *Atti del Convegno Paolo Giovio, Il Rinascimento e la Memoria* (Como, 1985), 31–50.

GIOMMI, E., *La monaca Arcangela Panigarola, madre spirituale di Denis Briçonnet, L'attesa del 'pastore engelico' annunciato dall' Apocalypsis Nova' del Beato Amadeo fra il 1514 e il 1520* (Università negli Studi di Firenze (Facoltà di Lettere e Filosofia) Ph.D. thesis).

GIOS, P., *L'Attività pastorale del vescovo Pietro Barozzi a Padova (1487–1507)* (Padua, 1977).

GIOVIO, P., *Le Vite di Leon Decimo et d'Adriano Sesto Sommi Pontefici et del Cardinal Pompeo Colonna*, trans. L. Domenichi (Florence, 1549).

——*Elogia virorum literis illustrium* (Basle, 1577).

——*Lettere*, ed. G. Ferrero (Rome, 1956).

GIUSTINIANI, A., *Castigatissimi annali* . . . *della ecclesia et illustrissima Repubblica di Genoa* (Genoa, 1537).

——and QUERINI, P., 'Libellus ad Leonem X Pontificem Maximum', *Annales Camaldulenses Ord. Sancti Benedicti*, ed. G. Mittarelli and A. Costadoni (Venice,

1763), ix. 613–719.

—— 'Trattati, lettere e frammenti dai manoscritti originali dell'Archivio dei Camaldolesi di Monte Corona nell'eremo di Frascati', ed. E. Massa, in E. Massa, *I manoscritti originali custoditi nell'eremo di Frascati* (Rome, 1967).

GLAY, M. le (ed.), *Négotiations diplomatiques entre la France et l'Autriche* (Paris, 1845).

GLIOZZI, G., *Adamo e il nuovo mondo. La nascita dell'antropologia come ideologia coloniale: dalle genealogie bibliche alle teorie razziale* (1500–1700) (Florence, 1976).

GNOLI, D., *La Roma di Leone X* (Milan, 1938).

GÖLLER, E., *Die päpstliche Poenitentiaria* (Rome, 1907–11).

GOLZIO, V., *Raffaello nei documenti nelle testimonianze dei contemporanei e nella letteratura del suo secolo* (Vatican City, 1936).

GOMBRICH, E., 'The Ecclesiastical Significance of Raphael's *Transfiguration:* "Ars auro prior"', in *Essays in Honour of Jan Bialostocki* (Warsaw, 1981), 241–3.

GOTTLIEB, T., *Büchersammlung Kaiser Maximilians* (Leipzig, 1900).

GOULD, C., *The Sixteenth-Century Venetian School* (National Gallery Catalogues) (London, 1959).

GRABSKI, J., 'Sul rapporto fra ritratto e simbolo nella ritrattistica del Lotto', in P. Zampetti and V. Sgarbi (eds.), *Lorenzo Lotto: Atti del Convegno internazionale di studi per il V centenario della nascita* (Asolo, 1980), 383–91.

GRECO, M. SCUVICINI, *Miniature riccardiane* (Florence, 1958).

GREGOROVIUS, F., *History of the City of Rome in the Middle Ages*, trans. A. Hamilton (London, 1902; Ital. trans., Rome, 1980).

GUICCIARDINI, F., *Storia d'Italia*, ed. C. Panigada (Bari, 1929).

—— *Scritti autobiografici e rari di Francesco Guicciardini*, ed. R. Palmarocchi (Bari, 1936).

GUTIÉRREZ, D., *The History of the Order of St. Augustine*, trans. J. Kelly (Villanova, Pa., 1979).

HALLMAN, B., *Italian Cardinals, Reform, and the Church as Property* (Berkeley, Calif., 1985).

HANE, P., *Historia sacrorum* (Kiel, 1728).

HEADLEY, J., 'Verso il recupero storico del grancancelliere di Carlo V: Problemi, progressi, prospettive', *Atti del Convegno di Studi Storici* (Vercelli, 1982), 69–104.

—— 'Germany, the Empire and *Monarchia* in the Thought and Policy of Gattinara', in H. Lutz (ed.), *Das römischdeutsche Reich im politischen System Karls V* (Munich and Vienna, 1982), 15–34.

—— *The Emperor and his Chancellor; A Study of the Imperial Chancellery under Gattinara* (Cambridge, 1983).

HERBORN, N., OM, *De insulis nuper inventis Ferdinandi Cortesii ad Carolum V Imperatorem narrationes . . . Item epitome de inventis nuper Indiae populis idolatris ad fidem Christi, atque adeo ad Ecclesiam Catholicam convertendis* (Cologne, 1532).

HEREDIA, P. de, *Corona regis* (Rome, 1486).

HILDEGARDE, ST, *Scivias*, ed. A. Führkötter (Turnhout, 1978).

HIRST, M., *Sebastiano del Piombo* (Oxford, 1981).

HOCHSTRATEN, J., *Ad sanctissimum dominum nostrum Leonem papam decimum. Ac divum Maximilianum Imperatorem semper augustum, Apologia . . . Contra dialogum Georgio Benigno Archiepiscopo Nazareno in causa Joannis Reuchlin . . . Opus novum, Anno MCCCCCXVIII Coloniae foelicter edictum.*

HOLLÄNDER, E., *Wunder, Wundergeburt und Wundergestalt in Einblattdrucken des 15.–18. Jahrhunderts* (Stuttgart, 1922).

HUDDLESTON, L., *Origins of the American Indians: European Concepts 1492–1729* (Austin, Tx., 1967).

HÜGEL, F. VON, *The Mystical Element of Religion as Studied in Saint Catherine of Genoa and her Friends* (London, 1923).

HURTER, H., *Nomenclator literarius theologiae catholicae*, 2: 1109–1563 (Innsbruck, 1906).

INFESSURA, STEFANO, *Diario della città di Roma*, ed. O. Tommasini (Rome, 1890).

ISOLANI, ISIDORO, *De imperio militantis Ecclesiae libri quattuor* (Milan, 1516).

——*Ex humana divinaque sapientia tractatus de futura nova mundi mutatione* (Bologna, 1523).

IUSTINIANI, A., *Psalterium Hebraeum, Graecum, Arabicum et Chaldaeum, cum tribus latinis interpretationibus et glossis* (Genoa, 1516).

JACOB, L., *Symbols for the Divine in the Kabbaláh* (London, 1984).

JAVARY, G., 'Panorama de la kabbale chrétienne en France au XVIe et au XVIIe siècles', in *Kabbalistes chrétiens* (Cahiers de l'Hermétisme, dir. A. Favre and F. Tristan; Paris, 1979), 69 ff.

JEDIN, H., *Papal Legate at the Council of Trent: Cardinal Seripando*, trans. F. Eckhoff (St Louis, 1947).

——*The History of the Council of Trent*, trans. D. Graf (London, 1957).

JOACHIM OF FIORE, *Liber Concordie Novi et Veteris Testamenti* (Venice, 1519).

——*Liber de Concordia Novi et Veteris Testamenti*, bks. 1–4, ed. E. R. Daniel, *Transactions of the American Philosophical Society*, 73, pt. 8 (Philadelphia, 1983).

——*Expositio in Apocalypsim* (Venice, 1527).

——*Il Libro delle Figure dell'Abate Gioacchino da Fiore*, ii, ed. L. Tondelli, M. Reeves, and B. Hirsch-Reich, 2nd edn. (Turin, 1954).

——*Psalterium decem chordarum* (Venice, 1527).

JOACHIM, pseudo, *Super Hieremiam Prophetam* (Venice, 1516).

——*Vaticinia de summis pontificibus* (in many 16th and 17th-cent. edns.).

——*Expositio magni prophete Joachim in librum beati Cyrilli*, see TELESPHORUS OF COSENZA, *Liber de magnis tribulationibus*.

——*Vaticinium Sibillae Erithreae*, ed. O. Holder-Egger, in 'Italienische Prophetien des 13. Jahrhunderts', *Neues Archiv der Gesellschaft für ältere deutsche Geschichtskunde*, 5 (1889), 155–73.

——*Oraculum Cyrilli*, ed. P. Piur, in K. Burdach and P. Piur, *Vom Mittelalter zur Reformation*, 2, pt. 4, Appendix (Berlin, 1912).

KRISTELLER, P., *Iter italicum*, 4 (London and Leiden, 1989).

KVAČALA, F., *Postelliana; Urkundliche Beiträge zur Geschichte der Mystik in Reformationszeitalter* (Jurjew, 1915).

LABOWSKY, L., 'Il cardinale Bessarione e le origini della Biblioteca Marciana', in A. Pertusi (ed.), *Venezia e l'Oriente tra tardo Medio Evo e Rinascimento* (Venice, 1966), 159–82.

LACTANTIUS, *Divinarum Institutionum Liber Septimus*, PL 6 (Paris, 1844).

LADNER, G., *Images and Ideas in the Middle Ages: Selected Studies in History and Art* (Rome, 1983).

LANDUCCI, L., *Diario fiorentino dal 1450 al 1516*, ed. I. Del Badia (Florence, 1883).

LANGASCO, C. DA, *Gli Ospedali degli incurabili* (Genoa, 1938).

——'Il Giustiniani e le tensioni ecclesiali del suo tempo', *Agostino Giustiniani annalista genovese ed i suo tempi: Atti del Convegno di studi Genova* (Genoa, 1984), 113–18.

LANGE, K., *Der Papstesel* (Göttingen, 1891).

LECLERCQ, J., *Un humaniste ermite: le bienheureux Paul Giustiniani (1476–1528)* (Rome, 1951).

LEHMANN, P., *Das Pisaner Concil von 1511* (Breslau, 1874).

LERNER, R., and MOYNIHAN, R., *Weissagungen über die Papste. Vat. Ross. 374* (Zurich, 1985).

LEVI DELLA VIDA, G., *Ricerche sulla formazione del più antico fondo dei manoscritti orientali della Biblioteca Vaticana* (Vatican City, 1939).

LEVIN, H., *The Myth of the Golden Age in the Renaissance* (New York, 1969).

LICHTENBERGER, J., *Prognosticatio* (Strasbourg, 1488).

LIGHTBOWN, R., *Botticelli* (Berkeley, Calif., 1978).

LIMBORCH, P. a, *Hist. Inquisitionis, cui subjungitur Liber Sententiarum Inquisitionis Tholosanae 1307–1323* (Amsterdam, 1692).

LITTA, P., *Famiglie celebri italiane* (Milan, 1820–56).

LOLLIS, C. de (ed.) *Scritti di Cristoforo Colombo: Raccolta di documenti e studi pubblicati dalla Reale Commissione Columbiana pel quarto centenario dalla scoperta dell'America* (Rome, 1892–4).

LONGOLIUS, C., *Epistolarum liber quartus* (Basle, 1558).

LOPEZ SANTIDRIAN, S., 'Quiñones, Francisco de', DS 12/2 (Paris, 1986), 2852–3.

LOYOLA IGNATIUS, St, SJ, *Iudicium de quibusdam opinionibus quae falso revelationes credebantur. Monumenta ignatiana* (Madrid, 1911), I. xii.

——*Letters*, ed. W. Young (Chicago, 1959).

LUMBROSO, M., and MARTINI, A., *Le Confraternite Romane nelle loro chiese* (Rome, 1963).

LUMMEN (LUMNIUS), J., *De extremo Dei iudicio et Indorum vocatione* (Antwerp, 1567; repr. Venice, 1569; Antwerp, 1594, with title *De vicinitate extremi iudicii Dei et consummationis saeculi libri duo*).

LUTHER, M., *Werke* (Weimar, 1883-).

LUTZ, H. (ed.), *Nuntiaturberichte aus Deutschland, erste Abteilung 1533–1559*, 15: *Friedenslegation des Reginald Pole zu Kaiser Karl V und König Heinrich VIII (1533–1556)* (Tübingen, 1981).

McGINN, B., *Visions of the End* (New York, 1979).

——*The Calabrian Abbot: Joachim of Fiore in the History of Western Thought* (New York and London, 1985).

——'Joachim of Fiore's *Tertius Status*: Some Theological Appraisals', in A. Crocco (ed.), *L'Età dello Spirito e la fine del tempi in Gioacchino da Fiore e nel Gioachinismo medievale—Atti del II Congresso Internazionale di Studi Gioachimiti* (S. Giovanni in Fiore, 1986), 219–36.

MACHIAVELLI, N., *The Prince*.

MACKAY, A., 'Averroistas y marginadas', *Actas del III Coloquio de Historia Medieval Andaluza* (Jaén, 1984), 247–61.

McKILLOP, S., *Franciabigio* (Berkeley, Calif., 1974).

MAIO, R. DE, *Savonarola e la curia romana* (Rome, 1969).

MANSELLI, R., 'Il problema del doppio Anticristo in Gioacchino da Fiore', in K. Hauck and H. Mordek (eds.), *Geschichtsschreibung u. geistiges Leben im Mittelalter* (Cologne and Vienna, 1978), 427–49.

MANSI, G., *Ss. Conciliorum Nova et Amplissima Collectio*, 32 (1438–1549) (Paris and Leipzig, 1902).

MANUTIUS, P., *Epistolarum libri IV* (Venice, 1560).

MARAVALL, J., 'La utopia politico-religiosa de los franciscanos en la Nueva Espana', *Estudios Americanos*, 1 (1948–9), 199–227.

——an. edn. in *Utopia y reformismo en la España de los Austrias* (Madrid, 1982), 78–110.

MARTELLI, M., 'I Medici e lettere', in C. Vasoli (ed.), *Idee, istituzioni, scienze e arti nella Firenze dei Medici* (Florence, 1980).

MARTÈNE, E., and DURAND, V., *Thesaurus novus anecdotorum* (Paris, 1717).

——*Veterum Scriptorum et Monumentorum . . . Amplissima Collectio* (Paris, 1724–33).

MARTIN, F. X., OSA, *The Problem of Giles of Viterbo* (Hénerlé-Louvain, 1960).

——'Giles of Viterbo as Scripture Scholar', *Egidio da Viterbo, O.S.A., e il suo tempo; Atti del V convegno dell'Istituto Storico Agostiniano* (Rome, 1983), 191–222.

MARTÍNEZ KLEISER, L., *Refranero general ideológico español* (Madrid, 1953).

MARTIRE, PIETRO D'ANGHIERA, *Opus epistolarum Petri Martiris Anglerii* (Amsterdam, 1670).

MARZI, D., *La questione della riforma del calendario nel Quinto Concilio Lateranense* (1512–1517) (Florence, 1896).

MASSA, E., 'Egidio da Viterbo e la metodologia del sapere nel cinquecento', in H. Bédarida (ed.), *Pensée humaniste et tradition chrétienne au XVe et XVIe siècles* (Paris, 1950), 185–239.

——*I fondamenti metafisici della 'dignitas hominis', e testi inediti di Egidio da Viterbo* (Turin, 1954).

——*I manoscritti originali custoditi nell'erèmo di Frascati* (Rome, 1967).

MATTIA OF MILAN, *Repertorium, seu interrogatorium, sive confessionale...* (Milan, 1516).

MATUCCI, A., 'Piero Parenti nella storiografia fiorentina', *Studi di filologia e critica offerti dagli allievi a Lanfranco Caretti* (Rome, 1985), i. 149–93.

Memoria delli novi segni e spaventevoli prodigii comparsi in più loci de Italia: et in varie parti del mondo: l'anno mille cinquecento undese (n.p., n.d.).

MENDIETA, G. DE, OM, *Historia ecclésiastica indiana*, ed. J. Garcia (Icazbalceta, Mexico, 1945).

MERCATI, A., *Prescrizioni pel culto divino nella diocesi di Reggio Emilia del vescovo Cardinale Marcello Cervini* (Reggio Emilia, 1933).

——'Paolo Pompilio e la scoperta del cadavere intatto sull'Appia nel 1485', *Opera minora*, iv (Vatican City, 1937).

MICCOLI, G., 'La storia religiosa', in G. Einaudi (ed.), *Storia d'Italia* (Turin, 1974), II. i. 431–1079.

MILHOU, A., *Colón y su mentalidad mesiánica en el ambiente franciscanista español* (Valladolid, 1983).

MILLS, J., 'The Coming of the Carpet to the West', in D. King and D. Sylvester (eds.), *The Eastern Carpet in the Western World from the 15th to the 17th Century* (London, 1983), 11–23.

Mirabilis Liber qui prophetias Revelationesque necnon res mirandas preteritas presentes et futuras aperte demonstrat (Rome recte Lyons, 1524).

MIRBT, K., and ALAND, C., *Quellen zur Geschichte des Papsttums und des Römanischen Katholizismus* (Tübingen, 1967).

MITTARELLI, G., and COSTADONI, A., *Annales Camaldulenses*, IX (Venice, 1763).

MITTARELLI, J., *Bibliotheca codicum manuscriptorum monasterii S. Michaelis Venetiarum prope Murianum* (Venice, 1779).

MOHLER, L., *Die Wiederbelebung des Platons: Studium in der Zeit der Renaissance durch Kardinal Bessarion* (Cologne, 1921).

——*Kardinal Bessarion als Theolog, Humanist und Staatsmann*, i: *Darstellung* (Paderborn, 1923); ii: *Bessarionis in calumniatorem Platonis* (Paderborn, 1927); iii: *Aus Bessarions Gelehrtenkreis. Quellen und Forschungen aus dem Gebeite der Geschichte* (Paderborn, 1942).

MONACO, M., *Il 'De Officio collectoris in regno Angliae' di Pietro Griffi da Pisa (1469–1516)* (Rome, 1973).

MONFASANI, J., *George of Trebizond; A Biography and a Study of his Rhetoric and Logic* (Leiden, 1976).

—— 'Sermons of Giles of Viterbo as Bishop', in *Egidio da Viterbo, O.S.A., e il suo tempo* (Atti del V Convegno dell'Istituto Storico Agostiniano (Rome, 1983), 137–89.

MONTE, A. DEL (ed.), *Concilium Lateranense V generale novissimum sub Julio II et Leone X celebratum* (Rome, 1521; repr. *Mansi*, 32).

MONUMENTA PERUANA, ed. A. de Egaña, SJ (Rome, 1961).

MOORMAN, J., *A History of the Franciscan Order: From its Origins to the Year 1517* (Oxford, 1968).

MORE, SIR THOMAS, *Utopia*, ed. J. Lupton (Oxford, 1895).

MORENI, D., *Memorie istoriche dell'ambrosiana Basilica di S. Lorenzo di Firenze* (Florence, 1816–18).

MORISI, A., *Apocalypsis Nova: Richerche sull'origine e la formazione del testo dello pseudo-Amadeo* (Istituto Storico Italiano per il Medio Evo, Studi Storici, 77; Rome, 1970).

—— 'Galatino et la Kabbale chrétienne', in *Kabbalistes chrétiens* (Cahiers de l'Hermétisme, dir. A. Favre and F. Tristan; Paris, 1979), 212–31.

MORONI, G., *Dizionario di erudizione storico-ecclesiastica* (Venice, 1840–79).

MORRISON, K., 'History malgré lui: A Neglected Bolognese Account of Charles V's Coronation in Aachen 1520', *Studia Gratiana, XV: Post Scripta* (Rome, 1972), 681–4.

MORSOLIN, B., *L'abate di Monte Subasio e il concilio di Pisa* (Venice, 1893).

MOTOLINIA, TORIBIO DE BENAVENTE, OM, *Memoriales e Historia de los Indios de la Nueva España*, ed. F. de Lejarza (Madrid, 1970).

—— an. edn., ed. E. O'Gorman (Mexico, 1973).

MOTTU, H., *La Manifestation de L'Esprit selon Joachim de Fiore* (Neuchatel and Paris, 1977).

NAEF, H., *Les origins de la Réforme à Genève* (Paris, 1936).

NARDUCCI, H., *Catalogus codicum manuscriptorum praeter Graecos et orientales in Bibliotheca Angelica olim Coenobii Sancti Augustini de Urbe* (Rome, 1893).

NAZARI, G. BATTISTA, *Discorso della futura et sperata vittoria contra il Turco estratto da i sacri profeti et da altre profetie, prodigii et pronostici et di nuovo dato in luce* (Venice, 1570).

NERI, TOMMASO, *Apologia . . . in difesa della dottrina del R.P.F. Girolamo Savonarola da Ferrara* (Florence, 1564).

NESI, GIOVANNI, *Oracolo de novo saeculo* (Florence, 1497).

NICCOLI, O., 'Il diluvio del 1524 fra panico collettivo e irrisione carnevalesca', *Scienze, credenze occulte, livelli di cultura* (Florence, 1982), 369–92.

—— 'Il mostro di Ravenna: teratologia e propaganda nei fogli volanti del primo Cinquecento', in D. Bolognesi (ed.), *Ravenna in età Veneziana* (Ravenna, 1986), 245–77.

—— *Profeti e popolo nell'Italia del Rinascimento* (Rome and Bari, 1987).

Niccolò delle Meraviglie (Nicolaus de Mirabilibus), *Finis quaestionis disputatae in Domo magnifici Laurentii Medices ultimo die Juni MCCCC-LXXXVIII* (Florence, 1488).

Nijhoff, W., and Kronenberg, M., *Nederlandsche Bibliographie van 1500 tot 1540* ('s-Gravenhage, 1923–61).

Oberhuber, K., 'Style and Meaning', in his *A Masterpiece Close-up: The Transfiguration by Raphael* (Cambridge, Mass., 1981).

Olin, J., *The Catholic Reformation: Savonarola to Ignatius Loyola* (New York, 1969).

O'Malley, J., SJ, *Giles of Viterbo on Church and Reform* (Leiden, 1968).

——*Praise and Blame in Renaissance Rome: Rhetoric, Doctrine and Reform in the Sacred Orators of the Papal Court c.1450–1521* (Durham, NC, 1979).

——*Rome and the Renaissance: Studies in Culture and Religion* (London, 1981).

Pallucchini, R., *Sebastian Viniziano* (Milan, 1944).

Pantera, G., *Monarchia del Nostro Signor Iesu Christo* (Venice, 1545; other edns. 1552, 1573).

Panvinio, O., *Historia della vite de i summi pontefici* (Venice, 1594).

Partridge, L., and Starn, R., *A Renaissance Likeness* (Berkeley, Calif., 1980).

Paschini, P., *Domenico Grimani cardinale di S. Marco (+1523)* (Storia e letteratura, 4; Rome, 1943).

——*La beneficenza in Italia e le Compagnie del Divino Amore nei primi decenni del Cinquecento* (Rome, 1925; repr. in *Tre Ricerche sulla storia della chiesa nel Cinquecento* (Rome, 1945), 3–88).

Pastor, L. von, *History of the Popes*, vols. i–vi, trans. and ed. E. Antrobus (London, 1891–8); vols. vii–xxiv, trans. and ed. R. Kerr (London, 1908–33).

Pastore, A., *Marcantonio Flaminio* (Milan, 1981).

Pecci, G., *Notizie storico-critiche sulla vita di Bartolomeo da Petrojo chiamato Brandano* (Lucca, 1763).

Pegolotto, A., *Una littera . . . de una prophetia di Santa Brigide* (n.p., 1542).

Pélissier, L., *De opere historico Aegidii Cardinalis Viterbiensis* (Montpellier, 1896).

——'Pour la biographie du Cardinal Gilles de Viterbe', *Misc. di studi critici editi in onore di Arturo Graf* (Bergamo, 1903), 789–815.

Perrone, B., 'Il *De re publica christiana* nel pensiero filosofico e politico di Pietro Galatino', *Studi storie pugliese in onore di Giuseppe Chiarelli* (Galatina, 1968), ii. 524–9.

Pesci, U., *I vescovi di Gubbio* (Perugia, 1919).

Petrarca, F., *Rime*, ed. G. Bezzola (Milan, 1976).

Phelan J., *The Milennial Kingdom of the Franciscans in the New World: A Study of the Writings of Gironimo de Mendieta (1525–1604)* (Berkeley, Calif., 1956).

Piatti, G., *Storia critico-cronologica de' Romani Pontefici e de' generali e provinciali concili* (Naples, 1765–8).

PIAZZA, C., *La Gerarchia cardinalizia* (Rome, 1702).

PICCOLOMINI, AENEAS SILVIUS (Pope Pius II), *Historia rerum Friderici III imperatoris*, ed. J. Schilter, *Scriptores rerum Germanicarum* (Strasbourg, 1687).

——*Pii II Commentarii rerum memorabilium que temporibus suis contigerunt*, ed. A. van Heck (Vatican City, 1984).

Platon et Aristote à la Renaissance. XVIe Colloque international de Tours (Paris, 1976).

PLATO, *Omnia opera* (Venice, 1513).

POGIANUS, J., *Epistolae et orationes* (Rome, 1762–8).

POLLIDORI, P., *De vita gestis et moribus Marcelli II* (Rome, 1744).

POMIAN, K., 'Astrology as a Naturalistic Theology of History', in P. Zambelli (ed.), *'Astrologi hallucinati': Stars and the End of the World in Luther's Time* (Berlin and New York, 1986), 29–43.

PONTANI, G., *Diario romano*, ed. D. Toni, *RIS* vol. iii, pt. 2.

PORRO, G., *Catalogo del manoscritti della Trivulziana* (Turin, 1884).

POSNER, K., *Leonardo and Central Italian Art: 1515–1550* (New York, 1974).

POUNCEY, P., and GERE, J., *Italian Drawings in the Department of Prints and Drawings in the British Museum* (London, 1962).

PREMOLI, O., *Storia dei Barnabiti nel Cinquecento* (Rome, 1913).

PRIGENT, A., *Apocalypse 12: Histoire de l'exégèse* (Tübingen, 1958).

PRODI, P., *Il sovrano pontifice: Un corpo e due anime: la monarchia papale nella prima età moderna* (Bologna, 1982).

——*The Papal Prince, One Body and Two Souls: The Papal Monarchy in Early Modern Europe*, trans. S. Haskins (New York, 1987).

Prognosticon super Monstrum ex Felsinea urbe oriundum (Ravenna, 1514).

Prophetia trovata in Roma: Intagliato in marmore in versi latini. Tratta in vulgar sentimento (Rome, [c.1510]).

PROSPERI, A., '"Otras Indias": missionari della Controriforma tra contadini e selvaggi', in G. Garfagnini (ed.), *Scienze, credenze occulte, livelli di cultura* (Florence, 1982), 205–34.

PULLAPILLY, C., *Caesar Baronius, Counter-Reformation Historian* (Notre Dame, Ind., 1975).

Questa è la vera prophetia prophetizata dal gloriosa Santo Anselmo: la quale declara la venuta de uno Imperatore: el qual mettera pace tra li christiani: et conquistara li infideli trovata in Roma (?Ferrara c.1510?).

RAINALDI, O., *Annales ecclesiastici ab anno MCXCVIII ubi desinit Cardinalis Baronius*, rev. D. Mansi (Lucca, 1747–56).

RAMUSIO, G., *Navigazioni e viaggi*, ed. M. Milanesi (Turin, 1979).

RANKE, L. VON, *History of the Popes* (New York, 1966).

RÉAU, L., *Iconographie de l'art chrétien* (Paris, 1955–9).

REDIG DE CAMPOS, D., *Le stanze di Raffaello* (Rome, 1950).

REEVES, M., *Influence of Prophecy in the Later Middle Ages: A Study in Joachimism* (Oxford, 1969).

——'Some Popular Prophecies from the Fourteenth to the Seventeenth Centuries', in G. Cuming and D. Baker (eds.), *Popular Belief and Practice* (Cambridge, 1972), 107–34.

——*Joachim of Fiore and the Prophetic Future* (London, 1976).

—— 'The Abbot Joachim's Sense of History', in *1274 Année Charnière: Mutations et Continuités, Colloques Internationaux du Centre National de la Recherche Scientifique*, 558 (Paris, 1977), 781–96.

—— 'The Development of Apocalyptic Thought: Medieval Attitudes', in C. Patrides and J. Wittreich (eds.), *The Apocalypse in English Renaissance Thought and Literature* (Manchester, 1984), 40–72.

—— 'Roma profetica', in F. Troncarelli (ed.), *La città dei segreti: Magia, astrologia e cultura esoterica a Roma* (Milan, 1985), 277–97.

—— 'The Third Age: Dante's Debt to Gioacchino da Fiore', in A. Crocco (ed.), *L'Età dello spirito e la fine dei tempi in Gioacchino da Fiore, Atti del II Congresso internazionale di studi Gioachimiti* (S. Giovanni in Fiore, 1986), 127–39.

REEVES, M., and HIRSCH-REICH, B., *The Figurae of Joachim of Fiore* (Oxford, 1972).

REGLÀ CAMPISTOL, J., *Introducció a la historia de la Corona d'Arago: dels origens a la Nova Planta* (Palma de Mallorca, 1969).

RENAUDET, A., *Le concile gallican de Pise-Milan: Documents florentins, 1510–1512* (Paris, 1922).

——*Préréforme et humanisme à Paris pendant les premières guerres d'Italie (1494–1517)* (Paris, 1916; 2nd edn. 1953).

RIDOLFI, R., *La vita di Girolamo Savonarola*, 2nd edn. (Rome, 1952).

——*La Stampa in Firenze nel secolo XV* (Florence, 1958).

ROBERTI, G., *S. Francesco di Paòlo, fondatore dell'Ordine dei Minimi (1416–1507): storia della sua vita*, 2nd edn. rev. (Rome, 1963).

ROCA, COUNT DE LA, *Epitome de la vida y echos del emperados Carlo* (Madrid, 1646).

ROCOCIOLO, F., *Ad illustrissimum ac excelentissimum principem divum Herculum Estensem Francisci Rococioli Mutinensis libellus de Monstro Romae in Tyberi reperto anno Domini MCCCCLXXXXVI* (n.p., n.d.).

RODERIGUEZ-VILLA, *Memorias para la historia del asalto y saco de Roma en 1527* (Madrid, 1875).

RODOCANACHI, E., *Una Cronica di Santa Sabina sull'Aventino* (Rome, 1898).

——*Histoire de Rome: le pontificat de Léon X 1513–1521* (Rome, 1931).

ROOVER, R. DE, 'Cardinal Cajetan on "Cambium" or Exchange Dealings', in E. Mahoney (ed.), *Philosophy and Humanism: Renaissance Essays in Honor of Paul Oskar Kristeller* (Leiden, 1976), 423–33.

ROQUETAILLADE, JEAN DE (RUPESCISSA), *Vade mecum in tribulatione, fasciculus rerum expetendarum et fugiendarum, prout ab Orthuino Gratio . . . editus est Coloniae . . . una cum appendice . . . scriptorum veterum . . . qui Ecclesiae romanae errores et abusus, detegunt et damnant . . . opera et studio Edwardi Brown* (London, 1690), ii. 496–507.

ROSCOE, W., *The Life and Pontificate of Leo the Tenth*, 2nd rev. edn. (London, 1806; 3rd rev. edn. 1846).

ROSE, P., *Bodin and the Great God of Nature* (Geneva, 1980).

ROSSBACH, H., *Das Leben und die politisch-kirchliche Wirksamkeit des Bernardino Lopez de Caraval, Kardinals von Santa Croce in Gierusalemme in Rom, und das schismatische Concilium Pisanum* (Breslau, 1892).

RUBERTI, J. BAPTISTE, *Monstrum apud urbem natum* [Rome, after 11 March 1513].

RUSCONI, R., 'Escatologia e povertà nella predicazione di Bernardino da Siena', in *Bernardino predicatore nella società del suo tempo* (Convegno del Centro di studi sulla spiritualità medievale, 16; Todi, 1976), 213–50.

——*L'Attesa della fine; Crisi della società, profezia ed Apocalisse in Italia al tempo del' grande scisma d'Occidente (1378–1417)* (Istituto Storico Italiano per il Medioevo, Studi Storici, 115–18; Rome, 1979).

—— ' "Ex quodam antiquissimo libello": La tradizione manoscritta della profezie nell'Italia tardo medievale: dalle collezioni profetiche alle prime edizioni a stampa', in W. Verbeke, D. Verhelst, and A. Welkenhuysen (eds.), *The Use and Abuse of Eschatology in the Middle Ages* (Louvain, 1988), 441–72.

—— 'Circolazione di testi profetici tra '400 e '500: La figura di Pietro Galatino', *Continuità e diversità nel profetismo gioachimita tra '400 e '500; Atti del 3° Congresso internazionale di studi gioachimiti, San Giovanni in Fiore, 17–21 sett. 1989* (forthcoming).

RUSSO, F., 'Francesco di Paola', *Bibliotheca Sanctorum* (Rome, 1965), v. 116–1175.

SACCANI, C., *I vescovi di Reggio Emilia* (Reggio Emilia, 1902).

SAFARIK, E. (ed.), *Catalogo sommario della Galleria Colonna in Roma* (Rome, 1981).

SAFFREY, H., 'Aristoteles, Proclus, Bessarion', *Atti dell' XI Congresso internazionale di filosofia* (Florence, 1960), 153–8.

SALIMBENE, OM, *Cronica*, *MGHS* 32.

SALVIATI, GIORGIO BENIGNO (DRAGIŠIĆ), *Opus de Natura Angelica Impressum cum maxima diligentia* Florentiae XIII Kalendas Augusti M.CCCCXXXXIX.

——*Propheticae Solutiones impressae per Ser Laurentium de Morgianis*, VI idus Aprillis M. CCCC.LXXXXVII.

——*De libertate et immutabilitate Dei sententias ad R. patrem et dominum Bessarionem Patriarcam Constantinopolitanum* (n.p., n.d.).

——*Mirabilia septem et septuaginto in opusculo Magistri Nicolai de Mirabilibus reperta mirabili praesenti opera annotavit* (see Ch. 7 n. 23).

——*Georgii Benigni Fratris Seraphice religionis ad Virum Magnanimum Laurentium Petri Cosmi Patriae Patri In opus septem quaestionum ab ipso propositarum* (see Ch. 7 n. 24).

——*Dialectica nova secundum mentem Doctoris subtilis. Et beati Thomae Aquinatis aliorumque realistarum* . . . Impressum Florentiae . . . Die Nono XVIII mensio Martii M.CCCC LXXXVIII.

——*Defensio praestantissimi viri Joannis Reuchlin LL. Doctoris a Reverendo patre Georgio Benigno Nazareno archiepiscopo Romae per modum dialogi edita* Anno nativitatis Dei, M.D. XVII . . .

SANCHEZ MONTEZ, J., *Franceses, Protestantes, Turcos: Los españoles ante la política internacional de Carlos V* (Pamplona, 1951).

SANUTO, M., *Diarii di Marino Sanuto* (Venice, 1879–1903).

SARTIO, M., *De episcopis eugubinis* (Pesaro, 1755).

SAVONAROLA, Fra Girolamo, OP, *Prediche italiane ai Fiorentini*, ed. R. Palmarocchi (Florence, 1930).

SBARALEA, J., *Supplementum* . . . *ad Scriptores Ordinis Minorum* (Rome, 1806; new edn. Rome, 1908).

——an. edn., *Supplementum* . . . *ad Scriptores trium ordinum S. Francisci* (Rome, 1921).

SCHARD, S., *Historicum opus. 2, Rerum Germanicarum scriptores* (Basel, 1574).

SCHLECHT, J. (ed.), *Briefmappe*, 2 (Reformationsgeschichtliche Studien und Texte, 40; Münster, 1922).

SCHMITT, J.-C., *Le Saint-Lévrier, Guinefort, guérisseur d'enfants depuis le XIIIe siècle* (Paris, 1979).

SCHNITZER, J., *Quellen und Forschungen zur Geschichte Savonarolas* (Leipzig, 1910).

——Savonarola, trans. E. Rutili (Milan, 1931).

SCHOLEM, G., *Major Trends in Jewish Mysticism*, 3rd edn. (London, 1955).

——'Considérations sur l'histoire des débuts de la kabbale chrétienne', in *Kabbalistes chrétiens* (Cahiers de l'Hermetisme, dir. A. Favre and F. Tristan; Paris, 1979), 34–45.

SCHUTTE, A., *Printed Italian Vernacular Religious Books, 1465–1550: A Finding List* (Geneva, 1983).

SCHWINEKÖPER, B., *Der Handschuh im Recht, Ämterwesen, Brauch und Volksglauben* (Berlin, 1938).

SECRET, F., 'L'émithologie de Guillaume Postel', in E. Castelli (ed.), *Umanesimo e esoterismo. Atti del V Convegno internazionale di studi umanistici, Oberhofen, 16–17 sett., 1960* (Padua, 1960), 381–437.

——*Le Zohār dans les Kabbalistes chrétiens de la Renaissance* (Paris and The Hague, 1964).

——*Les Kabbalistes chrétiens de la Renaissance* (Paris, 1964).

——*Introduction à Guillaume Postel, Le Thrésor des prophéties de l'univers; Manuscrit publié avec une introduction et des notes* (The Hague, 1969).

SEIDEL MENCHI, S., *Erasmo in Italia, 1520–1580* (Turin, 1987).

SEMERIA G., *Storia ecclesiastica di Genova e della Liguria dai tempi apostolici sino all'anno 1838* (Turin, 1838).

——*Secoli Cristiani della Liguria* (Turin, 1843).

SERAFINO DA FERMO, *Breve dichiaratione sopra l'apocalisse de Giovanni, dove si prova esser venuto il precursor de Antichristo et avicinarsi la per cossa da lui predetta nel sesto sigillo* (Venice, 1541).

SERIPANDO, G., *Commentari de vita sua, Concilium Tridentinum* (Freiburg, 1901–30), ii. 397–488.

SEVESI, P., 'Il B. Menez de Silva dei Frati Minori, Fondatore degli Amadeiti: Vita inedita di Fra' Mariano da Firenze e documenti inediti', *Luce e Amore*, 8 (1911), 529–42, 586–605, 681–710 (repr. *B. Amadeo Menez de Sylva dei Frati minori, Fondatore degli Amadeiti (Vita inedita di Fra Mariano da Firenze e documenti inediti)* (Florence, 1912).

SHAPLEY, F., *Paintings from the S. H. Kress Collection, Italian Schools, XV–XVI Centuries* (London, 1968).

SHUMAKER, W., *The Occult Sciences in the Renaissance: A Study in Intellectual Patterns* (Berkeley, Calif. 1972).

SIMONCELLI, P., *Evangelismo italiano del Cinquecento* (Rome, 1979).

SMITH, W., *Gloves, Past and Present* (New York, 1918).

SOJAT, P., OFM, *De voluntate hominis eiusque praeminentia et dominatione in anima secundum Georgium Dragišić (c.1448–1520). Studium historico-doctrinale et editio Tractatus: 'Friedericus, De animae regni principe'* (Rome, 1972).

SOUSA COSTA, A. DE, 'Studio critico e documenti inediti sulla vita del beato Amadeo da Silva nel quinto centenario della morte', in I. Vasquez Janeiro (ed.), *Nascere sancta: Miscellanea in memoria di Agostino Amore, OFM. (†1982)* (Rome, 1985), 101–360.

STEPHENS, J., *The Fall of the Florentine Republic 1512–1530* (Oxford, 1983).

STINGER, C., *The Renaissance in Rome* (Bloomington, Ind., 1985).

STORNAJOLO, C., *Alcune ricerche sulla vita del Cardinale Bessarione* (Siena, 1897).

TAURISANO, P., OP, *S. Sabina (Le Chiese di Roma illustrate)*, II (Rome, n.d.).

TAYLOR, R., *The Political Prophecy in England* (New York, 1900).

TEDALLINI, S. DI BRANCA, *Diario romano dal 3 maggio 1485 al 6 giugno 1524*, ed. P. Piccolomini, *RIS* 23, pt. 3.

TELESPHORUS OF COSENZA, *Liber de magnis tribulationibus... compilatus a... Theolosphori de Cusentia* (Venice, 1516).

THOMAS, K., *Religion and the Decline of Magic* (London, 1971).

THORNDIKE, L., *A History of Magic and Experimental Science*, I (New York, 1934).

TIRABOSCHI, G., *Storia della letteratura italiana* (Milan, 1824).

TOCCO, F., *Studi Francescani: Nuova biblioteca di letteratura, storia ed arte*, iii (Naples, 1909).

TOLNAY, C. DE, *Michelangelo II, The Sistine Ceiling* (Princeton, NJ, 1945).

TOMMASO DI SILVESTRO, *Diario*, ed. L. Fumi, *RIS* 15, V/2.

TONDELLI, L., REEVES, M., and HIRSCH-REICH, B. (eds.), *Il Libro delle Figure dell'Abate Gioacchino da Fiore*, vol. ii, 2nd edn. (Turin, 1954).

TORMO, E., *Monumentos de Españoles en Roma y de Portugueses e Hispano-Americanos*, vol. i (Rome, 1940).

TORRE, A. DELLA, *Storia dell'Accademia platonica fiorentina* (Florence, 1902; repr. Turin, 1960).

TRAVI, E., 'Paolo Giovio nel suo Tempo', *Atti del Convegno Paolo Giovio, Il Rinascimento e la Memoria* (Como, 1985), 313–30.

TREXLER, R., *Synodal Law in Florence and Fiesole, 1306–1518* (Studi e Testi, 268, Vatican City, 1971).

TRITHEMIUS, J., *Chronicon Hirsaugiense* (St Gall, 1690).

TROMBETTA, A., 'Quaestio super articulos impositos domino Gabriele sacerdoti', *Opus in Metaphysicam Aristotelis* (Venice, 1502).

UBALDINI, F., *Vita di Mons. Angelo Colucci: Edizione del testo originale italiano (Barb. Lat. 4882)*, ed. V. Fanelli (Vatican City, 1969).

UGHELLI, F., *Italia Sacra sive de episcopis Italiae et insularum adjacentium* (Venice, 1719).

UGURGIERI DELLA BERARDENGA, C., *Gli Acciaiuoli di Firenze nella luce dei loro tempi (1160–1834)* (Florence, 1962).

ULLMANN, W., 'Julius II and the Schismatic Cardinals', in G. Cuming and D. Baker (eds.), *Schism, Heresy and Religious Protest* (Cambridge, 1972), 177–93.

VALDÉS, *Dialogo de las cosas ocurridas en Roma* (Madrid, 1969).

VALLONE, G., 'Pietro S. detto il Galatino', *Studi offerti ad Aldo Vallone*, 1: *Letteratura e storia meridionale* (Florence, 1989), 87–105.

VANNICELLI, P., *S. Pietro in Montorio e il tempietto del Bramante* (Rome, 1971).

VARCHI, B., *Storia fiorentina* (Florence, 1858).

VASARI, G., *Vite de più eccellenti pittori, scultori e architetti*, ed. G. Bottari (Rome, 1759).

—— an. edn., ed. G. Milanesi (Florence, 1878–85).

VASCHO, A. DI, *Il diario della città di Roma dall'anno 1480 all'anno 1492*, ed. G. Chiesa, *RIS* 23, 3.

VASOLI, C., *Studi sulla cultura del Rinascimento* (Manduria, 1968).

—— 'Notizie su Dragišić', *Studi storici in onore di Gabriele Pepe* (Bari, 1969), 429–98; repr. with revisions in *Profezia e Ragione*.

—— *Profezia e ragione* (Naples, 1974).

—— 'Profezie e profeti nella vita religiosa e politica fiorentina', in *Magia, Astrologia e Religione nel Rinascimento* (Wrocław, Warsaw, Krakov, and Gdansk, 1974), 16–29.

—— *I miti e gli astri* (Naples, 1977).

—— *Religione e civiltà* (Bari, 1982).

—— *Immagini umanistiche* (Naples, 1983).

—— 'Postel, Galatino e l'*Apocalypsis nova*', in *Guillaume Postel 1581–1981. Actes du Colloque International d'Avranches 5–9 sept. 1981* (Paris, 1985), 97–108.

—— 'Un notaio fiorentino del Cinquecento: Ser Lorenzo Violi', in *Il notariato nella civiltà toscana. Atti di un Convegno (maggio 1981)* (Rome, 1985), 391–418.

——*Filosofia e religione nella cultura del Rinascimento* (Naples, 1988).

VÁZQUEZ JANEIRO, I. (ed.), *Noscere sancta: Miscellanea in memoria di Agostino Amore, OFM (†1982)* (Rome, 1985).

VEISSIÈRE, M., 'Guillaume Briçonnet et les courants spirituels italiens au début du XVe siècle', in M. Maccarrone and A. Vauchez (eds.), *Échanges religieux entre la France et l'Italie du Moyen Age à l'époque moderne* (Geneva, 1987), 215–28.

VERDE, A., *Lo studio fiorentino (1473–1503): Ricerche e documenti* (Florence, 1973).

VERNAZZA, B., *Lettera biografica del padre e della madre*, in *Delle Opere spirituali della reverende et divotiss. Vergine di Christo donna Battista da Genova* (Verona, 1602).

VEZIN, G., *Saint François de Paule fondateur des Minimes et la France* (Paris, 1971).

VILLARI, P., *Life and Times of Girolamo Savonarola*, trans. L. Villari (London, 1838).

VIO, T. DE (CAJETAN), *Sancti Thomas Aquinatis doctoris angelici opera omnia iussu impensaque Leonis XIII P.M. edita*, vol. ix: *Secunda secundae Summae theologiae . . . cum commentariis Thomae de Vio Caietani Ord. Praed. S.R.E. Cardinalis* (Rome, 1897).

——*Tractatus duodecimus De Maleficiis, Opuscula omnia* (Lyons, 1567).

VIOLI, L., *Le giornate*, ed. G. Garfagnini (Florence, 1986).

VITALE, G., *De monstro nato, Erfordiae per Mattheum Pictorium Anno novi seculi XIJ Mensis Iulio* (Erfurt, 1512).

——*Teratorizion* (Rome, 1514).

VOLPE, C., and LUCCO, M., *L'opera completa di Sebastiano del Piombo* (Milan, 1980).

WACHTEL, N., *La visione dei vinti: Gli Indianos del Perù di fronte alle conquista spagnola*, Italian trans. (Turin, 1977).

WADDING, L., *Annales Minorum* (Rome, 1731–6).

WAEL, P. DE, *Collectanea rerum gestarum et Eventuum Cartusiae Bruxellensis*, i (1625).

WALKER, D., *Spiritual and Demonic Magic from Ficino to Campanella* (Notre Dame, Ind., 1975).

WEIL, G., *Elie Lévita, humaniste et massorète (1469–1549)* (Leiden, 1963).

WEINSTEIN, D., *Savonarola and Florence: Prophecy and Patriotism in the Renaissance* (Princeton, NJ, 1970).

——'The Apocalypse in Sixteenth-Century Florence: The Vision of Albert of Trent', in A. Molho and J. Tedeschi (eds.), *Renaissance Studies in Honor of Hans Baron* (Florence, 1971), 313–31.

WICKS, J., *Cajetan und die Anfänge der Reformation*, trans. B. Hallensleben (Münster, 1983).

WIND, E., 'Maccabean Histories in the Sistine Ceiling', in E. Jacob (ed.), *Italian Renaissance Studies* (London, 1960), 312–27.

WINSPEARE, F., *La Congiura dei Cardinali contro Leone X* (Florence, 1957).

ZAMBELLI, P., 'Fine del mondo o inizio della propaganda? Astrologia, filosofia della storia e propaganda politico-religiosa nel dibattito sulla congiunzione del 1524', in G. Garfagnini (ed.), *Scienze, credenze occulte, livelli di cultura* (Florence, 1982), 291–368.

——(ed.), *'Astrologi hallucinati'. Stars and the End of the World in Luther's Time* (Berlin and New York, 1986).

ZAMBRINI, F., *Le opere volgari a stampa dei secoli XIII e XIV*, 4th edn. (Bologna, 1884).

ZANNONI, G., *Scrittori cortigiani del Montefeltro* (Rome, 1894).

ZIMMERMANN, T., 'Giovio e la Crisi del Cinquecento', *Atti del Convegno Paolo Giovio, Il Rinascimento e la Memoria* (Como, 1985), 9–18.

B. PERIODICALS AND WORKS OF REFERENCE

AMICO, J. D', 'A Humanist Response to Martin Luther: Raffaele Maffei's *Apologeticus*', *Sixteenth-Century Journal*, 6 (1975), 39–56.

ALESSANDRINI, A., 'Angelo da Vallombrosa', *DBI* 3 (Rome, 1961), 238–40.

BARBERI, F., 'Calvo, Francesco Giulio', *DBI* 17 (1974), 38–41.

BATTLORI, M., 'Bernardino Lopez de Carvajal, legado de Alejandro VI en Anagni, 1494', *Saggi storici intorno al Papato* (*Miscellanea Historiae Pontificiae*, 21; Rome, 1959), 171–88.

BAUMGARTNER, F., 'Henry II and the Papal Conclave of 1549', *Sixteenth-Century Journal*, 16 (1985), 301–14.

BELMON, C., 'Un evêque français à l'assemblée de Tours et au Concile de Pise: François d'Estaing', *Revue des études historiques*, 89 (1923), 300–20.

BENELLI, G., 'Di alcune lettere del Gaetano', *AFP* 5 (1935), 363–75.

BERNARDINO OF SIENA, St, 'Sermo intorno a S. Giuseppe, sposo della Vergine Maria', *Misc. Franc.* 311/312 (1883), 1–25.

BERTHIER, A., 'Un Maître orientaliste du XIIIe siècle: Raymond Martin, O.P.', *AFP* 6 (1936), 267–311.

BERTOLOTTO, G., '"Genua", Poemetto di Giovanni Maria Cataneo', *Atti della Società Ligure di Storia Patria*, 24 (1891–2), 721–881.

BERTONI, G., 'Intorno a tre letterati cinquecentisti modenesi. Francesco Roccocciolo', *Giornale storico della letteratura italiana*, 85 (1925), 376–7.

BEZOLD, F. VON, 'Jean Bodin als Okkultist und seine "Demonomanie"', *HZ* 105 (1910), 1–64.

BIGNAMI-ODIER, J., 'Les Visions de Robert d'Uzès, O.P. (†1296)', *AFP* 25 (1955), 258–310.

BIHEL, P., 'S. Franciscus Fuitne Angelus Sexti Sigilli? (Apoc. 7: 2)', *Antonianum*, 2, ser. 2 (1927), 59–60.

BIONDI, A., and PROSPERI, A., 'Il processo al medico Basilio Albrisio, Reggio 1559', *Contributi*, 4 (1976), pp. v–xlviii, 1–110.

BONGI, S., 'Francesco da Meleto, una profeta fiorentino a' tempi del Machiavelli', *ASI* 3 (1889), 27–38, 62–70.

BORGEAUD, P., 'La mort du Grand Pan. Problèmes d'interpretation', *Revue de l'histoire des Religions*, 200 (1983), 3–39.

BORNATE, C. (ed.), 'Historia, vite et gestorum per dominum magnum cancellarium . . . con note, aggiunte e documenti', *Miscellanea di storia Italiana*, 48 (1915), 233–568.

BRITNELL, J., and STUBBS, D., 'The *Mirabilis Liber*: Its Compilation and Influence', *JWCI* 49 (1986), 126–49.

BULL, M., 'The Iconography of the Sistine Chapel Ceiling', *Burlington Magazine*, 130 (August 1988), 597–605.

BURR, D., 'The Persecution of Peter Olivi', *Transactions of the American Philosophical Society*, NS 66, pt. 5 (1976).

—— 'Bonaventure, Olivi and Franciscan Eschatology', *Collectanea Franciscana*, 53 (1983), 23–40.

CANNAROZZI, C., 'Ricerche sulla vita di fra Mariano da Firenze', *Studi francescani*, 27 (1930), 31–71.

CANTERA BURGOS, F., 'Fernando de Pulgar y los conversos', *Sefarad*, 4 (1944), 295–348.

CENTI, T., 'L'attività letteraria di Santi Pagnini (1470–1536) nel campo delle scienze bibliche', *AFP* 15 (1945), 5–51.

CHASTEL, A., 'A Fly in the Pigment', *FMR* 4, ed. Franco Maria Ricci (1986), 62–81.

CLOUGH, C., 'Cardinal Bessarion and Greek at the Court of Urbino', *Manuscripta*, 8 (1964), 160–71.

CRIVELLI, I., 'Fieschi di Lavagna (Nicola) (*c.*1456–1524)', *DHGE* 16 (Paris, 1967), cols. 1440–1.

DANIEL, E. R., 'A Re-examination of the Origins of Franciscan Joachitism', *Speculum*, 43 (1968), 671–6.

DAVIS, C., 'Un appunto per Sebastiano del Piombo Ritrattista', *Mitteilungen des Kunsthistorischen Institutes in Florenz* (1982), 383–8.

DICTIONNAIRE, DE SPIRITUALITÉ, ed. C. Baumgartner, SJ, (Paris, 1953).

DIONISOTTI, C., '*Umanisti dimenticati?*', *Italia medioevale e umanistica*, 3–4 (1960–1), 287–321.

DONCKEL, E., 'Studien über die Prophezeiung des Fr. Telesforus von Cosenza, OFM (1365–1386)', *AFH* 26 (1933), 29–104, 282–314.

DOTSON, E., 'An Augustinian Interpretation of Michelangelo's Sistine Ceiling', *Art Bulletin*, 61 (1979), 250–5.

DYKMANS, M., 'Le cinquième Concile du Latran d'après le Diarie de Paris de Grassi', *Annuarium*, 14 (1982), 271–369.

EINEM, H. VON, 'Die "Verklärung Christi" und "Die Heilung des Besessenen" von Raffael', *Abhandlungen der Akademie der Wissenschaften und der Literatur in Mainz*, 5 (1966), 299–327.

ESPADAS BURGOS, M., 'Aspectos socioreligiosos de la alimentación española', *Hispania*, 131 (1975), 537–65.

FELICIANGELI B., 'Le proposte per la guerra contro i Turchi presentate de Stefano Taleazzi vescovo di Torcello a papa Alessandro VI', *Archivio della Reale Società Romana di Storia Patria*, 40 (1917), 5–63.

FONZO, P. DI, OFM, 'La Famosa Bolla di Leone X "Ite vos" non "Ite et vos" (29 Maggio 1517)', *Misc. Franc.* 44 (1944), 164–71.

FRAGNITO, G., articles on Carvajal, Bernardino López de', *DBI* 21 (Rome, 1978), 28–34; Castellesi, Adriano, *DBI* 21 (Rome, 1978); Capizucchi, Paolo, *DBI* 18 (1975), 571–2.

FRATI, L., 'Il Cardinale Francesco Alidosi e Francesco Maria Della Rovere', *ASI* 47 (1911), 114–58.

FUMAGALLI, E., 'Anecdoti della vita di Annio da Viterbo O.P.', *AFP* 50 (1980), 189–99.

GARFAGNINI, G., 'Ser Lorenzo Violi e le prediche del Savonarola', *Lettere italiane*, 37 (1986), 312–37.

GIACONE, F., and BEDOUELLE, G., 'Une lettre de Gilles de Viterbe (1469–1532) à Jacques Lefèvre d'Étaples (c.1460–1536) au sujet de l'affaire Reuchlin', *Bibliothèque d'Humanisme et Renaissance*, 36 (1974), 335–45.

GILBERT, F., 'Contarini on Savonarola: An Unknown Document of 1516', *AR* 59 (1968), 145–50.

GINZBURG, C., 'Vom finstern Mittelalter bis zum Blackout von New York—und zurück', *Freibeuter*, 18 (1983), 25–34.

GIORGETTI, A., 'Fra Luca Bettini e la sua difesa del Savonarola', *ASI*, 77 (1919), 164–231.

GLEASON, E., 'On the Nature of Sixteenth-Century Italian Evangelism', *Sixteenth-Century Journal*, 9 (1978), 3–25.

GLIOZZI, G., 'The Apostles in the New World: Monotheism and Idolatry between Revelation and Fetishism', *History and Anthropology*, 3 (1987), 123–48.

GOULD, C., 'Raphael's Papal Patrons', *Apollo*, 117 (1983), 358–61.

GRAGNITO, G., 'Cultura umanistica e riforma religiosa: Il *De officio viri ac probi episcopi* di Gasparo Contarini', *Studi veneziani*, 2 (1969), 167–89.

GRUNDMANN, H., 'Die Papstprophezien des Mittelalters', *AK* 19 (1929), 77–159.

—— 'Die Liber de Flore', *HJ* 40 (1929), 33–91.

HEADLEY, J., 'The Habsburg World Empire and the Revival of Ghibellinism', *Medieval and Renaissance Studies*, 7, ed. S. Wenzel (1978), 93–127.

—— 'The Conflict between Nobles and Magistrates in Franche-Comté 1508–1518', *Journal of Medieval and Renaissance Studies*, 9 (1979), 49–80.

—— 'Gattinara, Erasmus and the Imperial Configurations of Humanism', *Archiv für Reformationsgeschichte*, 71 (1980), 64–98.

—— 'L'imperatione e il suo cancelliere: discussioni sull'impero, l'amministrazione e il papa', *Rassegna degli Archivi di Stato (Roma)*, 46, 3 (1986), 534–51.

HELLER, H., 'Marguerite of Navarre and the Reformers of Meaux', *Bibliothèque d'Humanisme et Renaissance*, 33 (1971), 271–310.

—— 'The Briçonnet Case Reconsidered', *Journal of Medieval and Renaissance Studies*, 2 (1972), 223–58.

HÖFLER, C., 'Exemplum literarum domini Stephani Rosin Caesareae Majestatis apud S. Sedem Sollicitatoris ad Reverendum principem D. Carolum Gurcensem', *Analecten zur Geschichte Deutschlands und Italien, Abhandlungen der hist. Cl. del K. Bayer. Akademie der Wissenschaften*, 4 (1846), 1–59.

HOLDER-EGGER, O., 'Italienische Prophetien des 13. Jahrhunderts', *Neues Archiv der Gesellschaft für ältere deutsche Geschichtskunde*, 15 (1889), 155–73.

HUDON, E., 'Catholic Reform and Counter-Reform or Tridentine Reformation? Two Instructions to Preachers from the Tridentine Reformation', *Sixteenth-Century Journal*, 20 (1989), 457–70.

JEDIN, H., 'Ein Vorschlag für die Amerika-Mission aus dem Jahre 1513', *Neue Zeitschrift für Missionenwissenschaft*, 2 (1946), 81–4.

KAGAN, R., 'Lucrezia de Leon: per una valutazione dei sogni e delle visioni nella Spagna del Cinquecento', *Quaderni storici*, NS 68 (1988), 595–607.

KALKOFF, P., 'Zu Luthers römischen Prozess', *ZKG* 25 (1904), 90–147.

KAPPLER, C., 'La vocation messianique de Christophe Colomb: voyage, quête, pèlerinage dans la littérature et la civilisation médiévale', *Senéfiance* (Cahiers du Cuerma; Aix-en-Provence, 1976), 255–71.

KING, C., 'The Liturgical and Commemorative Allusions in Raphael's *Transfiguration* and *Failure to Heal*', *JWCI* 45 (1982), 148–59.

KLEINHANS, A., 'De vita et operibus Petri Galatini, OFM scientiarum Biblicarum cultoris', *Antonianum*, 1 (1926), 145–79, 327–56.

KLOR DE ALVA, J., 'The Aztec–Spanish Dialogues of 1524', *Alcheringa: Ethnopoetics*, 4 (1980), 52–193.

KOKOT, I., 'La fonte ispiratrice nei capolavori delle aquile (II): La cappella Borgherini a S. Pietro in Montorio di Sebastiano del Piombo', *Fede e Arte, Rivista internazionale di Arte Sacra*, 2, 4 (1954), 97–107.

LABOWSKY, L., 'Bessarion Studies, 1', *Medieval and Renaissance Studies*, 5 (1961), 108–62.

—— 'Bessarione, Basilio', *DBI* 9 (Rome, 1967), 686–96.

LEACH, E., 'Michelangelo's *Genesis*: Structuralist Comments on the Paintings on the Sistine Chapel Ceiling', *Times Literary Supplement* (18 March 1977), 311–13.

LEIJENHORST, C. VAN, 'Paul of Middelburg', *Contemporaries of Erasmus: A Biographical Register of the Renaissance and Reformation*, 3 (Toronto, 1987), 57–8.

LERNER, R., 'Refreshment of the Saints: The Time after Antichrist as a Station for Earthly Progress in Medieval Thought', *Traditio*, 32 (1976), 97–144.

—— 'Antichrists and Antichrist in Joachim of Fiore', *Speculum*, 60 (1985), 553–70.

—— 'On the Origins of the Earliest Latin Pope Prophecies: A Reconsideration', *MGHS* 33 (Hanover, 1988), 611–35.

LOERNETZ, P., 'Pour la biographie du Cardinal Bessarion', *Orientalia Christiana Periodica*, 10 (1944), 116–49.

LONGHI, R., 'Cartella tizianesca', *Vita artistica*, 2, 11–12 (1926–7), 216–26.

LOPEZ, A., 'Descriptio codicum francescanorum Bibliothecae Riccardianae Florentiae', *AFH* 1, (1908), 116–25, 433–42.

McGINN, B., 'The Abbot and the Doctors', *Church History*, 40 (1971), 30–47.

—— 'Angel Pope and Papal Antichrist', *Church History*, 47 (1978), 155–73.

—— 'Circoli gioachimiti veneziani (1450–1530)', *Cristianesimo nella storia*, 7 (1986), 19–39.

McMANAMON, J., 'The Ideal Renaissance Pope: Funeral Oratory from the Papal Court', *AHP* 14 (1976), 16–53.

McNALLY, R., 'Pope Adrian VI (1522–23) and Church Reform', *AHP* 7 (1969), 260–72.

MALIPIERO, D., 'Annali veneti dall'anno *1457 ad 1500*', *ASI* 7 (1843–4) 1–720.

MANUCCI, G., 'Il conclave di papa Marcello', *Bullettino senese di storia patria*, 27 (1920), 94–103.

MÁRQUEZ VILLANUEVA, F., 'El mundo converso de *La Lozana Andaluza*', *Archivo Hispalense*, 56 (1973), 87–97.

MARTELLI, M., 'La politica culturale dello ultimo Lorenzo', *Il Ponte*, 36 (1980), 923–50, 1040–69.

—— 'Benigno o Belcari?', *Interpres*, 7 (1987), 206–13.

MARTIN, F. X., OSA, 'The Writings of Giles of Viterbo', *Augustiniana*, 29, 1–2 (1979), 141–80.

MASSA, E., 'L'anima e l'uomo in Egidio da Viterbo e nelle fonti classiche e medioevale', *Archivio di Filosofia* (1951), 37–138.

MESEGUER FERNANDEZ, P., OFM, 'Breves de Clemente VII en favor de la Provincia de S. Pedro in Montorio y de su confessor Juan Antonio Tomas de Locarno, OFM', *AFH* 44 (1951), 161–90.

MESSINI, A., 'Profetismo e profezie ritmiche d'inspirazione gioachimito-franciscana nei secoli XIII, XIV e XV', *Misc. Franc.* 37 (1937), 39–54; 39 (1939), 109–30.

MILLS, J., '"Small Pattern Holbein" Carpets in Western Paintings', *Hali*, 1, 4 (1978), 326–34.

—— '"Lotto" Carpets in Western Paintings', *Hali*, 3, 4 (1981), 278–89.

MINNICH, N., 'Concepts of Reform Proposed at the Fifth Lateran Council', *AHP* 7 (1969), 163–251.

—— 'The Participants at the Fifth Lateran Council', *AHP* 12 (1974), 157–206.

—— 'Paride de Grassi's Diary of the Fifth Lateran Council', *Annuarium*, 14 (1982), 380–433.

—— 'The Healing of the Pisan Schism (1511–1513)', *Annuarium*, 16 (1984), 59–192.

—— 'Alexios Celadenus: A Disciple of Bessarion in Renaissance Italy', *Historical Reflections*, 15 (1988), 51–63.

MIRUS, J., 'On the Deposition of the Pope for Heresy', *AHP* 13 (1975), 231–48.

MORISI, A., 'Vangeli apocrifi e leggende nella cultura religiosa del tardo Medioevo. Ricerche sul pensiero teologico di Giorgio Benigno', *Bull. ISI* 85 (1974–5), 151–77.

MORISON, S., 'Marcello Cervini, Pope Marcellus II: Bibliography's Patron Saint', *Italia medioevale e umanistica*, 5 (1962), 301–18.

MOYNIHAN, R., 'The Development of the "Pseudo-Joachim" Commentary "Super Hieremiam": New Manuscript Evidence', *Mélanges de l'École Française de Rome, Série Moyen Âge et Temps Modernes*, 90, 1 (1986), 109–42.

NICCOLI, O., 'Profezia in piazza: Note sul profetismo popolare nell'Italia del primo Cinquecento', *Quaderni storici*, 41 (1979), 500–39.

NOBILE, B., ' "Romiti" e vita religiosa nella cronachistica italiana fra '400 e '500', *CS* 5 (1984), 303–40.

OBERHUBER, K., 'Vorzeichnungen zu Raffaels *Transfiguration*', *Jahrbuch der Berliner Museen*, 4 (1962), 116 ff.

—— 'Raphael and the State Portrait I: The Portrait of Julius II', *Burlington Magazine*, 113 (1971), 124–30.

O'GORMAN, J., 'An Interpretation of Andrea del Sarto's *Borgherini Holy Family*', *Art Bulletin*, 47 (1965), 502–4.

O'MALLEY, J., SJ, 'Giles of Viterbo: A Sixteenth-Century Text on Doctrinal Development', *Traditio*, 22 (1966), 445–50.

—— 'Fulfillment of the Christian Golden Age under Pope Julius II: Text of a Discourse by Giles of Viterbo, 1507', *Traditio*, 25 (1969), 265–338.

O'REILLY, C., ' "Maximus Caesar et Pontifex Maximus": Giles of Viterbo Proclaims the Alliance between Emperor Maximilian I and Pope Julius II', *Augustiniana*, 22 (1972), 80–117.

—— ' "Without Councils we cannot be saved . . .": Giles of Viterbo addresses the Fifth Lateran Council', *Augustiniana*, 27 (1977), 166–204.

PALL, F., 'Marino Barlezio. Uno storico umanista', *Mélanges d'histoire générale* (dell' Università di Cluj), 2 (1938), 135–515.

PANDZIC, B., 'Vida y obra de Jorge Dragišić, un humanista filosofo y teologo croato en al Renacimento italiano', *Studia Croatica*, 11 (1970), 114–31.

PASCHINI, P., 'Una predica inefficace (propositi di riforma ecclesiastica alla fine del sec. XV)', *Studi romani*, 1 (1953), 31–8.

—— 'Tre illustri prelati del Rinascimento', *Lateranum*, NS 23 (1957), 1–130.

PÉLISSIER, L., 'Manuscrits de Gilles de Viterbo à la Bibliothèque Angélique',

Revue des bibliothèques, 2 (1892), 228–40.

PERRONE, B., 'I Frati Minori di Puglia della Serafica Provincia di S. Niccolò (1590–1835)', *Archivi e Biblioteche*, 2 (1977).

PICOTTI, G., 'Un episodio di politica ecclesiastica medicea', *Annali delle Università toscane*, NS 14 (1930), 86 ff.

—— 'La Congiura dei Cardinali contro Leone X', *Rivista storica italiana*, 40–1 (1923–4), 249–67.

POLIZZOTTO, L., 'Confraternities, Conventicles and Political Dissent: The Case of the Savonarolian "Capi rossi"', *Memorie domenicane*, NS 16 (1985), 235–83.

—— 'Documents', *Memorie domenicane*, NS 17 (1986), 285–300.

POÚ Y MARTI, OFM, 'El Libro perdido de las platicas o coloquios de los doce primeros misioneros de México', in *Miscellanea Francesco Ehrle* (Rome, 1924), iii. 282–333.

PRATESI, R., 'De Silva Menesez, Amadeo, beato', *Bibliotheca Sanctorum* (Rome, 1969), iv. 587–8.

PRICE, M., 'The Origins of Lateran V's *Apostolici Regiminis*', *Annuarium*, 17 (1985), 464–72.

PROSPERI, A., 'Il monaco Teodoro: note su un processo fiorentino del 1515', *Critica storica*, 11 (1975), 71–101.

—— 'America e Apocalisse. Note sulla "conquista spirituale" del Nuovo Mondo', *Critica Storica*, 13 (1976), 1–61.

PUGLIESE, O., 'Apocalyptic and Dantesque Elements in a Franciscan Prophecy of the Renaissance', *Proceedings of the Patristic, Mediaeval and Renaissance Conference*, 10 (1985), 127–35.

REEVES, M., 'The Originality and Influence of Joachim of Fiore', *Traditio*, 36 (1980), 269–316.

—— and HIRSCH-REICH, B., 'The Seven Seals in the Writings of Joachim of Fiore', *RTAM* 21 (1954), 211–47.

RINALDIS, A. DE, 'Una interpretazione della *Trasfiguratione* de Raffaello in Vaticano', *L'illustrazione vaticana*, 6 (1935), 295 ff.

RISTORI, R., 'Cassini, Samuele', *DBI* 21 (Rome, 1978), 487–9.

ROSENTHAL, E., 'The Invention of the Columnar Device of Emperor Charles V at the Court of Burgundy in Flanders in 1516', *JWCI* 36 (1973), 198–230.

ROTONDÒ, A., 'Pellegrino Prisciani', *Rinascimento*, 11 (1960), 69–110.

RUSCONI, R., 'Apocalittica ed escatologia nella predicazione di Bernardino da Siena', *Studi Medievali*, 3rd ser. 22 (1981), 85–128.

—— 'Millenarismo e centenarismo: tra due fuochi', *Annali della Facoltà di Lettere e Filosofia della Università degli Studi di Perugia*, 22, NS 8 (1984/5), 2; *Studi Storici Antropologici*, 51–64.

—— 'Il collezionismo profetico in Italia alla fine del Medioevo ad agli inizi dell'età moderna (annotazioni a proposito di alcuni manoscritti italiani conservati nelle biblioteche parigine)', *Florensia*, 2 (1988), 61–90.

SAULNIER, V., 'Marguerite de Navarre au temps de Briçonnet. Étude de la correspondence générale', *Bibliothèque d'Humanisme et Renaissance*, 39 (1977), 437–78; 40 (1978), 7–47, 193–237.

SAXL, F., 'A Spiritual Encyclopaedia of the Later Middle Ages', *JWCI* 5 (1942), 82–134.

SCHENDA, R., 'Das Monstrum von Ravenna: Eine Studie zur Prodigien-Literatur', *Zeitschrift für Volkskunde*, 56 (1960), 209–25.

SCHMITT, C., 'Gianfrancesco Pico della Mirandola and the Fifth Lateran Council', *AR* 61 (1970), 161–78.

SCHNEIDER, F., 'Theologisches in Raffaels *Disputa* und *Trasfigurazione*', *Katholik*, 3 (1896), 11 ff.

SCHUTTE, A., 'The *Lettere Volgare* and the Crisis of Evangelism in Italy', *Renaissance Quarterly*, 28 (1975), 639–88.

SECRET, F., 'Guillaume Postel et les courants prophétiques de la Renaissance', *Studi Francesi*, 3 (1957), 375–95.

—— 'Les Dominicains et la Kabbale chrétienne à la Renaissance', *AFP* 27 (1957), 319–36.

—— 'Paralipomènes de la vie de François I par Guillaume Postel', *Studi Francesi*, 4 (1958), 50–62.

—— 'Paulus Angelus descendant des empéreurs de Byzance et la prophétie du Pape Angélique', *Rinascimento*, 2nd ser. 2 (1962), 211–14.

—— 'L'*Ensis Pauli* di Paulus de Heredia', *Sefarad*, 26 (1966), 79–102, 253–72.

—— 'Aspects oubliés des courants prophétiques au début du XVIe siècle', *Revue de l'histoire des religions*, 173 (1968), 178–201.

—— 'Flaminio, Antonio', *DHGE* 17 (Paris, 1971), 354.

SEIDEL MENCHI, S., 'Passione Civile e Aneliti Erasmiani di Riforma nel Patriziato Genovese del Primo Cinquecento: Ludovico Spinola', *Rinascimento*, 2nd ser. 18 (1978), 87–134.

SETTON, K., 'Pope Leo X and the Turkish Peril', *Proceedings of the American Philosophical Society*, 113 (1969), 367–424.

SEVESI, P., 'Il beato Amadeo Menezes de Silva e documenti inediti', *Misc. franc.* 31 (1931), 227–32.

SHEARMAN, J., 'The Vatican Stanze: Function and Decoration', *Proceedings of the British Academy*, 57 (1971), 363–424.

SHERR, R., 'A New Document Concerning Raphael's Portrait of Leo X', *Burlington Magazine*, 125 (1983), 31–2.

SIMONCELLI, P., 'Inquisizione romana e riforma in Italia', *Rivista storica italiana*, 100 (1988), 1–125.

SINDING-LARSEN, S., 'A Re-Reading of the Sistine Ceiling', *Acta Instituti Romani Norvegiae*, 4 (1969), 143–57.

SOULIÉ, M., 'Prophétisme et visions d'Apocalypse dans *Les Tragiques* d'Agrippa d'Aubigné', *Réforme, Humanisme, Renaissance*, 12 (1986), n. 22, pp. 5–10.

SPINI, G., 'Introduzione al Savonarola', *Belfagor*, 3, 4 (1948), 414–28.

STELZER, W., 'Neue Beiträge zur Frage des Kaiser-Papstplane Maximilians I im Jahre 1511', *Mitteilungen des Instituts für österreichische Geschichtsforschung*, 71 (1963), 311–32.

TOGNETTI, G., 'Sul "romito" e profeta Brandano da Petroio', *Rivista storica italiana*, 72 (1970), 20–44.

—— 'Un episodio inedito di repressione della predicazione post-Savonaroliana (Firenze, 1509)', *Bibliothèque d'Humanisme et Renaissance*, 24 (1962), 190–9.

—— 'Le fortune della pretesa profezia di S. Cataldo', *Bull. ISI* 80 (1968), 273–317.

—— 'Note sul profetismo nel Rinascimento e la letteratura relativa', *Bull. ISI* 82 (1970), 12–54, 129–57.

TRAMONTIN, S., 'I Teatini e l'Oratorio del Divino Amore a Venezia', *Regnum Dei*, 29 (1973), 53–76.

TUMMINELLO, G., 'Giano Vitale umanista del sec. XVI', *Archivio storico siciliano*, NS 8 (1883), 1–94.

ULIANICH, B., 'Accolti, Pietro (1455–1532)', *DBI* i (Rome, 1960), 106–10.

VARISCHI, C., 'Catalogo dei codici della biblioteca dei Minori Capuccini in Milan', *Aevum*, 11 (1937), 237–74; 461–503.

VASOLI, C., 'Le profezia di Francesco da Meleto', *Archivio di Filosofia*, 3 (1963), 27–38.

—— 'Sul probabile autore di una "profezia" Cinquecentesca', *Il Pensiero Politico*, 2 (1969), 464–72.

—— 'Ancora su Giorgio Benigno Salviati (Juraj Dragišić) e la "profezia" dello pseudo-Amadeo', *Il Pensiero Politico*, 3 (1970), 417–21.

—— 'Profezia e astrologia in un testo di Annio da Viterbo', *Studi sul Medioevo Cristiano*, 2 (1974), 1029–34.

—— 'Sisto IV professore di teologia e teologo', *L'età dei Della Rovere. Atti e memorie della Società Savonese di Storia Patria*, NS 34 (1988), 177–207.

—— 'Giorgio Benigno Salviati e la tensione profetica di fine '400', *Rinascimento*, 2nd ser. 29 (1989), 53–78.

VÁZANEZ JANEIRO, I., OM, 'Anticristo mixto, Anticristo mistico, varia fortuna de dos expressiones escatologicas medievale, *Antonianum*, 63 (1988), 522–50.

VEISSIÈRE, M., and TARDIF, H., 'L'emploi de l'écriture par Guillaume Briçonnet évêque de Meaux entre 1519 et 1524', *Revue de sciences philoso-phiques et théologiques*, 63 (1979), 345–64.

VOLPATO, A., 'La predicazione penitenziale-apocalittica nell'attività di due predicatori del 1473', *Bull. ISI* 82 (1970 [1974]), 113–28.

WEISS, R., 'Traccia per una biografia di Annio da Viterbo', *Italia Medioevale e Umanistica*, 5 (1962), 425–41.

WIND, E., 'The Crucifixion of Haman', *JWCI* I (1937–8), 245–8.

ZARRI, G., 'Le sante vive. Per una tipologia della santità femminile nel primo Cinquecento', *Annali dell'istituto storico italo-germanico in Trento*, 6 (1980), 371–445.

GENERAL INDEX

INDEX OF MANUSCRIPTS